# Innovative Economic Symposium (IES2019)

## Potential of Eurasian Economic Union

SHS Web of Conferences Volume 73 (2020)

Ceske Budejovice, Czech Republic
7 November 2019

**Editors:**

**Jakub Horak**
**Jaromir Vrbka**
**Zuzana Rowland**

ISBN: 978-1-7138-0293-8

**Printed from e-media with permission by:**

Curran Associates, Inc.
57 Morehouse Lane
Red Hook, NY 12571

CURRAN ASSOCIATES INC. proceedings.com

**Some format issues inherent in the e-media version may also appear in this print version.**



For additional information, please contact EDP Sciences – Web of Conferences at the address below.

EDP Sciences – Web of Conferences
17, Avenue du Hoggar
Parc d'Activité de Courtabœuf
BP 112
F-91944 Les Ulis Cedex A
France

Phone: +33 (0) 1 69 18 75 75
Fax: +33 (0) 1 69 28 84 91

contact-edps@webofconferences.org

**Additional copies of this publication are available from:**

Curran Associates, Inc.
57 Morehouse Lane
Red Hook, NY 12571 USA
Phone: 845-758-0400
Fax: 845-758-2633
Email: curran@proceedings.com
Web: www.proceedings.com

# TABLE OF CONTENTS

# Innovative Economic Symposium 2019

Potential of the Eurasian Economic Union

Jihočeský kraj

# Innovative Economic Symposium 2019

## Potential of the Eurasian Economic Union

http://ies.vstecb.cz/

Editors: Jakub Horák, Jaromír Vrbka, Zuzana Rowland

# About the conference IES 2019

### Date of the conference: 2019-11-07

**The conference will be held under the auspices of President of the Czech Republic Ing. Miloš Zeman, CSc.**

## Motto

*"Coming together is a beginning; keeping together is progress; working together is success."* Henry Ford

## Target

The IES 2019 conference brings together scientists and practitioners in economy, international relations, finance and marketing from Asia and Europe. The objective of the conference is to share knowledge and discuss the issues relating to economic potential, international business, Industry 4.0 and use of artificial intelligence in corporate practice and economy and corporate strategic decision-making in the global world.

## Thematic Areas

**Use of artificial intelligence in corporate practice and economy**

- Prediction of future corporate development.
- Prediction of future development of macroeconomic parameters.
- Analysis of businesses in sectors of the national economy.
- Machine learning, large data.

**International business**

- Entrepreneurship, international business and economic growth.
- Strategy in international business.
- Intellectual capital, knowledge and innovations in international business development.
- Direct foreign investments.

**Business administration view of the topic + Industry 4.0**

- Digitizing, automation of the production.
- Sector and corporate adaptation to the conditions of Industry 4.0.
- Digital economy.
- Quantitative and qualitative methods of management decision-making

- Economic, social and environmental dimensions of strategic management.
- Decision-making methods and optimization, econometric modelling of corporate strategies.
- Risk management and portfolio optimization.
- Financial sustainability of small and medium-sized enterprises.
- Controlling and reporting.
- Social media marketing.

**Macroeconomic view of the topic**

- Economic consequences related to the defined economic area.
- Influence of economic parameters on the consumer.
- Consumer behaviour, international marketing, marketing modelling.
- Currency policy, theory of currency area.
- Renewable sources of energy, environmental legislation.

## About organizers

**Institute of Technology and Business in České Budějovice (VŠTE)**

The primary focus of VŠTE activities is the implementation of full-time and part-time studies for professional bachelor's and master's degree programmes. The professional study programmes are supported by cooperation with regional entrepreneurs. VŠTE´s other priorities include applied research, the implementation of innovative technologies and knowledge transfer into areas that are current and crucial with regards to the regions needs. More information about VŠTE at www.vstecb.cz.

**North China University of Technology (NCUT)**

North China University of Technology (NCUT), located on the western side of Beijing, is a multi-disciplinary university that combines the natural sciences and engineering with liberal arts, economics, management and law. NCUT targets the development of qualified, high-level, application-oriented professionals with solid basic theoretical knowledge and strong practical skills and creative abilities. A quality education is the university's lifeline. Teaching is a primary focus at NCUT and the center of its activities, while the university also emphasizes the importance of research and the contribution research can make both to the educational environment and to address local and national needs. The under graduate program offers a full range of undergraduate degrees; there are also numerous graduate degree programs, as well as continuing education programs. The academic education is complemented with moral, intellectual, physical and aesthetic training in order to build well-rounded students prepared to meet the challenges of modern society. More information about NCUT at www.ncut.ciss.org.cn.

**Kookmin University**

Kookmin University, the first private university after Korea's independence, was established by Mr. Shin, Ik Hee and other main members of the Shanghai interim government of Korea. Deeply rooted in the four founding principles, humanism, nationalism, culturalism, and industrialism that Mr. Kim, Sung Gon had emphasized, Kookmin University has grown into a fixture of the academia of Korea with 24,000 students enrolled. Now universities of a knowledge based society have entered into a competition that is more severe than ever before. KMU is trying to solidify its educational foundation to maintain its role as a leading player by employing operative goals. The three operative goals are attracting outside funding, establishing the standards of reasonable distribution of resources, and building an efficient structural system. More information about Kookmin University at https://english.kookmin.ac.kr/.

## Conference report

This year's event took place on 7 November 2019 in the Spa Hotel Vita in České Budějovice, Czech Republic. The thematic focus of this year was Potential of the Eurasian Economic Union. Report on the symposium was also handed over to the President of the Czech Republic Miloš Zeman. The whole conference started with opening speech of prof. Ing. Marek Vochozka, MBA, Ph.D., Rector of the Institute of Technology and Business in České Budějovice (VŠTE). The participants of the conference were academics, officials and entrepreneurs from China, Russia, Korea, Ukraine, Mongolia, and the Czech Republic. The conference program consisted of four plenary sections and it is listed in the appendix. The topic Potential of the Eurasian Economic Union has been repeatedly mentioned as one of the promising opportunities for similar strategic partnerships at the Czech-Chinese level.

## Under the auspices of IES2019

Within the framework of the project "Stabilization and development of SMEs in rural areas", reg. No. TL01000349, the TACR Éta programme, an expert conference was organized under the auspices of the Innovative Economic Symposium 2019. The expert conference was organized in the cooperation with the Association of SMEs, EC and LAG, and focused on defining the issue of SMEs in rural areas and the identification of the main factors that limit the development of rural areas. At the conference, partial results of the solutions for individual areas were presented, and the participants were acquainted with the solution procedures and expected outputs including the possibilities of the ir application in business practice. At the same time, the conference was enable exchange of experience with the representatives of SMEs of all sectors and sizes, as well as the foreign participants in the conference. The conference was held on

27 November 2019 at Institute of Technology and Business in České Budějovice, Czech Republic, Building D, room 615.

## Expert guarantor

*prof. Marek Vochozka*
Rector of the Institute of Technology and Business (CZE)

*prof. Taehee Lee*
Chair of International School, Kookmin University (ROK)

*prof. Zhao Jixin*
Dean of the School of Economics and Management, North China University of Technology (CHN)

## The Organizing Committee

*prof. Marek Vochozka*, Rector of ITB, School of Expertness and Valuation, The Institute of Technology and Business in Ceske Budejovice (CZE)

*prof. Zhao Jixin*, Dean of the School of Economics and Management, North China University of Technology (CHN)

*prof. Taehee Lee*, Chair of International School, Kookmin University (ROK)

*Jakub Horák, MSc.*, ITB in Ceske Budejovice (CZE)

*Veronika Machová, MSc., MBA*, ITB in Ceske Budejovice (CZE)

*Pavel Rousek, Ph.D.*, ITB in Ceske Budejovice (CZE)

*Simona Hašková, Ph.D., MBA* ITB in Ceske Budejovice (CZE)

*Jaromír Vrbka, PhD., MBA*, ITB in Ceske Budejovice (CZE)

*Zuzana Rowland, MSc., MBA*, ITB in Ceske Budejovice (CZE)

*Tomáš Krulický, MSc.*, ITB in Ceske Budejovice (CZE)

*Petr Šuleř, PhD.*, ITB in Ceske Budejovice (CZE)

## The Scientific Committee

# *Conference program*

## Innovative Economic Symposium 2019

Potential of the Eurasian Economic Union

Spa Hotel Vita, November 7, 2019

***8.30 – 9.30***
***Registration***

***9.30 – 10.00***
***Opening ceremony and Introduction***

**prof. Ing. Marek Vochozka, MBA, Ph.D.**
Rector
Institute of Technology and Business in České Budějovice

***10.00 – 11.10***
***Plenary session I.***
Host: Pavel Rousek, Petr Šuleř

**10.00 – 10.30 Petr Šuleř**
ČEZ, a.s.
*Nuclear Communication in the Global World*

**10.30 – 10.50 Pavel Rousek**
Institute of Technology and Business
*EU policy vs. government functions*

**10.50 – 11.10 Radim Dušek**
Institute of Technology and Business
*A decision-making method based on consumer spending data*

Organizer

Co-organizers

Sponsor

Jihočeský kraj

***11.10 – 12.40***
***Plenary session II.***
Host: Pavel Rousek

**Inter-institutional cooperation – Universities, Faculties, Research**

- ✓ Kookmin University
- ✓ Peter the Great St. Petersburg Polytechnic University
- ✓ Kremenchuk Mykhailo Ostrohradskyi National University
- ✓ Henan University
- ✓ Samara State University of Economics

***12.40 – 13.40***
***Lunch***

***13.40 – 15.25***
***Plenary session III. (with discussion)***
Host: Pavel Rousek

**13.40 – 14.00 Taehee Lee**
Kookmin University
*Tax base erosion, Google tax and platform market competition in Korea*

**14.00 – 14.20 Pengfei Sheng**
Henan University
*The nexus between household wastewater emissions and urbanization in China*

**14.20 – 14.40 Natalia Pletneva**
Peter the Great St. Petersburg Polytechnic University
*Problems of transformation in the tourism industry in the digital economy*

**14.40 – 15.00 Iryna Trunina**
Kremenchuk Mykhailo Ostrohradskyi National University
*Globalization impact on the world travel market development*

**15.00 – 15.20 Olga Kalchenko, Svetlana Evseeva, Oksana Evseeva**
Peter the Great St. Petersburg Polytechnic University
*Strategy of economic and social development of St. Petersburg until 2030 achievement*

***15.25 – 15.55***
***Coffee break***

Organizer VŠTE

Co-organizers

Sponsor Jihočeský kraj

***15.55 – 17.15***

***Plenary session IV. (with discussion)***
Host: Pavel Rousek

**15.55 – 16.10 Zanabazar Altanchimeg, Jigjiddorj Sarantuya (*VIDEO*)**
National University of Mongolia
*The factors effecting entrepreneurial intention of university students: case of Mongolia*

**16.10 – 16.30 Eva Kalinová**
Institute of Technology and Business
*Everything you know and need to know about neural networks*

**16.30 – 16.50 Maxim Kiselev**
Kremenchuk Mykhailo Ostrohradskyi National University
*Predicting and decision making in complex systems*

**16.50 – 17.10 Radka Vaníčková**
Institute of Technology and Business
*The human factors errors and identification of causes of successful / unsuccessful completion of business projects*

***17.15 – 17.30***

***Closing ceremony***
Pavel Rousek, others

# Innovative Economic Symposium 2019

## Potential of the Eurasian Economic Union

Under the auspices of Ing. Miloš Zeman, President of the Czech Republic

### Motto

"Coming together is a beginning; keeping together is progress; working together is success.,, (Henry Ford)

### Target

The IES 2019 conference brings together scientists and practitioners in economy, international relations, finance and marketing from Asia and Europe. The objective of the conference is to share knowledge and discuss the issues relating to economic potential, international business, Industry 4.0 and use of artificial intelligence in corporate practice and economy and corporate strategic decision-making in the global world.

### Thematic Areas

1. Use of artificial intelligence in corporate practice and economy
   - Prediction of future corporate development
   - Prediction of future development of macroeconomic parameters
   - Analysis of businesses in sectors of the national economy
   - Machine learning, large data
2. International business
   - Entrepreneurship, international business and economic growth
   - Strategy in international business
   - Intellectual capital, knowledge and innovations in international business development
   - Direct foreign investments
3. Business administration view of the topic + Industry 4.0
   - Digitizing, automation of the production
   - Sector and corporate adaptation to the conditions of Industry 4.0
   - Digital economy
   - Quantitative and qualitative methods of management decision-making
   - Economic, social and environmental dimensions of strategic management
   - Decision-making methods and optimization, econometric modelling of corporate strategies
   - Risk management and portfolio optimization
   - Financial sustainability of small and medium-sized enterprises
   - Controlling and reporting
   - Social media marketing
4. Macroeconomic view of the topic
   - Economic consequences related to the defined economic area
   - Influence of economic parameters on the consumer
   - Consumer behaviour, international marketing, marketing modelling
   - Currency policy, theory of currency area
   - Renewable sources of energy, environmental legislation

### Number of participants and papers

Participants from around the world (mainly from Czech Republic, Russia, Ukraine, Poland, China, South Korea). Approximately 40 scientific papers. Full Paper length shall not exceed 10 pages. Expected Proceeding range is 420 pages (preface, content, 40 papers with 10 pages). A maximum of two articles per author and per volume is accepted.

**http://ies.vstecb.cz/**

**7th November 2019**

**Institute of Technology and Business**
**České Budějovice, Czech Republic**

### Sponsors and Partners

**Jihočeský kraj**

### Expert Guarantors

prof. Marek Vochozka,
Rector of the Institute of Technology and Business (CZE)

prof. Taehee Lee,
Chair of International School, Kookmin University (ROK)

prof. Zhao Jixin
Dean of the School of Economics and Management, North China University of Technology (CHN)

### Scientific Committee

prof. Dohyeon Kim,
Kookmin University (ROK)

prof. Patrick Saßmannshausen,
Regensburg University of Applied Sciences (GER)

prof. Byounggu Choi,
Kookmin University (ROK)

Dr. sc. Mario Bogdanović,
Senior research associate, high school professor (CRO)

prof. Svetlana I. Ashmarina,
Samara State University of Economics (RUS)

Dr. Katarína Valašková,
University of Žilina (SVK)

Dr. Simona Hašková,
Institute of Technology and Business (CZE)

prof. Ding Hui,
North China University of Technology (CHN)

prof. Shen Zhili,
North China University of Technology (CHN)

Dr. Vojtěch Stehel,
Institute of Technology and Business (CZE)

prof. Mateusz Grzesiak,
WSB University (POL)

prof. Alla Kasych,
Kyiv National University of Technologies and Design (UKR)

prof. Gregorz Slusarz,
University of Rzeszow (POL)

prof. Katarzyna Bilinska,
University of Economics in Katowice (POL)

Dr. Katarína Janošková,
University of Žilina (SVK)

prof. Izabela Barbara Sztangret
University of Economics in Katowice (POL)

**Organizing Committee:** **prof. Marek Vochozka; prof. Zhao Jixin; prof. Taehee Lee, Jakub Horák, MSc**
Pavel Rousek, Ph.D.; Simona Hašková, Ph.D.; Veronika Machová, MSc., MBA; Jaromír Vrbka, MSc., MBA; Zuzana Rowland, MSc., MBA; Tomáš Krulický, MSc.; Petr Šuleř, MSc.

# Statement of Peer review

In submitting conference proceedings to *Web of Conferences*, the editors of the proceedings certify to the Publisher that

1. They adhere to its **Policy on Publishing Integrity** in order to safeguard good scientific practice in publishing.
2. All articles have been subjected to peer review administered by the proceedings editors.
3. Reviews have been conducted by expert referees, who have been requested to provide unbiased and constructive comments aimed, whenever possible, at improving the work.
4. Proceedings editors have taken all reasonable steps to ensure the quality of the materials they publish and their decision to accept or reject a paper for publication has been based only on the merits of the work and the relevance to the journal.

Title, date and place of the conference

Innovative Economic Symposium 2019 - Potential of the Eurasian Economic Union

7th November 2019

Spa Hotel Vita, Skuherskeho 4, 37001 Ceske Budejovice, Czech Republic

Proceedings editor(s):

Ing. Jakub Horák

Ing. Jaromír Vrbka, PhD., MBA

Ing. Zuzana Rowland, MBA

Date and editor's signature

# The Peculiarities of the Budgetary Policy of Georgia and the Directions of Improvement in Association with EU

*George* Abuselidze[1*], *Linda* Mamuladze[1]

[1]Batumi Shota Rustaveli State University, Ninoshvili street 35, 6000-6010 Batumi, Georgia

**Abstract.** The work deals with the role of budgetary policy on socio-economic development of the country, the practical significance of budgetary policy improvement, the peculiarities of program financing, the problems of rationalization of budget resources, the existing reality and the main directions of improvement the effectiveness. The aspects of research are compared to the advanced international practice. The research is focused on the problems and the anti-rational usage of budget funds in the budgetary process, the peculiarities and efficiency of forming the program budget, the problems in the formation of expected results and budget performance indicators of the programs/sub-programs.
**Keywords:** budget policy, budget planning, program budget, effectiveness, Georgia, EU

## 1 Introduction

Budget planning and project preparation in the program format based on the priorities, which are defined according to the budgetary policy are the most important stages of the budgetary process. Although the preparation of program budget started since 2012 it is still not approved by the Budget Law. Also, the information presented in the program budget needs improvement. Systemic gaps can be noted in the stages of budget planning and reporting, which makes it difficult to compare the planned and achieved results and evaluate the efficiency/rationality of the funds spent from the budget.

The aim of the work is to study and analyze the problems in the process of planning the program budget and implementation of the programs/sub-programs and to elaborate the recommendations.

The objectives of the study were to set the following tasks:

- To study the peculiarities of program financing and evaluate the efficiency of distribution the budget funds;
- To study the existing gaps in the formation the program budget, identify the ways of solving the problems and elaborate the necessary recommendations;
- To analyze the current state of execution the state budget and reveal the modern tendency;

* Corresponding author: abuseri@mail.ru / george.abuselidze@bsu.edu.ge

•To compare the aspects of the research to the advanced international practice.

## 2 Data and methods

This research is based on the methods: quantitative, qualitative, statistical, component and comparative analysis and practical approaches, which are in practice and generalized in the Georgian and foreign economic writings and specially touch on the matters of budgetary policy and its effectiveness. The present article is based on the official data of Ministry of Finance of Georgia [21-26], Government of Georgia [37], State Audit Office of Georgia [31-35], The Swedbank [36] and etc.

In the process of writing of the article the scholarly articles by author [2-4] regarding budgetary policies in the regions of Georgia, Staehr [29, 30], Kajaks [10] concerning the socio-economic situation and fiscal policy in the Baltic States during the economic crisis period and Chakvetadze and Khetaguri [7] regarding the result-oriented budgeting have been used. The study of the existing literature on the development of the territorial units as a result is considered that they address the peculiarities and problems associated with the use of the budget policy were mainly considered in the context of accelerating economic growth (Kline and Moretti [11]; Gottlieb and Glaeser [9]; Busso, Gregory and Kline [5]), as well as from the point of view of the development of interbudgetary relations, regional budget or budget policy development in foreign countries (Ehrlich and Seidel [8]; Rodriguez-Pose [28]).

## 3 Discussion

### 3.1 The Peculiarities of the Budgetary Policy in the Baltic States

After gaining the independence Baltic States focused on the market and reducing the role of the government in the economy. The budgetary policy was providing the implementation of private sector supporting programs, the state-limited interventions in various activities and social security issues. This distinguishes the Baltic States economy from other European countries, where the government plays an important role in managing and coordinating the ongoing processes in the economy.

It should be noted that the Baltic States are different from their equal Western European countries and EU15 member states with the number of expenses incurred by the government in relation to GDP (Figure 1).

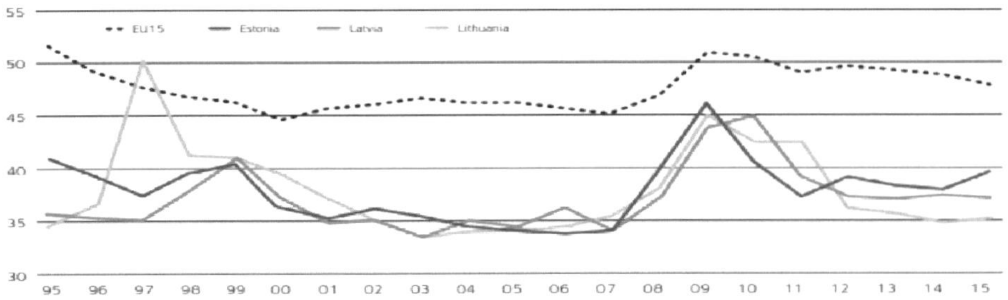


**Fig. 1.** Government's expenditure in GDP in 1995-2015

Source: Staehr [29].

https://doi.org/10.1051/shsconf/20207301001

According to Figure 1, a high level of expenses incurred in 1997 in Lithuania is related to compensation of the lost savings by the Government due to the extreme inflation since the beginning of 1990. Also, the government's expenditures in GDP have been significantly increased during the global financial crisis in all three Baltic countries due to the decrease of GDP and in the post-crisis period, as it is shown in the Figure 1, the share of expenditures in GDP is declining rapidly in parallel to GDP growth. Unlike Estonia and Latvia, the share of government's expenditure in GDP in Lithuania is lower (2015), which indicates Lithuania's high economic growth in this period in contrast to other Baltic countries.

The welfare policy of Baltic States is aimed at ensuring children's care, education, health care and the effective functioning of the pension system for people with disabilities and retirement age. A special unemployment insurance system is created in the countries. However, social policy is limited in terms of minimum protection.

Estonia was always adopting a prudent, reasonable fiscal policy while Lithuania and Latvia were facing a significant budget deficit (Figure 2).

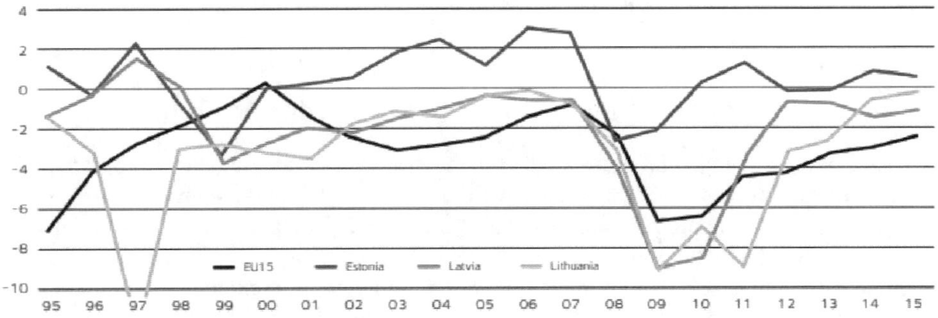


**Fig. 2.** Fiscal Balance 1995-2015

Source: Staehr [29].

According to Figure 2, for the Baltic States the period 2000-2010 can be divided into two periods: 2000-2007 years – economic boom period and 2008-2010 – economic crisis period. As we can see from the Figure 2, Estonia had a less deficient budget in 2008-2009, but the financial crisis period affected badly on the Lithuanian and Latvian budget, which was characterized by high levels of deficiency.

Despite the global financial crisis, it should be noted the strict policy taken by the Baltic States to get out of the crisis and economic consequences of this policy.

A large part of the budgetary policy was mainly aimed at reducing costs, which meant a significant reduction in employment and wages in public sectors, as well as reductions in social programs and investments [27] and certain structural reforms, such as the combination of schools and hospitals. For example, in Latvia the administrative office of the Ministry of Health decreased by 44%, many hospitals and 115 schools were closed. 29% of employees of public services have been released.

However, despite the crisis, all three Baltic states soon got out of the crisis and continued the economic growth process (Figure 3).

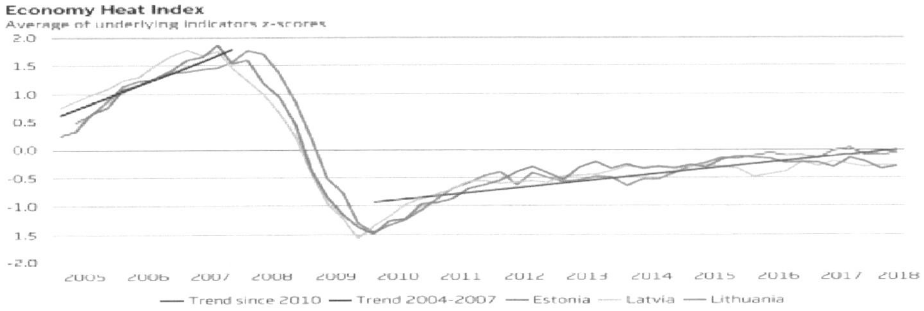


**Fig. 3.** Economy Heat Index 2005-2018

Source: Swedbank research [36].

From the experience of these countries, the main point of view is that macroeconomic stability can only be achieved by avoiding the excessive economic development and allocation of financial resources and adopting the strict fiscal policy during the economic boom.

## 3.2 The Peculiarities of Program Financing in the Budgetary Policy

The main purpose of the program budget is to implement programs/sub-programs and define their effectiveness by estimating the following correlation "resource → product → result". Although, it should be noted that increasing resources/expenditures does not always mean an effective budgetary policy due to the gaps during the budget execution process.
By 2014-2017 and 2018 (8 months) the average program implementation rate exceeds 100% every year (Figure 4).

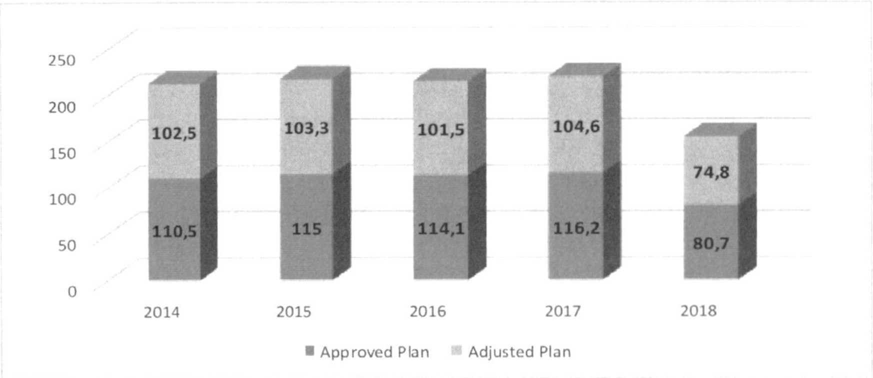


**Fig. 4.** The Average Program Implementation Rate, 2014-2018

Source: Prepared by author based on the data available in the State Audit Office report, 2014-2018 [31-35].

It must be noted that this data is partially contributed by the direct investment transfers and target grants that do not have a planned performance and are reflected directly in cash expenses.

Hence, we consider it to analyze the program implementation rate with the exception of the unplanned target grants and direct investment transfers, which ultimately changes the original picture (Figure 5).

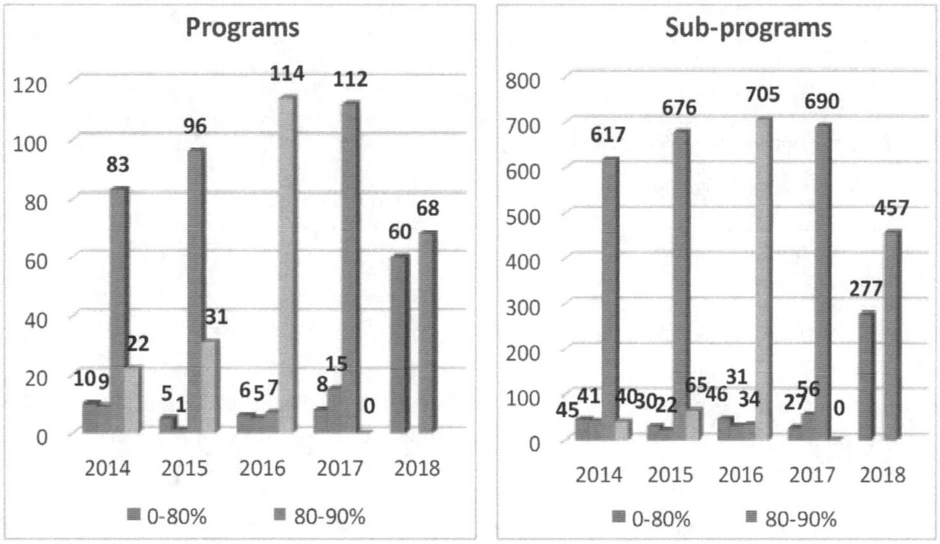


**Fig. 5.** Programs / Sub-Programs Implementation Rates Regarding to the Adjusted Plan, Excluding the Funds Spent on Target Grants and Direct Investment Transfers

Source: Prepared by author based on the data available in the State Audit Office report, 2014-2018 (2018 – 9 months) [31-35].

As it is shown in Figure 5, the high implementation rate of programs and sub-programs (>100%) is growing regarding to the adjusted plan in 2014-2016, in 2017 – it is between

90-100% and in 2018 – the majority of programs and sub-programs have low implementation rate, but the tendency of growth is in 90-100% interval rate.

However, in case of certain programs/sub-programs despite the high rate of implementation there are still some gaps and low rate of implementation.

In particular, some programs/sub-programs are characterized by the high rate of implementation regarding to the adjusted plan, whereas they have a low rate concerning to the budget plan. This is due to the distribution of funds allocated to other programs that indicates the gaps in the budget planning process. It should also be noted that despite some specification, some of the programs are characterized by low cash flow performance.

For example: on the program "Modernization of irrigation systems" was allocated 71,100 thousand GEL in 2014. However, during the year 21,000 GEL of unused funds were transferred to another program and finally allocation for this program amounted to 50,100 thousand GEL according to the adjusted plan. In this case actual performance (50,100 thousand GEL) is 100% according to the adjusted plan and 70.5% - according to the budget plan [18]. According to the Article 56 of the "Budget Code of Georgia" [6] the annual report should contain Clarification on the inconsistencies between the approved and adjusted budget allocations for spending institutions, if such inconsistencies exceed 30%. But in this case, the inconsistency is about 29.5%.

A similar tendency is revealed in the following years (Table 1). The table below shows the programs / sub-programs with low rate of performance.

**Table 1.** Low Rate Implementation Programs / Sub-Programs

| | Budget Law (1) | Adjusted Plan (2) | Cash Expenditure (3) | 3/1 % | 3/2 % |
|---|---|---|---|---|---|
| Provide educational institutions with information-communication technologies | 2,500 | 41 | 41 | 1.6% | 100% |
| Reconstruction and rehabilitation of Bagdati-Abastumani road | 15,000 | 1,680 | 1,679 | 11.2% | 99.9% |
| Development of Public Schools and Maintenance System | 2,000 | 257 | 245 | 12.3% | 95.5% |

Source: Prepared by author based on the data available in the State Audit Office report, (2017) [34].

In some cases, resources are not spent at all which is another problem, as the credit resources consider the payment of a commitment fee on the unused part of the loan and increase the non-purpose expenditures. In order to avoid a similar flow, it is necessary and appropriate to take effective measures by project implementers in order to facilitate the project implementation process.

Especially in 2018 within about 12 programs [19], no funds were spent during 8 months (Table 2). In addition, the total implementation rate of 8 months is less than 60% for 17 programs and 138 sub-programs.

**Table 2.** Zero Implementation Programs / Sub-Programs

| | Budget Law | 9 months adjusted plan | 8 months fact |
|---|---|---|---|
| Supporting agricultural cooperatives of beekeeping | 200 | 200 | 0 |
| Promoting pre-school education | 140 | | 0 |
| Forest Registration-Inventory Measures | 145 | | 0 |

Source: Prepared by author based on the data available in the State Audit Office report, (2018) [35].

From 2014-2018 programs "Adjara Solid Waste Project" is especially notable. The credit agreement was signed in 2010, but the implementation of the project and the use of funds has not been started yet (Table 3) and this project is still planned to implement according to 2019 budget.

**Table 3.** Adjara Solid Waste Project

| | Plan | Expenditure |
|---|---|---|
| 2014 | 2,548,400.0 | 0 |
| 2015 | 5,000,000.0 | 0 |
| 2016 | 7,600.0 | 0 |
| 2017 | 4,000,000.0 | 0 |
| 2018 | 4,000.0 | 0 |

Source: Prepared by the author, based on the reports of the State Audit Office and the reports of the state budget execution [31-35].

As a result, each year commitment fee (0.5%) is payed and by the end of 2017 the amount of 154.6 thousand euros was spent, while the credit resource has not yet been applied to the project.

### 3.3 Budget Spending Rationality Problems and the Main Directions of Increasing the Efficiency

Since the program budget shows the results achieved and not only the allocations used by spending institutions and their budgetary organizations, so during the evaluation process it is necessary to focus on the differences between planned and received results and not only the differences between approved and adjusted budget allocations and cash execution.

In this regard, if we consider the situation in the Autonomous Republic of Adjara in 2015-2018 we will see that the information presented in the Republican Budget Law [16-19] of the Autonomous Republic of Adjara and in the Annual Reporting [20, 22, 23, 25] on State Budget Execution is characterized by a number of gaps that makes it difficult to estimate the effectiveness of the program implementation.

The main focus is on identifying the flaws in the process of planning and implementation of programs by the three spending institutions of the region – The Ministry of Agriculture, Department of Tourism and Resorts and Department of Roads and Amelioration Systems Management.

The Figure below provides the information on the number of programs implemented by the Ministry of Agriculture of the Autonomous Republic of Adjara and by its organizations in 2015-2018 (Figure 6).

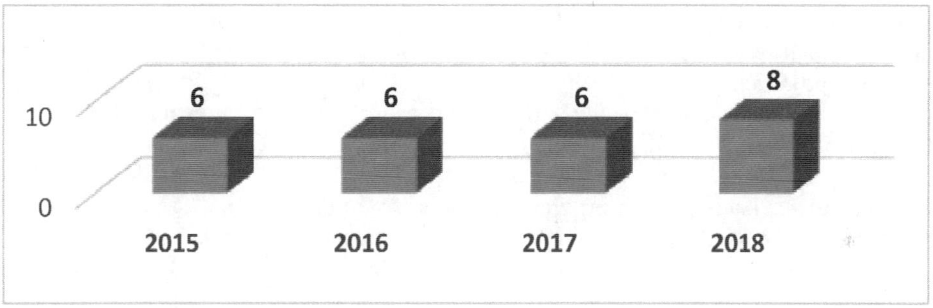


**Fig. 6.** Programs of the Ministry of Agriculture of the Autonomous Republic of Adjara 2015-2018

Source: Prepared by the author based on the program budget data, 2015-2018 [12-15].

As it is shown on the Figure 6, the Ministry of Agriculture of the Autonomous Republic of Adjara planned 6 projects in 2015-2016, which is confirmed by the program budget of 2015-2016 and the cash execution performed in the Budget Law (Table 4). However, it is noteworthy that the Annual Reporting on State Budget Execution does not include the information about all programs defined by the program budget.

**Table 4.** Cash Execution of the Programs in 2015-2016, Which are not Available in the Annual Reporting on State Budget Execution

| Name of the program | Fact of 2015 | Fact of 2016 |
|---|---|---|
| Development of agroservice in the region | 1,601.3 | 2,325.0 |
| Laboratory research services | 338.2 | 1,250.3 |

Source: Prepared by the author based on the Republican Budget Law of the Autonomous Republic of Adjara [16, 17].

It should be noted that in the Annual Reporting on State Budget Execution of 2017-2018 the information about the results achieved in the program "Development of agro service in the Region" is available, but in the case of "Laboratory Research Services" program, the information about the achieved results and the progress of the studies is not available.

The similar flaw on the one hand, contradicts the principle of Accountability and Transparency mentioned in Article 4 of the "Budget Code of Georgia" and also the Article 55 and 56 according to which the spending institutions should present the final report to the Ministry of Finance within the established timeframe. The final report must include the information about the programs/sub-programs implemented by spending institutions and the achieved results according to the Article 56. On the other hand, it prevents the effective management of budget resources.

Also, there are number of gaps in the formation of performance indicators for expected and final outcomes. In some cases, the indicators are irrelevant, incorrectly formulated or they are not a measure of effectiveness. Some programs and sub-programs require additional indicators such as quantitative, qualitative, efficient, and productivity indicators in order to ensure that the expected results are correctly evaluated.

One of the expected outcomes of the "Promotion of Livestock Development" is the production of high quality honey and the indicators are – "Transfer the honey extractor to beekeepers and agri-entrepreneur" and "Squeezed honey". The second expected result is the production of "Improved Milk and Dairy Products" and the performance indicators are - "Milk separator transferred to farmers and agro-entrepreneurs", "Butter making machine for farmers and agro- entrepreneurs", "Processed milk" and "Produced butter".

In both cases, the indicators are the program implementation plan and do not serve to evaluate its final result, except for the following indicators – "squeezed honey", "processed milk" and "produced butter", which represent quantitative performance indicators, but it should also be noted that only this information cannot be used to estimate the quality improvement rate and therefore the efficiency of the program.

The expected outcome and the performance indicator of the "Milk production-promotion of realization" sub-program is – "Milk produced by farmers and agro-entrepreneurs". This indicator repeats only the outcome and does not constitute a measure that would ensure the evaluation the efficiency of the results achieved within the sub-program.

The similar gaps are also in the formation of performance indicators in the Department of Tourism and Resorts of Autonomous Republic of Adjara.

In particular, in the sub-program "Marketing of the region as a tourist destination place on the international markets" measures to be taken to achieve the final result and the performance indicators are mixed. For example, the sub-program evaluation indicators are – "the amount of advertising/information materials" and "number of purchased promotional materials". These indicators do not allow to estimate the efficiency of the results achieved, as they are part of the action plan required to achieve the objectives set out within the sub-program, while "the number of unique online visitors in the target countries as a result of online advertising activities" is a quantitative evaluation indicator. Also the expected outcome of the program is formed incorrectly – "Advertising-marketing campaigns about the touristic potential of the region have been conducted in the target countries" - which is also the part of the action plan to achieve the objectives set out within the subprogram.

It must be noted that the program in 2015-2018 is characterized by the similar flaws, which shows the ignorance of the problem in previous years by the spending institutions.

The Department of Roads and Amelioration Systems Management of the Autonomous Republic of Adjara is implementing 2 main programs in 2015-2018 – "Roads and Amelioration Systems Management", which represents administrative costs and "Construction, rehabilitation and maintenance of road and amelioration infrastructure" which contains 5 sub-programs.

In the process of evaluation of the programs revealed that the Department of Roads and Amelioration Systems Management does not determine the usefulness of the domestic roads, whereas does not have the opportunity to prove whether the best option is chosen during the periodic maintenance and rehabilitation works. Therefore, in order to increase the efficiency of the results achieved within each program/sub program, we think it is desirable that the Department of Roads and Amelioration Systems Management should work out and elaborate the evaluation mechanisms for the usefulness of roads before starting the work.

"Unfortunately, in some regions (Kakheti, Imereti, Samegrelo-Zemo Svaneti, Shida Kartli, Guria, Mtskheta-Mtianeti) the facts of budget failure and/or excessive performance are very common, which clearly indicates that incorrect forecasts are used while forming the plans. Frequently, The Basic Data and Directions Document does not correspond to the actual potential of the country's economy and the performance indicators in the program budget are general and do not estimate the effectiveness of the results" [1].

The discussion of the result-oriented budget is also interesting based on the international experience. Let's discuss several countries.

**United Kingdom** – The Public Service Agreements (PSA System), used since 1998, is designed to provide efficient financing. The main purpose of PSA system is to ensure transparency and accountability. Accordingly, the spending institutions are obliged to publish an annual plan of duties on the unified web-portal and the measures to be taken to achieve results indicating the funds allocated to them. So, if since 1990 the spending institutions were obliged to submit these reports once a year, since 2002 they are obliged to submit reports twice a year.

**Australia** – As a result of 1999-2000 reforms, Australia has moved into accrual budgeting, which significantly increased the quality of reporting on efficiency. The existing system of budgeting is oriented towards evaluation of achieved results and its goal is to improve the efficiency of resource spending and public accountability. Every institution is obliged to provide accurate and detailed information on the expected results and the performance indicators within the programs and sub-programs that estimate the results achieved with quantitative, qualitative, economic and efficiency characteristics.

**Denmark** – Denmark has the Effectiveness Contracts Policy according to which spending institutions are obliged to publish an annual report where will be presented the comparison between actual and expected results.

## 4 Results

The following results have been obtained during the research process:

1. In case of certain programs/sub-programs, despite the high rate of implementation there are still some flaws and low performance tendency. Specifically, some programs/sub-programs are characterized by low performance regarding to the adjusted plan, while concerning to the budget plan and cash execution they have a low implementation rate.
2. In some cases, some programs/sub-programs that are not planned by the Budget Law are being implemented.
3. Some programs/sub-programs have low performance regarding to the adjusted and approved plan and in some cases, resources are not spent at all which is another problem, as the credit resources consider the payment of a commitment fee on the unused part of the loan and increase the non-purpose expenditures.
4. The Annual Reporting on State Budget Execution does not include the information about all programs defined by the program budget. Therefore, it is impossible to

evaluate how the goals set within the programs were achieved, which finally prevents to estimate the effectiveness of the program.

5. There are number of gaps in the formation of performance indicators for expected and final outcomes. In some cases, the indicators are irrelevant, incorrectly formulated or they are not a measure of effectiveness. Some programs and sub-programs require additional indicators such as quantitative, qualitative, efficient, and productivity indicators in order to ensure that the expected results are correctly evaluated.

## 5 Conclusions

Based on the study and analysis of budgetary policy of Georgia, we have developed the following conclusions and recommendations.

Depending on the above mentioned flaws, we consider that budgetary organizations should pay attention to the distribution of funds between programs and subprograms in the budget planning process during the year and take into consideration the experience of previous years in order to avoid frequent, massive distribution of funds and zero or low performance of the programs/sub-programs. At the same time, it is important to maximize the effective use of credit resources with minimum expenses and avoiding the improper expenditures on the loan, for which we believe that it is appropriate to identify the timely interruptions of the project by the relevant agencies and timely implementation of effective measures to eliminate them.

Also, we consider that The Annual Reporting on State Budget Execution should include the information about all the programs defined by the program budget to evaluate the results achieved of the program within the budget allocations.

In our opinion, it is appropriate to correctly determine the final and intermediate results in the process of planning/implementation of programs/sub-programs, through which the budget performance indicators will be formed. In addition, it is important to use such indicators as quantitative, qualitative, efficient, and productivity in order to estimate correctly the final/intermediate results achieved within the programs/sub-programs.

## References

1. G. Abuselidze, European Integration of Georgia and Financial-Economic Condition: Achievements and Challenges. *European Journal of Sustainable Development,* **8**(1), 53-68 (2019)
2. G. Abuselidze, Analysis of the formation and use of budgetary policies ensuring the socio-economic development of territorial units. *Economic Archive*, **2** (2019)
3. G. Abuselidze, Areas of Improvement of Finance Policy in the Process of Advancing of Upgrading the Territorial Economic Activity and Financial Capacity. *International Business Research,* **6**(5) (2013)
4. G. Abuselidze, Peculiarities of formation and functioning of Georgian budget system in transient period. *Author's abstract of dissertation* (2005)
5. M. Busso, J. Gregory, P. Kline, Assessing the incidence and efficiency of a prominent place-based policy. *American Economic Review*, **103**(2), 897-947 (2013)
6. Budget Code of Georgia. Part III, Chapter V, Article 34; Clause 1 [online], Available at: https://matsne.gov.ge/ka/document/view/91006?publication=39 (2009)
7. G. Chakvetadze, T. Khetaguri, Result-oriented budgeting [online], Available at: https://sao.ge/files/kanonmdebloba/pubnication/biujetireba-2016.pdf (2015)

8. M.V. Ehrlich, T. Seidel, The persistence effects of place-based policies. Evidence from the West-German Zonenrandgebiet. *American Economic Journal: Economic Policy, forthcoming* (2018)
9. E. Glaeser, J. Gottlieb, The economics of place-making policies. *Brookings Papers on Economic Activity* Spring, 155-253 (2008)
10. J. Kajaks, Economic and social situation in the Baltic States: Latvia. *European Economic* (2013)
11. P. Kline, E. Moretti, Local economic development, agglomeration economies and the big push: 100 years of evidence from the Tennessee Valley Authority. *Quarterly Journal of Economics,* **129**(1), 275-331 (2014)
12. Ministry of Finance and Economy of Autonomous Republic of Adjara. The Program Budget [online], Available on: https://matsne.gov.ge/document/view/2605063?publication=2 (2015).
13. Ministry of Finance and Economy of Autonomous Republic of Adjara. *The Program Budget* [online], Available at: https://matsne.gov.ge/document/view/3108436?publication=2 (2016)
14. Ministry of Finance and Economy of Autonomous Republic of Adjara. *The Program Budget* [online], Available at: https://matsne.gov.ge/document/view/3515842?publication=2 (2017)
15. Ministry of Finance and Economy of Autonomous Republic of Adjara. *The Program Budget* [online], Available at: https://matsne.gov.ge/document/view/3950679?publication=2 (2018)
16. Ministry of Finance and Economy of Autonomous Republic of Adjara. *The Republican Budget* [online], Available at: http://adjara.gov.ge/uploads/Docs/a3c4b20dbe5a4113982b4b3eb550.pdf (2015)
17. Ministry of Finance and Economy of Autonomous Republic of Adjara. *The Republican Budget* [online], Available at: http://adjara.gov.ge/uploads/Docs/bfedcf60982a4e6ab9dda85f739a.pdf (2016)
18. Ministry of Finance and Economy of Autonomous Republic of Adjara. *The Republican Budget* [online], Available at: http://adjara.gov.ge/uploads/Docs/cef43c2608eb4b2dbd749b810162.pdf (2017)
19. Ministry of Finance and Economy of Autonomous Republic of Adjara. *The Republican Budget* [online], Available at: http://adjara.gov.ge/uploads/Docs/0e4093bbbf75430fb867666f1690.pdf (2018)
20. Ministry of Finance and Economy of Autonomous Republic of Adjara. *Annual Execution Report of Republican Budget* [online], Available at: http://adjara.gov.ge/branches/listpage.aspx?pid=174&gid=5&ppid=162 (2015-2018)
21. Ministry of Finance of Georgia. *Approval of the methodology for drawing up the program budget*, Order N385. [online], Available at: https://matsne.gov.ge/ka/document/view/1400751?publication=0 (2011)
22. Ministry of Finance of Georgia. *Annual Report on the State Budget of Georgia* [online], Available at: http://www.parliament.ge/ge/ajax/downloadFile/32787/%E1%83%90%E1%83%9C%E1%83%92%E1%83%90%E1%83%A0%E1%83%98%E1%83%A8%E1%83%98 (2014)
23. Ministry of Finance of Georgia. *Annual Report on the State Budget of Georgia* [online], Available at: http://parliament.ge/ge/ajax/downloadFile/46937/07-

2.494_%E1%83%90%E1%83%9C%E1%83%92%E1%83%90%E1%83%A0%E1%83%98%E1%83%A8%E1%83%98_ (2015)

24. Ministry of Finance of Georgia. *Investment projects of the state budget of 2016 financed by donor organizations* [online], Available at: https://mof.ge/shesrulebis_angarishi_2016_12tve (2016)
25. Ministry of Finance of Georgia. *Annual Report on the State Budget of Georgia* [online], Available at: http://parliament.ge/ge/ajax/downloadFile/95078/%E1%83%90%E1%83%9C%E1%83%92%E1%83%90%E1%83%A0%E1%83%98%E1%83%A8%E1%83%98_2017_%E1%83%91%E1%83%98%E1%83%A3%E1%83%AF%E1%83%94%E1%83%A2%E1%83%98 (2017)
26. Ministry of Finance of Georgia. *Investment projects of the state budget of 2016 financed by donor organizations* [online], Available at: https://mof.ge/images/File/biujeti-2018-12tve/sainvesticio.pdf (2018)
27. Real Estate Investment Blog. *The economy in the Baltic states keeps growing and here's why* [online], Available at: https://blog.reinvest24.com/2019/the-economy-in-the-baltic-states-keeps-growing-and-heres-why/ (2019)
28. A. Rodríguez-Pose, The revenge of the places that don't matter (and what to do about it). *Cambridge Journal of Regions, Economy and Society,* **11**(1), 189-209 (2018)
29. K. Staehr, Public Finances and Fiscal Policy in the Baltic States 1991-2015. *Research in Economics and Business: Central and Eastern Europe*, **8**(1), 26-41 [online], Available at: http://rebcee.eu/index.php/REB/article/viewFile/73/59 (2016)
30. K. Staehr, Austerity in the Baltic states during the global financial crisis [online], Available at: https://archive.intereconomics.eu/year/2013/5/austerity-in-the-baltic-states-during-the-global-financial-crisis/ (2013)
31. State Audit Office Report on "2014 Georgia's State Budget Execution Annual Report." [online], Available at: https://sao.ge/files/auditi/2014-biujetis-cliuri-shesrulebis-shesaxeb-moxseneba-new.pdf (2014)
32. State Audit Office Report on "2015 Georgia's State Budget Execution Annual Report." [online], Available at: https://sao.ge/files/auditi/2015-clis-sax-biujet-shesrulebis-wliuri-angarishi.pdf (2015)
33. State Audit Office Report on "2016 Georgia's State Budget Execution Annual Report." [online], Available at: https://sao.ge/files/auditi/moxseneba-2016-biujetis-shesrulebis-cliuri-angarishis-shesaxeb.pdf (2016)
34. State Audit Office Report on "2017 Georgia's State Budget Execution Annual Report." [online], Available at: https://sao.ge/files/auditi/moxseneba-sakartvelos-2017-clis-saxelmcio-biujet.pdf (2017)
35. State Audit Office Report on "2018 Georgia's State Budget Execution Annual Report." [online], Available at: https://sao.ge/files/auditi/mokhseneba-2018-biujetis%20shesruleba.pdf (2018)
36. Swedbank. Macro Focus, *The temperature is slowly rising in the Baltic States* [online], Available at: https://www.swedbank-research.com/english/macro_focus/2018/18-10-01/macro_focus_economy_heat_index_final.pdf (2018)
37. The Government of Georgia. *Basic Data and Directions Document for 2018-2021* [online], Available at: https://mof.ge/images/File/BDD/2018-2021/saboloo/BDD-2018-2021-29.01.2018-saboloo.pdf (2018)

# Organizational Business Model of State Franchising

*Julia* Baldina[1], *Natalya* Masyuk[1*]

[1]Vladivostok State University of Economics and Service, Ulitsa Gogolya 41, 690014 Vladivostok, Russia

**Abstract**. The development of entrepreneurship in any country is inextricably linked with the participation of the state, without the intervention and support of which it is impossible to create a friendly institutional environment conducive to the development of entrepreneurial initiatives and entrepreneurial activity of the population. Small and medium entrepreneurship is the main driving force of the country's economic development throughout the world. The development strategy of the Russian Federation for 2018-2024 includes seven strategic priorities, some of which relate to the development of entrepreneurship and support for civic entrepreneurial initiative, the creation of a competitive economy in which government policy will be aimed at partnership with private business and reducing the direct participation of the state in the economy, and support of such socially important areas as health, education, social sphere, etc. In this connection, there is a need for s forms of the partnership of state and business structures in socially important sectors in order to attract small and medium-sized businesses. At the same time, a mechanism is needed to implement such a partnership, which should be based on a business model that will allow you to clearly understand the logical links and relationships. This paper describes the author's version of the configuration (template) of the business model of state franchising as a form of partnership between business and government. The basis is the canvas of the classic business models of Osterwalder-Pigneur and Ash Maurya. Special attention is paid to unfair (non-market) competitive advantage, which business receives. The described business model makes it possible to cover the partnership in general, concentrate on the main things, discover bottlenecks and interdependencies, contradictions and alternatives. Work within the framework of this business model gives a chance to significantly increase the efficiency of partners and to support socially significant areas of business.
**Key words:** entrepreneurship, partnership, business model, state franchising, socially significant industry, public-private partnership

* Corresponding author: masyukn@gmail.com

## 1 Introduction

The concept of a business model has become widespread. Currently, the term "business model" is used in the works of authors such as Guo, Zhao and Tang [1] to describe the process of creating value by an organization. Katkalo [2], Chesbrough and Rosenbloom [3], Stewart and Zhao [4], Drucker [5], use the same term to describe how to generate income. Amit and Zott [6], emphasize the importance of the concept of "value", which is central to the analysis of a business model. Mizik and Jacobson [7], say that value creation from activities occur when, after deducting all the opportunity costs of resources used, the utility for a particular member of society increases, and the value appropriation from activities occurs when a particular entity carrying out activity, able to assign a portion of the value created after taking into account all the costs associated with the resources used. The works of Santos [8] and Tucci et al. [10] reflect that value creation is measured at the level of society, the system, and the assignment of value at the level of a specific organization or another unit of analysis.

From the above reasoning, we can conclude that a business model is a new unit of business analysis; a business model can be viewed as a system that consists of components, links between them, and dynamic processes. Activity is the focus of consideration, and the main focus is value. Practically every company follows this or that option, chosen consciously or established spontaneously. The development of the economy and new technologies require original business models that would allow companies to be successful. The success or failure of a corporate strategy often depends on a business model. This predetermines the research interest in creating new business models and adapting existing ones to a dynamically changing market environment. The concept of "business models" in the Russian economy is a relatively new concept of modern business, and therefore contains quite a lot of unresolved issues and problem areas.

Since the 1990s, one can trace a certain trajectory of research on the concept of a business model – from the definition of the very concept of "business model" to the application of the concept of business model to study different types of organizations. So, in the works of Osterwalder and Pigneur [9]. Tucci et al. [10] and Sommerrock [11]. It was suggested, to use approaches to the definition of the stages of evolution of the concept of "business model", identify five stages of evolution of knowledge of the concept of a business model. At the first stage, researchers tried to find a definition of the concept of "business model", to develop a classification of business models, mainly based on the example of firms engaged in the electronic industry; at the second stage, the authors studying business models attempted to isolate the main components of the business model, the research data were mainly reduced to the enumeration of these components; at the third stage, the analysis of the components of the business model was carried out, the study of each component as a part of the construction of a common system; at the fourth stage, academic works were mainly devoted to the development of basic models and categories; and at the fifth stage, specific tools were developed for the construction and analysis of business models, the concept of business models was widely used in analyzing firms from various industries.

## 2 Data and methods

Further, we consider it expedient to clarify the conceptual apparatus. At the moment there is no single definition of the concept of "business model". The earliest works give the following definitions. Prahalad and Krishnan [12] defines the business model of an organization as the logic of combining resources and abilities in order to consistently achieve their fundamental goals and carry out business activities. Chesborough and

Rosenbloom [3] say that a business model is a way a company uses to create value and make a profit. Sliwotsky [13] gives a detailed definition, from which it follows that a business model is how a company chooses a consumer, formulates and delimits its proposals, allocates resources, determines which tasks it can fulfill on its own and for what will involve specialists from the outside, enters the market, creates value for the consumer and receives from this profit. Companies may offer products, services or technologies, but this proposal is based on an integrated system of actions and relationships, which is the company's business model [13]. Tucci et al. [10] interpret a business model as an idea of how an organization makes (or intends to make) money. The business model describes the value that an organization offers to different customers, reflects the organization's capabilities, the list of partners required to create, promote and deliver this value to customers, the capital ratios that are required to generate sustainable revenue streams [10].

When determining the essence of the concept of a business model, two approaches are distinguished. The first approach is focused on business processes/roles (approach directed inside the enterprise). As part of this approach, Kotelnikov [14] defines the business model as a mapping of a real (future) or planned business (enterprise) in a form that can clearly demonstrate all the significant properties and characteristics of an enterprise related to its ability to make a profit and be competitive. Shingarev [15] treats the business model as the presentation of a set of related model elements that define the internal and external environment in companies within a single system. Coolyatte [16] says that a business model is a look at an enterprise that focuses on the most relevant parameters for a business.

It should be noted here that the most interesting format of state franchising is for socially important industries in which the state is interested in reasonable prices and high quality.

The second approach implies a focus on value/customer (an approach aimed at the external environment of the enterprise). As part of this approach, Fedorenko [17] says that the business model logically describes how the organization creates, delivers to customers and acquires value - economic, social and other forms of value. A business model is a way in which a company creates value for customers and makes profits from it. Business models as an innovation in the activities of organizations are considered in the works of Gavrilova, Alsufiev and Janson, [18], Masyuk, Bushueva and Kotovshchikova, [19]; Petrichev, Masyuk and Bushueva [20], Strekalova [21] and others.

The methodology for constructing the author's business model of state franchising is based on using the methodological apparatus of business modeling and consists in modifying and integrating existing business models and detailing the resulting author's model in accordance with the content of government-business partnership in the context of state franchising.

## 3 Results and Discussion

The first result of our research is the update of the conceptual apparatus. We introduce the author's definition of state franchising. Under the state franchise, we offer to understand a special form of private-state entrepreneurship that is implemented between subjects of state and business structures with public and extra-budgetary financing in order to ensure the development of public goods markets, balanced interaction of state and non-state sectors of the economy, large business and small business based on products having special characteristics, exclusive right, company style and marketing support.

The next research results are the business model of state franchising, which helps to understand the logic underlying the activities of all stakeholders in partnership interaction in the format of state franchising.

The authors have developed an organizational business model of state franchising, which includes the architecture (template) and its meaningful content or detail. A distinctive feature of the proposed business model is its modified structure based on the integration of the Osterwalder business models – Pigneur and Ash Maurya.

First, we briefly discuss the description of the template of the classic business model of the Osterwalder-Pigneur business model, which is currently universal (Figure 1).

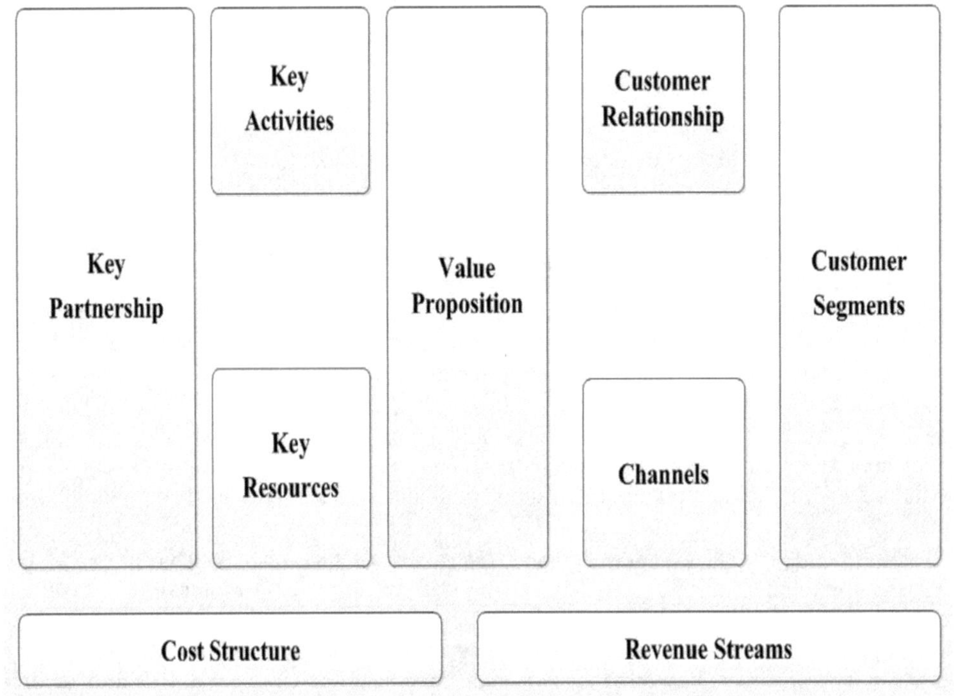


**Fig. 1.** The outline of the business model

Source: [9].

Its advantage is the original graphical solution (architecture), namely the matrix template, which the authors called canvas Business Model Canvas.

Canvas – a schematic display on one A4 page of the most important when detailing business processes, development schemes, plans, projects, startups. Canvas is a powerful tool for business modeling and innovation at the level of senior managers. Its potential lies in the fact that it can capture the full set of the most important elements of business modeling. Simply put, it expresses them with the help of visualization and simple schemes, allowing managers to create a business model on a single A4 sheet.

The business model of the Osterwalder-Pigneur consists of nine main blocks (components). Imagine them in the order adopted by the authors Tucci et al. [10]: Customer segments, Value proposition, Channels of promotion (Channels), Customer relationship technologies (Customer relationship), Revenue streams, Key resources, Key processes (Key activities), Key partners (Key partnership), Cost structure. These units are grouped into four basic areas: Consumers (Customers), Offer, Infrastructure and Profit Formula. It is the presence of a dedicated client direction consisting of three components - Target consumer groups, Promotional channels, and Client Relationship Technology - is, according to experts, the main advantage of this model in the context of the company's objectives for managing the client portfolio and client relationships.

Let us now consider the business model of Ash Maurya – the outline of the business model of a lean startup (Lin Canvas). Lean startup (lean startup) is a modern concept of entrepreneurship, first formulated by Rice [22] In various sources, the names of this concept are also found, such as “start-up on the principles of lean manufacturing”, “economical start-up”, “flexible start-up” and “ling-start-up”. Rice's theory has adopted the ideas of lean manufacturing concepts and the resulting concept of lean consumption for startups. In this case, we are talking about the organization of production, which will allow the company to produce effectively:

- what the consumer needs;
- when the consumer needs;
- as much as the consumer needs;
- at the price that suits the consumer;
- a quality that suits the consumer or exceeds his expectations.

From this, it follows that the focus of this concept is the consumer, his expectations and impressions of interaction with the company.

The outline of the business model of a “lean” startup Maurya [23] is presented in Figure 2.

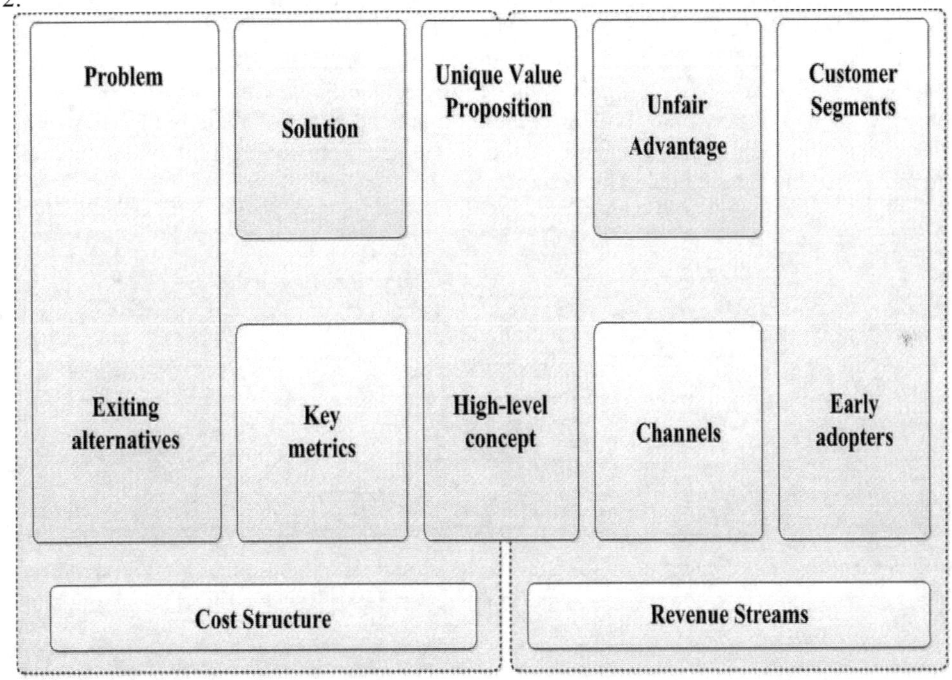


**Fig. 2.** The outline of the business model of a “lean” startup Esch Maurya

Source: [23].

The outline of the business model of a lean startup Maurya [23] includes nine components: Customer segments (Customer segments) and Promotion Channels (Channels), Problem, Solutions (means - consumer problems) (Solution), Unique value proposition, Unfair Advantage, Key metrics, Cost structure, Revenue streams. Ash Mauryi made changes to the traditional start-up business model. In the structure of the canvas, three elements are distinguished (subcomponent). The “Problems” component includes the subcomponents “Alternative proposals” (Exiting Alternatives), similar to the component of the same name, the canvas of the business model of Gladshtein [24] The element "Early Adopters" has been added to the structure of the "Target groups of consumers" component.

"Early users" means consumers who are starting to acquire a new product or use a new technology soon after the appearance of this product /technology on the market. The third «Motto» subcomponent (High-Level Concept or High-Concept Pitch) in the structure of the Unique Value Proposal component is a short phrase that should include a clear description of the company's value proposition, including using metaphors and comparisons with other famous products/services. It is believed that early followers and potential investors will accept the proposed description of the value proposition and will use it in internal and external communications, thus promoting the company's value proposition with word-of-mouth marketing methods. The main innovation of Ash Maurya was the replacement of the Key Processes component by the Key Metrics component. In the context of customer portfolio management and customer relations, an example of the most important metrics is considered to be the Customer Acquisition Cost (CAC), the customer Life Value (LTV). Startups use metrics such as vanilla metrics or vanity metrics. The outline of the business model of a lean- startup by Ash Maurya has gained wide popularity, resulting in numerous attempts to improve it.

Despite a detailed study of existing business models, the results of which are described in detail in the works of Masyuk and Baldina [25] and remained outside the scope of this article, analysis of business models showed that none of the existing business models are suitable for characterizing state franchising. because does not fully reflect the state-franchise interaction. Therefore, the authors chose to modify existing business models and integrate them. The most suitable for combining are the business models of Osterwalder (Pigneurur) and Ash Maurya. Figure 3 shows the architecture (pattern) of the author's organizational business model of state franchising.

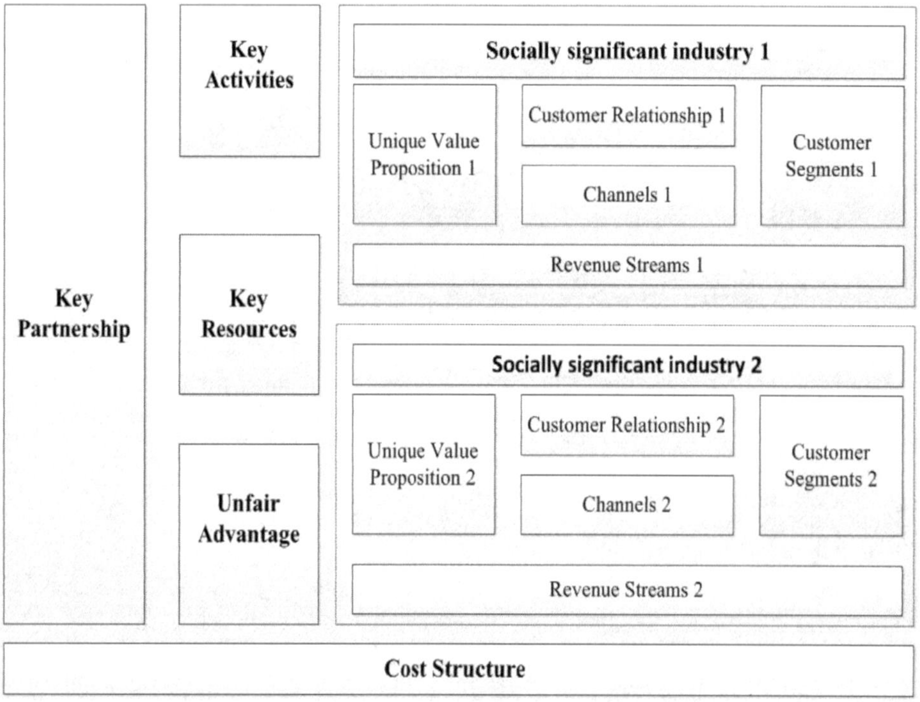


**Fig. 3.** Organizational business model of state franchising

Source: Authors.

In the modified template, the architecture of the classic business model of the Osterwalder-Pigneur is changed, namely such elements as "Key Partners", "Key Processes", "Key Resources", "Cost Structure" are common. Added the item "Unfair (non-market) competitive advantage" from the business model of Esch Maurya. This element is also common. The elements of "Socially significant industry 1", "Socially significant industry 2", for which their elements "Unique value proposition 1,2", "Technology of client relations 1,2", "Promotional channels 1,2", "Target consumer groups 1,2", "Revenue flows 1,2".

Table 1 shows the substantive content (detail) of the developed modified template of an organizational business model of state franchising.

**Table 1.** The content (detail) of the organizational business model of state franchising

<table>
<tr><td rowspan="8">Key Partnership<br><br>Franchiser (public (state) sector)<br><br>Franchisee (business sector)<br><br>Co-investor (business, funds, state)</td><td>Key Activities<br>Business processes of a core activity</td><td colspan="3">Socially significant industry 1</td></tr>
<tr><td rowspan="4">Key Resources<br>Federal, regional, local budgets. Private investment.</td><td rowspan="2">Unique Value Proposition 1<br>Franchise product 1/<br>Franchise service 1</td><td>Customer Relationship 1<br>CRM technology 1</td><td rowspan="2">Customer Segments 1<br>Socially significant industry 1<br>Socially unprotected groups of the population</td></tr>
<tr><td>Channels 1</td></tr>
<tr><td colspan="3">Revenue Streams 1</td></tr>
<tr><td colspan="3">Socially significant industry 2</td></tr>
<tr><td rowspan="3">Unfair Advantage<br>Exclusive right to produce a product/service</td><td rowspan="2">Unique Value Proposition 2<br>Franchise product 2/<br>Franchise service 2</td><td>Customer Relationship 2<br>CRM technology 2</td><td rowspan="2">Customer Segments 2<br>Socially significant industry 2<br>Post office sphere</td></tr>
<tr><td>Channels 2</td></tr>
<tr><td colspan="3">Revenue Streams 2</td></tr>
<tr><td colspan="5">Cost Structure</td></tr>
</table>

Source: Author's modified organizational business model of state franchising contains 17 blocks.

Block 1. Key Partnership. These are the subjects of state franchising: Franchiser (public (state) sector); Franchisee (business sector); Co-investor (business, funds, state). This partnership, mutually beneficial partnership, in which the Franchiser (public sector) gets the development of socially important industries and the ability to control, in contrast to outsourcing, the Franchisee (business sector) - the possibility of business development in the long term, Co-investor state) makes investments in socially important sectors and acts as a guarantor of transactions. At the same time, such cooperation reduces the risks associated with the conduct of competition policy and also makes it possible to attract business sector resources to the public sector for the development of socially important industries.

Block 2 "Key Activities". Implementation of state franchising in socially significant industries.

Block 3 Key Resources. Resources are necessary for successful operation. Resources can be intellectual (intellectual property, training, patents), labor (human) - highly qualified specialists, financial (federal, regional, local budgets, private investment from the business sector).

Block 4 "Unfair Advantage". In the organizational business model of state franchising, this component is expressed through the presence of an exclusive right to produce a product/service in the public sector (Franchiser), which it under certain conditions transfers to the business sector in order to attract resources, support entrepreneurship and develop socially important industries.

Blocks 5 and 11 "Socially significant industry 1, 2": 1 - Catering (catering baby food); 2 - Post office sphere.

Blocks 6 and 12 "Value Proposition 1, 2". The value proposition is a set of the most significant advantages for the consumer that the company offers. In this case, this component contains the Franchise product 1, 2 / Franchise service 1, 2.

Blocks 7 and 13 "Customer Relationship 1, 2". This component characterizes customer relationships, effective ways to attract, preserve and motivate customers to make a purchase. In this business model, technologies differ in customer segments, it is possible to use CRM technology, and this includes the relationship technology between the Franchiser and Franchisee in the interaction in the form of state franchising.

Blocks 8 and 14 "Channels 1, 2". Sales channels (promotion) solve the problem of informing customers about the goods and services of the company, give customers the opportunity to purchase specific goods and services, form an idea of the company's value offers from the consumer, and provide an opportunity to carry out service. In this business model, depending on the consumer segment, the industry and the Franchise product/Franchise service itself, sales channels, web sales, and customer service offices can be used.

Blocks 9, 15 "Customer Segments 1, 2". Consumers differ in their needs and preferences, communication with them is carried out through different distribution channels, different technologies of customer relations are needed, each segment has its own unique value proposition. In this business model, the target group of consumers depends on the socially significant industry in which the state catering (catering baby food), Post office sphere is implemented.

Blocks 10, 16 "Revenue Streams 1, 2". The formation of income/profit can occur in various ways. In this business model, this is the income of the franchisor in the form of royalties paid to the franchisee for using the exclusive right to the franchise product/service. And the Franchisee, in turn, receives income from the sale of the franchise product/service to the consumer.

Block 17 "Cost Structure". Any business model should minimize costs. The cost structure will depend on whether the focus is on spending or on value (utility). The proposed organizational business model of state franchising deals with activities in socially significant sectors, namely, in the field of baby food, which needs the low level of prices, high responsibility, and the high quality at the same time.

## 4 Conclusion

In conclusion, it can be noted that State franchising is a new form of the strategic partnership between the state and business for the Russian economy. A distinctive feature of the phenomenon of state franchising is its innovative nature, associated with the absence in the business circulation of this format of partnership interaction of government and business structures. State franchising involves a wide network of small and medium-sized businesses, as well as municipal, regional, federal authorities and other institutions. The organizational business model of state franchising can be considered as an alternative to the model of interaction between state authorities and organizations in terms of procurement in socially important sectors. The organizational business model of state franchising can be an alternative to outsourcing in the interaction of government and business structures in socially significant sectors. At the same time, the state will maintain control in this area, guarantee the quality, will attract resources, support and develop socially important industries, as well as the business sector. Successful interaction and involvement of the business sector in socially important industries is possible if there is a clear implementation mechanism based on the organizational business model of state franchising, the distinctive feature of which is a modified structure based on the Osterwalder-Pigneur and Asha Maurie models, which makes it possible to justify the occurrence of an unfair (non-market) competitive advantages in this form of partnership. From the point of view of institutional support in the Russian legislation, there are no legal acts in accordance with which it is possible to carry out this type of state-business activity, therefore there is a field for further research and legislative initiative.

## References

1. H. Guo, J. Zhao, J. Tang, The role of top managers' human and social capital in business model innovation. *Chinese Management Studies*, **7**(3), 447-469 (2013)
2. V. Katkalo, *The evolution of strategic management, Theory*. St. Peterburg: St: Peterburg University Publishing House (2010)
3. H. Chesbrough, R. Rosenbloom, The role of the business model in capturing value from innovation: evidence from Xerox Corporation's technology spin-off companies. *Industrial and Corporate Change*, **11**(3), 529-555 (2002)
4. D.W. Stewart, Q. Zhao, Internet Marketing, Business Models, and Public Policy. *Journal of Public Policy and Marketing*, 19-2, 287-296 (2000)
5. P. Drucker, The Theory of the Business. *Harvard Business Review* (1994)
6. R. Amit, C. Zott, Value Creation in Ebusiness. *Strategic Management Journal*, 6-722, 493-520 (2001)
7. N. Mizik, R. Jacobson, Trading off between value creation and value appropriation: the financial implications shifts in strategic emphasis. *Journal of Marketing*, **67**(1), 63-76 (2003)
8. F.M. Santos, A positive theory of social entrepreneurship. *Journal of Business Ethics,* **111**(3), 335-351 (2012)
9. A. Osterwalder, Y. Pigneur, *Business Model Generation: A Handbook for Visionaries, Game Changers, and Challengers*. John Wiley & Sons (2010)
10. C. Tucci, C. Chesbrough, F. Piller, J. West, When do firms undertake open, collaborative activities? Introduction to the special section on open innovation and open business models. *Industrial & Corporate Change*, **25**, 283-288 (2016)

11. K. Sommerrock, *Social entrepreneurship business models*. Houdsmill (etc.): Palgrave Macmillan (2010)
12. K.K. Prahalad, M.S. Krishnan, *Business Innovation Space: Creating Value Together with the Consumer.* Alpina Publishers: Yurait (2011)
13. A.J. Slywotsky, Миграция ценности. Что будет с вашим бизнесом послезавтра [Migration of value. What will happen to your budiness the day after tomorrow]. *Mann, Ivanov and Ferber* (2006)
14. V.Y. Kotelnikov, *New business models for a new era of rapid change driven by innovation*. M.: Eksmo (2007)
15. A.V. Shingarev, Building a business model of a development-oriented corporation. [online], Available at: http://www.aup.ru/books/m74/3.htm (2002)
16. A.Y. Soolyatte, A business model is the key to business development based on innovation. *Innovation Management*, **1**(9), 6-15 (2010)
17. A. Fedorenko, Enterprise Strategy vs Enterprise Business Model [online], Available at: http://socium.com.ua/2011/02/business-strategy-vsbusiness-model/#more-1190 (2011)
18. T. Gavrilova, A. Alsufyev, A. Yanson, Modern Notation of Business Models: Visual Trend. *Foresight-Russia*, **8**(2), 56-70 (2014)
19. N.N. Masyuk, M.A. Bushueva, E.A. Kotovshchikova, Business model of a company when looking for an investor and launching a startup. *Azimuth of Scientific Research: Economics and Administration*, **5**, 3-16, 122-125 (2016)
20. P.P. Petrichev, N.N. Masyuk, M.A. Bushueva, Typology of the University Business Model for Integration into the International Strategic Alliances. *Proceedings of the 32nd International Business Information Management Association Conference (IBIMA),* pp. 2839-2845 (2018)
21. N.D. Strekalova, Business model concept: system analysis methodology. *Proceedings of the A. Gertsen Russian State Pedagogical University*, pp. 92-96 (2009)
22. E. Rice, Business from scratch. Lean Startup method for quick testing of ideas and selection of a business model. *Alpina Publisher* (2013)
23. A. Maurya, Running Lean. How to Iterate from Plan A to a Plan That Works. *Published by O'Reilly Media,* 3-6 (2012)
24. B. Gladstein, Canvas DUOistic business model [online], Available at: http://bizmodelgu.ru/tag/brian-gladstein/ (2013)
25. N.N. Masyuk, Y.V. Baldina, Strategic partnership of state and business: global outsourcing and state franchising. *Economy & Entreprenership*, 12-3, 453-456 (2014)

# Problems of transformation in the tourism industry in the digital economy

*Svetlana* Bozhuk[1*], *Natalya* Pletneva[1], *Tatiana* Maslova[2] and *Konstantin* Evdokimov[3]

[1]Peter the Great St. Petersburg Polytechnic University, Polytechnicheskaya st., 29, 194064 St. Petersburg, Russia
[2]Federal State Budget Educational Institution of Higher Education "Saint-Petersburg State Economic University", Sadovaya st., 21, 191023 St. Petersburg, Russia
[3]Saint-Petersburg State Agrarian University, Peterburgskoe shosse, 2A, Pushkin, 196601 Saint-Petersburg, Russia

**Abstract.** The article is devoted to researching of current conditions of the tourism industry from consumers behavior position. In this work, the consumer is considered a legitimate participant in the creation of a tourist product. The aim of the work is to identify the impact of digital technology on the transformation of the decision-making process by tourists. The tasks are to assess the state of the digital tourism infrastructure, classify the available services to identify unfilled niches; to find out the attitude of consumers to various kinds of travel applications and their place in travel planning. The empirical basis of the work has been the research data by Google, PayPal, Booking.com which allowed to analyze tourists' behavior in the Internet and to identify the main travel trends for 2019. For achieve the research aim, authors applied an online survey of 100 consumers aged 18-35 too. As a result, the main directions of tourism transformation in the digital economy were determine and classification of applications in the category of "travel" were created.
**Key words:** Sustainable development, consumer behavior, tourism industry, digital economy, travel applications, tourism transformation

## 1 Introduction

The development of domestic and inbound tourism is one of the promising areas of import substitution. The federal target program defined development targets taking into account global trends and existing problems of tourism infrastructure.

The World Tourism and Travel Council (WTTC), in its annual report, provided a qualitative assessment of the impact of tourism on the global economy and employment in 185 countries and 25 regions. The study shows that this sector accounted for 10.4% of global GDP and 319 million jobs, or 10% of total employment in 2018. Most of the expenses of travelers accounted for the leisure market that is 78.5% of the total expenses compared with 21.5% of business expenses. The tourism sector accounted for 6.5% of total world exports and 27.2% of total world exports of services [1].

* Corresponding author: bojuk.svetlana@yandex.ru

According to various estimates in 2018 the volume of the Russian domestic tourism market in 2018 increased by 8-10% and reached about 900 billion rubles, more than 43 million Russians traveled outside their region of residence, that is 13 million more than in 2017. A similar trend is expected in 2019.

According to the Association of Tour Operators of Russia (ATOR), in 2018 there was an increase in the number of foreign tourists by 20% compared to 2017. The outbound tourism market significantly reduced its growth rate in 2018. The outbound tourism market finished the year of 2018 with a humble (compared to growth rates of 2017) growth in the number of tourists. At year-end 2018, the outbound tourist flow grew only by 6.1% (in 2017 – by 35.9%) [2].

The presented facts indicate the need for the development of domestic tourism infrastructure in Russia. According to the ATOR Analytical Service [3], the most promising areas are real-time dynamic packaging systems, new booking services for the b2b sphere, as well as online services for direct sales of travel packages and individual services to tourists on the Internet, including the possibility of self-designing a tour.

In our opinion ATOR's view is fully consistent with the digitalization of business and takes into account the changing distribution of tourist flows in favor of unorganized tourism.

According to the information portal Turazbuka [4] every third Russian organized leisure on his own (29%) in 2018. Young Russians aged 18 to 34 years old (35-39 %), plan trips on their own more often than others. According to various estimates, the share of unorganized tourists in the total inbound flow is 70-90%. The development of digital technology is significantly changing the process of planning and implementing of travel. The consumer becomes a full participant of the creation of a tourism product.

The purpose of the work is to identify the impact of digital technology on the transformation of the marketing environment of the tourism industry. In the environment of the tourism industry consumers and competitors (travel applications as the goods-substitutes) are present which the subject of study. The tasks are to assess the state of the digital tourism infrastructure, classify the available services to identify unfilled niches; to find out the attitude of consumers to various kinds of travel applications and their place in travel planning.

Considering the object of study, the authors took into account the classification of tourism depending on the method of organization into organized and unorganized ones. Organized tourism involves a pre-designed travel program and a travel organized by a company for tourist/tourists according to their wishes and budget. A travel company books and pays for all tourist services along the route, prepares all the necessary travel documents. Most of the work of researchers focuses on the features and development paths of organized tourism. However, the advent of digital services and travel applications significantly changes the ecosystem of the tourism business.

Unorganized tourism suggests that tourists organize their trip themselves without the participation and / or mediation of a travel agency. They independently develop a travel route, determine objects for excursion visits, plan overnight stays, do travel documents, etc. It is this category of tourists that remains beyond attention when analyzing the requirements for the digital transformation of tourist services.

There are few studies of unorganized (independent) tourism. Theorists and practitioners (association of travel agencies, individual tour operators) focus on the estimates of volume of unorganized tourism [5, 6]. The estimation of volumes is important for the tourism industry, as it characterizes the additional load on tourism facilities [7-9]. Strategies of interaction with clients under broad digitalization conditions are of interest [10, 11], especially with customers-representatives of generations X and Y [12]. The factor of digitalization has a close connection with the problem of sustainable development, as it provides optimization of resource consumption [13, 14]. S. Bozhuk and N. Pletneva observed current conditions of ecotourism in Russia, opportunities and threats for implementation of ecostrategies in tourism enterprises practice [15]. Under broad digitalization conditions specialists pay attention to the problems of service and staff training in the hospitality industry [16, 17]. However, many factors of consumer behavior remain outside the scope of research. It impedes the development of the industry.

## 2 Data and methods

The empirical sources of the study were reports and analytical articles characterizing the state of the tourism industry, as well as consumer reviews and the results of marketing research.

An important direction in the study of unorganized tourism is the development of digital technologies for collecting and processing data on the actions of Internet users [18]. While tour operators underestimate the segment of unorganized tourists, Internet companies are actively developing their service. The work analyzes data from the researches by Google, PayPal, Booking.com [19, 20].

In 2017 Google conducted a study of the needs and actions of Internet users in the field of vacation planning.

Booking.com analyzed over 163 million guest reviews and conducted a survey among 21,500 travelers from 29 countries to identify the main travel trends for 2019.

The payment service PayPal and the Russian research agency Data Insight investigate the habits of Russians with online payment, including travel services, on a sample of 4,000 people. The results indicate that more than half of tourists from the Russian Federation prefer to book and pay for trips via the Internet. In Moscow the share of online travel services purchases has grown to 61%.

Studying the characteristics of digital travel services, the authors applied an online survey of 100 consumers aged 18-35. Respondents were asked about the used travel applications, the desired benefits and the degree of customer satisfaction with an available service. The indicators of trust, willingness to pay for content and the search for new applications were considered as indicators of satisfaction.

The travel applications were evaluated as a weighted average. Every estimation is counting with the equation:

$$O_i = \frac{5*n_{5i}+4*n_{4i}+3*n_{3i}+2*n_{2i}+1*n_{1i}}{n_{5i}+ n_{4i}+n_{3i}+n_{2i}+n_{1i}}, \quad (1)$$

where Oi – the rating of the i-th travel application.

n5i – the number of respondents who gave the travel application 5 out of 5 points;
n4i – the number of respondents who gave the travel application 4 out of 5 points;
n3i – the number of respondents who gave the travel application 3 out of 5 points;
n2i – the number of respondents who gave the travel application 2 out of 5 points;
n1i – the number of respondents who gave the travel application 1 out of 5 points.

## 3 Results

The awareness of barriers to restructuring the decision-making process by tourists depends on the vector of consideration of the phenomenon of tourism. An analysis of various points of view made it possible to identify three vectors that differ in key elements and in the scope of the phenomenon. The general structure is presented in Table 1.

**Table 1.** The main directions of tourism transformation in the digital economy

| Vector of tourism consideration | The main function of digital technology | Digital technologies aimed at meeting needs / improving the quality of service |
|---|---|---|
| Short-term movement of people with different goals | It is associated with methods and tools for informing about tourist resources and infrastructure, the formation of the need for new experiences | Sites, blogs, groups in social networks, online advertising |
| Complex social environmental and economic system | Improving the comfort and attractiveness of recreational activities<br>Safety and environmental monitoring of the tourist center | Online business registration systems, online banking, digital systems for controlling payments and tax deductions, scanners for screening, video surveillance systems and patency accounting, baggage tracking, visa processing, including electronic ones |
| Economic sector | It is associated with the purchase of goods and services to meet the recreational and other needs of a consumer. | Online shops and mobile applications of travel business organizations for online purchase / booking of excursions and special equipment, accommodation, tickets for long-distance transport, booking a table, taxi, car rental, etc. |
| Product | Planning a unique experience, forecasting needs based on analysis of prior experience | A multifunctional approach to design of traveler's experience, VR and AR technologies, online translators, experience accumulation services |

Source: Own processing.

The consideration of tourism as a short-term movement of people outside their permanent place of residence for various purposes focuses on types of tourism. Depending on the purpose of the movement (health, cognitive, professional, business, sports, religious, educational ones and etc.), a tourist needs to get information about the relevant objects in the tourist center. The focus of attention with this approach is focused on communication in order to influence the choice of the place of visit. Digital technologies in this area have significant advantages. In this case marketing promotion tools are used for territories, cities, individual places and attractions, events and events, as well as other objects that meet the purpose of visiting (medical, educational, business ones and etc.). Currently, such technologies are actively used by both tour operators and consumers themselves. The main purpose is to increase awareness of the tourist and recreational value of objects, the formation of interest and desire to visit a promoted object. According to the NAFI Analytical Center [7], 97% of online shoppers choose a place to visit on the Internet. Travel goals of online shoppers are mainly related to recreational activities. For domestic tourism, a seaside vacation was chosen by 19% of online shoppers; 17% chose visiting attractions;

15% chose trips to nature, for sports. Business travel in Russia was carried out by 17% of online shoppers.

For outbound tourism to non-CIS countries, the leading purpose of travel is also sea vacations (18%); 13% of online shoppers chose sightseeing; 3% chose trips to nature, for sports, Business trips to non-CIS countries were chosen by 5%.

Sources of information about travel destinations are diverse. According to Google, 39% of Russian Internet users study online reviews, 21% study video reviews, 18% study YouTube, 15% use opinions on social networks, 11% use opinions on forums. Thus, we can conclude that the Internet generates travel expectations and creates motivation for visiting places.

The second direction of the study of tourism as a complex social ecological and economic system determines the focus on the stages and patterns of development of tourism infrastructure. The geographical component, tourist flows, and tourism industry enterprises are considered as the elements of the system. In its development the system goes through three stages (formation, maturity and degradation), which are characterized by different conditions and resources for the formation of a common and special infrastructure, the emergence and specialization of enterprises serving tourists. The digital economy accelerates the processes of formation and development of a business by digitalizing the registration of enterprises, the interaction of individual entities, conducting transactions and mutual settlements, etc. Digitalization provides the necessary level of control over the reliability of the functioning of natural and cultural complexes, technical systems, and statistically estimates the volume of specific tourist flows. At the time of preparing the publication, an experiment with a free electronic visa for citizens of 53 countries began in Russia. The experiment involved the Kaliningrad region, the regions of the Far East, St. Petersburg and the Leningrad region.

The third vector of research on tourism as a sector of the economy determines the use of digital technologies for the sale and purchase of services for moving, accommodation, meals, etc. Most organizations serving the tourism sector currently have sites with the function of booking and buying tickets and mobile applications. Digital technologies allow online travel agencies (online travel agencies) to interact with potential customers on the selection, payment and support of the tour completely online. The desire to expand sales through online consumers has led to the creation of online services owned by transport companies (for example, Aeroflot, Russian Railways), hotels and other tourist infrastructure organizations.

However, the tourism industry is saturated with various Internet and mobile applications that act as a metasearch engine or accumulate reviews and other useful information and are not directly related to the production of services. Having collected the results, the metasearch engine gives them in a convenient form for perception (taking into account the client's filters), which facilitate the search and planning of the tour. Having gained popularity, information portals often also begin to monetize their reputation by adding the functionality of a metasearch engine. Table 2 presents the classification and general characteristics of travel applications.

The classification of applications is given by the authors in terms of their specialization for the provision of tourist package services:

- a means of transportation to a place of rest (purchase of tickets for trunk transportation),
- accommodation (reservation of housing),
- a place of rest (selection of attractions and excursions, buying tickets to museums, theaters, stadiums, etc.),
- moving within the place of rest (maps, navigators, local public transport schemes, buying tickets for public transport, ordering a taxi, renting a car),

- catering (location of shops and catering points, booking a table in a restaurant).

Classification allows you to adequately assess the field of competition of travel applications and monitor changes in competitors' strategies. For example, Google consistently combines its travel services in maps and becomes a competitor to Booking. Competition is escalating, this means a dominant approach to tourism as a product. The data in Table 2 shows examples of an integrated approach to reviewing the product and meeting the needs of travelers. In this case, an adequate proposal to meet the need is the integration of all-in-one services. This trend is confirmed by a historical analysis of the functioning and diversification of applications and Internet services provided by the largest players in the online travel market. For example, Booking.com started as an online hotel reservation site, currently the service offers alternative accommodation reservations, travel guides and preferential conditions for buying tickets to museums, transfers from main lines to a place of accommodation and car rental services.

**Table 2.** Classification of applications in the category of "travel"

| Type of service | Examples | Rating | Main functions |
|---|---|---|---|
| Buying tickets to a vacation spot | Aviasales | 4.7 | the first meta-search engine in Runet with setting filters and sorting options, a calendar and a price map, saving favorites and search history |
| | OneTwoTrip | 4.7 | Meta search engine for tickets from 800 airlines, railway tickets, tickets for intercity buses, booking 2,000,000, car rental in 174 countries, filter settings |
| | Skyscanner | 4.6 | Search for cheap flights, hotels and car rental, sorting search results, price notification |
| | Kayak | 4.5 | search for airline tickets, hotels and cars, but also the ability to connect all this in a visual planner, forecasting ticket prices, travel planner, notification of flight status, time for inspection at the airport, terminal layouts |
| | Hopper | 4.5 | search for airline tickets, filters, price tracker |
| | Momondo | 4.3 | meta-search engine for flights, hotels and car rental, price calendar, filters, sorting options, saving favorites and search history |
| Buying a stay | Booking | 4.7 | online hotel reservations worldwide, online travel guides, transfer bookings and car rent |
| | Ostrovok | 4.5 | Russian metasearch engine for booking hotels online, with filters, offline mode, omnichannel mobile application, partner of Aeroflot and Pobeda airlines |
| | Airbnb | 4.5 | online site for renting and renting private housing |
| | Hotel Tonight | 4.4 | last minute reservations with big discounts for a maximum of a week in advance and up to 5 nights |
| | Hostelworld | 4.4 | hostel reservation service worldwide |
| Organization of leisure, choice of excursions and entertainment | izi.TRAVEL | 4.5 | an audio guide, allows you to download all the necessary routes and select the most interesting sights. There is paid content |
| | Tripadvisor | 4.4 | guide to attractions and entertainment, restaurants and hotels |
| | World Travel Guide by Triposo | 4.4 | guide in English about attractions, restaurants, hotels and various leisure options, personalizes the choice, there is paid content, advertising, online shopping |
| | Redigo | 4.2 | general information about the country (visa, time zone, currency, language, how to get there, etc.), offline maps with the sights of the city, photo gallery, phrasebook, schedule of current events |
| Moving within a vacation spot, | Offline maps Maps.me | 4.5 | shows attractions, shops, hotels and other objects; routes, current location, does not allow saving routes, there is |

| | | | |
|---|---|---|---|
| searching for places (restaurants, shops, etc.) | | | paid content |
| | Citymapper | 4.5 | builds multimodal routes taking into account the schedule of all types of transport (metro, buses, trams, trolleybuses, electric trains, bike rental, taxi / Uber) |
| | Google Maps | 4.3 | 220 countries and regions, as well as hundreds of millions of different places and companies, GPS navigation, traffic data, public transport routes, search for nearby restaurants, bars and attractions |
| | City Maps 2GO | 4.3 | search by address, location, menu in English and paid content, does not get directions |
| | Yandex maps | 4.2 | search for places, GPS navigation, traffic data, public transport routes, by car or on foot, search for nearby restaurants, bars and attractions |

Source: Own processing.

Functions of existing services, the emergence of new integrated services when the application contains training and entertainment elements. For example, using the augmented reality technology, the Momondo metasearch engine helps you determine the size of your hand luggage and its suitability for your carrier. Ozone Travel and AVIASALES practice humorous comments to smooth out delays in finding the right options and keep the consumer waiting for the result. The learning function has prospects for expansion, since the ability to share knowledge causes a positive attitude among independent travelers. Training of independent travelers also offer travel planning and gathering services, searching for opportunities to communicate with local residents, translators. Travel applications stimulate the creation of as many varied and enjoyable memories as possible and offer to share them with other travelers.

## 4 Discussion

Consumers of travel applications already have experience in interacting with different applications and their requirements become the basis for improving the functions of travel applications. Various ratings by consumer ratings signal the popularity of the application, but do not allow us to understand which properties are most valuable to consumers. At the same time, the saturation of the virtual market with travel applications forces us to differentiate our offer in order to avoid direct rivalry or surpass competitors.

An online survey conducted by the authors shows that consumers are actively using applications, while they are searching for new, more convenient or profitable options, based on the advice of the immediate environment and reviews on the Internet. The desired benefits when considering travel applications are free and useful content, an intuitive interface. The possibility of booking and paying online is considered by consumers as a standard of service. As a bonus, audio guides are considered; special places popular with locals; events and activities nearby.

More than a third of consumers read reviews of online applications. These data do not contradict the results of Google, where the Internet also acts as the main source of information. The popularity of other sources is estimated not so unambiguously. However, the results of the online survey provide grounds for hypotheses regarding the formation of loyalty of travel service users.

Adherents of one application use the service for 1.5-2 years. They do not want to change the service because they are used to it, its interface and method of use, they are not ready to spend time retraining in other services. They show interest in advertising places of visit in traditional media to the same extent as in online media.

Active travelers are constantly on the lookout for new travel services. They are interested in completeness of content, local life and authenticity of impressions. Their requirements are diverse, so the lack of filters for targeted search is a reason to change the application.

The research didn't reveal any significant dependencies. Researches of unorganized tourists, features of their choice of travel design should be studied further. There are no significant links between the fact of independent travel planning and socio-demographic characteristics of respondents. Socio-demographic characteristics of respondents don't have a significant impact on the assessment of travel applications. The authors are planning to continue the research to establish the initial momentum for actualizing the needs of independent travelers. There is a confidence that independent travelers are open for new impressions and choose at first a place of rest (the country, the city), and then how to carry out this rest (excursions, a beach, actions, etc.).

## 5 Conclusions

Digital technology has significantly changed the tourism infrastructure, which has made tourism a popular leisure activity. Travel applications for many consumers are an integral part of travel planning and implementation. The presence of many options on the market indicates the absence of obvious favorites among consumers.

The market of travel applications is saturated, there are no free niches, the competition for customer loyalty is high. The most promising approach to study the saturation of online applications is to consider tourism in terms of the proposed product to meet the needs. This approach is more in line with the trend of expanding the category of unorganized tourists who do an excellent job of organizing travel under the control of digital assistants. And if the question of the cost of the tour remains debatable, then the nature of the impressions clearly contributes to the further growth of this category of tourists.

Classification of applications in terms of their specialization for the provision of tourist package services allows not only to clarify the circle of competitors, but also to identify the features of travel applications that can be added to increase the value provided.

In the fight for the consumer, travel applications consolidate many functions in one package, provide reservations and pay online, and work on the convenience of the interface. However, for consumers this is a kind of standard, below which they do not consider the use of applications. The search for ways to develop digital solutions for unorganized travelers should be based on identifying the significance of new features and the time of use of the application.

## References

1. E. Zhilina, Tourism industry in 2018: key figures [online], Available at: https://rusunion.com/industriya-turizma-v-2018-godu-osnovnye-cifry/ (2018)
2. Rmaa Group, Blog about successful marketing strategies in Russia [online], Available at: https://russia-promo.com/blog/russia-in-the-world-travel-market (2019)
3. Tourism market in 2019: trends, risks, opportunities. *Herald of Ator* [online], Available at: https://www.atorus.ru/news/press-centre/new/45571.html (2019)
4. Tourism information portal Tur azbuka NEWS, Analysis of Russian domestic and foreign tourism in 2018 [online], Available at: https://ekec.ru/analiz-vnutrennego-i-vneshnego-turizma-rossiyan-v-2018-godu/ (2019)
5. I.P. Kulgachev, Global education monitoring report 2016. *Science and education in*

*global processes*, **1,** 82-87 (2016)

6. A.N. Buzni, A.A. Trenihina, On the institutional organization of active tourism. *Scientific Almanac*, **1**(15), 48-53 (2016)
7. NAFI Analytical Center, The market of tourist services. *Sociology. Statistics. Publications. Sector Report,* **1**(5), 79 (2017)
8. I. Krasyuk, Y. Medvedeva, V. Baharev, G. Chargaziya, Evolution of strategies of retail and technological systems under broad digitalization conditions. *International Scientific Conference "Digital Transformation on Manufacturing, Infrastructure and Service". IOP Conference Series: Materials Science and Engineering* (2019)
9. I. Aleksandrov, M. Fedorova, Strategic planning of the tourism development in small cities and rural territories as a tool for the development of the regional economy. *MATEC Web of Conferences*: *International Science Conference SPbWOSCE-2017 "Business Technologies for Sustainable Urban Development"*, **170** (2018)
10. Y. Klochkov, E. Klochkova, I. Krasyuk, S. Krymov, D. Gasyuk, N. Akobiya, Infocom Technologies and Optimization: Trends and Future Directions. *International Conference on Reliability,* pp. 133-142 (2018)
11. T. Lavrova, V. Plotnikov, The development of tourism: the experience of Russia. *MATEC Web of Conferences*: *International Science Conference SPbWOSCE-2017 "Business Technologies for Sustainable Urban Development"*, **170** (2018)
12. A. Thanamit, A. Kozlov, A. Teslya, Thai perceptions of Russia as a touristic destination between Thai tourists in generations X and Y. *31st IBIMA Conference,* pp. 2163-2183 (2018)
13. E.A. Konnikov, O.A. Konnikova, T.V. Bogacheva, O.U. Yuldasheva, Regional development sustainability to global economic growth, *Innovation Management and Education Excellence through Vision 2020*, pp. 5226-5232 (2018)
14. I.V. Rozdolskaya, M.E. Ledovskaya, V.V. Lysenko, K.V. Evdokimov. The Mediating Role of the Scientific Elite. *International Scientific Journal,* **4**, 48-55 (2017)
15. S. Bozhuk, N. Krasnostavskaia, T. Maslova, N. Pletneva, The problems of innovative merchandise trade in the context of digital environment. *IOP Conference Series: Materials Science and Engineering*, **497** (2019)
16. N.R. Saenko, V.V. Prokhorova, O.V. Ilyina, E.V. Ivanova, Service management in the tourism and hospitality industry. *International Journal of Applied Business and Economic Research,* **15**(11), 207-217 (2017)
17. E. Klochkova, K. Evdokimov, Y. Klochkov, V. Samorukov, Methodology for reducing risk of underperformance of personnel functions. *Engineering for Rural Development 17th International Scientific Conference*, pp. 1213-1222 (2018)
18. S.G. Bozhuk, A.S. Krasnov, Methodics of research of cunsumers psychographic characteristics in the internet. *Proceedings of the 2017 International Conference "Quality Management, Transport and Information Security, Information Technologies", IT and QM and IS,* pp. 166-172 (2017)
19. The market for online bookings: key trends in the industry [online], Available at: https://www.trn-news.ru/news/82827 (2019)
20. Official site of Data Insight [online], Available at: http://datainsight.ru/PaymentEvolution2018 (2018)

# Machine learning forecasting of CR and PRC balance of trade

*Tomáš* Brabenec[1], *Petr* Šuleř[2*]

[1]University of Economics, Faculty of Finance and Accounting, nám. W. Churchilla 1938/4, 13067 Prague, Czech Republic
[2]Institute of Technology and Business, School of Expertness and Valuation, Okružní 517/10, 37001 České Budějovice, Czech Republic

**Abstract.** International trade is an important factor of economic growth. While foreign trade has existed throughout the history, its political, economic and social importance has grown significantly in the last centuries. The objective of the contribution is to use machine learning forecasting for predicting the balance of trade of the Czech Republic (CR) and the People´s Republic of China (PRC) through analysing and machine learning forecasting of the CR import from the PRC and the CR export to the PRC. The data set includes monthly trade balance intervals from January 2000 to June 2019. The contribution investigates and subsequently smooths two time series: the CR import from the PRC; the CR export to the PRC. The balance of trade of both countries in the entire monitored period is negative from the perspective of the CR. A total of 10,000 neural networks are generated. 5 neural structures with the best characteristics are retained. Neural networks are able to capture both the trend of the entire time series and its seasonal fluctuations, but it is necessary to work with time series lag. The CR import from the PRC is growing and it is expected to grow in the future. The CR export to the PRC is growing and it is expected to grow in the future, but its increase in absolute values will be slower than the increase of the CR import from the PRC.
**Key words:** international trade, artificial neural networks, prediction, trade balance, seasonal fluctuations, turbulent relations

## 1 Introduction

According to Jánošová [1], international trade is an important factor of economic growth. The definition of international trade in the study of Fürst and Pleschová [2] is that the term "international trade" refers to exchange of goods, services and capital across international borders. The authors believe that carrying out this type of trade is much more complicated process than at the national level. Bernard [3] states that in most countries, international trade constitutes a significant share of gross domestic product. While foreign trade has existed throughout the history, its political, economic and social importance has grown significantly

* Corresponding author: petr.suler@cez.cz

in the last centuries. International trade has always been considered very important and has been given great attention to [4]. Drábek, Lipková a Gress [5] claim that balance of payment is considered a tool that provides one of the best analyses of eternal economic relations of a given country, thus serving as a monetary expression of economic transactions between the given country and its foreign partners over a specific period of time. Li et al. [6] state that international trade is a very important part of regional economic development. It can be said that the regions with faster developing international trade are also the regions showing the highest level of economic development. The reasons why given countries decide to open up to international trade vary depending on their national resources and conditions, as well as on the very different consumer tastes [7].

According to Sheikhan and Mohammadi [8], time series are characterized as a sequence of spatially and factually comparable observations arranged in time. De Baets and Harvey [9] define time series as arranged sequence of variables values at the same time intervals. León-Álvarez et al. [10] believe, time series analysis is a technique that involves the study of groups or individuals observed in consecutive moments. These are certain series of data points arranged in time order. Using time series analysis, it is possible to obtain very important statistics and other necessary data characteristics. Through time series forecast, it is possible to predict future values based on previously observed values [11]. An accurate and clear time series forecast is important for many areas, such as economics, transport, finance, energy, etc. [12]. Therefore, the most significant task of the time series analysis is forecasting. The main step in forecasting time series is to specify a concrete forecasting problem. According to Rostan and Rostan [13], when forecasting, it is necessary to determine the method, time horizon, and to have the most accurate data on the variables, data accessibility and characteristic. Pesaran and Smith [14] claim that the method of time series regression is a statistical method that enables to forecast future responses, and it is based on the response history and transmission of dynamics from relevant predictors. Using this method, it is possible to understand and forecast the behaviour of dynamic systems from experimental or observational data. Time series regression is often used for forecasting and modelling economic, financial, and biological systems [15].

Klieštik [16] states in his work that standard statistical methods are not suitable for forecasting the development of trade (or general predicting of macroeconomic indicators) due to the fact that most of them cannot capture both hidden and non-linear dependencies. On the contrary, neural networks (ANNs) could be used for this. According to [8], ANNs try to capture the behaviour of time series and best forecast individual data points. In order to forecast the outputs of systems as fast and as accurately as possible, it is possible to design models of time series based on the ANNs processes. Hu and Hwang [17] state that ANNs need to be trained properly to be able to work with time series correctly. According to Chen, Yang and Dong [18], the effectiveness of the designed time series and learning process appears to be a most suitable tool for forecasting a complex non-linear time series.

According to Sánchez and Melin [19], ANNs can be widely applied in various fields. Time series analysis is the field where ANNs can be applied a lot. As Rowland and Vrbka [20] say, the advantage of using time series compared to other methods consists in the precision of results and their ability to work with big data. Vrbka and Rowland [21] see another ANNs in a relatively easy application for solving complex problems and predictions. Santin [22] adds that in terms of application, ANNs are really flexible. However, Hossain et al. [23] see their disadvantage in the method of creating individual ANNs models or the requirement for big sample data. Tealab [24] states that ANNs models can be used for functions approximation with high precision, and contain a hidden layer of neurons, which uses non-linear activation for forecasting financial trends. Sloboda [25] claims that in using regression for forecasting, he considers time series, thus trying to forecast the future. According to Horák and Krulický [26] there can be a number of problems with time series data. When using regression models with time series data, it is necessary to distinguish between two types of forecast, that is, ex ante and ex post. Ex-ante forecasts are those that are only made using data available in advance, and ex-post forecasts are those using later predictor data [27]. Yu, Wang and Lai [28] in their study report that the difficulty of forecasting the volume of foreign trade is often due to the limitations of many conventional forecasting models. To improve forecasting performance, it is necessary to propose an approach that would hybridize artificial intelligence models and econometric models.

According to Cai, Chen and Fang [29], CNY exchange rates can be considered financial time series characterized by high non-linearity, uncertainty and behaviour changing over time. Liu et al. [30] made exchange rates forecasts using ANNs RBF (radial basis function). Their results have shown that the efficiency of ANNs RBF in forecasting CNY exchange rates if acceptable. Chang and Vochozka [31] deal with ANNs, decision trees, and hybrid model of ANNs. They compare the models of stock prices forecasting models created by means of these three methods. Their study also found that compared to other two methods, ANNs is a more stable forecasting method. Weijin and Yuhui [32] present a model used for forecasting import and export within a single sector. The authors have found that non-linear forecasts can deal not only with the combination of data and improving the accuracy but can also reflect non-linear characteristics of the forecasting system.

De Souza, Freitas and De Almeida [33] introduced a new predictor of neural time series weight, which uses virtual generalized ANN weighting with a random approach to forecasts of future incomes from shares. The predictor was evaluated based on predicting future weekly returns on 46 shares from Brazilian stock market. The results have shown that the predictors of weighted ANN with a random-access weight generate forecasts of future revenues with the same error level and characteristics of autoregressive ANNs basic predictors 5,000 times faster. Dongdong and Wenhong [34] note that financial time series are non-linear, non-stationary, and stochastic, which can be seen as difficult. In their work, the authors used a specific method based on wavelet analysis and artificial intelligence for forecasting the A300 index in China and NASDAQ index in the USA. Their model shows superiority in performance forecasting. The results show that these methods are suitable only for short-term forecasting. The prediction difference between A300 and NASDAQ indicates that the Chinese stock market is less efficient than the US market. The aim of Tkacz [35] is to improve the accuracy of financial forecasting of Canadian production growth using the most significant models of artificial neural networks (ANNs). In his study, the author has found that ANNs show statistically smaller forecasting errors in terms of year-on-year growth of real GDP compared to linear and univariate models. However, such forecasting improvements are less significant in forecasting quarterly growth of real GDP.

Copper prices in international trade markets are volatile. Accurate forecasting of copper price can result in commodity trading and fixed profits in the copper industry. For predicting copper prices, Wang et al. [36] developed a hybrid forecasting techniques combining complex networks and traditional ANNs techniques. This technique consists in transformation of the original price time series into price volatility network (PVN) and extracts the volatility characteristics from the PVN topological structure. For original data reconstruction, three ANNs techniques are widely used for future copper prices forecasting: BPNN, RBFNN and ELM. For investigating forecasting performance of proposed PVN-ANNs techniques, published data on copper spot prices from the New York Commodity Exchange (COMEX) are used. Empirical results show that the proposed hybrid PVN-ANNs techniques can achieve favourable forecasting effect compared to traditional ANNs techniques. This result clearly indicates the efficiency of the proposed hybrid forecasting techniques in detecting non-linear patterns of international copper prices. Vochozka and Horák [37] in their work aim to carry out a regression analysis of the copper price development on New York Stock Exchange using non-linear regression and ANNs. First, linear regression is carried out. Subsequently, ANNs are used for regression analysis. From linear regression, the curve obtained by means of spline function appears to be the most suitable one. All ANNs have proven applicable in practice.

Vijayalakshmi and Girish [38] provide an overview of literature dealing with short-term forecasting of spot electricity prices in deregulated competitive energy markets using ANNs. Accurate prices forecasts enable the participants in energy markets to maximize their profit and meet their commitments in the area of energy using the right combination of energy purchase agreements, bilateral trade, and purchase/sale of electricity through energy exchanges in a reasonable, efficient, and effective manner. ANNs models have proven to be the tool for short-term forecasts of spot electricity prices.

The foreign exchange market is one of the most invested markets in the world, with an average daily volume of trade of USD 1.8 trillion. Due to extreme volatility and uncertainty associated with foreign currency fluctuations, exchange rate forecasting is one of the most demanding and difficult tasks both for researchers and experts from practice. Traditionally, exchange rate is forecast for some technical indicators which just followed past price trends without considering a number of other factors. However, if the foreign exchange market is influenced by an accidental event, these indictors will provide misleading forecasts. This means that traditional forecasting techniques, such as linear trend analysis, would not work for forecasting future foreign currency fluctuations. To overcome this lack of traditional forecasting techniques, Emam and Min [39] propose ANN, which have proven useful for forecasting volatile financial time series, such as exchange rates. After applying ANN to actual data, the authors have found that the proposed ANN is very effective for forecasting daily fluctuations of exchange rates.

## 2 Data and methods

The current world trade is significantly affected by the trade policy of the United States of America (the USA), which gradually imposed sanctions on the Russian Federation, Turkey, and other states. They started trade war with the PRC, the European Union, and other countries, thus creating a new framework of the trade relationships between countries directly involved in the international politics game, but also the countries which are not directly affected, threatened or influenced by the US trade policy.

It can thus be stated that the new relations will definitely be hard to forecast (also due to the way the USA changes them. Stability on international markets is also derived from the prediction of the future development of the key macroeconomic indicators. The question therefore is, whether at present, to a certain extent turbulent time, it is possible to forecast the key macroeconomic indicators using standard statistical methods. As the research suggests, artificial neural networks are supposed to be successful in solving such tasks. Some authors already proved that the artificial multilayer perceptron neural networks are able to find the correlation even between very non-linear variables. At the same time, they are also able to find certain trend in year-on-year, month-on-month and week-on-week development of time series. They can therefore identify seasonal fluctuations quite accurately.

The aim of the contribution is to investigate and subsequently smooth two time series:

1. The CR import from the PRC.
2. The CR export to the PRC.

On the basis of the results obtained, it will be possible to predict the future development of the balance of trade of two trade partners.

The data set shows the course of the time series at monthly intervals starting from January 2000 to June 2019. The course of both time series is shown in Figure 1.

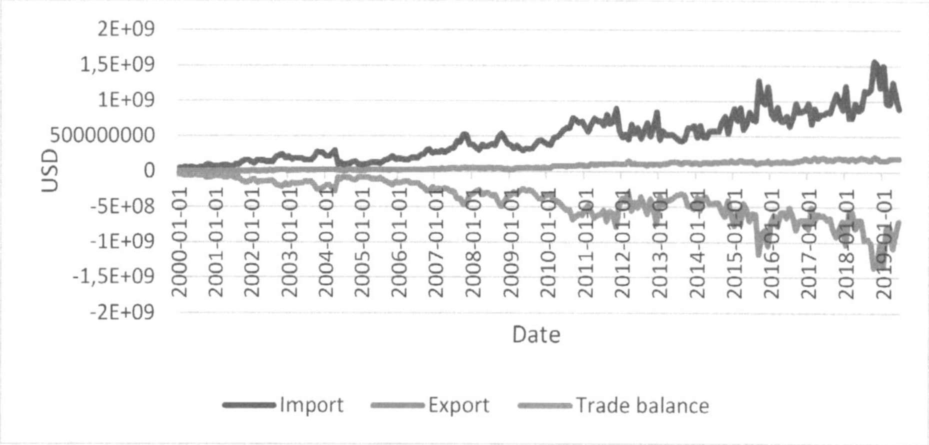


**Fig. 1.** Development of the USA and the PRC balance of trade between January 2000 and June 2019

Note: The values are given in American dollars (USD).

Source: [40], edited by authors

The figure does not show seasonal fluctuations in the time series development within the individual years of the monitored period. However, there is a clear increase in import to almost 20-time higher volume since 2000. There was also an increase in the CR export to the PRC during the monitored period (more than 50 times higher), but the difference between import and export is almost five times in favour of the CR import from China. Nevertheless, it can be assumed that the time series development is not random but subject to seasonal fluctuations, although they are not obvious at first sight. As evident from the figure, the balance of trade of both countries in the entire monitored period is negative from the perspective of the CR. The shape of the curve is very similar to the inverse curve of the import. The export development is less noticeable on the curve of the balance of trade. Table 1 shows the basic statistical characteristics of the data set.

**Table 1.** Basic statistical characteristics of monitored data set

| Statistics | Date (Input variable) | Month (Input variable) | Year (Input variable) | Import (Output (target)) | Export (Output (target)) |
|---|---|---|---|---|---|
| Minimum (Training) | 36556.00 | 1.00000 | 2000.000 | 6375.60 | 863.10 |
| Maximum (Training) | 43646.00 | 12.00000 | 2019.000 | 48127.80 | 13147.80 |
| Average (Training) | 40050.47 | 6.45455 | 2009.115 | 27572.76 | 6415.19 |
| Standard deviation (Training) | 2001.16 | 3.50688 | 5.473 | 12039.78 | 3382.42 |
| Minimum (Testing) | 36585.00 | 1.00000 | 2000.000 | 6584.40 | 972.70 |
| Maximum (Testing) | 43677.00 | 12.00000 | 2019.000 | 52202.30 | 12382.10 |
| Average (Testing) | 40130.09 | 6.00000 | 2009.371 | 26146.36 | 5918.11 |
| Standard deviation (Testing) | 2418.96 | 3.38683 | 6.691 | 12282.23 | 3492.76 |
| Minimum (Validation) | 36646.00 | 1.00000 | 2000.000 | 7070.50 | 1227.50 |
| Maximum (Validation) | 43373.00 | 12.00000 | 2018.000 | 50015.00 | 13644.80 |
| Average (Validation) | 40412.86 | 6.71429 | 2010.086 | 29966.20 | 7025.04 |
| Standard deviation (Validation) | 3403.28 | 3.44159 | 9.284 | 21320.19 | 5528.71 |
| Minimum (Overall) | 36556.00 | 1.00000 | 2000.000 | 6375.60 | 863.10 |
| Maximum (Overall) | 43677.00 | 12.00000 | 2019.000 | 52202.30 | 13644.80 |
| Average (Overall) | 40116.30 | 6.42553 | 2009.298 | 27716.79 | 6431.99 |
| Standard deviation (Overall) | 2069.19 | 3.45140 | 5.669 | 12057.45 | 3429.19 |

Source: Authors.

Regression will be carried out using multilayer perceptron networks. Smoothing of the time series will be performed separately for import and export time series. The balance of trade will be the difference between the two variables over time. The procedure chosen corresponds to the situation where the two time series seasonal fluctuations are different. Smoothing the time series separately and subsequent calculation of the forecast development of balance of trade using the most successful neural networks help achieve more accurate results.

Multilayer perceptron networks (MLP) will be generated. The time series lag will be 5 months. This means that the target (forecast) variable will be calculated using the data from the 5 previous months. This way a possible time series fluctuation caused by the latest data (in this case, this refers to the time series of the USA import from the PRC). A smaller lag can thus result in extreme fluctuations of smoothed time series, which would be a significant decrease in the case of import and increase in the case of export. Larger time series lag can mean averaging of the values. However, each time series lag brings higher demands on the complexity of the artificial neural structure, more specifically the neurons in the input layer (in the case of this calculation, it will be 15 neurons I the input layer of the artificial neural structure).

The independent continuous variable will time. Seasonal fluctuations will be represented by categorical variable in the form of the month and the year in which the value was measured. We will thus work with possible monthly seasonal fluctuations of the time series. However, it should also be possible to capture the overall trend of the time series. They will be entered as continuous variables in the "Integer" regime. The dependent variable will be the CR import from the PRC or the export from the CR to the PRC.

The time series will be divided into three sets – Training, Testing, and Validation. The first group will contain 70% of the input data. The training data set will be used for generating neural structures. The remaining two sets will contain 15% of the input data each. Both sets will be used for verifying the reliability of the neural structure or the model found. A total of 10,000 neural networks will be generated, out of which 5 with the best characteristics† will be retained. The hidden layer will contain at least 2 and at most 9 neurons. The following distribution functions in the hidden layer and the output layer will be considered: Linear, Logistic, Atanh, Exponential, Sinus.

Other settings will be default (according to the ANN tool – automated neural networks). Finally, the prediction of both time series until December 2020 will be carried out. At the same time, the balance of trade will be calculated through the export and import difference. However, this implies that from the retained networks, the neural network able to forecast the concrete time series development most accurately.

## 3 Results

***Import***

Table 2 shows the overview of the neural networks retained in smoothing the import time series at the time series lag of 5 months.

† Least squares method will be used. Generating of networks will be finished if there is no improvement, that is, if there is no reduction in the sum of the squares. We will thus retain the neural networks whose sum of the residuals squares to the actual CR import from the PRC or the CR export to the PRC is as low as possible (zero in ideal case).

**Table 2.** Retained neural networks for import time series smoothing

| Index | Network | Training perform. | Testing perform. | Validation perform. | Training error | Testing error | Validation error | Training algorithm | Error function | Activ. of hidden layer | Output activation function |
|---|---|---|---|---|---|---|---|---|---|---|---|
| 1 | MLP 15-5-1 | 0.944810 | 0.931109 | 0.940005 | 5.380788E+15 | 8.664971E+15 | 6.583494E+15 | BFGS (Quasi-Newton) 59 | Sum of squares | Tanh | Tanh |
| 2 | MLP 15-7-1 | 0.950087 | 0.941681 | 0.941613 | 4.880342E+15 | 7.504740E+15 | 6.307628E+15 | BFGS (Quasi-Newton) 59 | Sum of squares | Tanh | Identity |
| 3 | MLP 15-3-1 | 0.951657 | 0.951374 | 0.941369 | 4.734524E+15 | 6.118437E+15 | 6.399299E+15 | BFGS (Quasi-Newton) 86 | Sum of squares | Tanh | Exponential |
| 4 | MLP 15-4-1 | 0.948516 | 0.949282 | 0.940584 | 5.025787E+15 | 6.288411E+15 | 5.796231E+15 | BFGS (Quasi-Newton) 73 | Sum of squares | Logistic | Identity |
| 5 | MLP 15-10-1 | 0.970359 | 0.953558 | 0.942601 | 2.920810E+15 | 5.761694E+15 | 5.662860E+15 | BFGS (Quasi-Newton) 182 | Sum of squares | Tanh | Exponential |

Source: Authors.

Due to the time series lag, the input layer contains 15 neurons. The table shows that the neural networks with 4-10 neurons in the hidden layer were retained. For the activation of the hidden layer, the function of hyperbolic tangent and logistic function are used. For the activation of the output layer, exponential, identity, and hyperbolic tangent functions are used. The correlation coefficient of all neural networks data sets achieves very high values, always from more than 0.93 to more than 0.97, which indicates high degree of direct dependence. The retained neural networks thus shall be able to forecast the future development of the CR import from the PRC very accurately. The error was set by means of the sum of the least squares. The error is approximately the same as the error calculated in the first experiment. The error is rather large but acceptable with regard to the high values of the variable (in all data sets and networks).

Figure 2 shows the comparison of the actual course of the time series and the smoothed time series.

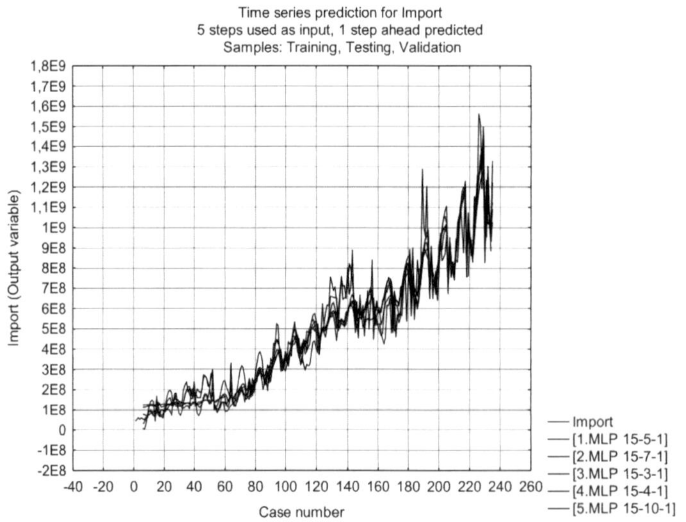


**Fig. 2.** Smoothed import time series

Source: Authors.

When comparing with Figure 2 (Experiment 1, it can be stated that it shows better results. Neural networks retained in Experiment 2 are more accurate in smoothing the time series. The only significant deviation is identified around the case No. 190, where there is a huge jump increase, then slight decrease, a further increase in the volume of import and subsequently a significant decrease in the CR import (this was between September 2015 and December 2015. The value is not stabilized until January 2016).

In order to determine whether the networks are applicable, we will deal with their application for forecasting. The development of the CR import from the PRC for the period of July 2019-December 2020 will be forecast. Concrete data is shown in Table 3.

**Table 3.** Development of predictions for the period August 2019-January 2020 by networks retained for smoothing import time series

| Date | MLP 15-5-1 | MLP 15-7-1 | MLP 15-3-1 | MLP 15-4-1 | MLP 15-10-1 |
|---|---|---|---|---|---|
| 31 July 2019 | 1.028362E+09 | 1.123861E+09 | 1.329162E+09 | 1.092625E+09 | 1.221804E+09 |
| 31 August 2019 | 1.118063E+09 | 1.202537E+09 | 1.326695E+09 | 1.227136E+09 | 1.340655E+09 |
| 30 September 2019 | 1.196789E+09 | 1.301656E+09 | 2.406627E+09 | 1.357579E+09 | 1.449726E+09 |
| 31 October 2019 | 1.261490E+09 | 1.415406E+09 | 1.466920E+09 | 1.463281E+09 | 1.621533E+09 |
| 30 November 2019 | 1.312878E+09 | 1.524919E+09 | 1.239474E+08 | 1.568247E+09 | 1.929219E+09 |
| 31December 2019 | 1.351882E+09 | 1.611339E+09 | 1.239474E+08 | 1.761729E+09 | 2.516401E+09 |
| 31 January 2020 | 1.380977E+09 | 1.659723E+09 | 1.239474E+08 | 2.089505E+09 | 3.692948E+09 |
| 29 February 2020 | 1.000428E+09 | 1.330634E+09 | 1.239474E+08 | 1.189211E+09 | 1.070819E+09 |
| 31 March 2020 | 1.332371E+09 | 1.577977E+09 | 1.239474E+08 | 1.731316E+09 | 1.178964E+09 |
| 30 April 2020 | 1.263561E+09 | 1.470854E+09 | 1.239474E+08 | 1.641207E+09 | 8.136747E+09 |
| 31 May 2020 | 1.300011E+09 | 1.500345E+09 | 1.239474E+08 | 1.752477E+09 | 1.246006E+09 |
| 30 June 2020 | 1.195967E+09 | 1.523270E+09 | 1.239474E+08 | 1.709404E+09 | 5.236437E+09 |
| 31 July 2020 | 1.289576E+09 | 1.626052E+09 | 1.239474E+08 | 2.023368E+09 | 8.511753E+09 |
| 31 August 2020 | 1.357345E+09 | 1.737773E+09 | 1.239474E+08 | 2.277100E+09 | 1.187036E+10 |
| 30 September 2020 | 1.403060E+09 | 1.861599E+09 | 1.239474E+08 | 2.459162E+09 | 1.497943E+10 |
| 31 October 2020 | 1.432786E+09 | 1.995227E+09 | 1.239474E+08 | 2.647926E+09 | 1.737552E+10 |
| 30 November 2020 | 1.451778E+09 | 2.127699E+09 | 1.239474E+08 | 2.882422E+09 | 1.899166E+10 |
| 31 December 2020 | 1.463856E+09 | 2.241309E+09 | 1.239474E+08 | 3.123546E+09 | 1.999224E+10 |

Source: Authors.

The results shown in the table represent a possible development of the time series. I tis evident that the network 3. MLP 15-31 will not be considered for the application, since it achieves very low values during the monitored period, and the network 5. MLP 15-10-1, which achieves extremely high values. The detailed course of predictions can be seen in Figure 3.

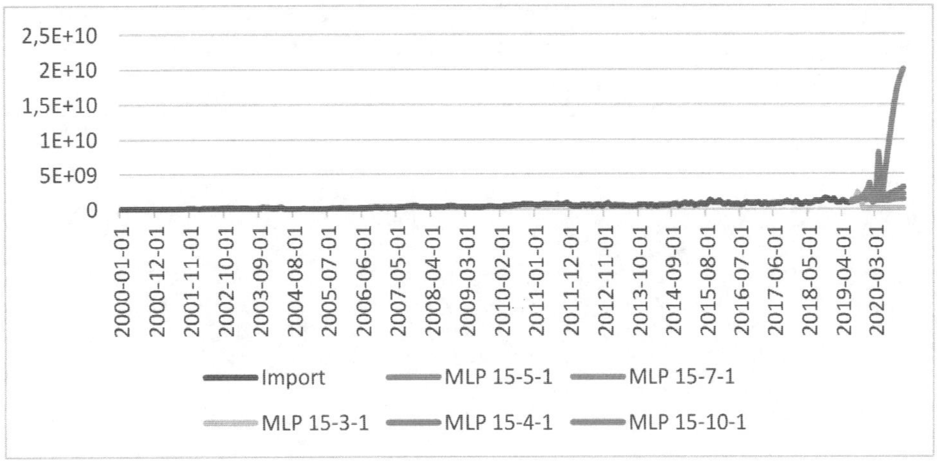


**Fig. 3.** Development of CR import from PRC and predictions until December 2020

Source: Authors.

Comparing the actual development of the CR import from the PRC and the forecast development of the variable over time forecast using artificial neural networks retained in Experiment 2, it can be concluded (namely with regard to the table of statistics in Annex 2) that the network 1. MLP 15-5-1 can be seen as the best artificial neural network with a potential of forecasting further development of the CR import from the PRC. The network is also able to capture the seasonal fluctuations and the overall trend of the time series.

***Export***

Also in the case of smoothing the export time series, 10,000 artificial neural networks were generated, out of which 5 with the best characteristics were retained (see Table 4).

**Table 4.** Retained neural networks for export time series

| Index | Network | Training perform. | Testing perform. | Validation perform. | Training error | Testing error | Validation error | Training algorithm | Error function | Activat. of hidden layer | Output activation function |
|---|---|---|---|---|---|---|---|---|---|---|---|
| 1 | MLP 15-4-1 | 0.978543 | 0.989979 | 0.983787 | 7.177847 E+13 | 3.906779 E+13 | 5.464654E +13 | BFGS (Quasi-Newton) 12 | Sum of squares | Tanh | Exponent. |
| 2 | MLP 15-10-1 | 0.977797 | 0.991030 | 0.983598 | 7.848493 E+13 | 4.622919 E+13 | 5.225459E +13 | BFGS (Quasi-Newton) 25 | Sum of squares | Sinus | Exponent. |
| 3 | MLP 15-6-1 | 0.976574 | 0.988310 | 0.983421 | 8.178466 E+13 | 5.142567 E+13 | 5.367858E +13 | BFGS (Quasi-Newton) 33 | Sum of squares | Sinus | Sinus |
| 4 | MLP 15-7-1 | 0.973193 | 0.986142 | 0.983858 | 8.913029 E+13 | 5.477165 E+13 | 5.132736E +13 | BFGS (Quasi-Newton) 49 | Sum of squares | Sinus | Sinus |
| 5 | MLP 15-8-1 | 0.978088 | 0.987068 | 0.983508 | 7.363016 E+13 | 5.048309 E+13 | 5.231443E +13 | BFGS (Quasi-Newton) 32 | Sum of squares | Sinus | Identity |

Source: Authors.

In the case of export, the input layer also contains 15 neurons. It results from the table that the neural networks with 4-10 neurons in the hidden layer were retained. For the activation of the hidden layer, the neural networks use the functions of hyperbolic tangent and sinus. The output layer of neurons is activated by means of the exponential, sinus, and identity functions. The performance of the artificial neural networks measured by correlation coefficient achieves the values of more than 0.97 in the training data set, more than 0.98 in the testing data set (above 0.99 in the case of the 2. MLP 15-10-1) and more than 0.98 in the validation data set. This indicates direct dependence in almost all cases. Based on this parameter, it can be stated that the networks (if not suffering from overfitting) are able to forecast the further development of the CR export in the PRC very accurately (since they were smoothing the time series in the past very accurately). For the calculation of the error, the least squares method was used. In the case of export, the error calculated achieves quite high values; however, since we work with the parameter given in USD and with regard to the individual values, the error is acceptable.

Figure 4 shows comparison of the actual course of the time series and smoothed time series.

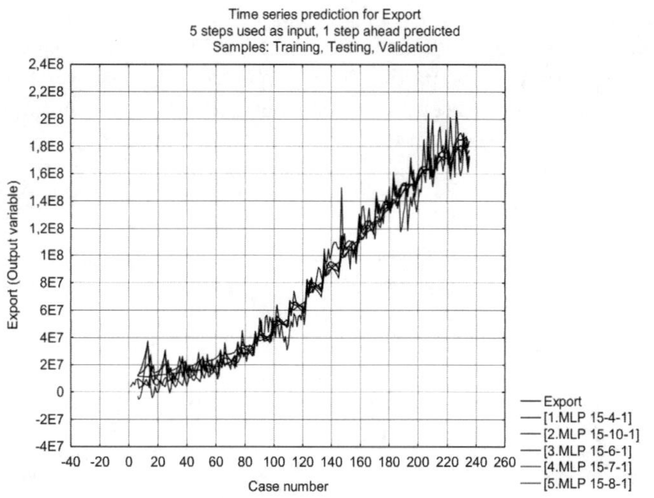


**Fig. 4.** Smoothed time series of CR export in PRC

Source: Authors.

It results from Figure 4 that all smoothed time series are able to follow the actual development of the CR export to the PRC relatively well. However, the actual time series is so disordered that the models are not able to capture all local extremes, especially in the last quarter of the monitored period, which shows extreme fluctuations. Nevertheless, the parameters of the neural structures acquired provide a precondition of quality monitored time series forecasting (although they are not able to fully capture the local extremes of the time series, they react to the fluctuations of the neural network). Also in the case of export, we will deal with forecasting the further development of the variable monitored, predicting the development of the CR export to the PRC for the period of July 2019-December 2020. Concrete data is shown in Table 5.

**Table 5.** Export forecast for July 2019-January 2020 according to retained artificial neural networks

| Date | MLP 15-4-1 | MLP 15-10-1 | MLP 15-6-1 | MLP 15-7-1 | MLP 15-8-1 |
|---|---|---|---|---|---|
| 31 July 2019 | 172085976 | 173383947 | 177267045 | 171188690 | 184032463 |
| 31 August 2019 | 178610635 | 177220223 | 179445382 | 178547110 | 188236320 |
| 30 September 2019 | 184708986 | 180211843 | 181876699 | 183572486 | 192644763 |
| 31 October 2019 | 189833524 | 182393335 | 184437226 | 186118002 | 197281886 |
| 30 November 2019 | 194080437 | 183239438 | 187140812 | 186030035 | 202122082 |
| 31 December 2019 | 197630682 | 182220837 | 189870505 | 182858384 | 207156320 |
| 31 January 2020 | 200919286 | 179080078 | 192458079 | 175808449 | 212219314 |
| 29 February 2020 | 201436611 | 172425100 | 192800101 | 186519165 | 210530052 |
| 31 March 2020 | 201813408 | 181984192 | 192941279 | 190441065 | 206889928 |
| 30 April 2020 | 190548337 | 182061308 | 193436480 | 184962130 | 209417703 |
| 31 May 2020 | 182585752 | 195542912 | 193308065 | 176301596 | 211552152 |
| 30June 2020 | 173288416 | 194206056 | 194780316 | 167197955 | 199586766 |
| 31 July 2020 | 178624641 | 195832010 | 196514700 | 177553626 | 205811991 |
| 31 August 2020 | 185180396 | 195075557 | 197926020 | 183973610 | 211240797 |
| 30 September 2020 | 192538221 | 191880039 | 199046611 | 186790961 | 215823010 |
| 31 October 2020 | 199952841 | 186343273 | 199917499 | 186322552 | 219518769 |
| 30 November 2020 | 206647795 | 178666616 | 200579953 | 182496166 | 222296204 |
| 31 December 2020 | 212125976 | 169174861 | 201064563 | 174926978 | 224137215 |

Source: Authors.

The data in the table indicate that none of the neural networks suffer from overfitting. Predictions made using the retained neural networks shall thus be applicable. To choose the most suitable structure for forecasting the CR export to the PRC, the graphical illustration of the course of predictions will be used (for more details, see the graph in Figure 5).

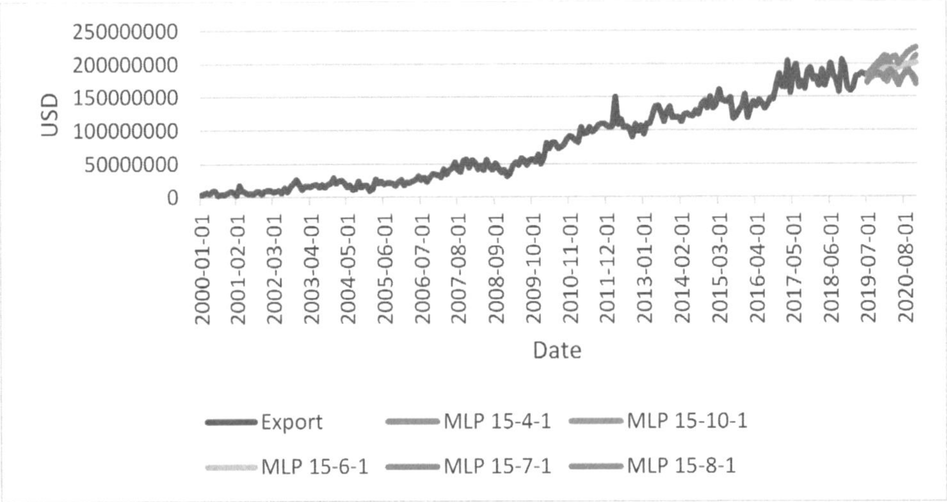


**Fig. 5.** Development of export time series and predictions using retained artificial neural networks

Source: Authors.

Comparing the actual development of the CR export to the PRC with the forecast development of the variable over time forecast for the period of July 2019-December 2020 using artificial neural networks retained in smoothing the export time series, it can be concluded (especially with regard to the table of statistics in Annex 2) that the network 1. MLP 15-4-1 shows the best predictions.

# 4 Results discussion

Within the research, smoothing of time series of the CR import from the PRC and the CR export to the PRC was carried out. Time series lag was set for 5 months in both cases. This eliminated potential extraordinary fluctuations caused by fluctuations of the time series in the individual months. First, it was observed how multilayer perceptron neural networks are able to smooth time series. For both time series, 10,000 artificial neural networks were generated, out of which five best structures were retained. In forecasting the future development of import, two neural structures had to be excluded since they apparently suffered from overfitting. In graphical comparison, another network suffering from overfitting had to be excluded. The two remaining networks were compared in terms of their basic statistical parameters and with regard to the current development of the time series. Subsequently, a network for calculating the balance of trade of both monitored countries was chosen. Also, artificial neural networks smoothing the time series of the CR export to the PRC were analysed. For this, all retained networks appeared to be applicable, but one of the captures the future development more accurately than others. It was thus used for calculating the CR and PRC balance of trade. It shall also be stated that the error identified is acceptable. The difference of predictions calculated using the most successful neural structures was used for calculating the balance of trade from the perspective of the Czech Republic (for more details, see Table 6).

**Table 6.** Comparison of most successful neural structures from all experiments

| Date | CR import from PRC | CR export to PRC | Balance of trade |
|---|---|---|---|
| | 1. MLP 15-5-1 | 1. MLP 15-4-1 | |
| 31 July 2019 | 1.028362E+09 | 172085976 | -8.562758E+08 |
| 31 August 2019 | 1.118063E+09 | 178610635 | -9.394527E+08 |
| 30 September 2019 | 1.196789E+09 | 184708986 | -1.012080E+09 |
| 31 October 2019 | 1.261490E+09 | 189833524 | -1.071657E+09 |
| 30 November 2019 | 1.312878E+09 | 194080437 | -1.118797E+09 |
| 31 December 2019 | 1.351882E+09 | 197630682 | -1.154251E+09 |
| 31 January 2020 | 1.380977E+09 | 200919286 | -1.180058E+09 |
| 29 February 2020 | 1.000428E+09 | 201436611 | -7.989909E+08 |
| 31 March 2020 | 1.332371E+09 | 201813408 | -1.130558E+09 |
| 30 April 2020 | 1.263561E+09 | 190548337 | -1.073013E+09 |
| 31 May 2020 | 1.300011E+09 | 182585752 | -1.117425E+09 |
| 30 June 2020 | 1.195967E+09 | 173288416 | -1.022678E+09 |
| 31 July 2020 | 1.289576E+09 | 178624641 | -1.110951E+09 |
| 31 August 2020 | 1.357345E+09 | 185180396 | -1.172164E+09 |
| 30 September 2020 | 1.403060E+09 | 192538221 | -1.210522E+09 |
| 31 October 2020 | 1.432786E+09 | 199952841 | -1.232833E+09 |
| 30 November 2020 | 1.451778E+09 | 206647795 | -1.245130E+09 |
| 31 December 2020 | 1.463856E+09 | 212125976 | -1.251730E+09 |

Source: Authors.

Table 6 demonstrates the potential development of the monitored variables and mainly the target variable – balance of trade. It assumes the growth in the CR import from the PRC, slight growth in the CR export to the PRC and the growth in negative balance of trade of the CR and the PRC. This situation is even better illustrated in Figure 6.

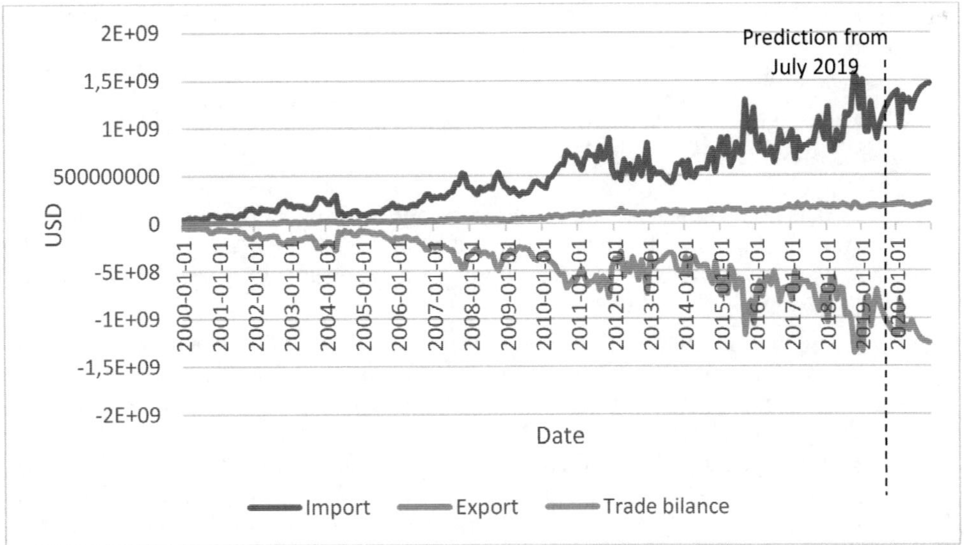


**Fig. 6.** Comparison of most successful neural structures of all experiments

Source: Authors.

The figure clearly shows that the actual course of all three time series and their predictions is very similar. It also better illustrates what is indicated in the table. It is very probable that the difference between the CR import from the PRC and the CR export to the PRC will increase. From the perspective of the Czech Republic, the balance of trade of both countries will be even more negative than now.

## 5 Conclusion

The objective of the contribution was to predict the development of the Czech Republic (the CR) and the People´s Republic of China (the PRC) balance of trade in analysing and machine learning forecasting of the CR import from the PRC and the CR export to the PRC.

Both monitored time series were smoothed. In both cases, the most successful neural network was chosen. From the difference of the CR export to the PRC and the CR import to the PRC, the balance of trade of both countries was calculated.

It can be stated that the objective of the contribution was achieved:

1. Multilayer perceptron networks are a suitable tool for forecasting the development of the time series with extraordinary fluctuations.
2. MLP networks are able to capture both the trend of the entire time series and its seasonal fluctuations.
3. It is necessary to work with time series lag.
4. Turbulent relations in the world trade currently do not significantly affect the result of machine learning forecasting.
5. The CR import from the PRC is growing and it is expected to grow in the future.
6. The CR export to the PRC is growing and it is expected to grow in the future, but its increase in absolute values will be slower than the increase of the CR import from the PRC.
7. By December 2020, the deficit in the balance of trade of both countries will grow if there are no interventions in terms of the state regulation of the mutual trade.

Further research shall be focused on an experiment that would help to identify what time series lag would be the most suitable for this type of tasks and variables. This can be different for each variable and also within the sub-intervals of the monitored time series. In any case, when identifying the most suitable time series lag, it should be possible to achieve better results in forecasting.

## References

1. D. Jánošová, Human Resources and Personnel Marketing Management in Business Practice from the Perspective of Globalization. *Globalization and its Socio-Economic Consequences, 16th International Scientific Conference Proceedings, PTS I-V.*, pp. 767-774 (2016)
2. R. Fürst, G. Pleschová. Czech and Slovak Relations with China: Contenders for China's Favour. *Europe-Asia Studies*, **62**(8), 1363-1381 (2010)
3. A.B. Bernard, Exporting and Productivity in the USA. *Oxford Review of Economic Policy*, **20**(3), 343-357 (2004)
4. M. Vochozka, J. Vrbka, Estimation of the development of the Euro to Chinese Yuan exchange rate using artificial neural networks. *SHS Web of Conferences: Innovative Economic Symposium 2018 - Milestones and Trends of World Economy (IES2018)*, **61** (2019)
5. J. Drábek, L. Lipková, M. Gress, Development of the balance of payments in the Czech Republic since 1993 with the emphasis on current account. *Economic Annals*-XXI, **7-8**(2), 4-7 (2015)
6. N. Li, L.P. Sun, X. Luo, R. Kang, M.D. Jia, Foreign Trade Structure, Opening Degree and Economic Growth in Western China. *Economies*, **7**(2) (2019)

7. P. Hnát, M. Tlapa, China-V4 Investment regime - a Czech perspective. *Trends and Perspectives in Development of China - V4 Trade and Investment*, pp. 82-95 (2014)
8. M. Sheikhan, N. Mohammadi, Time series prediction using PSO-optimized neural network and hybrid feature selection algorithm for IEEE load data. *Neural Computing and Applications* **23**(3-4), 1185-1194 (2013)
9. S. De Baets, N. Harvey, Forecasting from time series subject to sporadic perturbations: Effectiveness of different types of forecasting support. *International Journal of Forecasting*, **34**(2), 163-180 (2018)
10. A. León-Álvarez, J. Betancur-Gómez, F. Jaimes-Barragán, H. Grisales-Romero, Clinical and epidemiological round: Interrupted time series. *Iatreia*, **29**(3), 344-351 (2016)
11. L. Ashander, J. Klieštikova, P. Durana, J. Vrbka, The decision-making logic of big data algorithmic analytics. *Contemporary Readings in Law and Social Justice*, **11**(1), 57-62 (2019)
12. F. Rodrigues, I. Markou, F.C. Pereira, Combining time-series and textual data for taxi demand prediction in event areas: A deep learning approach. *Information Fusion*, **49**, 120-129 (2019)
13. P. Rostan, A. Rostan, The versatility of spectrum analysis for forecasting financial time series. *Journal of Forecasting*, **37**(3), 327-339 (2018)
14. M.H. Pesaran, R.P. Smith, Signs of impact effects in time series regression models. *Economics Letters*, **122**(2), 150-153 (2014)
15. K. Valášková, T. Klieštik, M. Kováčová, Management of financial risks in Slovak enterprises using regression analysis. *Oeconomia Copernicana*, **9**(1), 105-121 (2018)
16. T. Klieštik, Models of autoregression conditional heteroskedasticity garch and arch as a tool for modeling the volatility of financial time series. *Ekonomicko-manažerské spektrum*, **7**(1), 2-10 (2013)
17. Y.H. Hu, J.N. Hwang, *Handbook of neural network signal processing*. Boca Raton: CRC Press (2002)
18. Y. Chen, B. Yang, J. Dong, Time-series prediction using a local linear wavelet neural network. *Neurocomputing*, **69**(4-6), 449-465 (2006)
19. D. Sánchez, P. Melin, Modular Neural Networks for Time Series Prediction Using Type-1 Fuzzy Logic Integration. *Design of Intelligent Systems Based on Fuzzy Logic, Neural Networks and Nature-Inspired Optimization*, pp. 141-154 (2015)
20. Z. Rowland, J. Vrbka, Using artificial neural networks for prediction of key indicators of a company in global world. *16th International Scientific Conference on Globalization and its Socio-Economic Consequences*, pp. 1896-1903 (2016)
21. J. Vrbka, Z. Rowland, Stock price development forecasting using neural networks. *SHS Web of Conferences*: *Innovative Economic Symposium 2017 – Strategic Partnership in International Trade*, **39** (2017)
22. D. Santin, On the approximation of production functions: a comparison of artificial neural networks frontiers and efficiency techniques. *Applied Economics Letters*, **15**(8), 597-600 (2008)
23. M.S. Hossain, Z.C. Ong, Z. Ismail, S. Noroozi, S.Y. Khoo, Artificial neural networks for vibration based inverse parametric identifications: A review. *Applied Soft Computing*, **52**, 203-219 (2017)
24. A. Tealab, Time series forecasting using artificial neural networks methodologies: A systematic review. *Future Computing and Informatics Journal*, **3**(2), 334-340 (2018)
25. B. Sloboda, Book review: Forecasting, time series, and regression: An applied approach, 4th edition. *International Journal of Forecasting*, **21**(2), 391-392 (2005)

26. J. Horák, T. Krulický, Comparison of neural networks and regression time series in estimating US and China trade balance. *International Masaryk Conference for Ph.D. Students and Young Researchers*, pp. 244-253 (2018)
27. K. Valášková, T. Kliešik, L. Švábová, P. Adamko, Financial risk measurement and prediction modelling for sustainable development of business entities using regression analysis. *Sustainability*, **10**(7) (2018)
28. L. Yu, S. Wang, K. K. Lai, Forecasting China's Foreign Trade Volume with a Kernel-Based Hybrid Econometric-Ai Ensemble Learning Approach. *Journal of Systems Science and Complexity*, **21**(1), 1-19 (2008)
29. Z. Cai, L. Chen, Y. Fang, A New Forecasting Model for USD/CNY Exchange Rate. *Studies in Nonlinear Dynamics and Econometrics*, **16**(3) (2012)
30. Z. Liu, Z. Zheng, X. Liu, G. Wang, Modelling and Prediction of the CNY Exchange Rate Using RBF Neural Network. *2009 International Conference on Business Intelligence and Financial Engineering*, pp. 38-41 (2009)
31. T.S. Chang, M. Vochozka, A comparative study of artificial neural networks, and decision trees for digital game content stocks price prediction. *Expert Systems with Applications*, **38**(12), 14846-14851 (2019)
32. J. Weijin, X. Yuhui, A novel method for nonlinear time series forecasting of time-delay neural network. *Wuhan University Journal of Natural Sciences*, **11**(5), 1357-1361 (2006)
33. A.F. De Souza, F.D. Freitas, A.G.C. De Almeida, High performance prediction of stock returns with VG-RAM weightless neural networks. *Workshop on High Performance Computational Finance at SC10 (WHPCF),* pp. 1-8 (2010)
34. H. Dongdong, Z. Wenhong, The Forecast of Price Index Based on Wavelet Neural Network. *Fourth International Conference on Business Intelligence and Financial Engineering*, pp. 32-36 (2011)
35. G. Tkacz, Neural network forecasting of Canadian GDP growth. *International Journal of Forecasting*, **17**(1), 57-69 (2001)
36. C. Wang, X. Zhang, M. Wang, M.K. Lim, P. Ghadimi, Predictive analytics of the copper spot price by utilizing complex network and artificial neural network techniques. *Resources Policy*, **63** (2019)
37. M. Vochozka, J. Horák, Comparison of Neural Networks and Regression Time Series When Estimating the Copper Price Development. *Sustainable Growth and Development of Economic Systems*, pp. 169-181 (2019)
38. S. Vijayalakshmi, G.P. Girish, Artificial neural networks for spot electricity price forecasting: A review. *International Journal of Energy Economics and Policy*, **5**(4), 1092-1097 (2015)
39. A. Emam, H. Min, The artificial neural network for forecasting foreign exchange rates. *International Journal of Services and Operations Management*, **5**(6) (2009)
40. World Bank [online], Available at: https://data.worldbank.org/indicator/ (2019)

## *Annex 1*

**Basic statistics of smoothed import time series with 5-month lag**

| Statistics | 1.MLP 15-5-1 | 2.MLP 15-7-1 | 3.MLP 15-3-1 | 4.MLP 15-4-1 | 5.MLP 15-10-1 |
|---|---|---|---|---|---|
| Minimal prediction (Training) | 2.870538E+07 | 2.084195E+06 | 1.249895E+08 | 1.134611E+08 | 7.142744E+07 |
| Maximal prediction (Training) | 1.279052E+09 | 1.354799E+09 | 1.386515E+09 | 1.482664E+09 | 1.459338E+09 |
| Minimal prediction (Testing) | 9.253505E+07 | 9.400378E+07 | 1.278659E+08 | 1.229188E+08 | 1.065232E+08 |
| Maximal prediction (Testing) | 1.179155E+09 | 1.251712E+09 | 1.249326E+09 | 1.252984E+09 | 1.302836E+09 |
| Minimal prediction (Validation) | 1.019749E+08 | 6.793485E+07 | 1.272657E+08 | 1.183130E+08 | 9.380888E+07 |
| Maximal prediction (Validation) | 1.181279E+09 | 1.201416E+09 | 1.259741E+09 | 1.181576E+09 | 1.198775E+09 |
| Minimal residuals (Training) | -3.033300E+08 | -2.885160E+08 | -2.702194E+08 | -2.716622E+08 | -2.305341E+08 |
| Maximal residuals (Training) | 4.443217E+08 | 4.862127E+08 | 4.117377E+08 | 4.551818E+08 | 4.482359E+08 |
| Minimal residuals (Testing) | -2.821787E+08 | -3.106774E+08 | -2.805951E+08 | -2.616854E+08 | -2.627082E+08 |
| Maximal residua (Testing) | 3.845328E+08 | 3.119759E+08 | 3.143621E+08 | 3.107034E+08 | 3.160098E+08 |
| Minimal residuals (Validation) | -3.054264E+08 | -3.255626E+08 | -2.925264E+08 | -2.696388E+08 | -2.710663E+08 |
| Maximal residuals (Validation) | 2.292086E+08 | 2.769787E+08 | 2.457431E+08 | 3.414572E+08 | 3.138268E+08 |
| Minimal standard residua (Training) | -4.135161E+00 | -4.129949E+00 | -3.927158E+00 | -3.832015E+00 | -4.265633E+00 |
| Maximal standard residuals (Training) | 6.057237E+00 | 6.959870E+00 | 5.983875E+00 | 6.420707E+00 | 8.293826E+00 |
| Minimal standard residuals (Testing) | -3.031382E+00 | -3.586261E+00 | -3.587235E+00 | -3.299963E+00 | -3.460976E+00 |
| Maximal standard residuals (Testing) | 4.130949E+00 | 3.601250E+00 | 4.018926E+00 | 3.918101E+00 | 4.163183E+00 |
| Minimal standard residuals (Validation) | -3.764249E+00 | -4.099222E+00 | -3.656780E+00 | -3.541682E+00 | -3.602115E+00 |
| Maximal standard residuals (Validation) | 2.824898E+00 | 3.487492E+00 | 3.071957E+00 | 4.485010E+00 | 4.170346E+00 |

## *Annex 2*

**Basic statistics of smoothed export time series with 5-month lag**

| Statistics | 1.MLP 30-6-1 | 2.MLP 30-3-1 | 3.MLP 30-5-1 | 4.MLP 30-9-1 | 5.MLP 30-7-1 |
|---|---|---|---|---|---|
| Minimal prediction (Training) | 9.581047E+07 | 8.978954E+07 | 9.523406E+07 | 8.620605E+07 | 1.226069E+08 |
| Maximal prediction (Training) | 1.143141E+09 | 1.169006E+09 | 1.128150E+09 | 1.156245E+09 | 1.152624E+09 |
| Minimal prediction (Testing) | 1.275041E+08 | 1.225674E+08 | 9.857046E+07 | 1.188153E+08 | 1.261200E+08 |
| Maximal prediction (Testing) | 1.142398E+09 | 1.163636E+09 | 1.069460E+09 | 1.144003E+09 | 1.085287E+09 |
| Minimal prediction (Validation) | 1.274159E+08 | 1.139746E+08 | 9.620851E+07 | 1.102632E+08 | 1.244502E+08 |
| Maximal prediction (Validation) | 1.124260E+09 | 1.155792E+09 | 1.050281E+09 | 1.166400E+09 | 1.092989E+09 |
| Minimal residuals (Training) | -3.298498E+08 | -3.663393E+08 | -4.178500E+08 | -3.515643E+08 | -4.260971E+08 |
| Maximal residuals (Training) | 3.812593E+08 | 3.643545E+08 | 5.210952E+08 | 3.813481E+08 | 4.795802E+08 |
| Minimal residuals (Testing) | -2.326371E+08 | -2.291130E+08 | -2.144975E+08 | -2.476463E+08 | -2.300545E+08 |
| Maximal residua (Testing) | 4.212896E+08 | 4.000518E+08 | 4.942276E+08 | 4.196847E+08 | 4.784008E+08 |
| Minimal residuals (Validation) | -2.435260E+08 | -2.420521E+08 | -1.957826E+08 | -2.076164E+08 | -2.164618E+08 |
| Maximal residuals (Validation) | 1.888672E+08 | 1.813321E+08 | 3.057997E+08 | 1.845939E+08 | 2.816919E+08 |
| Minimal standard residua (Training) | -4.469026E+00 | -4.922313E+00 | -4.922472E+00 | -4.724518E+00 | -5.116134E+00 |
| Maximal standard residuals (Training) | 5.165556E+00 | 4.895644E+00 | 6.138750E+00 | 5.124770E+00 | 5.758303E+00 |
| Minimal standard residuals (Testing) | -2.791572E+00 | -2.759282E+00 | -2.331060E+00 | -2.998620E+00 | -2.583201E+00 |
| Maximal standard residuals (Testing) | 5.055342E+00 | 4.817954E+00 | 5.371037E+00 | 5.081741E+00 | 5.371794E+00 |
| Minimal standard residuals (Validation) | -3.274627E+00 | -3.307242E+00 | -2.799055E+00 | -3.043960E+00 | -3.104309E+00 |
| Maximal standard residuals (Validation) | 2.539645E+00 | 2.477603E+00 | 4.371944E+00 | 2.706417E+00 | 4.039783E+00 |

# Family and related link as a cradle of behavior of youth in political rent-seeking society

*Rais* Burganov[1*] *Emina* Altynbaeva[1], *Regina* Sinnurova[1]

[1]Kazan State Power Engineering University, Krasnoselskaya Str. 51, 420066 Kazan, Russia

**Abstract.** Among the pressing problems of any society, a special place is occupied by the desire of various categories of people to receive a political rent-seeking society. To start solving this sensitive problem, it is necessary to have baseline data, including on the views of young people on the possibility of obtaining political rent through the use of family-family ties in social development. This paper presents some of the research results on the basis of a survey of student youth. The results of the study are divided into three groups: students' perception of the modern political and economic atmosphere of society, assessment of opportunities for access to political rent through family and family ties, evaluation of measures for state regulation of the process of obtaining political rent by representatives of family and related groups of society. According to the author, it is necessary to consider in depth and wider regulations on the participation of family-related groups in the implementation of political rent.
**Key words:** family, young people, rent-seeking, society

## 1. Introduction

In modern conditions, among the difficult problems in the development of society and the economy, a special place is occupied by the desire of various participants in economic and political life to obtain various benefits. Ideally, such a desire characterizes the common human trait of their behavior. However, in reality, there are many contradictory and / or opposing interests of the state, social groups, and individuals about access to these benefits. In particular, if the benefits are acquired through the use of personal connections, patronage of individuals engaged in a higher position in society, then the tension in society can be cyclical, even explosive. The emergence of a multidimensional nature of relations in this area affects the interests of various social groups, among which the view of students can be especially distinguished. The age level of student youth is directly related to the stage of personality formation, which represents a specific family-related group of society. In fact, this person has information about the successive stages of the process of obtaining political rent by individual family-related groups in society. Also, the study of rent-oriented behavior of students young people is interested in the fact that they represent an active part

* Corresponding author: burraabr@gmail.com

of young people, are potential job fillers in government structures and institutions, and are part of the "Z" generation.

The theory of political rent itself occupies a special place in economic theory. Researchers are constantly working to reduce the level of tension in the distribution of political rents between participants in the social system. As you know, capital is accumulated and distributed through family channels. The concentration of family capital, economic mutual assistance and support of family group members is carried out mainly at local borders, within the framework of a family household. Concentrated family business groups in the world have a long historical tradition. Initially, financial groups were of a family-dynastic character: Morgan, Rockefellers. Dupons, Rothschilds, etc. Currently, such family groups are known LG Group (Family Co.), South Korea; BMW (Kundt Family), Germany; Motorola (Galvin Family), USA; Mars (Mars Family), USA.

In Russia, the process of concentration of family capital has gone through several stages. 1985-1989 - This is the initial stage of formalizing family capital. The transfer of a number of state means of production (buildings, structures, machinery, equipment, etc.) into the ownership of the cooperative contributed to the beginning of changes in the concentration of family capital. In the 1990s, after a 70-year hiatus, accelerated concentration and formalization of family capital begins. The history of the appearance of Russians in the lists of the richest people on the planet begins in 1997. Then there were 6 of our compatriots. Currently, the personal wealth of the Russian oligarchs is enormous.

The manifestation of political rent is a behavioral-economic problem. The political arena, access to the benefits of the budget is the market. The economic behavior of individual households in the market is similar to that of an entrepreneur. In conditions of limited resources, each member of the society faces the choice of one of the available alternatives. To describe human behavior in the political sphere, hypotheses of pursuing personal interest, completeness of preferences, rational maximization of the objective function are used. The process of identifying household preferences is most often understood in terms of market interaction: it is assumed that relations between people in the political sphere can be described in terms of mutually beneficial market exchange.

The attitude of student youth to the search for political rent is formed in the process of their stay in the most family-related group of society. That is, a student-family relationship – society and the state – emerges. All links have their own problems. The public oversight bodies could not develop conceptual principles for the participation of family-related groups and communities in the implementation of political rent. Political, economic, social and managerial actions of society should be based on taking into account the predictive behavior of young people, especially students. An analysis of their views on the possibility of obtaining political rent makes it possible to develop targeted programs to reduce the corruption of society. Brown J. notes that corruption is a syndrome of several generations [1]. According to some information, the opinion of a member of society towards the possible receipt of free benefits from the state begins to take shape in early childhood. Alfred Adler notes that the foundations of life style are laid by five years, after which fundamental changes no longer occur [2]. However, more conscious behavior regarding the possibility of obtaining political rent arises at about 20 years of age.

The purpose of the study is to determine the attitude of student youth towards the generation of political rent by family-related groups.

## 2. Data and Methods

The research methodology is based on the use of express questionnaire data, which was conducted among students of different courses of the university in an anonymous form. It was attended by more than a hundred full-time and part-time students in the field of economics. Of course, the representativeness of the study may cause some doubt. Perhaps the manifestation of the antisyndrome of Pavlik Morozov. However, there are certain trends in the responses of respondents. In addition, the methods used include such methods as comparing and summarizing, analyzing and synthesizing data on the state of problems in a selected area of research.

The need to study the complex relationship between the state and family-related groups of society in modern conditions is indicated in a number of scientific studies. Among contemporaries, we can highlight the work of R. Fismana, which analyzes interpersonal relations between family members and famous families in business or political circles in order to gain special access to resources [3]. Ibrahim Yilmaz provided an overview of the relationship between family business and the state [4]. The article by Fangjun Wang et al. Examined political connections, internal control, and firm value using the example of China's anti-corruption campaign [5]. Earlier, the author of this article considered some conceptual aspects of the participation of family-family relations in public management [6].

## 3 Results

The results of the study can be divided into three groups:

1. The perception of students of the modern political and economic atmosphere of society.

Great influence on the formation of rent-oriented behavior of students has a general socio-economic atmosphere, which is made up of many of its indicators. For example, on what basic values does the economy function? So, respondents to the question "Does everything have a price? (Is everything being sold and bought?) "-" yes ", 30% answered, the rest -" no ". At the same time, the question "Do you agree with the fact that" money rules the whole world "received such answers. The point "yes" was noted by 68.8% of respondents, 17.2 - "no", and the rest chose the item "I find it difficult to answer". The presence of market opportunities to solve pressing problems in a family-related group predetermines the behavior of each of its members.

According to sociologists, 41% of Russians do not approve of the moral state of modern society [7]. Among the reasons for this assessment is the strengthening of the role of family-related business groups in the life of society. Respondents assess the current situation in Russian society related to the influence of family-related business groups on political decision-making of the state differently (Table 1).

**Table 1.** Assessment of the situation of the influence of family-related business groups on the adoption of important political decisions of the state

| № | Questionnaire questions | Response rate in% |
|---|---|---|
| 1 | Yes, family-related business groups fully influence the political decisions of the state | 38.46 |
| 2 | Yes, but there is a slight effect | 41.7 |
| 3 | Do not affect | 18.68 |
| 4 | Own answer | 1.16 |

Source: Author.

However, to the question "Do you agree that in modern conditions, family-related business groups are the basis for the development of the political life of Russia?" - "yes" was answered - 17.6%, "no" - 45.88, "I find it difficult answer"- 36.47% of respondents.

If we analyze these answers with the indicators in Table 1, then we can see that pointing out the great influence of family-related groups on making politically important decisions does not mean that they are the basis for the development of political life.

2. Assessing the possibilities of access to political rent through family and family ties.

It is interesting for the respondents to understand the possibilities of using family and family ties to solve their particular issues. Twenty percent of the respondents would like to have a relative in senior positions in the state structure. At the same time, forty-seven percent of respondents to the question "Would you like to have a relative in leadership positions in the state structure?" Answered "no", the rest said "I don't care." That is, hypothetically 20 percent of respondents are potential agents of increasing the corruption of society.

When interviewing 35% of respondents to the question: "If your relative (a) has a high position in the state structure, will you wait for help from him?" Noted the item - "yes" - 18.5%; 23.4% - "no", 58.1% - "it all depends on the situation." This means that some rely on their personal potential or know their relatives as unreliable and distrustful people.

As you know, the harm from nepotism lies in the fact that a person is appointed to a new position without passing through professional selection and often not being a good specialist. To the question "Do you condemn if your acquaintances got a job not by skill level, but by family ties?" 9.7% answered in the affirmative, 48.9% answered "no", the rest said "I don't care."

Among the reasons for the search for political rent by family-related groups, one can point out the insufficient level of trust in the possibilities of formally solving top-priority problems. The proportion of Russians claiming to be able to provide for themselves and their families independently, without the support of the state, is slowly but steadily growing. In 2015, their number was 44% of respondents, today it is already approaching 50%. And among young people aged 18–30, their share is 62 against 38% who said that they and their families could not survive without state support [6].

Many parents want their children to be on the list of rich Russians. And the officials, taking advantage of their official position, punch various financial benefits and concessions for their children and relatives. Even the reproduction of individual professions and specialties is carried out only at the expense of the rich and middle layers of the population.

3. Evaluation of measures for state regulation of the process of obtaining political rent by representatives of family-related groups of society.

As is well known, within the framework of the anti-corruption policy of the state, a solid legislative base has been created for regulating family and family relations in the public service, in state institutions, etc. For example, if property benefits in the form of money, other values, the provision of material services are provided to the family and friends of an official with his consent or if he did not object to it and used his official powers in favor of the bribe giver, the official's actions should be qualified as accepting a bribe.

The question of the survey "Is there harm to society if representatives of the same family-related group work in government?" Received the following answers (Table 2).

**Table 2.** Estimation of harm to society, if representatives of the same family-related group work in state bodies

| № | Questionnaire questions | Response rate in% |
|---|---|---|
| 1 | Yes definitely | 33.3 |
| 2 | Yes, slightly | 26.3 |
| 3 | No, definitely | 3 |
| 4 | No, if the generally accepted norms of life are not violated | 37.4 |
| 5 | Not | 0 |

Source: Author.

As can be seen from the table, more than half of the respondents notice the negativity of the work of representatives of one family-related group of society in government bodies.

Interesting is the level of awareness of students' youth about the activities aimed at regulating relations between the state and family-related groups of society. As a result of the survey, it turned out that the question "Do you consider the measures affecting the decline in the role of family-related groups in the political life of society sufficient?" Respondents noted the following options (Table 3).

**Table 3.** On the adequacy of measures affecting the reduction of the role of family-related groups in the political life of society

| № | Questionnaire questions | Response rate in% |
|---|---|---|
| 1 | Sufficient measures | 16.9 |
| 2 | Measures insufficient | 20.6 |
| 3 | I do not know the details of these measures | 35.2 |
| 4 | Difficult to answer | 27.3 |

Source : Author.

More than 60% of respondents do not have information about the measures taken by the state in the field of anti-corruption policy, especially in terms of limiting the participation of representatives of one family-related groups of society in the public service.

## 4 Discussion and Conclusion

It should be noted that the above answers of the respondents only supplement the information about the presence of problems in the search for political rent associated with the formation of political and economic behavior of interested parties, that is, on the one hand a representative of an individual family, on the other - a representative of the state and society. In fact, the legislative principles contain the basic principles of interaction between these parties, such as maximizing the interests of the parties; creation of conditions for self-realization at the place of work of members of family-related groups of the company (in particular, through the realization of interests in the family business), responsibility for illegal actions (for example, increasing the responsibility of the entire family-related group of the company for a certain crime committed by one of the members of such groups), transparency of access to the resources of the state and society, etc. At the same time, the widespread penetration of the processes of market relations in the spiritual and moral sphere of society, in the upbringing and education requires interdisciplinary research. For example, at the junction of economics and education.

Of course, one cannot see in the mutual support of the members of one family-related group only the negative. Most of the family-related groups of society from childhood instill in their children the fundamental principles of honesty and decency, attitude to people, attitude to family, determination, hard work and, of course, respect for elders.

It is interesting to compare the data obtained with the results of studies in other countries. For this purpose, a considerable number of publications in well-known foreign databases were considered. However, no materials were found for comparison. One of the reasons for the lack of materials is the use of an author's analysis technique. Despite this, it can be stated with certainty that the children's principle "do you know who my dad is", that they can do everything and dad will save them in any case is a universal principle.

Thus, maintaining a balance of interests between the state and family-related groups of society regarding the realization of political rent is a necessary requirement for the stable development of society. Representatives of students mainly treat with understanding the existing problems in relations between the state and family-related groups as participants in the market of political rents.

The data obtained from the study can be used in compiling different models for observing the interests of the state and family-related groups of society based on the distribution of political rent and taking appropriate measures. Examples of interaction between the state and family-related groups of society in the search for political rent include:

1. The model of administrative and legal pressure to solve problems in the implementation of political rent. In the space of political rents, severe barriers are created for individual families to gain access to state and public goods. However, in the digital economy, there has been an increase in the share of infrastructure services provided by the private sector of the economy [8]. Possible contradictions.
2. A search model for reconciling the interests of individual families and state-public institutions regarding the implementation of political rents. So, one of the directions for coordinating rental interests can be the expansion of the system of public goods.

In general, a civilized solution to the acute and sensitive problems that arise between the state and individual family-related groups about obtaining political rent should be started in the early period of becoming a citizen of the country. A more in-depth discussion of the above models will be described in other publications of the authors.

## Acknowledgements

The reported study was funded by RFBR and EISR according to the research project № 19-011-31037.

## References

1. J. Brown, M. Loosemore, Behavioural Factors Influencing Corrupt Action in the Australian Construction Industry. *Engineering, Construction and Architectural Management,* **22**(4), 372-389 (2015)
2. S.R. Maddi, Teorii lichnosti: sravnitel'nyy analiz. Per.s ang. SPb: Izdatel'stvo Rech'. 539 (2002)
3. R. Fisman, Estimating the Value of Political Connections. *American Economic Review,* **91(**4), 1095-1102 (2001)
4. I. Yilmaz, Political connected Firm: a review. *International Journal of Management Economics and Business*, **14**(4), 1081-1101 (2019)
5. W. Fangjun et. el., Political connections, internal control and firm value: Evidence from China's anti-corruption campaign. *Journal of Business Research,* **86,** 53-67 (2018)
6. R.A. Burganov, On the question of the formation and development of family "capitalism" in Russia. *Economic science of modern Russia,* **1**(13), 72-74 (2008)
7. Uchonyye zapiski FNISTS RAN materialy zasedaniya Uchonogo soveta (Moskva, 27 fevralya 2019 g.). Vypusk vtoroy. M.: FNISTS RAN. p. 64 (2019)
8. V.V. Vishnever, R.A. Burganov, I.V. Nusratullin, Institutional transformations in the field of sevices in digital economy. *GCPMED 2018 – International scientic conference global challenges and prospects o the modern economic development*, **57**, pp. 1759-1768 (2019)

# A decision-making method based on consumer spending data

*Radim* Dušek[1*]

[1]Institute of Technology and Business, Faculty of Corporate Strategy, Okružní 517/10, 37001 České Budějovice, Czech Republic

**Abstract.** The aim of the paper is to introduce method which allows to calculate the estimation of the total consumer spending for 2019 in different regions of the Czech Republic. Presented method is based on combination of publicly available consumer spending data sets and geomarketing information. It can be applied to estimate consumer spending in total or for example only in one specific consumption expenditure group. Thus obtained findings, that can be visualized on a map with the use of GIS software, can be considered as a basis for more effective SMEs‘ marketing and management decision-making process concerning Czech Republic local consumer goods markets‘ size and characteristics as well as for forecasting future growth rate.
**Keywords:** decision-making method, geomarketing, consumer spending, GIS software, SMEs

## 1 Introduction

Household consumer spending in the Czech Republic is divided among many different businesses that include many international enterprises, but more often, small and medium sized enterprises (SMEs). These SMEs have to face many limitations and disadvantages in their decision-making process when compared to their bigger competitors.

The simple and low-cost methods based on combination of publicly available data sets and geomarketing tools can help to reduce mentioned disadvantages. They can support adequate information to manager's decision-making process concerning selected markets growth and size as well as new opportunities.

Geomarketing can extend typical marketing mix with the spatial view. It is a discipline that uses geographic data in the process of planning and implementation of marketing decisions, so when combined with consumer spending data sets, it can be used to estimate size of the selected markets [1, 2, 3, 4]. With the use of geographic information systems (GIS), which often allow using many special analysis modules, it is also possible to visualize results on a map instead of large tables, so it is easier to recognize correlations between market potential and characteristics of consumers' behaviour [5, 6, 7, 8].

* Corresponding author: dusek@mail.vstecb.cz

## 2 Data and methods

The aim of this paper is to introduce method which allows to calculate an estimation of the total consumer spending for 2019 in different regions of the Czech Republic. As this method is based on combination of consumer spending data sets and geomarketing information, the results will be visualized on a map with the use of GIS software. Obtained findings can be used as a basis for more effective marketing and management decision-making processes concerning Czech Republic local consumer goods markets‘ size and characteristics as well as for forecasting future growth rate.

The publicly available data about consumer spending (on the level of NUTS2 over the period from 2011 to 2016) [9] as well as the Czech Republic demographic and geographic information [10, 11] were used as a basis of this method. In the first step, data sets were decomposed to the level of Czech Republic districts and then the estimation of the local consumer spending in individual districts (77) from 2011 to 2016 was calculated with the following formula:

$$CS_{D,y} = AS_{NUTS2,y} \times I_{D,y} \quad (1)$$

where: CS = local consumer spending; D = district, Y = year; AS = average consumer spending; NUTS2 = district NUTS2 region; I = inhabitants.

The average annual consumer spending growth from 2011 to 2016 in all districts was calculated with the following formula:

$$CSG_D = \sqrt[5]{\frac{CS_{D,2016}}{CS_{D,2011}}} - 1 \quad (2)$$

where: $CSG_D$ = average annual consumer spending growth in specific district.

The average annual growth rate values over the period from 2011 to 2016 were used as a basis for the calculation of total consumer spending forecast for 2019. The results were visualized with geographic information system software QGIS.

## 3 Results

### 3.1 Estimated consumer spending for 2019

The first part of results chapter is devoted to the figures with estimated total consumer spending for 2019 in the individual Czech Republic districts. The structure of the total consumer spending consists of these consumption expenditure groups: food and non-alcoholic beverages; alcoholic beverages, tobacco; clothing and footwear; housing, water, electricity, gas and other fuels; furnishings, household equipment and routine household maintenance; health; transport; communication; recreation and culture; education; restaurants and hotels; miscellaneous goods and services. Presented method can be obviously used to calculate estimated consumer spending only in one or two selected consumption expenditure groups.

The highest estimated consumer spending for 2019 have been calculated in Capital city of Prague district. Brno-city district, Ostrava-city and Karviná. The results for 40 Czech districts (out of a total of 77 districts) with the biggest estimated consumer spending (in Czech crowns – CZK) can be seen in table 1.

**Table 1.** Czech districts with the biggest estimated consumer spending for 2019 (in billion CZK)

| District | Consumer spending estimation for 2019 | District | Consumer spending estimation for 2019 |
|---|---|---|---|
| Capital city of Prague | 192.19 | Vsetín | 18.11 |
| Brno-city | 47.19 | Uherské Hradiště | 18.04 |
| Ostrava-city | 39.29 | Děčín | 17.60 |
| Karviná | 30.06 | Teplice | 17.44 |
| Olomouc | 29.75 | Mladá Boleslav | 16.89 |
| Brno-countryside | 28.10 | Chomutov | 16.74 |
| Frýdek-Místek | 26.39 | Přerov | 16.48 |
| Praha-east | 24.33 | Trutnov | 16.38 |
| Zlín | 24.31 | Litoměřice | 16.21 |
| Liberec | 24.30 | Ústí nad Labem | 16.12 |
| Pardubice | 23.98 | Karlovy Vary | 15.42 |
| České Budějovice | 23.54 | Šumperk | 15.26 |
| Plzeň-city | 23.17 | Náchod | 15.21 |
| Hradec Králové | 22.71 | Most | 15.17 |
| Kladno | 21.72 | Příbram | 15.03 |
| Opava | 21.66 | Žďár nad Sázavou | 14.71 |
| Prague-west | 19.72 | Břeclav | 14.48 |
| Hodonín | 19.19 | Chrudim | 14.42 |
| Ústí nad Orlicí | 19.07 | Svitavy | 14.39 |
| Nový Jičín | 18.61 | Česká Lípa | 14.28 |

Source: Own processing.

For better differences visualization between total consumer spending estimations for 2019 in all individual districts, the results have been divided into six intervals, each with a different shade of grey on the map (see fig. 1).

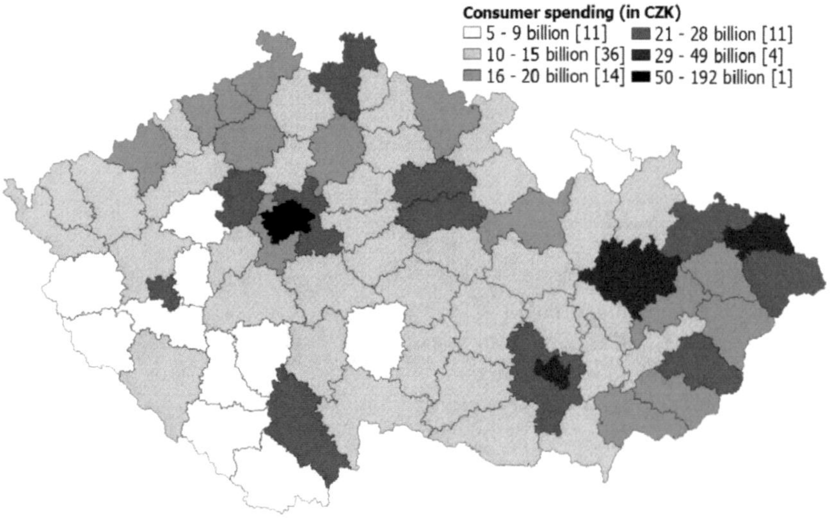


**Fig. 1.** Czech Republic districts' total consumer spending estimation for 2019 (in billion CZK)

Source: Own processing.

The darker shade means higher consumer spending in a specific districts. The quantity of particular districts that fall into each interval is indicated in square brackets.

The map clearly shows some interesting differences in consumer spending within the Czech Republic. Based on an estimated forecast results, it is possible to argue that residents of the northern, eastern and central districts of the Czech Republic will spend more money in 2019 on goods and services than households living in the southwest parts of the country.

### 3.2 Consumer spending change 2011-2019

For the effective decision-making process it is undoubtedly important to have information if the consumer spending in particular markets of interest is significantly increasing or decreasing within a certain period of time.

The answer to the question of how big are the differences in districts' estimated consumer spending when comparing years 2011 and 2019 can be seen in fig. 2. It shows percentage change in the estimated total consumer spending between 2011 and 2019 in all individual districts. From the map is evident that the biggest increase in consumer spending during the analysed period occurred especially in the west and east parts of Prague and in the northern and north-eastern districts as well.

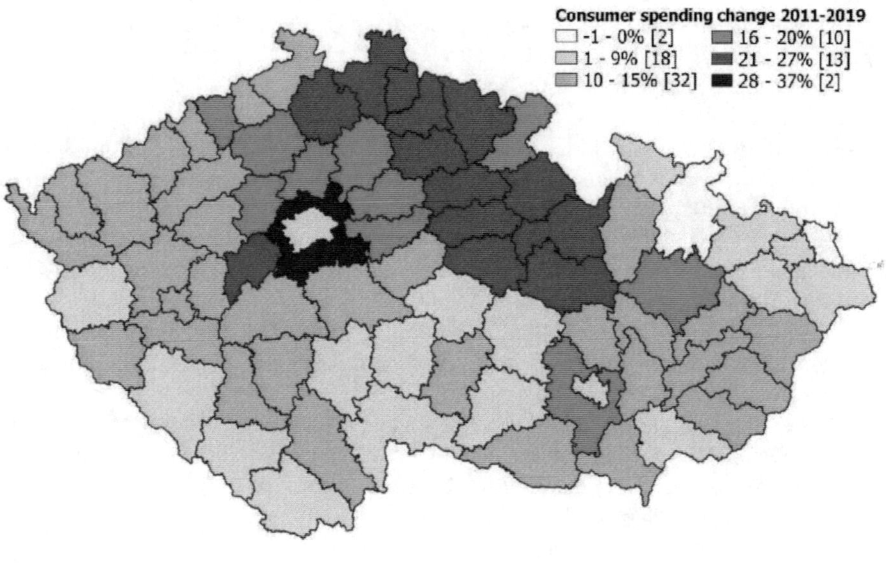


**Fig. 2.** Czech Republic districts' percentage change in the estimated consumer spending between years 2011 and 2019

Source: Own processing.

## 4 Conclusions

This study presented method which allows to forecast an estimation of the total consumer spending in individual districts of the Czech Republic. This low-cost tool enables to simply calculate estimated size and growth of a selected local market. As this method is based on combination of publicly available consumer spending datasets and geographic information system software, the total amount of money spent by Czech households could be visualized on a map at the district level.

Based on an estimated forecast results, it is possible to say that in 2019 residents of the northern, eastern and central districts of the Czech Republic will spend more money on goods and services than households living in the southwest parts of the country.

Presented method can be obviously applied to calculate estimated consumer spending only for a specific consumption expenditure group – for example food and non-alcoholic beverages. It is also possible to determine demographic profile of specific Czech district by including more detailed demographic data, for example the numbers municipalities, residents or the age structure of residents.

Thus obtained findings can be used as a basis for more effective marketing and management decision-making process concerning the Czech Republic local consumer goods and services markets‘ size, characteristics and growth, as well as potential opportunities or threats.

## References

1. T. Serajnik, S. Amaduzzi, G. Paulus, Geomarketing Analyses of the Città Fiera Mall, GI_Forum 2014 - *Geospatial Innovation for Society*, Herbert Wichmann Verlag, Berlin (2014)
2. J.C. Cross, T.J. Belich, W. Rudelius, How marketing managers use market segmentation: an exploratory study. *Proceedings of the 1990 Academy of Marketing Science Annual Conference* (1990)
3. A.C. Tynan, J. Drayton, Market segmentation. *Journal of Marketing Management*, **2**(3), 301-335 (1987)
4. S. Touzani, R. Van Buskirk, Estimating Sales and Sales Market Share from Sales Rank Data for Consumer Appliances. *Physica A: Statistical Mechanics and its Applications,* **451**, 266-276 (2016)
5. L. Wang, H. Fan, T. Gong, The Consumer Demand Estimating and Purchasing Strategies Optimizing of FMCG Retailers Based on Geographic Methods. *Sustainability,* **10**, 466-484 (2018)
6. O. Yarosh, Digital geomarketing methods for analyzing the development of the economy of modern urban space. *IOP Conf. Series: Materials Science and Engineering,* **497** (2019)
7. D. Baker, D. Bridges, R. Hunter, G. Johnson, J. Krupa, J. Murphy, K. Sorenson, *Guidebook to Decision Making Methods*. USA: Department of Energy (2001)
8. W. Tappert, *Geomarketing in der Praxis: Grundlagen – Einsatzmöglichkeiten –Nutzen,* Bernhard Harzer Verlag, Karlsruhe (2007)
9. Czech Statistical Office, *Expenditures and Consumption of Households Included in the Household Budget Survey – 2016* [online]. Available at: https://www.czso.cz/csu/czso/expenditures-and-consumption-of-households-included-in-the-household-budget-survey-2016 (2017)
10. Czech Statistical Office, *Database of Demographic Indicators for Selected Towns of the Czech Republic* [online]. Available at: https://www.czso.cz/csu/czso/database-of-demographic-indicators-for-selected-towns-of-the-czech-republic (2019)
11. State Administration of Land Surveying and Cadastre, *Registry of territorial identification, addresses and real estate* [online]. Available at: http://www.cuzk.cz/Uvod/Produkty-a-sluzby/RUIAN/2-Poskytovani-udaju-RUIAN-ISUI-VDP/Poskytovani-udaju-RUIAN-ISUI-VDP.aspx (2019)

# Fuzzy Approach to Short-Term Youth Unemployment Forecast

*Simona* Hašková[1]*

[1]The Institute of Technology and Business, School of Expertness and Valuation, Okružní 517/10, 37001 České Budějovice, Czech Republic

**Abstract.** The practical output of the paper is the prediction of the percentage of the short-term youth unemployment in the following two consecutive years. For this purpose, the non-traditional fuzzy model of the relevant prediction task is formulated in its theoretical-methodological part, which is subsequently applied in the application part. By means of the exact tools of the fuzzy approach (fuzzy sets and inference rules for manipulation with them), the synergy, uncertainty and complexity of the values (terms) of vaguely defined linguistic variables of an economic-psychological nature represented by sets of selected macroeconomic indicators is reflected. Despite the fact that the results of the fuzzy prediction are not significantly different from the results of conventional statistical prediction, the demonstration of the approach of the case of knowledgeable expertise to the choice of the relevant fuzzy sets and the formulation of the set of inference rules can be considered as a secondary benefit.
**Key words**: uncertainty, unemployment, fuzzy sets, fuzzy rules

## 1 Introduction to Youth Unemployment

In the last decade youth unemployment has been in the center of attention of economists and politicians, in particular in the countries severely affected by the latest finantial cricis [1]. Regadless of the economic recession it applies that the youth unemployment rate is generally higher than adult unemployment rate. The reasons are related to the lower level of human productivity, which, ceteris paribus, entitles employers to behave rationaly and prefer adults to young people [2].

Young people are generally far more hit by the employment crises than the adults with possible severe consequences that can result in "discouraged workers" effect and social

* Corresponding author: haskova@mail.vstecb.cz

exclusion from labour market. Thus, the integration of young people into the labour market is an essential objective of the European Employment Strategy [3].

In Tomić [4] analysed the key macroeconomic, microeconomic and sociological determinants of youth unemployment in 28 Member States of the European Union (EU). Its level was more pronounced in the countries with poor GDP growth, a low share of construction activity and high public debt. Futhermore, reduced mobility as a result of homeownership, perception of corruption, a high level of remittances and a lack of possibilities for young people to live outside parental homes appeared to be significant contributors as well.

The sensitivity of youth unemployment rate to economic crises differs accros the countries in dependence of e.g. the economic structure, the institutional framework, the policy-makers response at different levels, etc. [5, 6]. From the empirical point of view, the negative relationship was identified between the changes in unemployment rates and output growth in the short period known as "Okun's law". Many studies confirm this relationship for overall unemployment whereas research on youth unemployment in connection of Okun´s law is not common [7]. The exeption presents the study of Dunsch [8] that examined changes in youth unemployment in Germany and Poland by means of the Okun's law while having tested whether young employees are more vulnerable to the business cycle.

The purpose of this paper is to predict the youth unemployment rate expressed as a percentage further marked as YUNE (%), in two upcoming periods 2019 and 2021 in the Czech Republic on the basis of a historically known series of selected macroeconomic indicators of ΔIND (annual change in industrial production index in %), UNE (the total unemployment in %), ΔGDP (the GDP growth rate in %). The prediction tool is a fuzzy approach that, unlike standard statistical models, forms its predictive outputs within a given algorithm, which is capable to include the contribution of "soft factors" such as opinions and expertise of experts and economists.

## 2 Uncertainty in Economic Predictions of Youth Unemployment

The inaccuracy of the outcome of the prediction of the youth unemployment rate of any state is a consequence of the lack of information needed to the elimination of uncertainty we face in every non-deterministic environment [9]. This uncertainty is associated with both the inputs to the prediction model and its functioning. In the first case, we talk about "external" uncertainty resulting from the incomplete knowledge of the relevant values of *known* factors entering the prediction model [10]. They, together with the *unknown* values and therefore non-considered factors within the model, influence the future development of each macroeconomic prediction.

In the case of the uncertainty resulting from the model functioning, we talk about the "inner" uncertainty stemming from the approximate character of the formal description of the considered relationships between inputs and outputs of the prediction model [11]. In each of these cases, we can encounter the uncertainty of two different kinds. The uncertainty in terms of randomness, whose *objectively identified* basic statistical characteristics are known, and uncertainty in the sense of "*fuzziness*", which predominantly derives from the vagueness of the terms (e.g., few, little, approximately, a little, essentially, simply, complexly, significantly, analogously, etc.) occurring in the description of the situation and indicating the *subjective understanding* of intuitive concepts.

Econometrics often identifies uncertainty with randomness by considering the existing uncertain alternatives as equally probable in the context of the *indifference principle* and as a result building the rediction models solely on the basis of the probability theory and mathematical statistics (see e.g. [12]).

However, a number of system theory authorities (see, for example, [13]) call into question the effectiveness of decision-making and management procedures based on the approximation of uncertainty with randomness. In terms of uncertainty, their works operate with the terms of *linguistic variables* formalized by *fuzzy sets* instead of the numerical values of *random variables*. Zadeh´s work conception of terms as representatives of intuitive concepts is in line with Kahneman´s conception of the functioning of the human mind (see [14]).

The aim of the paper is to present the fuzzy algorithm of the short-term prediction of the youth unemployment rate in the Czech Republic operating under conditions of inner uncertainty and formulated within Zadeh´s fuzzy approach offering *subjectively* expected values as a viable alternative to *statistically* expected prediction. It deals with the fuzzy algorithm of the progression of the time series, which is preceded by a brief discussion of the basic principles of the fuzzy approach. In the application section, the fuzzy algorithm is used for the estimation of the youth unemployment in 2019 (the comparison with the published econometric forecast) and 2021 (a contribution to the paper).

## 3 Methodology: Brief Introduction to the Fuzzy Approach

The *fuzzy approach* can be traced in different versions of fuzzy logic, which was formed by adapting the binary numerical characteristics of the proposition operators to the interval $\langle 0,1 \rangle$ [15]. The fuzzy logic performes a tool for handling of fuzzy sets, the theory of which was published by Zadeh [16].

In brief let us describe the set theory as follows (for more details see e.g. [17]): Let the set $U$ be a field of consideration or discussion. Let $\mu_{\underline{A}}: U \rightarrow \langle 0,1 \rangle$ be a membership function and let $\underline{A} = \{(y, \mu_{\underline{A}}(y)): y \in U\}$ be a set of all pairs $(y, \mu_{\underline{A}}(y))$ in which the numbers $0 \leq \mu_{\underline{A}}(y) \leq 1$ assign to the given $y \in U$ a membership degree of the pair $(y, \mu_{\underline{A}}(y))$ to the set $\underline{A}$. Then $\underline{A}$ is a *fuzzy subset* on the universe $U$. The significant characteristic of fuzzy subset $\underline{A}$ is its support $U_{\underline{A}} = \{y: 0 < \mu_{\underline{A}}(y) \leq 1, y \in U\} \subset U$. In terms of fuzzy logic $\mu_{\underline{A}}(y) = \lfloor y \in U_{\underline{A}} \rfloor$. The element $y \in U$ with $\mu_{\underline{A}}(y) = 0.5$ is called the *crossover point* in $\underline{A}$. In the case of values greater than 0.5, the element $y$ rather belongs to $U_{\underline{A}}$, in the case of the smaller ones it rather does not belong to it [18].

In this paper, the numerical fuzzy sets are formal representations of terms of linguistic variables. For our purpose, the model with one internal and two border fuzzy sets for the terms *low* ($\underline{L}$), *middle* ($\underline{M}$), and *high* ($\underline{H}$) value is suitable. Interval $U$ is divided with the points a, b, c, d into five sections with the following membership functions (1):

$$
\begin{aligned}
(\underline{L}) \quad & \mu_{\underline{L}}(y) = 1 \text{ for } y < \text{a}, \\
& \mu_{\underline{L}}(y) = (\text{b} - y) \,/\, (\text{b} - \text{a}) \text{ for a} \leq y < \text{b}, \\
& \mu_{\underline{L}}(y) = 0 \text{ otherwise.} \\
(\underline{M}) \quad & \mu_{\underline{M}}(y) = (y - \text{a})/(\text{b} - \text{a}) \text{ for a} \leq y < \text{b}, \\
& \mu_{\underline{M}}(y) = 1 \text{ for b} \leq y < \text{c}, \\
& \mu_{\underline{M}}(y) = (\text{d} - y)/(\text{d} - \text{c}) \text{ for c} \leq y < \text{d}, \\
& \mu_{\underline{M}}(y) = 0 \text{ otherwise.}
\end{aligned}
\tag{1}
$$

($\underline{H}$) $\mu_{\underline{H}}(y) = 0$ for $y < c$,
$\mu_{\underline{H}}(y) = (y - c) / (d - c)$ for $c \le y < d$,
$\mu_{\underline{H}}(y) = 1$ otherwise.

The role of the expert is to determine the position of the points a, b, c and d in the given universe $U$. In the case of their regular distribution, the courses of the functions μ are given as shown in Fig. 1 in Part 5. From it we see that the linguistic variable acquires the values at two levels: at the level of the basal values $y$ in the universe $U$ and at the level of terms (intuitive concepts) as fuzzy subsets of $\underline{L}$, $\underline{M}$, $\underline{H}$ in the universe $U$. Each of these terms is defined by its membership function $\mu_{\underline{L}}$, $\mu_{\underline{M}}$, $\mu_{\underline{H}}$ over the field of its support, which is a subset of $U$.

Another important tool of the fuzzy set theory is the *rule*. In our considered model with n input linguistic variables and one output linguistic variable it is an element $((\underline{A}_1,\ldots,\underline{A}_n), \underline{C})$ of the relation $F \subset (\{\underline{L}_1, \underline{M}_1, \underline{H}_1\}\times\ldots\times\{\underline{L}_n, \underline{M}_n, \underline{H}_n\})\times\{\underline{L}, \underline{M}, \underline{H}\}$, which is a projection $F$: $(\{\underline{L}_1, \underline{M}_1, \underline{H}_1\}\times\ldots\times\{\underline{L}_n, \underline{M}_n, \underline{H}_n\}) \rightarrow \{\underline{L}, \underline{M}, \underline{H}\}$ in the form of $F(\underline{A}_1,\ldots,\underline{A}_n) = \underline{C}$, where $\underline{C} \in \{\underline{L}, \underline{M}, \underline{H}\}$ and $\underline{A}_i \in \{\underline{L}_i, \underline{M}_i, \underline{H}_i\}$, i = 1,…, n. The n-tuple of terms $(\underline{A}_1,\ldots,\underline{A}_n)$ is the left side of the rule, the term $F(\underline{A}_1,\ldots,\underline{A}_n)$ is the right side of the rule. The relation $F$ has a maximum of $n^3$ elements. We call it a set of *inferential* rules.

One of the basic concepts of the fuzzy set theory is the extension principle (in detail see [19]). Our modification of the extension principle is based on the following steps:

a) *Fuzzification* in which the input vector $x = (x_1,\ldots, x_n)$ converts each inference rule from the set $P$ into the *logical* notation mode.
b) *A set of partial results* is a set $B = \{\min\{\min\{\mu_1(x_1),\ldots, \mu_n(x_n)\}, \mu_{\underline{B}}\}: (\min\{\mu_1(x_1),\ldots, \mu_n(x_n)\}, \mu_{\underline{B}}) \in P^*\}$.
c) *Aggregation* or summation of functions of set $B$ into a compact unit and its aggregate $\mu_{agg}$ detection; this compact unit is a fuzzy subset on the universe $V$ with $\mu_{agg} = \max\{\min\{\min\{\mu_1(x_1),\ldots, \mu_n(x_n)\}, \mu_{\underline{B}}\}: (\min\{\mu_1(x_1),\ldots, \mu_n(x_n)\}, \mu_{\underline{B}}) \in P^*\}$.
d) *Defuzzification*, which transforms the result from the level of terms (the function $\mu_{\underline{AGG}}(y)$ of the fuzzy set $\underline{AGG}$) into $y_0 \in Y$ in the space of numerical values of the output linguistic variable. We ascertain the numeric value $y_0$ as the horizontal coordinate of the center of gravity of the area under the course of the function $\mu_{\underline{AGG}}(y)$. Thus:

$$y_0 = \int y \cdot \mu_{\underline{AGG}}(y)dy \,/ \int \mu_{\underline{AGG}}(y)dy \qquad (2)$$

where $\int$ is the symbol of a certain integral over the universe $Y$.

Since the resulting constant $y_0$ is largely the result of the subjective experience and opinions of experts who created the model, we call it a *subjectively* expected value.

## 4 The Set of Input Variables and Assumptions

The formulation of the fuzzy prediction model of the next time series member has its own specificity consisting of the fact that previous members of the resulting series are known (the historically measured values). Specifically, in the case of the YUNE rate prediction it is possible to evaluate the current YUNE development (decline, growth, stagnation, etc.). We also know the previous part of the baseline values of several linguistic variables, on which YUNE depends (albeit, largely vaguely). This reflects in the values of the extreme limits within which we look for the result of the prediction.

Table 1 lists the baseline values of ΔIND (annual change in industrial production index in %), UNE (the total unemployment in %), ΔGDP (the GDP growth rate in %) and the output linguistic variable YUNE (the YUNE rate in %) of the Czech Republic between 2010 and 2018 and their econometric forecasts in 2019 and 2020 [20, 21, 22, 23, 24] (highlighted columns in Tab. 1).

**Table 1.** Basic macroeconomic data for fuzzy predicting models of youth unemployment rate development in the Czech Republic

| Year | 2010 | 2011 | 2012 | 2013 | 2014 | 2015 | 2016 | 2017 | 2018 | 2019 | 2020 |
|---|---|---|---|---|---|---|---|---|---|---|---|
| ΔIND(%) | 25.7 | -2.5 | -6.3 | 0.7 | 5.1 | -0,7 | -0.9 | 3.0 | -3.1 | 0.3 | -0.6 |
| UNE(%) | 7.28 | 6.71 | 6.98 | 6.95 | 6.11 | 5.05 | 3.95 | 2.89 | 2.24 | 2.11 | 2.10 |
| ΔGDP(%) | 2.3 | 1.8 | -0.8 | -0.5 | 2.7 | 5.3 | 2.5 | 4.4 | 2.9 | 2.4 | 2.4 |
| **YUNE%** | **18.3** | **18.1** | **19.5** | **18.9** | **15.9** | **12.6** | **10.5** | **7.9** | **6.7** | **3.74** | **3.73** |

Source: Own processing.

The opinion of experts is an important specificity of the fuzzy model formulation; the experts take their knowledge and experience into account through qualified interventions in the model structure and the inference rules formulation.

The model works with the *dimensionless* values $u_{\Delta IND}$, $u_{UNE}$, $u_{\Delta GDP}$ of input universes $U_{\Delta IND}$, $U_{UNE}$, $U_{\Delta GDP}$, or respectively with *dimensionless* value $y$ of output universe $Y_{YUNE}$. These dimensionless input values are obtained by converting the given basal input values $x_{\Delta IND}$, $x_{UNE}$, $x_{\Delta GDP}$ of the respective input linguistic variables into intervals $\langle 0, 100\rangle$ by the formula $u = 100 \cdot (x - x_{min}) / (x_{max} - x_{min})$, where $x_{max}$, or respectively $x_{min}$ is the highest, or respectively the lowest given value $x$ within the monitored period.

In our case, the output linguistic variable is the growth rate of the YUNE(%) in the year immediately following the monitoring period. The conversion of the inserted basal output value $x_{YUNE}$ of the universe of the output linguistic variable to the dimensionless value $y \in Y_{YUNE} = \langle 0, 100\rangle$ (in the following text and figures $Y_{YUNE}$ is shortly marked as $Y$) is also given by the formula $y = 100 \cdot (x - x_{min}) / (x_{max} - x_{min})$, where $x_{max}$ is the value at the end of monitoring period and $x_{min}$ is so called "natural rate of unemployment" a priori (subjectively) expected at value of 3 %. For the conversion in the opposite direction $x = x_{min} + y \cdot (x_{max} - x_{min}) / 100$ applies.

## 5 The Results of the YUNE Prediction

The course of the YUNE(%) (see Table 1) during the period 2014-2018 shows a decreasing course. In the 2019 and 2020 further decline is expected. The formulation of *the predictive fuzzy model for the year 2019* is based on the characteristics of this period.

**Table 2.** The conversion of basal data of Table 1 relevant to the period 2014-2018 into dimensionless ones

| Year | 2014 | 2015 | 2016 | 2017 | 2018 |
|---|---|---|---|---|---|
| ΔIND(%) | 100 | 29.2 | 26.8 | 74.4 | 0 |
| UNE(%) | 100 | 72.6 | 44.2 | 16.8 | 0 |
| ΔGDP(%) | 7.1 | 100 | 0 | 67.8 | 14.3 |
| **YUNE(%)** | | | | | **100** |

Source: Own processing.

Thus, from last column of Table 2 the triple (0, 0, 14.3) $\in U_{\Delta IND} \times U_{UNE} \times U_{\Delta GDP}$ has been created. Because of the *external* and *internal* uncertainty of the model we choose the points a, b, c, d within the interval $\langle 0, 100 \rangle$ evenly distributed (a = 20, b = 40, c = 60, d = 80), (see formula (1) in Part 3). Therefore, the courses of the membership functions of the converted output and all the converted inputs are identical (see Fig. 1, in which these courses are plotted above the domain of universe *Y* of converted values of the output linguistic variable).

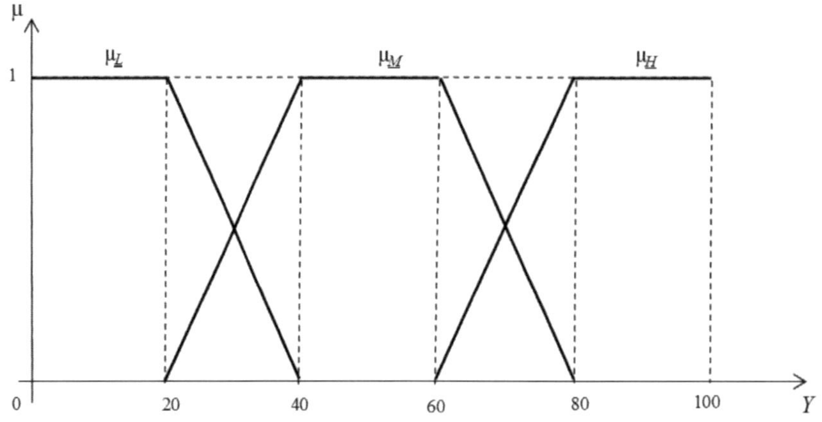


**Fig. 1.** Courses of the membership functions with even distribution of the points a, b, c, d within the interval values $y \in Y = \langle 0, 100 \rangle$

Source: Own processing.

The following fuzzification table valid for i = IND, UNE, GDP, whose elements are the values $\mu_{\underline{A}i}(u_i)$, where index $\underline{A} \in \{\underline{L}, \underline{M}, \underline{H}\}$, is derived from the above stated equations and inequalities:

**Table 3.** Fuzzification table valid for i = IND, UNE, GDP

| **Interval** | $u_i < 20$ | $20 \le u_i < 40$ | $40 \le u_i < 60$ | $60 \le u_i < 80$ | $u_i \ge 80$ |
|---|---|---|---|---|---|
| $\underline{L}_i$ | 1 | $(40 - u_i) / 20$ | 0 | 0 | 0 |
| $\underline{M}_i$ | 0 | $(u_i - 20) / 20$ | 1 | $(80 - u_i) / 20$ | 0 |
| $\underline{H}_i$ | 0 | 0 | 0 | $(u_i - 60) / 20$ | 1 |

Source: Own processing.

In the fuzzification table, only non-zero elements are taken into account, with the help of which the input vector (0, 0, 14.3) $\in U_{\Delta IND} \times U_{UNE} \times U_{\Delta GDP}$ generates the set $X = \{(\underline{L}_{\Delta IND}, 1), (\underline{L}_{UNE}, 1), (\underline{L}_{\Delta GDP}, 1)\}$. From its three elements only one element set $LF = \{(\underline{L}, \underline{L}, \underline{L})\}$ can be created. The triad $(\underline{L}, \underline{L}, \underline{L})$ of the input fuzzy sets the projection *F* assigns the output fuzzy set $\underline{L}$ according to the strategy of the predominant element and then applies $\mu_{AGG}(y) = \mu_{\underline{L}}(y)$ (see the highlighted course of line in Fig. 2.

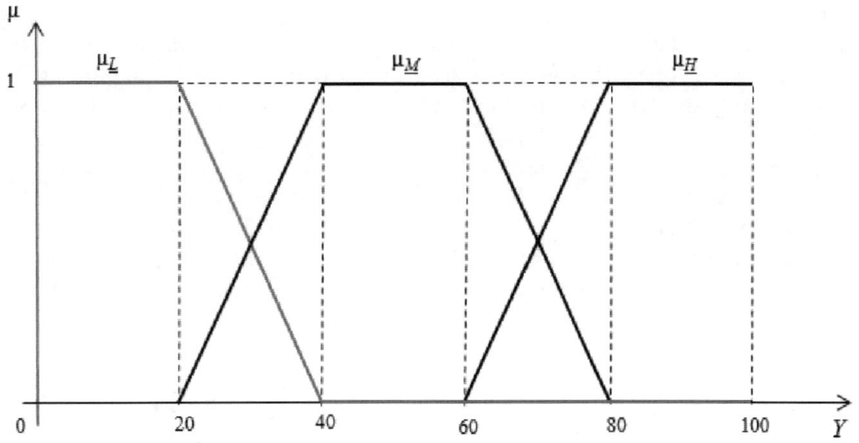


**Fig. 2.** The course of the membership function $\mu_{\underline{AGG}}$

Source: Own processing.

For dimensionless value $y_0$ given by formula (2) than holds:

$$y_0 = \int y \cdot \mu_{\underline{L}}(y)\mathrm{d}y \,/\, \int \mu_{\underline{L}}(y)\mathrm{d}y \approx 15.55$$

and predicted basal value $x_0$ of $x_{\mathrm{YUNE}}$ for year 2019 is $x_0 \approx 3 + 15.55 \cdot (6.7 - 3) / 100 = 3.57$.

The formulation of *the predictive fuzzy model for the year 2021* is based on the characteristics of the period 2016-2020.

**Table 4.** The conversion of basal data of Table 1 relevant to the period 2016-2020 into dimensionless ones

| **Year** | **2016** | **2017** | **2018** | **2019** | **2020** |
|---|---|---|---|---|---|
| ΔIND(%) | 36 | 100 | 0 | 55.7 | 41 |
| UNE(%) | 100 | 42.7 | 7.5 | 1 | 0 |
| ΔGDP(%) | 12 | 100 | 30.5 | 0 | 0 |
| **YUNE**(%) | | | | | **100** |

Source: Own processing.

Thus, from last column of Table 4 the triple (41, 0, 0) $\in U_{\Delta\mathrm{IND}} \times U_{\mathrm{UNE}} \times U_{\Delta\mathrm{GDP}}$ has been created. From Table 3 it is obvious that this input vector generates the set $X = \{(\underline{M}_{\Delta\mathrm{IND}}, 1), (\underline{L}_{\mathrm{UNE}}, 1), (\underline{L}_{\Delta\mathrm{GDP}}, 1)\}$. From its three elements only one element set $LF = \{(\underline{M}, \underline{L}, \underline{L})\}$ can be created. The triad $(\underline{M}, \underline{L}, \underline{L})$ of the input fuzzy sets the projection $F$ assigns the output fuzzy set $\underline{L}$ according to the strategy of the predominant element and then applies $\mu_{\underline{AGG}}(y) = \mu_{\underline{L}}(y)$ with the same highlighted course of line as in the previous case.

$$y_0 = \int y \cdot \mu_{\underline{L}}(y)\mathrm{d}y \,/\, \int \mu_{\underline{L}}(y)\mathrm{d}y \approx 15.55$$

and predicted basal value $x_0$ of $x_{\mathrm{YUNE}}$ for year 2021 is $x_0 \approx 3 + 15.55 \cdot (3.73 - 3) / 100 = 3.11$.

## 6 Results Discussion and Summary

The results of the statistical forecasts suggest that the Czech economy will appear near the state of "full employment" of young people. This refers to a combination of frictional, structural, and surplus unemployment that occurs among young people in a healthy economy.

This, therefore, suggest (and the statictical data confirm it) that Czech economy will operate above its potential in the next 2 years if similar market conditions remain.

The fuzzy prediction of YUNE (%) shows, in accordance to the development of the input variables, a significantly decreasing trend that does not fundamentally differs from statistical estimates.

In terms of the "quality" of the result, the fuzzy approach can be evaluated as a "more complex" method because, unlike the conventional statistical approach, it takes into account expert knowledge that projects into the selection of relevant fuzzy sets and formulation of a set of inference rules. This "quality" is a resultant consisting also of subjective attitudes and opinions, and as such can be considered as a method, which leads to a more complex informative value.

## References

1. G.M. Caporale, L. Gil-Alana, Youth unemployment in Europe: Persistence and macroeconomic determinants. *Comparative Economic Studies,* **56**(4), 581-591 (2014)
2. M. Choudhry, E. Marelli, M. Signorelli, The impact of financial crises on youth unemployment rate. *Quaderni del Dipartimento di Economia, Finanza e Statistica*, **79** (2010)
3. L. Grinevica, R. Kovalevs. Integration of young people into the Latvian labour market. *Economics and Business,* **27**(1), 64-68 (2015)
4. I. Tomić, What drives youth unemployment in Europe? Economic vs non-economic determinants. *International Labour Review,* **157**(3), 379-408 (2018)
5. M. Choudhry, E. Marelli, M. Signorelli, The impact of financial crises on youth unemployment rate. *Quaderni del Dipartimento di Economia, Finanza e Statistica*, **79** (2010)
6. J. O'Reilly, W. Eichhorst, A. Gábos, K. Hadjivassiliou, D. Lain, J. Leschke, S. McGuinness, L. Mýtna Kureková, T. Nazio, R. Ortlieb, H. Russell, P. Villa, Five characteristics of youth unemployment in Europe: Flexibility, education, migration, family legacies, and EU policy. *Sage Open,* **5**(1), 1-19 (2015)
7. A. Banerji, H.H. Lin, S. Saksonovs, Youth unemployment in advanced Europe: Okun's law and Beyond. *International Monetary Fund*, **15** (2015)
8. S. Dunsch, Okun's law and youth unemployment in Germany and Poland. *International Journal of Management and Economics,* **49**(1), 34-57 (2016)
9. R. Bayrak, H. Tatli, The Determinants of Youth Unemployment: A Panel Data Analysis of OECD Countries. *The European Journal of Comparative Economics,* **15**(2), 231-248 (2018)
10. C. López-Duarte, M.M. Vidal-Suárez, External uncertainty and entry mode choice: Cultural distance, political risk and language diversity. *International Business Review,* **19**(6), 575-588 (2010)
11. D. Caldara, C. Fuentes-Albero, S. Gilchrist, E. Zakrajšek, The macroeconomic impact of financial and uncertainty shocks. *European Economic Review,* **88**, 185-2017 (2016)
12. M. Timmermans, R. Heijmans, H. Daniels, Cyclical patterns in risk indicators based on financial market infrastructure transaction data. *De Nederlandsche Bank Working Paper* No. 558 (2017)

13. L.A. Zadeh, Fuzzy logic = computing with words. *Computing with Words in Information/Intelligent Systems 1*, 3-23 (1999)
14. D. Kahneman, Maps of bounded rationality: Psychology for behavioral economics. *American Economic Review,* **93**(5), 1449-1475 (2003)
15. L. Běhounek, P. Cintula, From fuzzy logic to fuzzy mathematics: A methodological manifesto. *Fuzzy Sets and Systems,* **157**(5), 642-646 (2006)
16. L.A. Zadeh, Outline of a new approach to the analysis of complex systems and decision processes. *IEEE Transactions on systems, Man, and Cybernetics,* **1**, 28-44 (1973)
17. S. Hašková, P. Fiala. A fuzzy approach for the estimation of foreign investment risk based on values of rating indices. *Risk Management,* **21**(3), 183-199 (2019)
18. D. Dubois, H. Prade, What are fuzzy rules and how to use them. *Fuzzy Sets and Systems,* **84**(2), 169-185 (1996)
19. C. Kahraman, *Fuzzy multi-criteria decision making: theory and applications with recent developments*. Springer Science & Business Media (2008)
20. Czech statistical office Industry - Time series, Industry: Index průmyslové produkce: meziroční indexy [Industrial production index: year-on-year indices] [online], Available at: https://www.czso.cz/csu/czso/pru_ts (2019)
21. Ministry of Finance of the Czech Republic [online], Available at: https://www.mfcr.cz/cs/verejny-sektor/makroekonomika/makroekonomicka-predikce/2019/47-kolokvium--setreni-prognoz-makroekono-35203 (2019)
22. Real GDP growth rate – volume Percentage change on previous year [online], Available at: https://ec.europa.eu/eurostat/data/database (2019)
23. Trading Economics Czech Republic Youth Unemployment Rate – Forecast [online], Available at: https://tradingeconomics.com/czech-republic/youth-unemployment-rate?embed/forecast (2019)
24. OECD [online], Available at: https://data.oecd.org/unemp/youth-unemployment-rate.htm (2019)

# Using RBF neural networks to identify relationship between development of oil prices in world market and value of Chinese currency

*Jakub* Horák[1*], *Jaromír* Vrbka[1], *Tomáš* Krulický[1]

[1]Institute of Technology and Business, School of Expertness and Valuation, Okružní 517/10, 37001 České Budějovice, Czech Republic

**Abstract.** The objective of the contribution is to identify a possible relationship between the development of the price of Brent oil (Brent in USD/barrel) and the CNY / USD Exchange rate by means of artificial neural networks. Understanding future fluctuation characteristics and the trend in oil prices is the basis for a deep understanding of systemic mechanisms and trends in related research areas. However, given the complexities of oil prices, it is very difficult to obtain accurate forecasts. Within the experiment, a total of 50,000 artificial RBF neural networks were generated. Was found the CNY / USD price will play a significant role in creating China's real product. Given that it was already proven that the CNY / USD exchange depends on Brent in USD / barrel, it is important to focus the further research on finding out the time lag with which the price of Brent in USD / barrel is actually reflected in the price of CNY / USD.

**Key words:** RBF neural networks, value, oil prices, exchange rate

## 1 Introduction

Oil is a vitally important energy source. The fluctuation of international oil prices influences all aspects of economy. The exchange rate is one of the important channels which demonstrate how the international shock on oil prices is reflected on the actual economy and financial markets. The change in the exchange rate of the US dollar will inevitably influence international oil prices. Quiang et al. [1] explore impacts of the fluctuation of oil prices on the national exchange rate in oil-importing countries. As a result of the detailed exploration the authors argue that a future research may be conducted in exploring inherent laws on the reserve of oil price volatility in different modes of exchange rates and in studying the way in which and to which extent the fluctuation of oil prices influences the effects of exchange rates.

The understanding of fluctuation characteristics and trends in oil prices gives grounds for a deep understanding of system mechanisms and gradual trends in related researches. In a view of the characteristics of oil prices being rather complex, it is very difficult to make accurate predictions. In order to increase the accuracy of international predictions of oil

* Corresponding author: horak@mail.vstecb.cz

prices, Chai et al. [2] developed a new prediction hybrid model of which the results show that the accuracy of the prediction of the designed hybrid model, based on fuzzy clustering algorithm with dynamic time deformation, is fairly good. In case of China, the main world's importer of oil, the slightest fluctuation of the oil price or the exchange rate will have a dramatic impact on the oil import [3]. China's dependence rate on foreign oil continuously increases. The fluctuation of the international oil price has an overwhelming influence on Chinese oil import, which continuously shocks the economic development of China. This article proves Markov's properties of the international fluctuation of oil prices and the strong probability of a state's transfer to Markov's chain is calculated. This document estimates that the fluctuation of the international oil price may cause a loss about 47,078 billion yuan on GNP in a single month [4].

Baek and Kim [5] examine the influence of the fluctuation of oil prices on exchange rates in selected sub Saharan African countries (SSA). In order for this issue to be thoroughly explored, the authors are, in contrast to previous studies, delayed in specifically describing asymmetrical effects of changes in oil prices in their modelling process using the nonlinear autoregressive model. The results demonstrate strong evidence that changes in oil prices will have long-term asymmetrical effects on actual exchange rates. The fluctuation of actual exchange rates thereby reflects the increase in oil prices rather than their drops. Changes in oil prices have various impacts on financial ratios of global markets and economies [6]. The aim the study of Kayalar, Küçüközmen and Selcuk-Kestel [7] is to explore the structure of the dependence between oil prices and indexes of stock markets and exchange rates in various economies categorized in regard to their position in developing markets and countries which import or export oil. The authors of the research found out that exchange rates and stock market indexes of most of the countries exporting oil demonstrate a higher dependence on the oil price; on the other hand, developing markets where oil imported are less sensitive to price fluctuations.

This article focuses on the use of artificial neural structures, RBF in particular, which are suggested in the next part. Artificial neural networks (ANNs) are considered as a highly efficient method of data collection, their analysis and prognosis; therefore, they may be applied while resolving various complex issues and situations including the prognosis of the development of stock market prices [8; 9]. ANNs may be used for the classification, regression or, for example, cluster analysis. Their big advantage is the ability to process a large amount of data, a simple application of the acquired neural network or the accuracy of results [10]. On the other hand, their disadvantage is the way in which individual models of artificial neural networks are created [11].

ANNs are currently a very popular method for the prognosis of key ratios in a company [12]. ANNs have gradually become very significant in universal domains Adhikari and Agrawal [13] and Sayadi et al. [14] argues that they are highly accurate in the prognosis. One of the crucial issues, though, is that there is no consensus about how to use them the best [15]. Šnorek [16] argues that the biggest advantages of ANN methods for the prognosis of key ratios of the company are non-linear approximate functions and the ability to learn and generalize. Rowland, Šuleř and Vochozka [17] on the other hand argue that disadvantages of ANNs are that variables should be carefully selected and prioritized (it requires highly accurate data); furthermore, there is a risk of illogical behaviour of the network and their architecture requires an exact definition.

The time series can be referred to as a specific data monitoring; these data are recorded horizontally with respect to time – from the past to present [18]. The analysis of time series is mainly used for the future prognosis [19]. Vrbka and Rowland [20] declare that the analysis of time series provides for ANNs a wide application. Rodrigues, Markou and Pereira [21] state that through the time series prognosis it is possible to predict future values from previously monitored values. The accurate prediction of time series is very important for a wide range of areas of competence [22]. The prediction requires a most accurate data on prognosis variables, data characteristics and data accessibility; moreover, the method of the prediction and its time horizon need to be devised [23].

Networks of radial basic functions (RBF) belong to ANNs called feedforward networks since the data in the network structure are processed in one direction – from input neurons to output neurons. These networks are different from other ANNs of this sort; they actually use radial basic functions as activation functions. This enables them learn faster while their sensitivity to the order of the presented data is reduced [24]. They usually consist of three layers: the input layer, the hidden layer with non-linear activation RBF function and the linear output layer. RBF can be applied in a lot of ways including approximation of a function, time series prediction and system classification and management. RBF are the best used to deal with problems of a smaller number of inputs [25].

## 2 Data and methods

Some researchers claim that exist relationship between the oil price in the world market and the value of Chinese Yuan. However, a question remains, whether the influence is somehow significant and clear from the comparison of the two-time series.

The data for analysis are available on the World Bank web pages [26]. For the purpose of the analysis, the information on (CNY/USD) exchange rate will be used. The second time series will be (Brent in USD/barrel). The use of the USD value for calculating the prices of CNY and Brent[†]oil enables to eliminate the delay that could occur due to the fluctuations of the USD value. These data are also available on the World Bank web pages [26]. The contribution will use the final value of both variables at the time interval of 1 September 2014-30 August 2019.

The descriptive characteristics of both time series are available in Table 1.

**Table 1.** Descriptive characteristics: Brent in USD/barrel and Brent in USD/barrel

| Samples | Brent in USD/barrel Input variable | CNY/USD Output (target) |
|---|---|---|
| Minimum (Training) | 27.8800 | 0.139600 |
| Maximum (Training) | 102.7900 | 0.163600 |
| Average (Training) | 58.5614 | 0.152175 |
| Standard deviation (Training) | 12.9949 | 0.006092 |
| Minimum (Testing) | 30.5000 | 0.139800 |
| Maximum (Testing) | 100.2000 | 0.163600 |
| Average (Testing) | 61.7022 | 0.152119 |
| Standard deviation (Testing) | 12.2878 | 0.006461 |
| Minimum (Validation) | 31.5500 | 0.139600 |
| Maximum (Validation) | 102.7700 | 0.163600 |
| Average (Validation) | 59.8710 | 0.152370 |
| Standard deviation (Validation) | 18.2674 | 0.009356 |
| Minimum (Overall) | 27.8800 | 0.139600 |
| Maximum (Overall) | 102.7900 | 0.163600 |
| Average (Overall) | 59.2272 | 0.152196 |
| Standard deviation (Overall) | 13.1782 | 0.006168 |

Source: Own processing.

[†] In this contribution, Brent oil price represents the price of oil in the world market, because it is used for valuation of two thirds of the world oil supply.

Course of Brent in USD/barrel time series is shown in Figure 1.

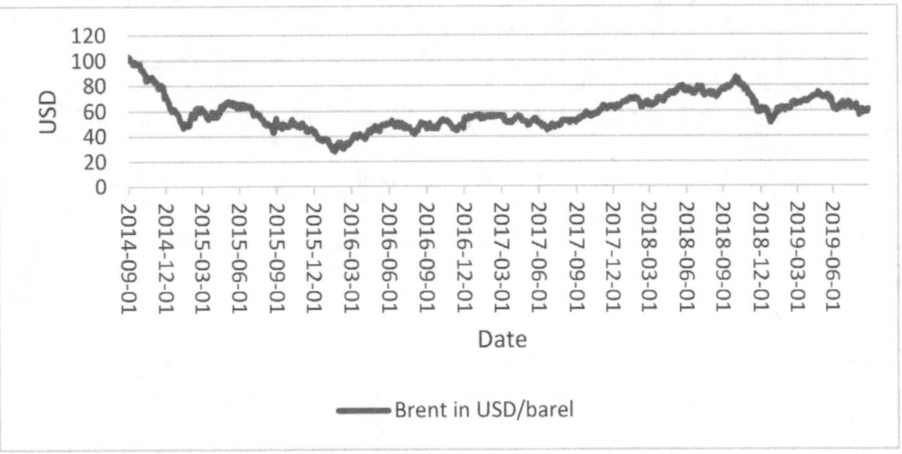


**Fig. 1.** Course of Brent in USD/barrel time series

Source: World Bank [26], own processing.

The course of the Brent in USD/barrel series can be seen in Figure 2.

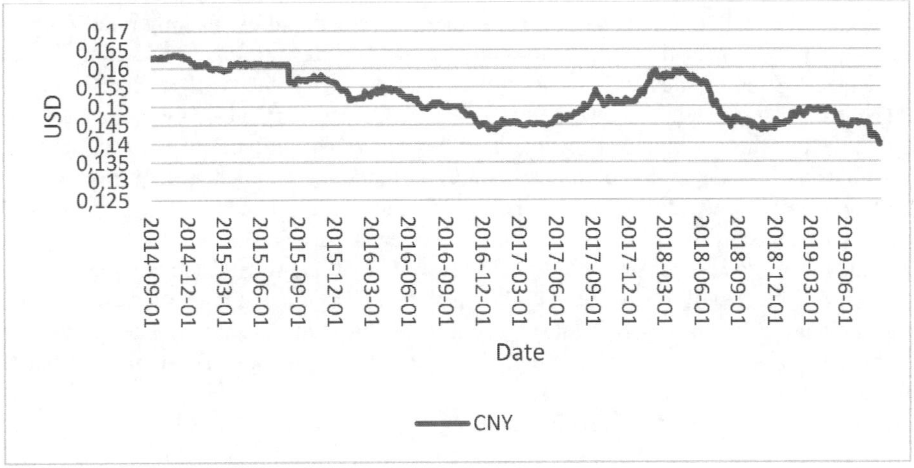


**Fig. 2.** Course of CNY/USD

Source: World Bank [26], own processing.

The influence of Brent in USD/barrel on the development of CNY/USD does not have to be immediate. It can be seen on the same day or with one-week or even one-month lag. Time lag cannot be calculated. It does not even have to be constant time lag throughout the monitored period of time. It is therefore necessary to carry out an experiment to find out whether there is a real dependence between the monitored time series and at which time lag this dependence is greatest. The following factors will be assumed:

1. Change in CNY/USD price shows on the same day. In such a case, CNY/USD would probably not react on Brent in USD/barrel, but the exchange rate would more likely reflect the cause resulting in the change of Brent in USD/barrel price. There can also be signs that the oil price will change.

2. Change in CNY/USD price shows with a one-day lag after the change of Brent in USD/barrel price. It would be a really small lag, which would indicate high sensitivity of CNY/USD to Brent in USD/barrel.
3. Change in CNY/USD price shows with a five-day lag after the change in the Brent in USD/barrel price.
4. Change in CNY/USD price shows with a 10-day lag after the change in Brent in USD/barrel price.
5. Change in CNY/USD price shows with a 30-day lag after the change in Brent in USD/barrel price.

Given that due to the complexity of the calculation we do not work with only one-day lag, there is a risk that if there is a real dependence between the two-time series, time lag will not be identified. Despite this potential risk, the objective of the contribution can be achieved.

Regression using neural structures will be carried out in DELL´s Statistica software, version 12. Radial basis function (RBF) neural networks will be generated. The independent variable will be Brent in USD/barrel. The dependent variable will be CNY/USD. The time series will be divided into three data sets – training, testing and validation. The first set will contain 70% of the input data. Based on the training data set, neural structures will be generated. The remaining data sets will contain 15% of the input data. Both sets will be used for the verification of the neural structure reliability. There will be generated 10,000 neural networks for each time lag. 5 calculations will thus be carried out. 5 artificial neural networks with the best characteristics[‡] will be calculated for each calculation. We will thus compare 25 artificial neural structures; and their performance expressed by the correlation coefficients of the training, testing, and validation data sets will be investigated. In ideal case, the correlation coefficient values will be significant (over 0.7). Statistical dependence is at the interval of 0.5-0.7. Nevertheless, the dependence can be identified if the correlation coefficient values are at the interval of 0.3-0.5. In such a case, however, the dependence is not strong. At the same time, it is important that the correlation coefficient value is approximately the same in all data sets. The error will also be assessed. The sensitivity analysis will also be interesting.

In the hidden layer of RBF there will be always 21-30 neurons. The activation function of the hidden layer will always be the Gaussian function. For the activation of the output layer, the Identity function will be used. As an error function, the sum of the least squares will be used. For the networks training, the RBFT algorithm integrated in the Statistica software will be used.

Other settings will remain default according to the ANN (Automated Neural Networks) tool. If needed, the corrections of the individual neurons weight will be corrected and tested using the VNN tool. It must be stated that improving artificial neural network using this tools is rather accidental than an exact process with a predictable outcome.

## 3 Results

### 3.1 Result without lag

First, regression analysis using RBF without a time lag was carried out (Table 2).

[‡] The least squares method will be used. Generating of networks will be finished when there is no improvement, that is, no reduction in the sum of squares. The retained networks will be those whose sum of residual squares to the development of CNY / USD will be as low as possible (zero in ideal case).

**Table 2.** Retained neural networks for time series without lag

| Index | Network | Training performance | Testing performance | Valid. performance | Training error | Testing error | Validation error |
|---|---|---|---|---|---|---|---|
| 1 | RBF 1-24-1 | 0.412751 | 0.372668 | 0.511868 | 0.000015 | 0.000018 | 0.000014 |
| 2 | RBF 1-29-1 | 0.402858 | 0.411735 | 0.501684 | 0.000016 | 0.000017 | 0.000015 |
| 3 | RBF 1-27-1 | 0.421069 | 0.363556 | 0.501947 | 0.000015 | 0.000018 | 0.000015 |
| 4 | RBF 1-29-1 | 0.416481 | 0.331440 | 0.505133 | 0.000015 | 0.000019 | 0.000015 |
| 5 | RBF 1-25-1 | 0.403294 | 0.314854 | 0.501389 | 0.000016 | 0.000019 | 0.000015 |

Source: Own processing.

The table clearly indicates that the retained networks contain 24-29 neurons in the hidden layer. The performance of all retained neural structures is above 0.3. The validation data sets always show the highest values; in their case, correlation coefficient value is always above 0.5, while the lowest values are in the case of the testing data set, namely between 0.31 (the 5. RBF 1-25-1) and 0.41 (the 2. RBF 1-29-1). The values of training performance are between 0.4 and 0.42. The error is always below 0.00002. It can thus be stated that compared to the absolute value of CNY/USD the value is minimal.

## 3.2 Results with one-day lag

The results of regression analysis with a one-day lag are shown in Table 3.

**Table 3.** Retained networks for time series with one-day lag

| Index | Network | Training performance | Testing performance | Validation performance | Training error | Testing error | Validation error |
|---|---|---|---|---|---|---|---|
| 1 | RBF 1-25-1 | 0.367709 | 0.265241 | 0.423495 | 0.000016 | 0.000019 | 0.000017 |
| 2 | RBF 1-27-1 | 0.412785 | 0.344563 | 0.432329 | 0.000015 | 0.000018 | 0.000017 |
| 3 | RBF 1-25-1 | 0.434149 | 0.292056 | 0.427362 | 0.000015 | 0.000019 | 0.000017 |
| 4 | RBF 1-30-1 | 0.448699 | 0.321214 | 0.421693 | 0.000015 | 0.000018 | 0.000017 |
| 5 | RBF 1-26-1 | 0.411882 | 0.336311 | 0.431904 | 0.000015 | 0.000018 | 0.000017 |

Source: Own processing.

It results from the table that the retained networks contain 25-30 neurons in the hidden layer. The highest values are shown by the validation data set, where the value of the correlation coefficient is always above 0.42, while the lowest average values of the correlation coefficients are shown by the testing data set, namely between the value higher than 0.26 (1. RBF 1-25-1) and almost 0.35 (2. RBF 1-27-1). The training performance achieves the values of almost 0.37 – 0.45. The error is always lower than 0.00002. It can be stated that compared with the absolute value of CNY/USD the value is minimal. In general, however, the performance in the case of the individual neural networks and data sets is uneven.

## 3.3. Results with five-day lag

Another investigated time lag was a five-day lag. The results are shown in Table 4.

**Table 4.** Retained neural networks for time series with 5-day lag

| Index | Network | Training performance | Testing performance | Validation performance | Training error | Testing error | Validation error |
|---|---|---|---|---|---|---|---|
| 1 | RBF 1-28-1 | 0.359699 | 0.428676 | 0.462018 | 0.000016 | 0.000018 | 0.000017 |
| 2 | RBF 1-21-1 | 0.375812 | 0.449748 | 0.483160 | 0.000015 | 0.000018 | 0.000016 |
| 3 | RBF 1-23-1 | 0.383388 | 0.387543 | 0.459824 | 0.000015 | 0.000019 | 0.000017 |
| 4 | RBF 1-25-1 | 0.369094 | 0.448173 | 0.468541 | 0.000015 | 0.000018 | 0.000016 |
| 5 | RBF 1-21-1 | 0.344893 | 0.422891 | 0.459610 | 0.000016 | 0.000018 | 0.000017 |

Source: Own processing.

It results from the table that the retained neural networks contain 21-28 neurons in the hidden layer. The performance of all retained neural structures is above 0.3. The highest values are shown by the validation data set, with the correlation coefficient value always above 0.46. The lowest average values are shown by the training data set correlation coefficient, namely the values between more than 0.34 (5. RBF 1-21-1) and more than 0.38 (3. RBF 1-23-1). The testing performance value is between more than 0.38 and almost 0.45. The error is always smaller than 0.00002. It can thus be stated that compared to the absolute value of CNY/USD the error is minimal.

### 3.4 Results with 10-day lag

The results of regression analysis of the Brent in USD/barrel and CNY/USD time series are shown in Table 5.

**Table 5.** Retained networks for time series with 10-day lag

| Index | Network | Training performance | Testing performance | Validation performance | Training error | Testing error | Validation error |
|---|---|---|---|---|---|---|---|
| 1 | RBF 1-23-1 | 0.422277 | 0.289404 | 0.411383 | 0.000015 | 0.000018 | 0.000017 |
| 2 | RBF 1-22-1 | 0.449527 | 0.335184 | 0.421136 | 0.000015 | 0.000018 | 0.000017 |
| 3 | RBF 1-25-1 | 0.446497 | 0.281631 | 0.413589 | 0.000015 | 0.000018 | 0.000017 |
| 4 | RBF 1-28-1 | 0.450363 | 0.285487 | 0.417970 | 0.000014 | 0.000018 | 0.000017 |
| 5 | RBF 1-21-1 | 0.412808 | 0.271661 | 0.417731 | 0.000015 | 0.000018 | 0.000017 |

Source: Own processing.

It results from the table that the retained networks contain 21-28 neurons in the hidden layer. The highest values are shown by the training data set, where the correlation coefficient values are at the interval of more than 0.41 to more than 0.45, and the lowest average values are achieved by the testing data set, namely from more than 0.27 (5. RBF 1-21-1) to 0.33 (2. RBF 1-22-1). In this case, the performance values of networks and data sets are very low. The validation data set performance achieves the values between more than 0.41 to more than 0.42. The error is always smaller than 0.00002.

### 3.5. Results with 30-day lag

The last calculation concerns a 30-day time lag (Table 6).

**Table 6.** Retained networks for time series with30-day lag

| Index | Network | Training performance | Testing performance | Validation performance | Training error | Testing error | Validation error |
|---|---|---|---|---|---|---|---|
| 1 | RBF 1-29-1 | 0.504936 | 0.417204 | 0.469858 | 0.000013 | 0.000015 | 0.000015 |
| 2 | RBF 1-25-1 | 0.492835 | 0.402539 | 0.467906 | 0.000013 | 0.000016 | 0.000015 |
| 3 | RBF 1-28-1 | 0.491749 | 0.426946 | 0.474171 | 0.000013 | 0.000015 | 0.000015 |
| 4 | RBF 1-24-1 | 0.503934 | 0.421252 | 0.477249 | 0.000013 | 0.000015 | 0.000015 |
| 5 | RBF 1-27-1 | 0.515606 | 0.481461 | 0.468058 | 0.000013 | 0.000014 | 0.000015 |

Source: Own processing.

The table shows that the retained networks contain 24-29 neurons in the hidden layer. All correlation coefficient values are above 0.40. The training data set coefficients achieve the values of more than 0.49 to almost 0.52. The lowest values are achieved by the testing data set correlation coefficients, namely from more than 0.40 (2. RBF 1-25-1) to more than 0.48 (5. RBF 1-27-1). The validation data set achieves the performance from almost 0.47 to almost 0.48. The error is never higher than 0.00002. In the case of the results for the time lag it must be noted that compared to other results, these achieve the lowest fluctuation in all three data sets.

### 3.6 Analysis of results

The analysis of the results given in Tables 2-6 clearly shows that, considering the overall performance of the retained neural networks, the correlation coefficient values, and all data sets error, the most successful is the set of results presented in Table 6. It declares certain (significant) dependence of CNY/USD exchange rate on the development of Brent in USD/barrel. This gives the response to the question whether Brent in USD/barrel affects the value of the CNY/USD, indicating that there is certain dependence. The influence is not decisive but it is not negligible. It can also be stated that the price of Brent in USD/barrel will be reflected in the CNY/USD exchange in approx. 30 days. Now it is necessary to assess which of the retained neural networks captures the relation between the Brent in USD/barrel and CNY/USD best.

Based on the correlation coefficients given in Table 6, where the absolute value in each set and ideally equalized values in all three data sets, the most successful appears to be the 5. RBF 1-27-1 neural network. Its suitability is mainly due to the performance of the testing data set, which is the highest one of all the retained networks. However, interesting data can be see also in Table 7 showing the basic forecasts statistics.

**Table 7.** Basic statistics of forecasts (30-day lag)

| Statistics | 1.RBF 1-29-1 | 2.RBF 1-25-1 | 3.RBF 1-28-1 | 4.RBF 1-24-1 | 5.RBF 1-27-1 |
|---|---|---|---|---|---|
| Minimal forecast (Training) | 0.14318 | 0.14680 | 0.14586 | 0.14487 | 0.14386 |
| Maximum forecast (Training) | 0.16677 | 0.16676 | 0.16564 | 0.16341 | 0.16400 |
| Minimal forecast (Testing) | 0.14683 | 0.14680 | 0.14594 | 0.14620 | 0.14510 |
| Maximum forecast (Testing) | 0.16676 | 0.16675 | 0.16496 | 0.16336 | 0.16400 |
| Minimal forecast (Validation) | 0.14625 | 0.14683 | 0.14599 | 0.14798 | 0.14671 |
| Maximum forecast (Validation) | 0.16669 | 0.16676 | 0.16534 | 0.16341 | 0.16400 |
| Minimal residuals (Training) | -0.01463 | -0.01658 | -0.01601 | -0.01592 | -0.01620 |
| Maximum residuals (Training) | 0.01449 | 0.01460 | 0.01280 | 0.01310 | 0.01403 |
| Minimal residuals (Testing) | -0.01449 | -0.01406 | -0.01344 | -0.01425 | -0.01405 |
| Maximum residua (Testing) | 0.01227 | 0.01228 | 0.01082 | 0.01312 | 0.01202 |
| Minimal residuals (Validation) | -0.01255 | -0.01428 | -0.01317 | -0.01454 | -0.01483 |
| Maximum residuals (Validation) | 0.01485 | 0.01246 | 0.01521 | 0.01308 | 0.01258 |
| Minimal standard residua (Training) | -4.01689 | -4.51769 | -4.35931 | -4.36985 | -4.48091 |
| Maximum standard residuals (Training) | 3.97793 | 3.97692 | 3.48442 | 3.59497 | 3.88114 |
| Minimal standard residuals (Testing) | -3.68314 | -3.54739 | -3.44145 | -3.62991 | -3.71718 |
| Maximum standard residuals (Testing) | 3.11746 | 3.09997 | 2.76965 | 3.34226 | 3.17966 |
| Minimal standard residuals (Validation) | -3.25771 | -3.70885 | -3.44086 | -3.80729 | -3.84626 |
| Maximum standard residuals (Validation) | 3.85619 | 3.23586 | 3.97297 | 3.42616 | 3.26253 |

Source: Own processing.

The basic characteristics (at least those whose absolute values are given) are compared with the basic characteristics in Table 1 (showing the basic characteristics of the examined time series). Also, the residuals analysis is important. In this case, the comparison is vertical (that is, the same characteristic is compared for all data sets of each retained networks). The only network that slightly differs (in a negative way) is the network 3. RBF 1-28-1. Other retained networks show approximately the same performance.

Furthermore, we will deal with the sensitivity analysis of the individual networks. We will examine the sensitivity analysis of CNY/USD to Brent in USD/barrel (Table 8).

**Table 8.** Sensitivity analysis of CNY/USD to Brent in USD/barrel

| Networks | Brent in USD/barrel |
|---|---|
| 1.RBF 1-29-1 | 1.551800 |
| 2.RBF 1-25-1 | 1.602996 |
| 3.RBF 1-28-1 | 1.320312 |
| 4.RBF 1-24-1 | 1.529973 |
| 5.RBF 1-27-1 | 1.543923 |
| Average | 1.509801 |

Source: Own processing.

Based on the results of the analysis, the network with the highest sensitivity of CNY/USD to changes in Brent in USD/barrel appears to be 2. RBF 1-25-1.
For better illustration, see Figure 3.

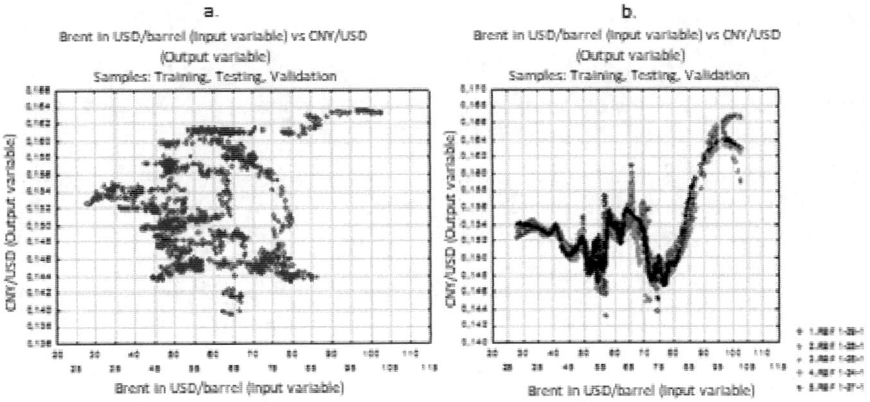


**Fig. 3.** Comparison of data distribution for real relationship between CNY/USD and Brent in USD/barrel and forecast

Source: Own processing.

Figure a. shows the actual distribution of time series data, where Brent in USD/Barrel is on the axis *X* and CNY/USD (without time sequence) on the axis *Y*. Figure b. shows the value of forecasts, which seems to indicate that the values follow the development over time but this is not true.

It follows from the figure that:

1. None of the neural networks is able to forecast the actual development of the CNY/USD exchange rate depending on Brent in USD/barrel completely.
2. It is clear that there is dependence between the CNY/USD time series and Brent in USD/barrel.

## 4 Conclusion

The objective of the contribution was to identify possible relationship between the price of Brent oil (Brent in USD/barrel) and the Chinese Yuan to American dollar (CNY/USD) exchange rate by means of artificial neural networks of the radial basis function (RBF).

Within the experiment, a total of 50,000 artificial RBF neural networks were generated. In each setting based on the time lag of the impact on the CNY/USD exchange rate (0, 1, 5, 10 and 30 days), five artificial neural structures with the best characteristics were retained. Subsequently, the results were analysed with regard to the objective of the contribution. It can be concluded as follows:

1. There is a dependence of CNY/USD on Brent in USD/barrel. It is not a majority one but is measurable.
2. None of the 50,000 generated RBF neural networks is capable of forecasting the actual development of the CNY/USD exchange rate depending on Brent in USD/barrel completely.
3. Time lag of the influence of Brent in USD/barrel on CNY/USD is almost 30 days. It is obvious that the time lag is not constant in the entire course of the monitored time interval.

It is thus evident that the CNY/USD price will play a significant role in creating China´s real product.

Given that it has already been proven that the CNY/USD exchange depends on Brent in USD/barrel, it is important to focus the further research on finding out the time lag with which the price of Brent in USD/barrel is actually reflected in the price of CNY/USD.

## References

1. W. Qiang, A. Lin, C. Zhao, Z. Liu, M. Liu, X. Wang, The impact of international crude oil price fluctuation on the exchange rate of petroleum-importing countries: a summary of recent studies. *Natural Hazards*, **95**(1-2), 227-239 (2019)
2. J. Chai, Y. Wang, S. Wang, Y. Wang, A decomposition – integration model with dynamic fuzzy reconstruction for crude oil price prediction and the implications for sustainable development. *Journal of Cleaner Production*, **229**, 775-786 (2019)
3. S. Liu, W. Fang, X. Gao, F. An, M. Jiang, Y. Li, Long-term memory dynamics of crude oil price spread in non-dollar countries under the influence of exchange rates. *Energy*, **182**, 753-764 (2019)
4. C.P. Zhang, X. Cao, Y.N. Zhao, J. Zhang, L. Yu, Research on the Impact of International Crude Oil Price Fluctuation and Oil Import to China's GDP. *2013 International Conference on Management Science and Engineering(ICMSE)*, pp. 1168-1173 (2013)
5. J. Baek, H.Y. Kim, On the relation between crude oil prices and exchange rates in sub-saharan African countries: A nonlinear ARDL approach. *Journal of International Trade & Economic Development*, 1-12 (2019)
6. F. Novotný, The Link Between the Brent Crude Oil Price and the US Dollar Exchange Rate. *Prague Economic Papers*, **21**(2), 220-232 (2012)
7. D.E. Kayalar, C.C. Küçüközmen, A.S. Selcuk-Kestel, The impact of crude oil prices on financial market indicators: copula approach. *Energy Economics*, **61**, 162-173 (2017)
8. M. Vochozka, J. Horák, P. Šuleř, Equalizing Seasonal Time Series Using Artificial Neural Networks in Predicting the Euro–Yuan Exchange Rate. *Journal of Risk and Financial Management*, **12**(2) (2019)
9. J. Horák, T. Krulický, Comparison of exponential time series alignment and time series alignment using artificial neural networks by example of prediction of future development of stock prices of a specific company. *SHS Web of Conferences: Innovative Economic Symposium 2018 – Milestones and Trends of World Economy*, **61** (2019)
10. P. Šuleř, Using Kohonen´s neural networks to identify the bankruptcy of enterprises: Case study based on construction companies in South Bohemian region. *Proceedings of the 5th International conference Innovation Management, Entrepreneurship and Sustainability*, pp. 985-995 (2017)
11. Z. Rowland, J. Vrbka, Using Artificial Neural Networks for Prediction of Key Indicators of a Company in Global World. *Globalization and Its Socio-Economic Consequences, 16TH International Scientific Conference Proceedings*, pp. 1896-1903 (2016)
12. V. Machová, Z. Rowland, Value generators of enterprises in the processing industry. *Proceedings of the 6th International conference Innovation Management, Entrepreneurship and Sustainability*, pp. 624-634 (2018)
13. R. Adhikari, R.K. Agrawal, A combination of artificial neural network and random walk models for financial time series forecasting. *Neural Computing and Applications*, **24**(6), 1441-1449 (2014)
14. A. R. Sayadi, S. M. M. Tavassoli, M. Monjeziand, M. Rezaei, Application of neural networks to predict net present value in mining projects. *Arabian Journal of Geosciences*, **7**(3), 1067–1072 (2012)

15. A. Hogenboom, W. Ketter, J. Van Dalen, U. Kaymak, J. Collins, A. Gupta, Adaptive Tactical Pricing in Multi-Agent Supply Chain Markets Using Economic Regimes. *Decision Sciences*, **46**(4), 791-818 (2015)
16. M. Šnorek, *Neural networks and neurocomputers*. 1st ed. Prague: Publishing CVUT (2002)
17. Z. Rowland, P. Šuleř, M. Vochozka, Comparison of neural networks and regression time series in estimating the Czech Republic and China trade balance. *SHS Web of Conferences: Innovative Economic Symposium 2018 – Milestones and Trends of World Economy*, **61** (2019)
18. K. Valášková, T. Klieštik, L. Švábová, P. Adamko, Financial risk measurement and prediction modelling for sustainable development of business entities using regression analysis. *Sustainability*, **10**(7) (2018)
19. Y. Mai, H. Chen, J.Z. Zou, Currency co-movement and network correlation structure of foreign Exchange market. *Physica A-Statistical Mechanics and its Applications*, **492**, 65-74 (2018)
20. J. Vrbka, Z. Rowland, Stock price development forecasting using neural networks. *SHS Web of Conferences: Innovative Economic Symposium 2018 – Strategic Partnerships in International Trade,* **39** (2017)
21. F. Rodrigues, I. Markou, F.C. Pereira, Combining time-series and textual data for taxi demand prediction in event areas: A deep learning approach. *Information Fusion*, **49**, 120-129 (2019)
22. T. Klieštik, M. Mišánková, K. Valášková, L. Švábová, Bankruptcy prevention: New effort to reflect on legal and social changes. *Science and Engineering Ethics*, **24**(2), 791-803 (2018)
23. P. Rostan, A. Rostan, The versatility of spectrum analysis for forecasting financial time series. *Journal of Forecasting*, **37**(3), 327-339 (2018)
24. M. Vochozka, Using the Radial Basic Function neural network for determining the financial plan of a company. *Management 2016: International Business and Management, Domestic Particularities and Emerging Markets in the Light of Research*, pp. 331-337 (2016)
25. J. Vrbka, Z. Rowland, Stock price development forecasting using neural networks. *SHS Web of Conferences: Innovative Economic Symposium 2018 – Strategic Partnerships in International Trade,* **39** (2017)
26. World Bank [online], Available at: https://www.worldbank.org/ (2019)

# Agricultural enterprises development strategy investment support in the field of animal husbandry in Ukraine

*Svitlana* Ishchenko[1*], *Kateryna* Skrypniuk[1], *Dmytro* Pyrogov[1], *Tetyana* Tkach[1]

[1]Kremenchuk Mykhailo Ostrohradskyi National University, Pershotravneva Street 20, 39600, Kremenchuk, Ukraine

**Abstract.** The article studies the theoretical and methodological principles of investment support for agricultural enterprises development strategy. Under modern conditions, when the Ukrainian producers are forced to revise the strategic guidelines for their own development as a result of the lost traditional markets for agricultural products, the investment support for implementing the long-term development programs becomes especially topical. Such elements of investment support for enterprise development as investment environment, investment potential and investment climate, have to be considered not only by the commodity producers but also influenced by the state on a regular basis. The enterprises investment support system is dynamic and flexible in terms of quite turbulent modern economic development of Ukraine, it is constantly transformed under the influence of a wide range of factors. This requires the formation and implementation of an effective national policy to attract foreign investment, to boost the investment attractiveness of the national economy as a whole, to ensure an adequate protection level of investments and to implement other system measures to create a favorable investment environment. The study of the livestock sector development trend in Ukraine shows that after years of neglect to the industry development by the state we face degradation, evidenced by the negative dynamics of both volumetric and qualitative indicators. In terms of the research the attention was paid to the development of tools for forecasting the investment support impact on the livestock enterprises development indicators in order to use the predictive model in strategic planning.
**Keywords:** investment support, development strategy, management, model, investment climate, investment environment

## 1 Introduction

Investments in Ukraine are characterized by insufficient investment volume, a significant difference between investment and investment needs, and the limited ability of funding from various sources. It is stipulated by the imperfect legislative framework, high tax pressure, and poorly developed stock market and financial and credit system, inefficient use of depreciation charges and undistributed profits. Under these conditions the problem of

* Corresponding author: tusya22111975@gmail.com

providing the enterprise investment resources is very topical, which is impossible without efficient investments.

Studying the problems of investment support to make any enterprise function and develop in general and the agricultural enterprises in particular has become one of the most popular issues in economic science. It could be explained with the fact that to ensure a stable market, to increase the level of enterprise competitiveness it is necessary to replace the outdated material and technical base, to reconstruct the production capacities, to install new technologies and to produce new types of products.

The formation and functioning of the investment support for the various enterprises of the Ukrainian economy was considered by many scholars, among which we may admit the contribution of I.V. Bezpiata, O.V. Bilotkach, I.M. Vakhovych, A.V. Vitiuk, V.O. Herasina O.I. Hutorov, N.M. Dashchenko, A.A. Dyskina, S.M. Didukh, Yu.O. Kapitanets, M.I. Kisil, M.Yu. Kodenska, V.M. Koshelnyk, O.V. Lytvyn, Yu.A. Malakhova, I.M. Manaienko, D.H. Molodchenko,M.A. Odnoroh, O.O. Okhrimenko, L.D. Pavlovska, O.M. Prosovych, V.I. St rashynskyi, S.M. Sudomyr, A.M. Tkachenko, O.V. Feier, V.V. Chornii, A.Ya. Shevtsova, I. A. Yurieva and others.

Such scientists as Th. P. Tomich, P. Lidder, M. Coley, D. Gollin, R. Meinzen-Dick, P.Webb, P.Carberry drvoted their scientific work to the issues of investment support for the development of griculture [1].

But despite a wide range of issues considered by the scholars, the problems of investment support for enterprises strategic development in dynamic environment still need further investigation.

## 2 Data and methods

The purpose of the research is to develop the theoretical and methodological principles how to provide the investment support for the enterprises development and to offer the tools for forecasting the impact of investment support on agricultural enterprises development indicators in the field of animal husbandry in Ukraine.

In the course of investigation we have used the methods of theoretical generalization, comparison and systematization when disclosing the essence of "investment support for the enterprise development strategy" concept; statistical and economic analysis when assessing the livestock development indicators in Ukraine; grouping and classification when systematizing the factors influencing the level of investment attractiveness, enterprise investment potential; graphics when presenting the research results; correlation analysis when modeling the investment support impact on the livestock enterprises development indicators.

The study is based on the legislative and normative acts of Ukraine, domestic and foreign scientific publications, the official annual publications of the State Statistics Committee of Ukraine, the results of authors' research.

## 3 Results

### 3.1 Terminological study of the investment support strategy for enterprise development

Modern economic conditions make any enterprise not only solve the problems of ensuring the effective functioning at a certain period of time, but also to understand, to target, to plan and to justify the resource usage and to ensure its long-term development, given the predictable and unpredictable changes in the external environment, which, in its turn, creates the need for strategy development and strategic planning and management.

Despite the fact that the problems of strategic development have been analysed in a number of both domestic and foreign scientific works, many questions are still debatable. There is a large variety of such definitions as the concepts of "strategy" and "enterprise development" as well as the concept of "enterprise development strategy" among the scholars.

Approaches to the interpretation of the term "strategy" can be conventionally united into several directions.

The first direction, followed by Ansoff [2], Shershnova [3], Boiarska [4], Martynenko and Ihnatieva [5], Vikhanskii [6], deals with interpreting the strategy as a concept, a system of views on the variable enterprise future development, taking into account the variable environment, the goals set, and resource constraints. M. Martynenko and I. Ignatieva define the strategy as "organization business concept with a particular strategic perspective in the form of a long-term business plan that can develop this concept and provide the competitive advantages in achieving the goals" [4].

The second direction, formed by such scientists as Porter [7], Blank [8], Shekhovtseva [9], is aimed at understanding the strategy as a certain model, a set of rules for desicion-making process in management and the ways of reacting to external influence to achieve the enterprise long-term goals.

The third direction, presented to the scientific community by such scientists as Vinokurov [10], Saienko [11], Dergousova [12], is focused on interpreting the essence of the strategy as a plan of action, a set of ways to achieve the goals of the enterprise by coordinating the available resources and their distributing in accordance with the prioritized goals.

The authors studying the questions of forming and implementing the enterprises strategies combine these approaches in different combinations, which results in a wide variety of this concept interpretation.

The concept of "development" in general and "enterprise development" in particular can be also the subject of ongoing scientific discussions.

Development as a whole is considered as a result, a new qualitative state of the object, or as a process, a dynamic system of transforming the states of the object.

The authors tend to consider the enterprise development as a process, a set of changes, but the consequences of these changes cause quite different the scientists' views. A number of scholars believe that development does not necessarily have to be connected with the improvement of the state of the object in relation to its current characteristics, but it involves only the acquisition of new qualities, different from the previous ones. What is more, these qualities may be desirable for the goals of enterprise functionning, as well as undesirable.

For example, Boyarska [4], Epifanova and Dziubko [13] refer all the following basic strategies to the development ones such as diversified growth strategies, integrated growth strategies, concentrated growth strategies and reduction strategies. They note that a development strategy should include the proper allocation of enterprise resources.

Another group of authors persists that the development involves the improvement, advanced state of the enterprise as a system, increasing its viability, developing the potential of confronting the negative impact of the external environment, improving the efficiency of functioning.

So, we believe that Tyukha and Denysiuk [14] have reasonably distinguished between the strategies with development features and the strategies without development features, noting that, for instance, a strategy aimed at halting production or eliminating an enterprise is not a development strategy, because it does not include progress or improvement, but, on the contrary, it is the result of decreasing working life, regression, decline, degradation, contradicting the very interpretation of the term "development".

According to Dergousova [12], the enterprise development strategy involves the growth of the enterprise, strengthening of its competitiveness and market position, improvement of the working life under unfavourable operating conditions.

Golovinov and Lytvynov [15] consider that the development strategy includes a set of managerial decisions resulting in the improvement of the enterprise state by strengthening its internal potential and ability to adapt to continuously changing environment. Vinichenko [16] states that the enterprise development is based on three interrelated processes such as changing, growing and improving.

Summarizing the results of researchers' works about the essence of the enterprise development strategy, it is advisable to identify a number of distinct characteristics indicating the signs of development (Fig. 1).

Investment support is a prerequisite and driving force of strategic development in a market economy. Without proper investment support the economic growth, strategic transformation and increased productivity of enterprises are impossible. Investment support acts as a compulsory tool for implementing the development strategy of an entity.

Investment support is a necessary component of any short-term or long-term development program.

Terminological study of the "investment support" concept in modern economic science suggests that the modern scholars' approaches to the interpretation of this term can be divided into two main groups. The first group, represented by Parakonnyi [17], Kapitanets [18], Bilotkach and Pavlova [19], Kodenska [20], Kisil [21] believes that the investment support is a set of investment resources, conditions and means of investment activities.

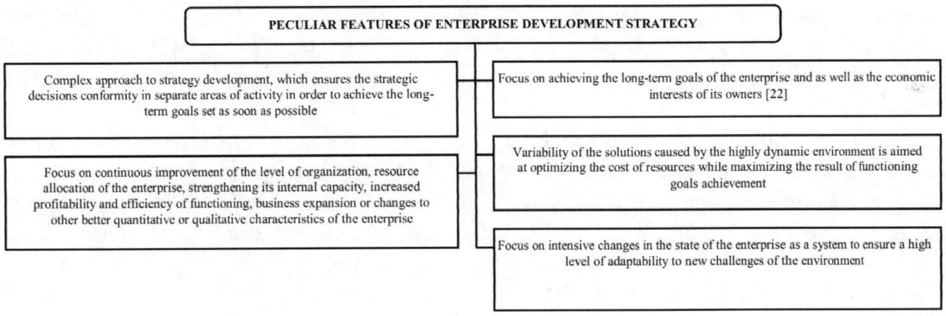


**Fig. 1.** Peculiar features of enterprise development strategy

Source: Own processing based on [20, 21].

The second group of scientists (Bezpiata [23], Tovmasian [24], Shevtsova [25], Vakhovych [26], Feier [27]) tends to consider the investment support as a system of economic relations that arise between the business entities and investment market participants, aimed at the accumulation of investment resources, their rational allocation and funding in order to realize perspective social economic goals.

The elements of the investment support system and its functions determine the structure how the mechanism operates, including such elements as resource support, organizational support and institutional provision of investment activity (Fig. 2).

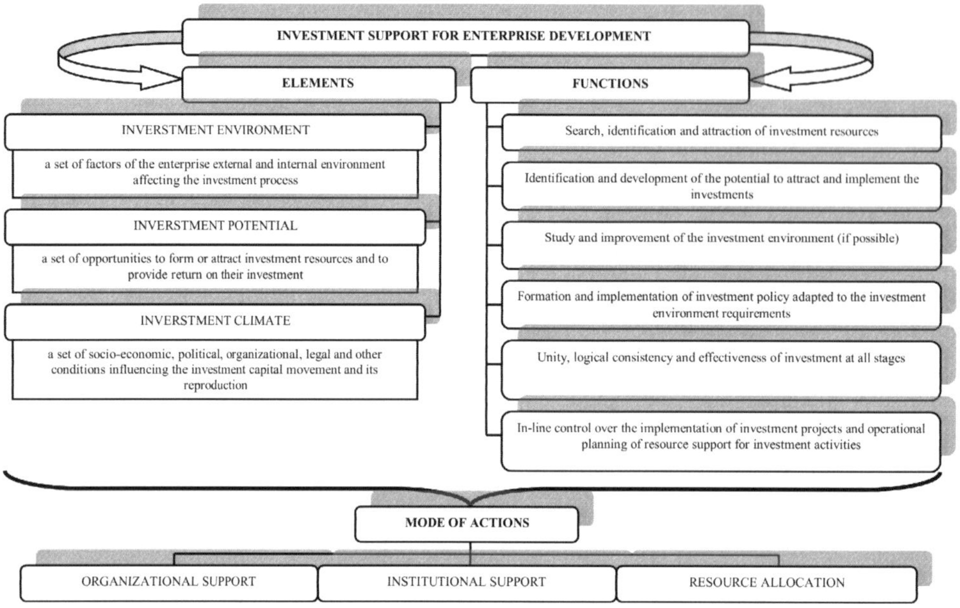


**Fig. 2.** Enterprise development investment support system

Source: Own processing.

Enterprise investment support system is dynamic being constantly transformed given the various changing factors, for example, investment resources availability level, enterprise innovative susceptibility level, the level of investment attractiveness, the market competitive position and the developed competitive advantages, the level of congenial investment climate, changing capacity of the company internal investment potential, and many others.

First of all, the effectiveness of investment support system is stipulated by adhering to the principles of systemicity, complexity and preventive adaptability to conjunctural changes. It should be noted that the impact on the investment support system is provided by the investment and financial markets as well as the market situation the company faces having impact on the level of its investment attractiveness. Moreover, the economic situation has cyclical fluctuations so the enterprise investment support system should have a sufficient degree of flexibility and strategic vision to ensure a biased response to negative market conditions.

After considering the theoretical meaning of the components of the "enterprise development strategy investment support" term we have come to the following vision of this category (Fig. 3).

The given terminological structure of this category has allowed systematizing, generalizing and contributing to the existing scientific views on its essential content and economic significance.

## 3.2 Trends and prospects of livestock development in Ukraine and its investment support

Traditionally, the livestock sector, with the production amounting to about 20% of the total volume of gross Ukrainian agricultural production, plays an important role in the agricultural complex development in Ukraine. Moreover, the industry is undoubtedly an essential element of the national food security. But, despite the strategic importance of the industry, its state could be described as ongoing stagnation.

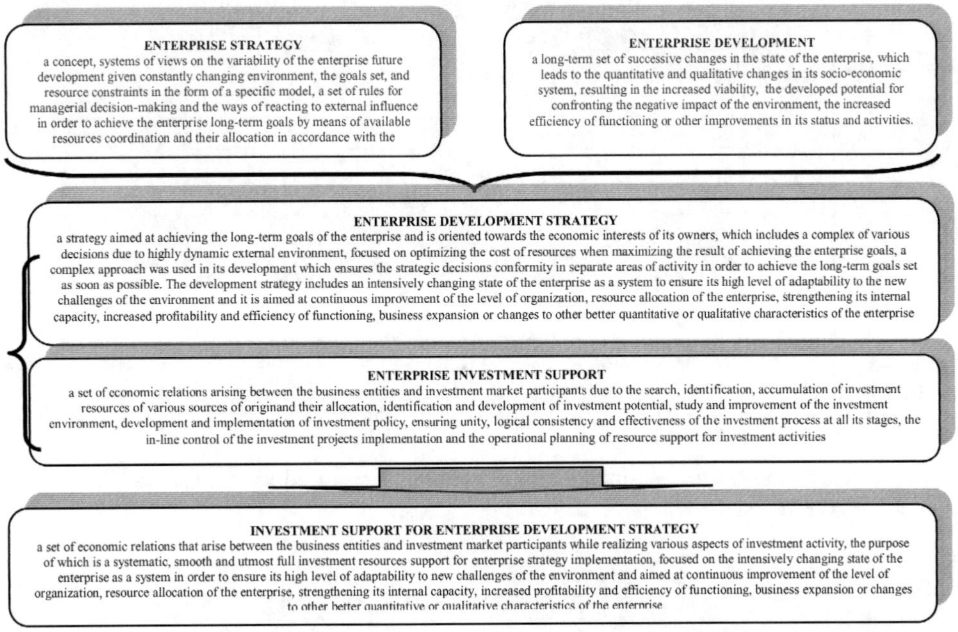


**Fig. 3.** Terminological structure of the category "investment support for enterprise development strategy"

Source: Own processing.

Apart from the overall negative impact of socioeconomic instability in Ukraine in the last few years, such livestock enterprises as dairy-and-meat producers have faced a new challenge being the loss of the traditional external market for livestock products in the Russian Federation, caused by the deteriorating political and economic relations with a neighboring country because of the aggression against Ukraine. So, Russia began to restrict the import of Ukrainian cheese in 2012. In July 2014 the Russian Federation resorted to a total ban on the import of dairy-and-meat products from Ukraine, and in 2016 it introduced a food embargo against Ukraine. In general, over the past five years the volume of Ukrainian dairy-and-meat products exported the Russian Federation, whose share in 2012 amounted to about 70% of the total export volume of these Ukrainian products, has decreased 117 times (Fig. 4), from $ 169 million per year to almost zero (Fig. 5) [28].

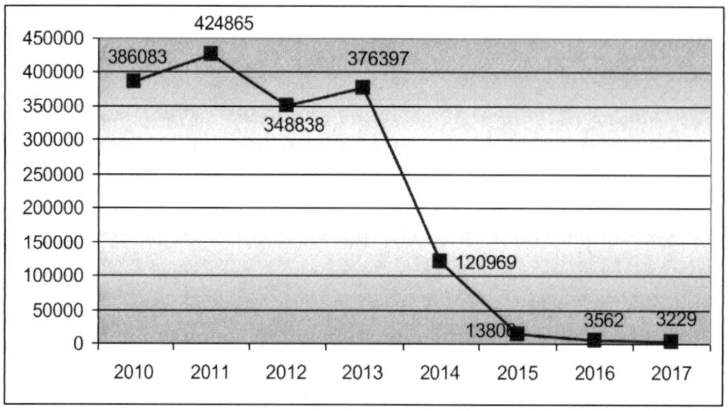


**Fig. 4.** Dynamics of the Ukrainian milk and dairy products exported to the Russian Federation (USD Thousand)

Source: Own processing based on [28].

Dynamics of the Ukrainian meat and meat products exported to the Russian Federation is presented on fig. 5.

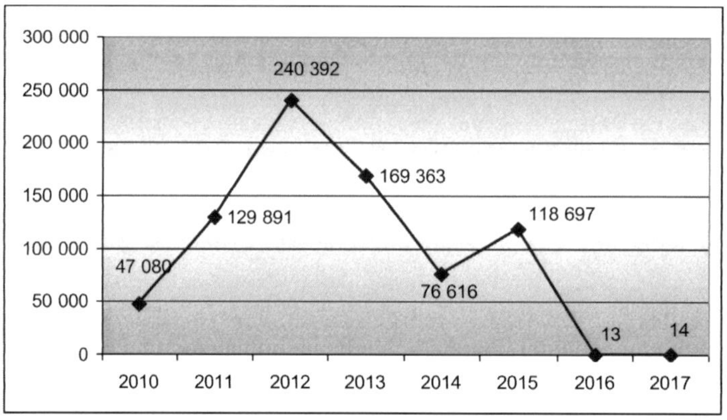


**Fig. 5.** Dynamics of the Ukrainian meat and meat products exported to the Russian Federation (USD Thousand)

Source: Own processing based on [28].

Negative external factors contributed to the deepening of the existing internal problems of the industry. Thus, compared with the indicators in 1990, at the beginning of 2018, the number of cattle decreased by almost 7 times, the pig population by 3 times, the poultry by 1.3 times, despite the fact that there are mostly positive trends of poultry farming development during last 15 years [28].

During 2005-2018, the livestock population in Ukraine decreased almost twice (Fig. 6), including the milking herd population.

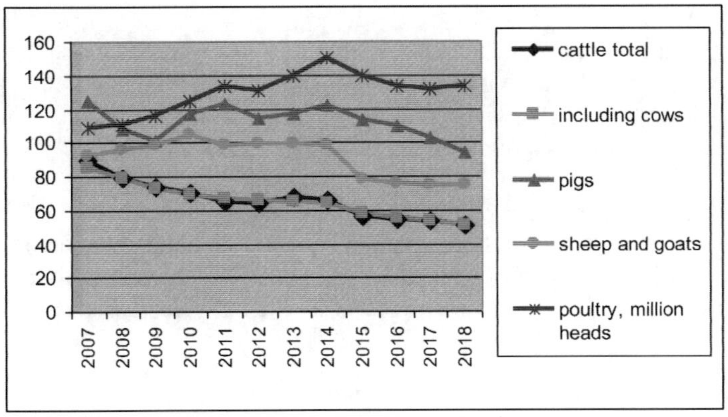


**Fig. 6.** Dynamics of livestock and poultry products growth rates as at January 1 (until 2005 base year), %

Source: Own processing based on [29].

In the last five years, sheep and goat population has been significantly shrinked as a result of the Crimea annexation being the place with traditionally developing sheep breeding.

Reductions in pig breeding as in other livestock sectors are partially caused by the loss of control over the temporarily occupied territories of Luhansk, Donetsk regions and the Crimea, but during the past three years, the dynamics of sector development also remains negative as a result of the African swine fever epidemic outbreaks in various regions of Ukraine.

During the analyzed period, the positive tendencies were observed only in poultry farming, but there are no steady development trends either.

During 2006-2017 the meat production volume in Ukraine grew by one and a half times (Fig. 7), but a significant part of this increase was the result of the total cattle slaughtering, the number of which for the same period decreased almost twice.

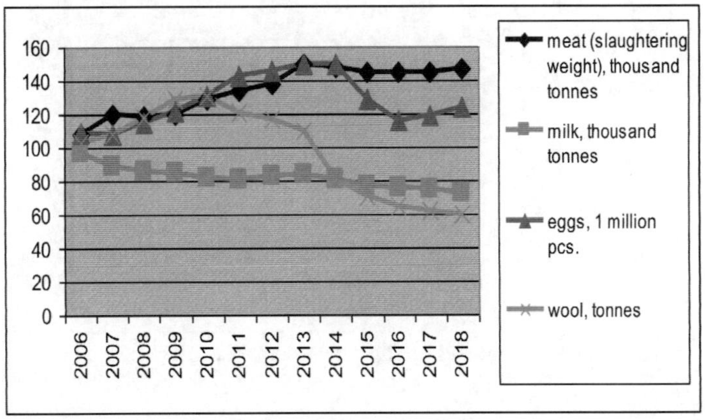


**Fig. 7.** Dynamics of main livestock production volume rates growth (until 2005 base year), %

Source: Own processing based on [29].

Due to the reductions of cow population, milk production declined at a significant pace despite the fact that the milking herd productivity has increased significantly. So, if in 2005 the annual productivity of one cow on average was 3.49 tons of milk per year, in 2017 this indicator reached the point of 4.87 tons. In general, over the last 12 years, the gross milk production has decreased by 25.04%

The livestock sector development is negatively affected by various factors. A significant negative effect is caused by the violation of the equivalence principle while setting prices for the products. The increase in procurement prices for livestock products does not cover the increase in costs for its production, caused by a significant increase in energy prices, inflationary processes, increase in the cost of feedstuff, the cost of imported veterinary drugs and grooming aids due to lowering hryvnia exchange rate, etc.

The decline in population purchasing power also has a negative impact on the industry development trends. For example, over the last three years, the consumption of meat and meat products, milk and dairy products, eggs has significantly decreased (Fig. 8).

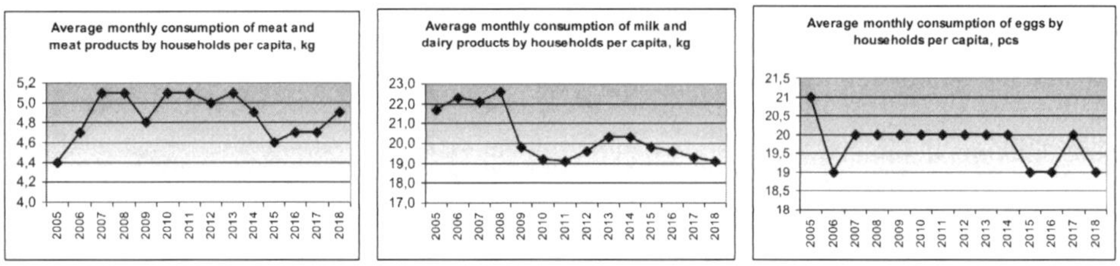


**Fig. 8.** Dynamics of average monthly consumption of livestock products by households in Ukraine per capita

Source: Own processing based on [29].

The problem of renovation of the production and technical base is also relevant to livestock enterprises, as well as to most sectors of the Ukraine's national economy.

Speaking about the prospects of the industry development, it is impossible to ignore the issue of implementing the export livestock potential.

During 2007-2017, Ukraine has gradually increased the meat and meat products exports (Fig. 9), which during the stated period increased almost 13 times.

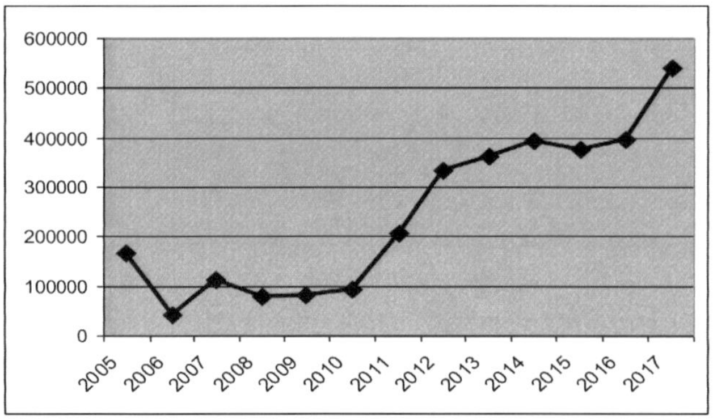


**Fig. 9.** Dynamics of Ukraine's meat and meat products exports, USD Thousand

Source: Own processing based on [28].

The positive tendencies of the industry development in this direction are evidenced by the gradual increase of Ukraine's share in meat products world export (Fig. 10).

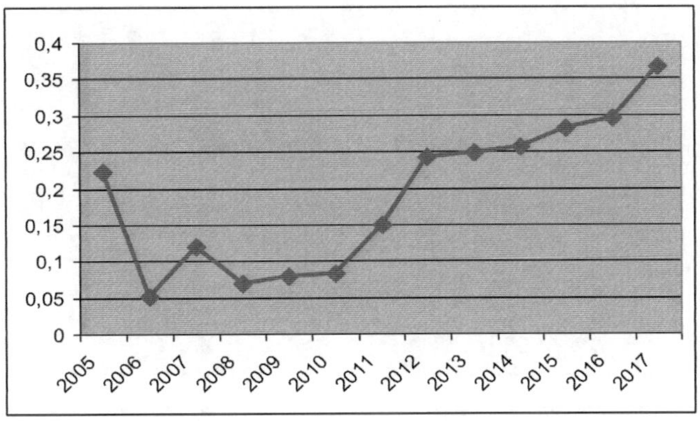


**Fig. 10.** Dynamics of Ukraine's share in meat and meat products world export, %

Source: Own processing based on [28].

Unlike meat and meat products export, the dynamics of Ukrainian milk and dairy products export during 2011-2016, on the contrary, showed a very negative trend. From 2005 to 2013, the volumes of milk export were subject to significant fluctuations, and from 2014 to 2016, they generally declined at a rather high pace (Fig. 11).

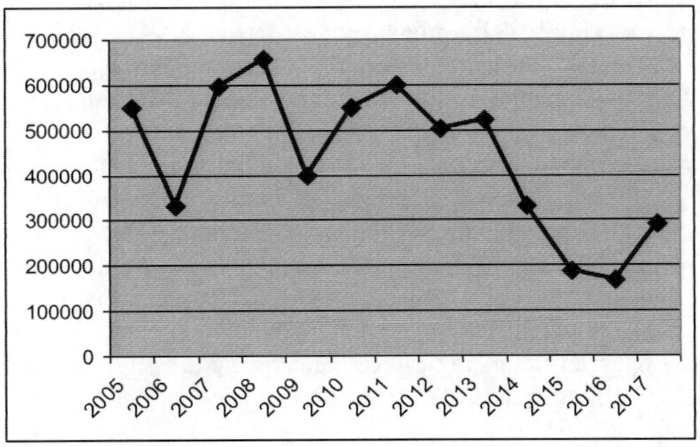


**Fig. 11.** Dynamics of Ukraine's milk and dairy products exports, USD Thousand

Source: Own processing based on [28].

As for the share of Ukrainian exports in the milk and dairy products world export, since 2005 there has been a general tendency of its reducing, from 1.22% in 2005 to 0.23% in 2016 (Fig. 12) [28].

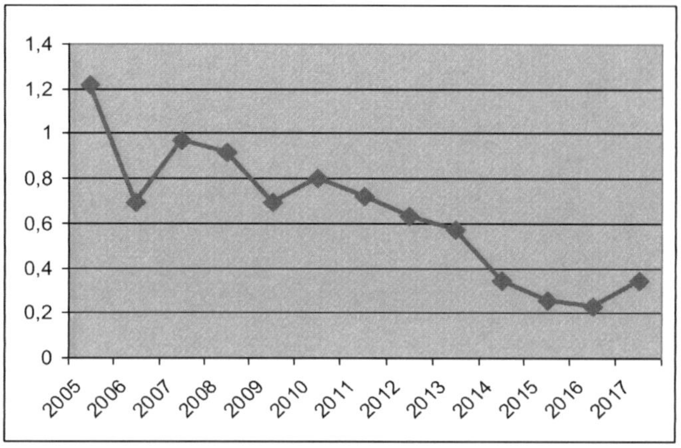


**Fig. 12.** Dynamics of Ukraine's share in milk and dairy products world export, %

Source: Own processing based on [28].

The main problems of Ukrainian milk and dairy products export development and its reorientation to more demanding European markets include dairy raw materials of low quality, low production capacity, the production failing to meet the European quality requirements, a lack of certification according to the European standards. It should be noted that Ukraine can export only dairy products made from extra-class dairy raw materials, the share of which from a total milk production barely reaches 16.5%.

The largest share of milk to be further processed comes from the private sector (about 32% of the total amount of processed dairy raw material), which is characterized by primitive conditions of animal welfare, far from the European standards, the lack of proper sanitary and hygienic conditions, and, as a result, the low quality of raw materials (87% of the milk procured from the population belongs to the second (lowest) grade).

Positive changes in Ukrainian exported milk and dairy products can be observed in 2017. Thus, compared to 2016, the total milk and dairy products export increased by 75.6% in 2017, being a significant increase.

The dairy farming development perspective can be considered in a positive way thanks to the fact that Ukraine significantly increased the dairy products export to the European developed countries in 2017, where the quality standards are much higher than those in the traditional post-soviet markets for Ukrainian livestock products (Fig. 13).

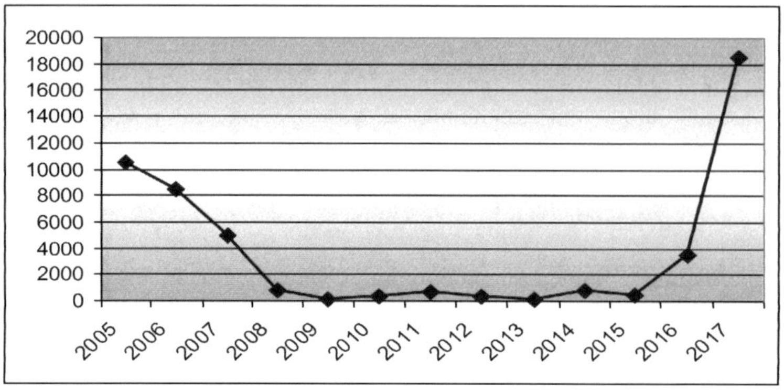


**Fig. 13.** Dynamics of Ukrainian milk and dairy products exported to the EU countries

Source: Own processing based on [28].

New export prospects create the conditions for attracting the agricultural producers' interest to develop the dairy farming. However, in order to increase the quality of milk production to the European standards, the industry needs to undertake a significant modernization, which, in its turn, raises the problem of investment support for the agricultural enterprises development.

The crisis in the country's economy, the complex geopolitical situation, internal political tensions, and military actions in the east of the country had significant negative consequences for the investment climate in Ukraine. It has led to the sharp decrease in the ability of agricultural enterprises to attract the investment resources in the amount necessary for active development, both at the expense of internal and external sources. The situation in the country forced even large enterprises to reconsider their current investment plans and curb their appetite for spending a lot, especially when faced with the lost sales markets in the Russian Federation, they had been working with for many years.

If we analyze the dynamics of capital investment in livestock production in hryvnia equivalent, the trend seems to be positive, the trend line indicates a gradual increase in capital investment (Fig. 14).

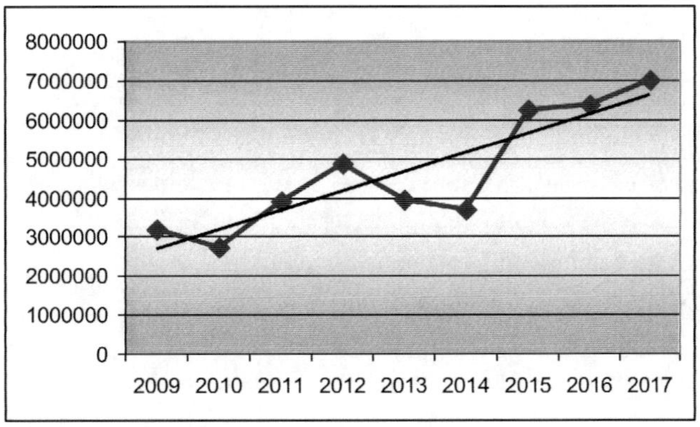


**Fig. 14.** Dynamics of capital investment volume in animal husbandry, UAH Thousand

Source: Own processing based on [29].

Calculation of the capital investment volume in dollar equivalent by the average annual rate significantly changes the picture (Fig. 15).

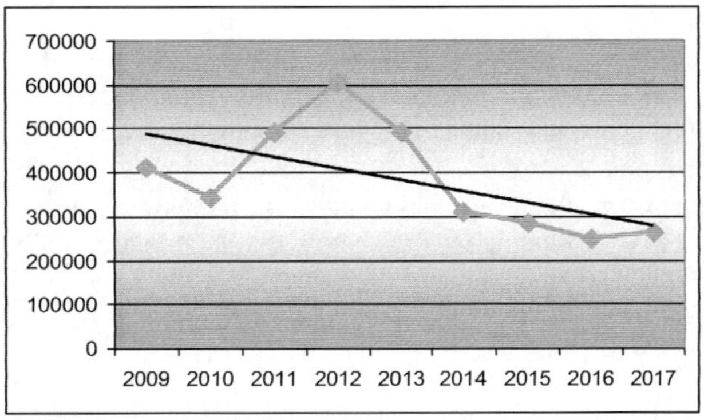


**Fig. 15.** Capital investment volume in animal husbandry, calculated in US dollars as to the average annual rate, USD Thousand

Source: Own processing based on [29].

The results of the calculations show that the increase of capital investments in livestock industry during 2009-2017 was quite nominal, while real investment volumes were gradually decreasing.

As to the share of capital investment in livestock production relative to the total volume of capital investment in Ukraine (Fig. 16), it remained relatively constant during 2009-2017, which indicates the impact on investment activity in the area of general trends in the national economy and the general deterioration of the investment climate.

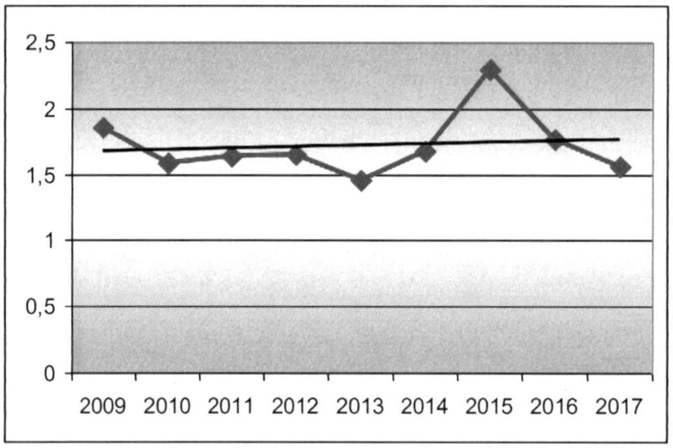


**Fig. 16.** Dynamics of the capital investment share in livestock production in the total volume of capital investments in the national economy of Ukraine, %

Source: Own processing based on [29].

We faced only one exception which is 2015, when the possibility of losing the main market of products forced the livestock enterprises to invest to increase the level of products compliance with the European quality standards despite the negative investment climate.

The maximum exploitation of all possible investment resources is the necessary condition for the further development of the livestock sector, strengthening and successful implementation of its export potential.

Today the main source of livestock enterprises investment resources is their own funds. But given the long-term crisis in the industry, it is difficult to find an enterprise having an internal potential for this.

High loan interest rates also do not contribute to the revitalization of investment activity.

At the beginning of 2018 the Cabinet of Ministers approved the agrarian support programs within the framework of state budget allocation, including financing in the amount of 4 billion UAH stipulated by "State support of the livestock sector" budget program. This program also provides the partial compensation for the cost of construction and reconstruction of livestock farms and complexes, milking rooms. In addition, the farm support program guarantees the partial compensation for the cost of purchased agricultural machinery and equipment of domestic production, the partial compensation of the interest rate on loans taken from the national banks in the national currency [30]. Undoubtedly, these are the positive steps towards stimulating the domestic livestock development, but, as practice shows, they are not sufficient to provide the urgent qualitative changes.

Nowadays, the dynamics of direct foreign investment in the agricultural, forestry and fishing sectors of Ukraine has a negative trend (Fig. 17), in spite of improvements observed in 2017. The main investors here are Germany, Denmark, Great Britain and Poland.

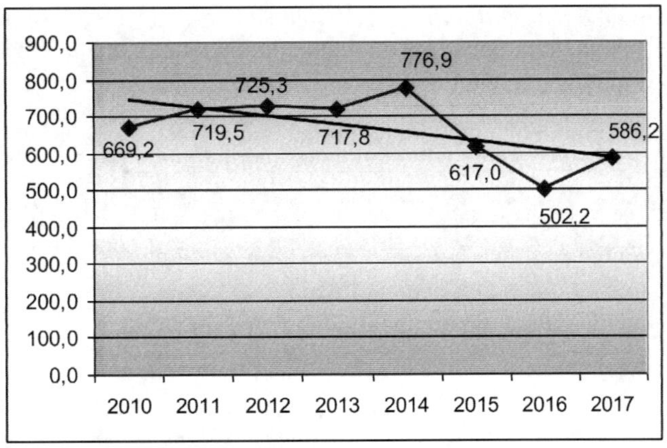


**Fig. 17.** Dynamics of direct foreign investments in agriculture, forestry and fishing industry of Ukraine, US Dollars Million

Source: Own processing based on [29].

Under the given conditions, the state is required to stimulate the investment activity.

In addition to ensuring the improvement of the overall investment climate in the country by means of stabilizing the economic and political conditions, the main elements for creating an attractive investment environment in agriculture by the state include:

- formation of a logical, understandable and predictable legislative framework;
- simplified administrative procedures for managerial decision making in the sphere of investment;
- ensuring equal working conditions and protection for different types of investors;
- providing access to budget agricultural producers financing of different forms of ownership and different production capacities without lobbying for the interests of large agricultural holdings, etc.

It is impossible to ensure both the Ukrainian economy systemic development as a whole and the livestock industry development in particular without implementing an effective national policy of attracting foreign investments, boosting investment attractiveness of the national economy as a whole, ensuring an adequate level of investments protection.

## 4 Discussion

Improved mechanism of investment support for animal husbandry strategic development involves a scientific footing of the prospects for changing indicators characterizing the industry development level, based on the predictive models.

Forecasting the impact of investment support on development indicators implies:

- selecting the development indicators which are really significant for the level of industry investment support;
- identifying the type of functional relationships between the factor (the level of investment support) and performance indicators;
- analysing the significance of the constructed models and the choice of the most suitable.

At the first stage of the analysis, the statistical data on the dynamics of livestock development indicators were collected from available open sources, which includes [31]:

- annual volumes of milk production in actual measurement;
- milk production profitability;
- annual milk and dairy products export volumes.

After calculating the pair correlation coefficients (Table 1) and determining the binding force between the factor and the resultant indicator according to the Chaddock scale we found out that the most significant correlation is between the volume of livestock capital investment and the volume of milk production in actual measurements (correlation coefficient 0.792) and the correlation between the volume of capital investment and the volumes of milk and dairy exports (correlation coefficient of 0.725).

The link between the volume of capital investment and the profitability of milk production is reversed and much less significant, so we do not consider this dependence in the course of our investigation.

**Table 1.** Pair correlation coefficients results

| | | (X)<br>Annual volume of livestock capital investments, USD Thousand | Binding force under the Chaddock scale |
|---|---|---|---|
| (Y) | Milk production volumes, tonnes | 0.728658 | high |
| | Profitability / unprofitability of milk production, % | -0.55508 | average |
| | Annual volumes of exported milk and dairy products, USD Thousand | 0.725038 | high |

Source: Own processing.

After identifying the connection between the investigated indicators, we set to evaluate the forms of dependencies between the productive and factor characteristics in two selected models.

Based on the available initial data on the volume of livestock production capital investments in 2009-2017 and the annual milk production volumes for the corresponding period, we construct a correlation field (Fig. 18).

The location of the points on Fig. 18 suggests the presence of a linear or logarithmic relationship between the studied variables.

| | Annual volume of livestock capital investment, USD Thousand (X) | Milk production volumes, tonnes (Y) | |
|---|---|---|---|
| | X | Y | Ln(X) |
| 2009 | 410444,09 | 11609,6 | 12,9249949988 |
| 2010 | 341320,25 | 11248,5 | 12,7405764651 |
| 2011 | 489404,09 | 11086 | 13,1009437871 |
| 2012 | 605584,66 | 11377,6 | 13,3139496505 |
| 2013 | 490368,2 | 11488,2 | 13,1029118165 |
| 2014 | 310419,63 | 11132,8 | 12,6456803063 |
| 2015 | 286541,77 | 10615,4 | 12,5656395986 |
| 2016 | 248791,57 | 10381,5 | 12,4243707566 |
| 2017 | 262634,77 | 10280,5 | 12,4785196388 |

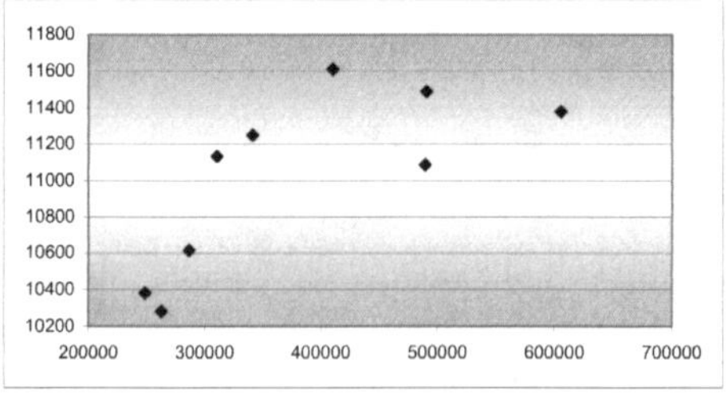


**Fig. 18.** Initial data and correlation field of the model of the milk production depending on the level of livestock enterprises investment support

Source: Own processing.

The results of constructing two variants of the model of the milk production depending on the level of livestock enterprises investment support are shown in Fig. 19-20.

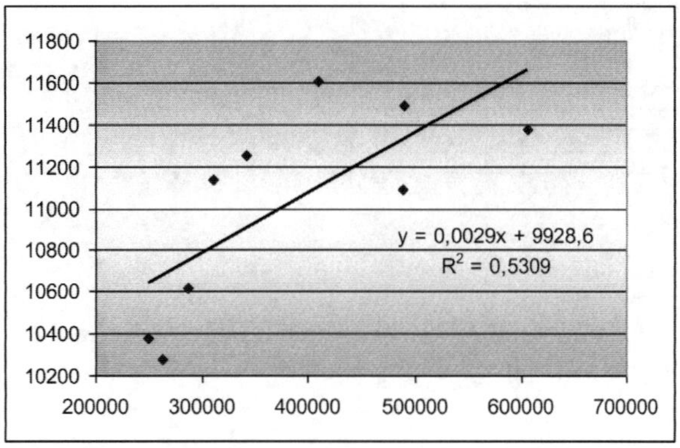


**Fig. 19.** Results of calculating the indicators of the linear model of the milk production depending on the level of livestock enterprises investment support

Source: Own processing.

Fig. 20 show the results of calculating the indicators of the logarithmic model of the milk production depending on the level of livestock enterprises investment support

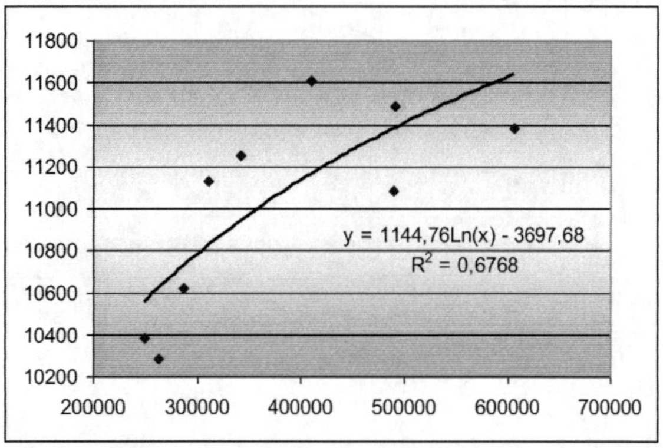


**Fig. 20.** Results of calculating the indicators of the logarithmic model of the milk production depending on the level of livestock enterprises investment support

Source: Own processing.

Figures 19-20 show that in order to evaluate the indicators of the model of milk production depending on the level of livestock enterprises investment support, it is necessary to apply the logarithmic form of communication, since the determination coefficient of this function (0.6768) is higher than in the linear form (0.5309).

Similarly, we build two variants of the model of the exported milk and dairy products depending on the level of livestock enterprises investment support. Output data is shown in Fig. 21.

| | Annual volume of livestock capital investment, USD Thousand (X) | Annual volume of exported milk and dairy products, USD Thousand (Y) | |
|---|---|---|---|
| | X | Y | Ln(X) |
| 2009 | 410444,09 | 400122,78 | 12,924995 |
| 2010 | 341320,25 | 548633,645 | 12,74057647 |
| 2011 | 489404,09 | 598674,076 | 13,10094379 |
| 2012 | 605584,66 | 503027,91 | 13,31394965 |
| 2013 | 490368,2 | 521915,7 | 13,10291182 |
| 2014 | 310419,63 | 330963,8 | 12,64568031 |
| 2015 | 286541,77 | 186550,01 | 12,5656396 |
| 2016 | 248791,57 | 151096,323 | 12,42437076 |
| 2017 | 262634,77 | 372857 | 12,47851964 |

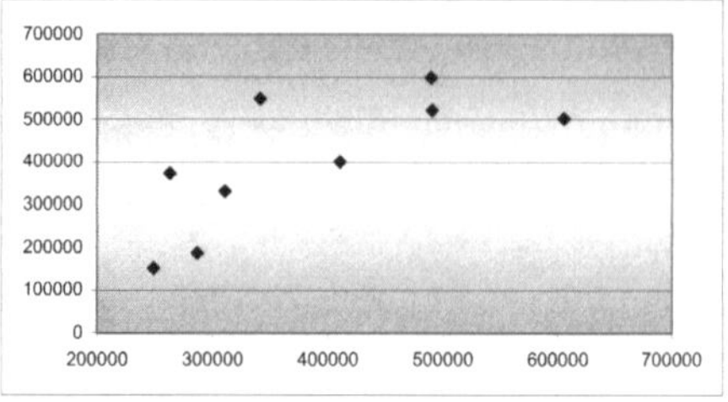


**Fig. 21.** Output data and correlation field for analysing the model of the exported milk and dairy products depending on the level of livestock enterprises investment support

Source: Own processing.

The results of the calculations are shown in Fig. 22-23.

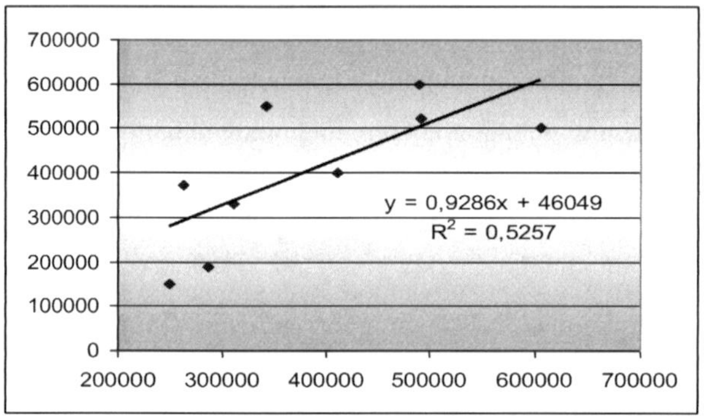


**Fig. 22.** Results of calculating the indicators of the linear model of the exported milk and dairy products depending on the level of livestock enterprises investment support

Source: Own processing.

Results of calculating the indicators of the logarithmic model of the exported milk and dairy products depending on the level of livestock enterprises investment support are presented in fig. 23.

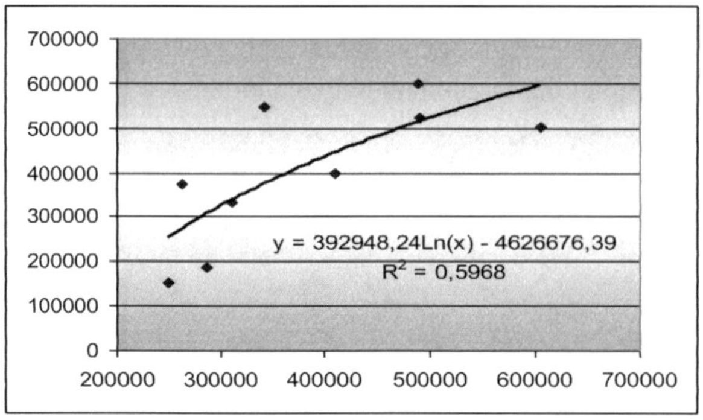


**Fig. 23.** Results of calculating the indicators of the logarithmic model of the exported milk and dairy products depending on the level of livestock enterprises investment support

Source: Own processing.

Figures 22-23 show that in order to evaluate the indicators of the model of dependence, it is necessary to apply the logarithmic form of communication, since the determination coefficient of this function (0.5968) is higher than in the linear form (0.5257).

Thus, the results of the analysis include two logarithmic models, one of which shows the level of milk production in actual meters depending on the level of livestock enterprises investment support, and the other one shows the exported milk and dairy products depending on the level of livestock enterprises investment support. Both models are reasonable and can be used to predict the impact of changing volumes of investment support on the indicators of milk production and the volume of exported products.

## 5 Conclusion

Thus, as a result of the research, it was identified that the constituent elements of the enterprise investment development include investment environment, investment potential and investment climate.

Depending on the development level of each given element, it is reasonable for the enterprise to attract certain financial resources for investment activity, to focus on particular elements of investment development to maximize the investment needs meeting for the implementation of strategic development programs.

For example, the level of the investment climate in the state determines the priorities and moods of potential investors, their willingness to invest in the development of enterprises of this or that industry, or to invest money or not at all.

The development level of the enterprise internal investment potential determines its ability to develop at the expense of internal sources of growth, their need for the resources involved.

The level of investment attractiveness determines the possibilities of the company as a capital consumer, its ability to draw the potential investor's attention and boost his/her interest to invest in the development of this particular enterprise.

A program of investment support is required for each variant of the given factors combination. It will enable efficient using both the existing internal capabilities of the enterprise and the external environment.

Functioning and development of agricultural enterprises as well as the enterprises of other national economy sectors depend on numerous factors of the internal and external environment.

Modern enterprises operate in rapidly changing environment, which increases the risks associated with the implementation of strategic development programs. It requires more focus on predicting and forecasting future economic processes, designing the development models based on assessing the changing enterprise environment.

The design of realistic strategic development plans based on forecasting the economic processes both within and outside the enterprise and the practical implementation of these plans is impossible without using the appropriate tools. Therefore, the use of the proposed predictive models will give a clearer idea how to implement the investment support projects, their results on indicators of agricultural enterprises strategic development in livestock sector.

## References

1. T.P. Tomich, P. Lidder, M. Coley, D. Gollin, R. Meinzen-Dick, P. Webb, P. Carberry, Food and agricultural innovation pathways for prosperity [online], Available at: https://www.sciencedirect.com/science/article/pii/S0308521X17305383 (2019)
2. H. Ansoff, *Strategic Management*. Moscow: Economics (1989)
3. Z.Y. Shershnova, *Strategic management.* Kyiv: KNEU (2004)

4. M.O. Boiarska, Analysis of development strategy of the enterprise. *Global and national problems of economics*, **5**, 306-310 (2015)
5. M. Martynenko, I. Ihnatieva, *Strategic Management*. Kyiv: Karvella (2006)
6. O.S. Vikhanskii, *Strategic Management*. Moscow: The Economist (2004)
7. M. Porter, *Competitive strategy: Techniques for analyzing industries and competitors*. Moscow: Albina Business Book (2006)
8. E.A. Blank, *Trade management*. Kyiv: Ukrainian-Finnish Institute of Management and Business (1997)
9. L.S. Shekhovtseva, About some concepts of strategic management. *Management in Russia and Abroad*, **6**, 34-41 (2002)
10. V.A. Vinokurov, *Organization of strategic management at the enterprise*. Moscow: Center for Economics and Marketing (1996)
11. M.G. Saienko, *Enterprise strategy*. Ternopil: Ekonomichna dymka (2006)
12. A.A. Dergousova, The issue of transport enterprise development strategy concept determination. *Economic and Management of Machine Building Enterprise: the Problems of Theory and Practice*, **4**(12), 88-101 (2010)
13. I.Y. Yepifanova, M.Y. Dzyubko, Formation of the enterprise development strategy in modern conditions. *Odesa National University herald*, **22**(55), 99-103 (2017)
14. I.V. Tiukha, I.O. Denysiuk, Strategy of development in the system of enterprise strategies. *Ekonomika kharchovoi promyslovosti*, **3**, 33-37 [online], Available at: http://nbuv.gov.ua/UJRN/echp_2013_3_9 (2013)
15. M.I. Holovinov, O.I. Lytvynov, Enterprise development strategy: nature and characteristics. *Scientific Bulletin of UNFU*, **21**(19), 224-228 (2019)
16. O.M. Vinichenko, Characteristics, types and essence of enterprise development. *Agrosvit*, **15**, 49-57 (2015)
17. S.V. Parakonnyi, L.V. Vojshvillo, I.V. Shpankovs'kyj, Areas of investment support for the development of the economic potential of the enterprise. *Visnyk Skhidnoukrains'koho natsional'noho universytetu imeni Volodymyra Dalia: nauk. zhurnal, Luhans'k*, **11**(153), 51-58 (2011)
18. Y.O. Kapitanets, Investment provision of farms. *Innovative Economy*, **5**, 141-144 (2010)
19. O.V.Bilotkach, G.Y. Pavlova, Theoretical and methodological foundations of investment support for the development of agricultural enterprises. *Naukovyj visnyk Uzhgorods`kogo nacional`nogo universytetu. Seriya: Mizhnarodni ekonomichni vidnosyny ta svitove gospodarstvo*, **15**, 29-32 (2017)
20. M.Y. Kodens'ka, Motivation factors of investment providing of agrarian and industrial production development. *Visnyk Akademiyi praci i social`nyh vidnosyn Federaciyi profspilok Ukrayiny*, **2**, 62-66 (2013)
21. M.I. Kisil, Investment support for the development of agrarian production and rural areas. *Menedzhment: Zbirnyk naukovyh prats*, **11**, 225 (2009)
22. M.V. Khatser, Strategy of development of the enterprise in conditions of instability of the state economy. *Scientific Papers of Tavria State Agrotechnological University (Economic Sciences)*, **3**(27), 109-112 (2014)
23. I.V. Bezpiata, *Investment provision of the agrarian sector of the economy*. Mykolayiv (2016)

24. V.P. Tovmasyan, The economic essence of the investment support and its role in the development of enterprises. *Investytsiyi: praktyka ta dosvid*, **2**, 60-63 (2015)
25. Y.A. Shevtsova, Investment support of the enterprises: the dissertation author's abstract for the degree of candidate of economic sciences. *Lugansk*, 20, (2003)
26. I.M. Vakhovych, G.L. Denysyuk, Conceptual grounds for investment support of region's innovative development. *Current problems of the economy*, **1**, 138-144 (2011)
27. O.V. Feyer, The investment support of national enterprises. *Scientific Bulletin of Mukachevo State University, Series «Economy»*, **2**(6), 112-117 (2016)
28. Merchandise trade matrix – detailed products, exports in thousands of dollars, annual, 1995-2017. *United Nations Conference of Trade and Development* [online], Available at: http://unctadstat.unctad.org/wds/TableViewer/tableView.aspx?ReportId=24738 (2019)
29. Agriculture of Ukraine. *Livestock breeding, Statistical Collection, State Statistics Service of Ukraine* [online], Available at: http://www.ukrstat.gov.ua (2019)
30. The state of financing of agro-industrial complex in 2018, *Ministry of Agrarian Policy and Food of Ukraine* [online], Available at: http://www.minagro.gov.ua/node/25302 (2019)
31. O. Maslak, P. Sokurenko, N. Grishko, M. Maslak, Formation of a marketing support complex for an export-oriented enterprise in conditions of risk and uncertainty. *SHS Web of Conferences: Innovative Economic Symposium 2018 – Milestones and Trends of World Economy,* **61** (2019)

# Strategy of economic and social development of St. Petersburg until 2030 achievement

*Olga* Kalchenko[1*], *Svetlana* Evseeva[2], *Oksana* Evseeva[1], and *Kristina* Plis[3]

[1]Graduate School of Management and Business, Institute of Industrial Management, Economics and Trade, Peter the Great St. Petersburg Polytechnic University, Polytechnicheskaya 29, 195251 St. Petersburg, Russia
[2]Graduate School of Service and Trade, Institute of Industrial Management, Economics and Trade, Peter the Great St. Petersburg Polytechnic University, Polytechnicheskaya 29, 195251, St. Petersburg, Russia
[3]International Graduate School of Management, St. Petersburg State Marine Technical University, Locmanskaya 3, 190121 St. Petersburg, Russia

**Abstract.** The paper focuses on the goals' achievements of the Strategy of economic and social development of St. Petersburg until 2030. Strategic directions of St. Petersburg development until 2030 (human capital development, improving urban environment, providing sustainable economic growth, ensuring effective governance and civil society development) are examined. The feasibility of economic and social development indicators, predicted in 2012 for 2020 and 2030, is verified on the basis of actual data for 2016. The indicators, predicted values of which are hard-to-reach and require revision, are revealed. Such indicators are: employed in economy, percent of people with higher education, workplaces with high labor productivity and foreign investment in economy of St. Petersburg. The paper examines the dynamics of the main indicators presented by the Committee for Economic Policy and Strategic Planning of St. Petersburg. The list of indicators is not exhaustive. The Strategy is based on the UN sustainable development goals and one of its scenarios has prospects for implementation.
**Key words:** strategy, economic, social, indicators, Russia, St. Petersburg

## 1 Introduction

St. Petersburg is one of the most developed Russian cities in terms of economy, innovation, industry, culture and social infrastructure [1, 2].

In 2015, St. Petersburg became the winner in the ranking of the Association of innovative regions of Russia in the framework of the "Open innovations" forum and also took the 48th place out of 442 in the ranking of innovative cities in the world according to the portal "Innovation cities Program" [3].

St. Petersburg economic and social development is of particular interest as a federal city in Russia.

* Corresponding author: kalchenko_oa@spbstu.ru

Research methods used: structural analysis, synthesis, comparison and generalization, deduction and induction.

The initial data are the statistical data from PetroStat, Committee for Economic Policy and Strategic Planning of St. Petersburg, government of St. Petersburg and The Strategy of economic and social development of St. Petersburg until 2030.

The paper is aimed at verifying the feasibility of the main economic and social development indicators, projected in 2012 for 2020 and 2030, on the basis of actual data for 2016.

## 2 The strategy of economic and social development of St. Petersburg until 2030

The Strategy of economic and social development of St. Petersburg until 2030 is based on the sustainable development goals, outlined in The United Nations report "Our Common Future" over 30 years ago [4, 5].

The Strategy was created with the assistance of leading international experts in economy and strategy planning and defines the future of the city for the next 15 years. The Strategy is the complex document, which contains justification of developmental scenario of St. Petersburg and description of priorities of a long-term development of the city on the basis of priorities of the Russian Federation. It also defines mission and purposes' systems of economic and social development of St. Petersburg for the period till 2030. The Strategy was adopted in May 2014 [6].

Initially, 3 scenarios of the Strategy were developed: conservative, moderately optimistic and innovative. As the main scenario, the innovative one is chosen.

Priority functions of St. Petersburg:

- a multifunctional city, providing comfortable conditions and high quality of life,
- the leader of the Baltic Sea region,
- an outpost of the Arctic exploration,
- the center of culture, science, education, tourism, and high-tech industry.

The general objective of the Strategy: Ensuring the stable improvement of the quality of life in the city and an increase of the global competitive position of St. Petersburg by achieving the national development priorities, ensuring a sustained economic growth, and using the results of innovation and technology development.

Strategic areas aimed at achieving the general objective of the Strategy and the main priorities of St. Petersburg development until 2030 are also consistent with the sustainable development goals. Strategic directions are: human capital development, improving urban environment, providing sustainable economic growth, ensuring effective governance and civil society development [7].

### 2.1 Human capital development

**Demographic development:**

- permanent population will increase to 5,9 million people (in 2012 – 5 million people);
- city population increase due to migration will range from 40-70 thousand people per year;
- mortality rate will decrease to 11.3 ppm (in 2012 – 12.4).

**Development of the public health system:**

- increase in life expectancy to 78 years;
- mortality reduction from all causes per 1,000 people up to 11.3 ppm;
- ratio of doctors and nurses 1:3,5.

**Development of education, including professional:**
• availability of pre – school education will be at least 100% (in 2012 – 95.65%);
• ratio of the unified state exam (USE) average score in 10% of schools with the best exam results to the average score in 10% of schools with the worst exam results in the main subjects will be 1,1 and 1,07 respectively (in 2012 – 1,42 and 1,82 respectively);
• value of the education index will be 0,984 (in 2012 – 0,971).
**Development of culture:**
• share of cultural heritage objects in satisfactory condition in the total number of objects of cultural heritage of federal, regional and local (municipal) importance will reach 100%;
• attendance of all types of cultural institutions will be at least 3 times per 1 inhabitant per year.
**Development of tourism:**
• number of tourists entered the city for various purposes will reach 10 million people a year;
• capacity of the city hotel complex will be 45 thousand rooms;
• share of sea, river cruise and resort tourism segments will increase by 15%;
• number of resort complex rooms will reach 15 thousand (will increase more than 2 times) and will create a modern infrastructure of outdoor activities.
**Development of physical culture and sports:**
• proportion of residents, systematically engaged in physical culture and sports, will reach 50%;
• number of sportsmen-dischargers and masters of sports will increase by 40%;
• proportion of people with disabilities and disabled people who are systematically engaged in physical training and sports will be 40 thousand people (in 2012 – 24.2 thousand people);
• up to 1.15 people for 10 thousand inhabitants will increase the number of athletes of high qualification, prepared for the reporting year.
**Development of social support system for citizens:**
• Giny coefficient, which characterizes the stratification of society by income, can drop to 0.440 (in 2012 – 0.442);
• proportion of the population living below the poverty line will not exceed 8% of the total population (in 2012 – 9.2%).
**Providing affordable housing for citizens:**
• indicator of the average level excess of the interest rate on the mortgage housing loan in relation to the consumer price index will be reduced to a level not exceeding 2.2 percentage points;
• number of issued mortgage housing loans will increase to 4,200 per year;
• share of communal apartments in the total number of apartments will not exceed 1%.

## 2.2 Improving urban environment

**Improvement of urban territories:**
• the main direction of development is the transition from the regulatory and distributive principle of management to the program-investment principle;
• within the framework of the project management system, it will be possible to plan the direct and indirect effects of budget investments.
**Improvement of city ecology:**
• reduction of air and water basins pollution by 20% compared to the level of 2013;
• air and water environment will be 100% compliant with the regulations of the Baltic region.

**Development of the transport system:**
• 18 new metro stations and 4 additional lobbies of existing stations will be built, which will ensure pedestrian accessibility of metro stations for 55% of the population;
• by 2030, the metro network will consist of 89 stations;
• at least 75% of all population movements will be carried out by public transport.
**Development of energy and engineering infrastructure:**
• transformer capacity of 110 kV network will increase to 21,356.0 MVA;
• installed heat capacity of heat supply sources will be at least 31003.2 Gcal/hour;
• capacity of head structures of the water supply system will be 2800.0 cubic meters/day;
• capacity of the drainage system will reach 2,444.5 m3/day
**Territorial development of St. Petersburg:**
• transition to complex development of territories of the city;
• clarification of the borders;
• adjustment of the functional zoning of territories associated with the placement of objects of engineering and transport infrastructure;
• clarification of the boundaries of protected natural areas.
**Development of St. Petersburg agglomeration:**
• strengthen the importance of small and medium-sized cities;
• reduce differences in the quality of life of the population by ensuring the availability of social, communal and household infrastructure, including through the creation of transport links;
• improve the environmental situation.

## 2.3 Providing sustainable economic growth

**Economic development, establishment of bases of knowledge economy:**
• share of St. Petersburg's economy in Russia's economy will increase to at least 8% (3.8% in 2012);
• share of knowledge economy in gross disposable income of St. Petersburg will be at least 35%;
• the demand for qualified personnel will be provided by at least 80% of highly qualified employees of St. Petersburg;
• number of high – performance jobs will increase to at least 400 thousand units (in 2012 – 102,7 thousand).
• degree of fixed assets depreciation will be reduced to 30% (2012 – 39.2 per cent).
**Innovation and technological development of industry:**
• share of high-tech industry in gross disposable income will be at least 25%;
• share of innovative products in the total volume of shipped products in St. Petersburg should be at least 30% (12.1% in 2012);
• output per industrial worker will increase by at least 2.5 times (in 2012 prices);
• number of advanced production technologies will increase to at least 450 (in 2012 – 259).
**Development of science and innovation activity:**
• share of high–tech and knowledge-intensive sectors in gross disposable income will be at least 25% (in 2012 – 15.6%);
• at least 17 thousand advanced production technologies per year will be used in economic activities (in 2012 – 6.5 thousand)
• number of personnel engaged in research and development will increase to 114 thousand people (in 2012 – 80.7 thousand people).
**Development of the system of continuous education, training and retraining of highly qualified staff for the knowledge economy:**
• a closer link between vocational education and labour market needs is required.

**Development of trade and consumer markets:**
• increase of St. Petersburg budget taxation revenues from implementation of trade activity and services by 20% in comparison with indicator of 2012;
• decrease of consumer goods price index level in comparison with the current period by 15%;
• reduction of three largest suppliers share in retail chains to 45%.
**Small business development:**
• number of small enterprises, including microenterprises, will reach 340 thousand units:
• share of employees of small and medium-sized enterprises in the total average number of employees will reach 50%;
• indicators will allow St. Petersburg to take the World Bank ranking position on favorable business conditions at least 10th.
**Development of the construction industry:**
• volume construction works will reach 1 480.0 billion rubles (in 2012 – 353.1 billion rubles.);
• total housing area input in St. Petersburg will amount to 4.1 million sq. m.;
• total housing stock area will increase to approximately 182 million square metres (a 52% increase compared to 2012).
**Development of the real estate market and support of development projects:**
• investments into fixed assets will increase to 3,000 billion rubles (in 2012 – 351.9 billion rubles);
• investments per capita will increase up to 511.1 thousand rubles. (in 2012 – 70.5 thousand rubles).
**Providing activity connected with Arctic zone development:**
• at least 50% of all purchases (goods, works and services) will be carried out by enterprises and organizations of St. Petersburg (Krylov State Research Centre, shipyard "Severnaya Verf", JSC "SPMBM" Malakhit", etc.);
• 60% of the total research amount on Arctic issues will be carried out by St. Petersburg scientific organizations;
• 15% of St. Petersburg employees will be involved in projects and programs related to the Arctic zone development.
**Development of creative industries:**
• contribution of creative industries to the formation of St. Petersburg gross disposable income will increase to 12%;
• ratio of professionals in creative industries in relation to the total number of able-bodied population in St. Petersburg will be 16%.

## 2.4 Ensuring effective governance and civil society development

**Improvement the quality of public management:**
• share of population positively assessing the executive authorities activities will be 60%;
• share of public services that the population can receive using modern technologies will be 100%.
**Providing personal and public safety of citizens:**
• reduction of deaths in road accidents per 100 thousand people by 40% compared to 2012;
• reduction of thefts and crimes connected with drug trafficking to 15,2 thousand and 8,9 thousand a year, respectively;
• increase in number of police districts to 632 units per 10 thousand people.
**Better integration of young people in modern society:**
• the proportion of young people enrolled in vocational guidance services will be at least 80% of the total number of students;

• the proportion of socially supported minors and young people from the total number of those registered will be 90% of the total number of registered persons;
• the proportion of minors and young people covered by prevention of drug addiction, will be 50% of the number of minors and youth.
**Development of interethnic and interfaith relations:**
• reduction of crimes committed by migrants – by 40% compared to 2013;
• reduction of crimes against migrants – by 40% compared to 2013;
• reduction of illegal migrants – by 80% compared to 2013.
Development of relations with social organizations:
• increase in the number of non-profit organizations – more than 3 times (compared with 2012);
• share of purchased goods, works and services for state and municipal needs realized non-profit organizations will increase by more than 5 times (in comparison with 2012);
• number of volunteers in the non-profit sector will increase by more than 4 times (compared to 2012).

The implementation mechanism of the Strategy involves the monitoring of its performance, which allows to control the achievement of the planned results during the social and economic development of St. Petersburg, based on a comprehensive assessment of targets and benchmarks of the Strategy and state programmes in St. Petersburg.

It is assumed, that combining the efforts of government, businesses, and society will allow St. Petersburg to succeed in the implementation of the Strategy and become a fully global city with a comfortable habitat.

**Table 1.** Resource background of the strategy

| Main indicators | 2012 | 2016 | 2020 | 2030 |
|---|---|---|---|---|
| | Actual data | | Predicted data | |
| Employed in economy (thousand people) | 2,530.4 | 2,970.5 | 2,966.9 | 3,174.2 |
| Percent of people with higher education | 44.7 % | 42.3% | 47.0 % | 50.0 % |
| Workplaces with high labor productivity(thousand units) | 102.7 | 783.1 | 280 | 400 |
| Water use (thousand cubic meter per day) | 1,625 | 1,416 | 1,500 | 1,500 |
| Gross Regional Product of Saint-Petersburg (billion roubles) | 2,530.4 | 3,742.2 | 5,310.5 | 14,000 |
| Fixed capital investment (billion roubles) | 351.9 | 582.3 | 982.7 | 3019 |
| Foreign investment in economy of Saint-Petersburg (billion USD) | 10.8 | 4.8 | 15.2 | 17 |
| Consolidated budget revenues (billion roubles) | 379.9 | 476.7 | 648.3 | 1401.4 |
| Consolidated budget expenditures (billion roubles) | 382.3 | 496.5 | 630.9 | 1401.4 |

Source: Own processing based on [7-9].

According to statistics for 2016, some indicators differ significantly from the predicted for 2020-2030. Particular attention should be paid to: employed in economy, percent of people with higher education, workplaces with high labor productivity and foreign investment in economy of St. Petersburg.

Fig. 1 clearly shows that in 2016 the number of employed in the economy exceeded the predicted data for 2020. The forecast for 2020 needs to be revised.

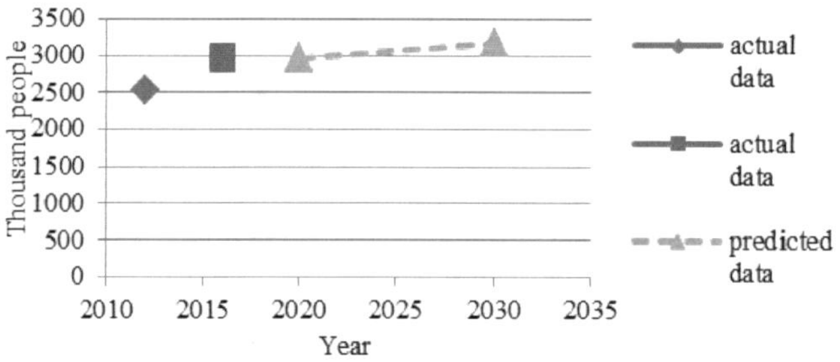


**Fig. 1.** Employed in economy.

Source: Own processing based on [7-9].

The percent of people with higher education, on the contrary, in 2016 decreased compared to 2012, as can be seen in Figure 2.

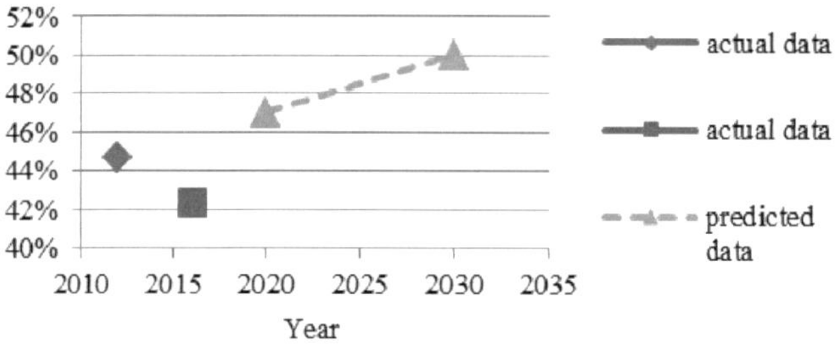


**Fig. 2.** Percent of people with higher education.

Source: Own processing based on [7-9].

Fig. 3 reflects a significant leap in the number of workplaces with high labor productivity in 2016, which exceeds forecasts for 2020 and 2030. Accordingly, the forecast also needs to be revised.

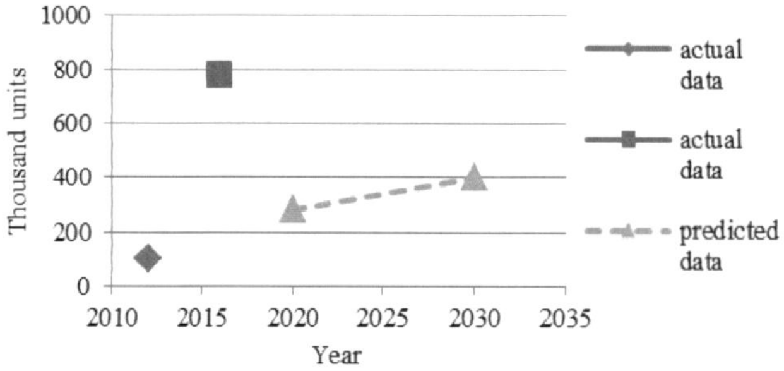


**Fig. 3.** Workplaces with high labor productivity.

Source: formed by author based on [7-9].

Foreign investment in economy of St. Petersburg in 2016 significantly decreased even compared to 2012. Forecasts for 2020 and 2030 are seen as weakly achievable and need to be revised.

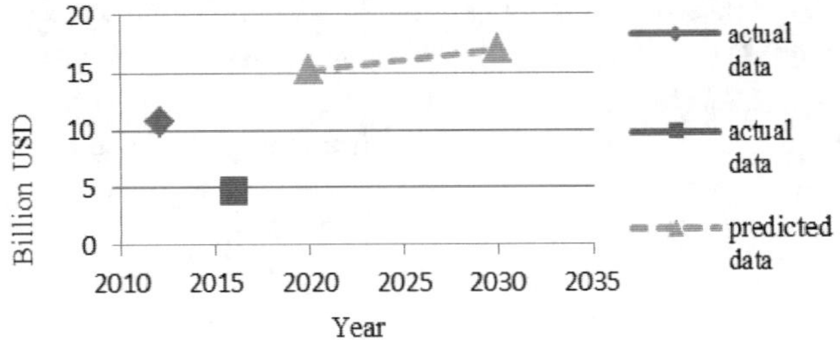


**Fig. 4.** Foreign investment in economy of Saint-Petersburg.

Source: formed by author based on [7-9].

Other indicators of the Table 1 are within the predicted ranges and are achievable.

## 3 Conclusions

An updated strategy until 2035 was recently presented by the Committee for Economic Policy and Strategic Planning of St. Petersburg (CEDIPT) [10]. Unfortunately, information on whether the predicted indicators highlighted in the paper have been revised is not yet available. This represents the limitation of the study and provides opportunities for future research.

During the paper preparation, a greater number of indicators were revealed on various information portals of the St. Petersburg government, committees, and statistical services and in policy documents [11, 12]. For the study, the main indicators were selected, according to CEDIPT.

Current strategy until 2030 and updated strategy until 2035 are fully consistent with the UN sustainable development goals and one of its scenarios has prospects for implementation [13, 14]. A number of pilot innovative projects in St. Petersburg are aimed at sustainable urban development. Major research universities take part in the implementation of most of those projects; some of them are supported by instruments of cooperation across borders through project funding [15, 16]. Cooperation, exchange of experience, joint financing make it possible to achieve sustainable development goals and indicators of economic and social development of such a large European city as St. Petersburg.

This paper was financially supported by the Ministry of Education and Science of the Russian Federation on the program to improve the competitiveness of Peter the Great St. Petersburg Polytechnic University (SPbPU) among the world's leading research and education centres in the 2016-2020.

## References

1. O. Golubchikov, World-city-entrepreneurialism: globalist imaginaries, neoliberal geographies, and the production of new St Petersburg. *Environment and Planning A*, **42**(3), 626 (2010)
2. J. O'Connor, Creative Exports: Taking Cultural Industries to ST Peterburg. *International Journal of Cultural Policy*, **11**(1), 45-60 (2005)

3. Association of innovative regions of Russia [online], Available at: http://i-regions.org/eng/ (2019)
4. Assembly UN General "The Future we want (Resolution adopted by the General Assembly on 27 July 2012) [online], Available at: http://www.un.org/disabilities/documents/rio20_outcome_document_complete.pdf. (2012)
5. Report of the World Commission on Environment and Development: Our Common Future [online], Available at: http://www.un-documents.net/our-common-future.pdf (2019)
6. Strategy of economic and social development of St. Petersburg until 2030 [online], Available at: http://spbstrategy2030.ru (2019)
7. Strategy of economic and social development of St. Petersburg until 2030. CEDIPT [online], Available at: http://cedipt.spb.ru/media/acts/2015/12/09/Strategy_of_economic_and_social_development_of_St._Petersburg_until_2030.pdf (2015)
8. PetroStat [online], Available at: http://petrostat.gks.ru/ (2019)
9. Government of St. Petersburg resolution [online], Available at: http://gov.spb.ru/law?d&nd=822403603&nh=1 (2019)
10. S. Evseeva, O. Kalchenko, O. Evseeva, Circular economy for the energy transition in Saint Petersbutg. *MATEC Web of Conferences: International Science Conference SPbWOSCE-2017 "Business Technologies for Sustainable Urban Development"*, **170** (2018)
11. A.R. Bril, O.V. Kalinina, I.V. Ilin, Sustainable Economic Growth, Education Excellence, and Innovation Management through Vision 2020. *Proceedings of the 29th IBIMA Conference*, pp. 2268-2277 (2017)
12. O.A. Kalchenko. *Proceed. of the ICTM 2016*, 46 (2016)
13. V.V. Okrepilov, V.L. Makarov, A.R. Bakhtizin, S.N. Kuzmina, Application of supercomputer technologies for simulation od socio-economic systems. *R-Economy*, **1**(2), 340-350 (2015)
14. V. Plotnikov, G. Fedotova, E. Popkova, A. Kastyurina, Harmonization of Strategic Planning Indicators of Territories' Socioeconomic Growth. *Regional and Sectoral Economic Studies*, **15**(2), 105-114 (2015)
15. I. Ilin, A. Levina, O. Iliashenko, Enterprise Architecture Analysis for Energy Efficiency of Saint-Petersburg Underground. *Advances in Intelligent Systems and Computing: International Scientific Conference Energy Management of Municipal Transportation Facilities and Transport*, **692**, pp. 1214-1223 (2018)

# The Level of Production Localization of Automotive Industry Enterprises in Russia

*E.A.* Kandrashina[1], *S.I.* Ashmarina[1], *D.V.* Aleshkova[1*], and *M.V.* Vorotnikova[1]

[1]Samara State University of Economics, Soviet Army Str. 141, 443090 Samara, Russia

**Abstract.** Currently, in Russia there is a need for inclusion of Russian companies in the global value chains. The international alliances created in the automotive industry are able to influence this situation to a certain extent. Increasing the level of localization of such enterprises is a possible solution to the problem of production value chains formation. The aim of this research is to compare the level of production localization of automotive industry enterprises in Russia. The main research methods are comparison, analysis and synthesis of the obtained information. The authors analyzed the degree of localization of the largest automobile enterprises included in international alliances.
**Keywords:** localization, automotive industry, value chains, international alliances

## 1 Introduction

The issues of production localization seem to be the most promising both from the point of view of foreign enterprises and domestic producers. Most often, the production localization is of interest to companies seeking to expand their sales markets. To do this, they need to take into account principles of the industrial policy of the localization country. These principles regulate measures of the state support for the national economy, as well as restrictions and even prohibitions on the import of certain goods. This approach is the most widespread in the Russian automotive industry.

This aspect is particularly relevant from the point of view of the inclusion of Russian producers in global production chains. The main problems interfering scientific and technological development of the country are marked in the Strategy of Scientific and Technological Development of the Russian Federation (article 11) [1]. One of them is the difficulty of forming value-added production chains. The production localization of the automotive industry is able to solve this problem to a certain extent.

## 2 Methods

The methodological basis of the study is the theoretical analysis and systematization of conceptual approaches to the research issue: production localization within the value chains. In this study, the authors applied the following methods: theoretical (analysis and

* Corresponding author: dashajuly343@gmail.com

synthesis, generalization and systematization), which allowed to identify the basic characteristics of the production localization and its level in the considered chains); empirical methods (the study of the practical experience of the production localization in the Russian automotive industry); systematic analysis of data on foreign counterparties in the compared chains, as well as generalization and structuring of the factual data obtained from open sources about the current state of the considered enterprises.

## 3 Findings

Ford and Renault are examples of international automobile alliances localizing their production in Russia. They are the first foreign car companies that concluded agreements with the Ministry of Economic Development of the Russian Federation on the work in the industrial assembly mode. Later, the contracts were activated by Volkswagen (also producing Skoda and Audi models), tandem Hyundai–Kia, GM (Opel, Chevrolet, Cadillac), Nissan, Toyota, PSA Peugeot Citroën in partnership with Mitsubishi Motors. JSC "Avtotor", experimenting in the Kaliningrad special economic zone with Chinese companies, organized the assembly of BMW, some models of Hyundai–Kia and GM.

The requirements of the first mode of industrial assembly were relatively loyal. This was due to the fact that at that time the government's goal was to attract numerous assembly platforms to Russia. The first level of industrial assembly required the production of 25 thousand cars per year and localization of automotive components at the level of 30%.

In exchange, foreign companies received from Russia the right of preferential customs import of components for the assembly of machines (it was a rate from 0 to 5% instead of 15% on the average), these companies were also able to participate in state programs for sales support (recycling, trade-in, preferential car loans and leasing), as well as received numerous subsidies and some tax reliefs (partial or complete zeroing of property tax, etc.).

In 2012, the main industry regulator in Russia – MinPromTorg – came to the second stage of the assembly project. To maintain the previous benefits, foreign manufacturers were offered to create capacities for 300-350 thousand cars per year, and it was required to bring the localization up to 60%. In addition, the car companies had to build factories in Russia for the production of motors, transmissions, as well as to create research centers. This process is similar to the formation of the automotive industry in China.

To solve this task, foreign companies began to unite in industrial unions with Russian companies. So we got a tandem of Volkswagen and GAZ Group, the Franco-Japanese alliance Renault-Nissan united around PJSC "AVTOVAZ", which came under the control of Renault. The company JSC "SOLLERS" (controlling public corporation "UAZ") united with "Mazda" and "Toyota". Trucks and commercial vehicles of Daimler Group appeared at the enterprises of KAMAZ and GAZ [2]. For foreign manufacturers, it ensured compliance with the localization requirements, and for Russian automakers and their suppliers' – inclusion in global production chains, which correlates with the problem identified in the Strategy of Scientific and Technological Development of the Russian Federation, approved by the President Decree of December 1, 2016 № 642 [1].

According to the amendments of 07.06.2009 in the Decree of the Government of the Russian Federation of July 17, 2019 №719, for the established automobile alliances, there will be a point system of evaluation of technological operations, depending on the amount of added value. Points will be awarded both for the actually produced cars models and for obligations assumed for this model or product and fixed in the special investment contract. According to the presentation of the Ministry, to participate in public procurement and measures to stimulate demand for its cars, the manufacturer should score at least 2,000 points since 2019, at least 3,200 points since 2021, at least 4,500 points since 2023, and at least 6,000 points since 2025. At the moment, the aim is not to increase the localization level, but only to obtain the required number of points [3].

Considering the level of production localization for 2018 in the largest alliances, we can see the following results:

- the localization level by Volkswagen production reaches 40%, while it is planned to increase it up to 75% in the next three years;
- the localization level of PSA Peugeot Citroën and Mitsubishi Motors is 35% and it is planned to increase it to 50%;
- the localization level of PJSC "AVTOVAZ" is about 50%. According to the development strategy of this the company, the localization level should grow up to 75%.

Next, we consider functioning of such large members of international alliances as PJSC "AVTOVAZ" and JSC "SOLLERS".

AVTOVAZ Group is a part of the Renault-Nissan-Mitsubishi Alliance and produces cars in the full production cycle for 4 brands: LADA, Renault, Nissan, Datsun.

After the procedure of tender purchase, Alliance Rostec Auto B.V. increased its stake in the share capital of PJSC "AVTOVAZ" up to 100%, the distribution of the shares in ARA BV: 67,61% – Renault, of 32.39% – State Corporation "Rostec".

PJSC "AVTOVAZ" actively cooperates with foreign partners. For example, Lada Xray is assembled from 1800 parts from 250 suppliers and is now localized by 50%. The main part of automotive components for Xray is now produced by foreign companies that have their own production in Russia. And in the future, the localization level of this car is planned to increase up to 75%.

It is known that the French company "Valeo" supplies radiators and heaters for Xray, "Faurecia" makes front seats, the German Bosch supplies a new hydraulic module ABS, "Eberspacher" produces the exhaust system. "Continental" supplies tires and washer system, "Brose" makes electric window regulators. Japanese "Takata" supplies a heated steering wheel as well as passenger airbags and a driver airbag module. AGS company provides the conveyor with side and rear windows, Turkish "Martur" makes rear seats, and another supplier of tires is the Italian "Pirelli".

Other imported components include door handles (U-Shin), mirrors with heating control unit (Magna), rear lights (Valeo), rear door lock (Kiekert) and seals (Standard Profile). Heated windshields are supplied by AGAR, Leoni and Tubor make ropes, joint venture Koito/Avtosvet makes headlights, and Russian Itelma supplies the ERA-GLONASS module. It can be concluded that the most high-tech components are imported for the use in the production process [4].

Engines, transmissions (including automatic ones), front suspension subframes, front and rear suspension arms, instrument panel, front and rear bumpers and radiator grille are produced by "AVTOVAZ" itself.

An important component of the company's activity is the interaction with customers. The company sells not only in Russia, but also for export (Belarus, Kazakhstan, Uzbekistan, Azerbaijan, Germany, Egypt, etc.).

Analyzing the revenue from sales for export, we can say that in 2018, compared to 2017, it has increased by more than 90% (2017 – 10,536 million rubles, 2018 – 20,030 million rubles).

At the same time, revenue from sales in Russia increased by 21.7% and was 271,743 million rubles in 2018. In 2018, PJSC "AVTOVAZ" had one main customer – the JSC "Renault Russia". The revenue received by JSC "Renault Russia" was 38,572 million rubles.

As for the place PJSC "AVTOVAZ" in the car production chain in general, the company is present in the three points of the car production chain:

- design (stage with high added value) is carried out by the engineering center of PJSC "AVTOVAZ";

- production of components (the stage with the added value above average): production and stamping of car bodies both for needs of PJSC "AVTOVAZ" and for other dependent enterprises (GM-AVTOVAZ); production of the engine and transmission is also carried out by the enterprise independently with help of the subsidiaries which are engaged in metal processing (for example, LLC LADA TOOL, LLC VMZ);

- final assembly (stage with average added value) – the last stage in which the company participates creating added value. After that, the cars are transferred to distributors for sale.

Like most manufacturers, PJSC "AVTOVAZ" is not a full-cycle enterprise. Therefore, we can see that the raw materials and materials for production are supplied by external suppliers, as well as a part of the components required for the car assembly.

The relationships with suppliers are based on supply contracts, according to which suppliers must produce and transfer raw materials or components for PJSC "AVTOVAZ" for further production at a previously agreed cost.

At the same time, PJSC "AVTOVAZ" acts as a consumer not in all cases, but sometimes as a manufacturer. Such examples can be the supply of car bodies for the needs of JSC "GM-AVTOVAZ" and the transfer of finished cars after the final assembly to distributors.

According to the reporting of PJSC "AVTOVAZ", the share of imported components in 2017 was about 34%, and in 2018 it increased by 36% and was about 46% (i.e. 107,752 million rubles). As for the exported products of PJSC "AVTOVAZ", it should be noted that in 2017 the share of exports was 4.5% of the total sales. In 2018, this share increased to 6.9%. Also in 2018, the share of exports increased by more than 90% compared to 2017 and amounted to 20,030 million rubles. We can conclude that the costs of imported components to a large extent exceed the amount of revenue received from the export of finished products (by more than 5 times) [5].

PJSC "SOLLERS"– the Russian company which is also the official representative of several car manufacturers with world names: Mazda, Toyota, Isuzu, SsangYong and Ford. The direct parent company of the Group is LLC "Erfix", its share in the share capital of PJSC "SOLLERS" is about 68%.

PJSC "SOLLERS" owns some production platforms in the Volga region, where there is the production of UAZ SUVs (UAZ), Japanese ISUZU trucks, as well as gasoline and diesel engines of ZMZ (ZMZ). In 2011, a joint venture Ford Sollers was launched, which is the exclusive manufacturer and distributor of a wide range of Ford cars in Russia.

In June 2019, PJSC "SOLLERS" bought 51% of the new joint venture from Ford for 75 million US dollars. In the new structure of Ford Sollers shareholders from Ford Motor Company and PJSC "SOLLERS" remain partners, while PJSC "SOLLERS" will own a share of 51% in the authorized capital of the joint venture.

PJSC "SOLLERS" has managed to bring the level of production localization up to 70%. Contracts have been signed with many Russian companies. Among them are Bor glass factory (glass), Vladimir "Avtopribor" (instrument cluster), Magneto (Kingisepp, wheels), Togliatti AKOM (battery), "Nizhnekamskshina". Also, Russian companies will supply noise isolators, seals, mirrors, etc.

About 30% of components are supplied by foreign companies. Among them are BASF Polyurethanes GmbH, who supplies engine cushions; Delphi Italia Automotive Systems r.l., producing air conditioning and heating; A. Raymond Jablonecs r.o. – plastic mounting brackets; UFI Filters S.p.A. – fine filters; LovatoGas – gas equipment; Eberspecher HYDRONIC – pre-heater; German Bosch – diesel system Common Rail CRS2-14.

It is clear that by analogy with PJSC "AVTOVAZ", mainly high-tech components are purchased by foreign companies. At the same time, both companies have common foreign contractors, for example, Eberspecher HYDRONIC and Bosch.

PJSC "SOLLERS" is also present in the three points of the car production chain. They are identical to the positions by PJSC "AVTOVAZ": design, production of components and final assembly. The difference is that most of the raw materials are supplied to PJSC "SOLLERS" directly by its subsidiaries, which not only increases the level of localization, but also adds an additional niche occupied in the car production chain.

According to the financial reporting statements of PJSC "SOLLERS", the share of imported components in recent years has remained at 30%. So, in 2018, the amount of the imported components was approximately 4 745 million rubles.

At the same time, the share of exported products in 2018 increased by about 11% and was on the average 22% of the total output. The revenue from export sales amounted to approximately 8,685 million rubles. In the case of PJSC "SOLLERS", export volumes exceed import volumes by approximately 83% [6].

## 4 Discussion

The importance of enterprises' participation in the global value chains and increase in the localization level is noted by many domestic and foreign authors. V.E. Dement'ev, E.S. Novikova and E.V. Ustyuzhanina consider global value chains as a new form of both cooperation and competition [7]. This opinion is supported by T.S. Naumova and E.I. Shakirzyanova. In their work, they write that global value chains open new competitive advantages for corporate participants [8]. E.D. Suslin notes that the integration of the country's economy into global production processes determines some trends towards the production localization of certain territories [9].

Speaking about the production localization, M. Hannibal and G. Knight mention that its level depends on such key factors as technological complexity of production, logistics network, demand, availability and technical competence of enterprises [10]. Based on this, we can conclude that not every value chain has the ability to localize production by 100%. At the same time, it remains important that production should be localized near the main end users, as it is noted by A.O. Laplume, B. Petersen, and J. M. Pearce [11]. J.S. Arlbjørn and T. Lüthje believe that the production localization in global value chains has an impact not only on the economy of the country, but also on the economy of individual chain participants [12]. As a consequence, after the production localization, there is an increase in productivity at the enterprise. We can say that all the authors (whose works were analyzed within this research) note the importance of inclusion in global value chains and increasing the localization level, as this has a positive impact on the economy of individual enterprises, industries and the country as a whole. However, at enterprises there are subjective factors that can limit the degree of production localization.

## 5 Conclusion

As part of the study, it was revealed that the enterprises of the same industry, that take part of the global production chains within the Alliance, may have different levels of the localization. As part of the Renault-Nissan-Mitsubishi Alliance, PJSC "AVTOVAZ" achieved a localization level of 50%, and its competitors in the face of PJSC "SOLLERS" were able to localize production by 70%. This difference is caused by the backlog of PJSC "AVTOVAZ" in the production of high-tech components, which is planned to be overcome in the near future.

Many authors note the importance of localization of production processes in certain areas. At the same time, they note the importance of overcoming the obstacles associated with internal processes at enterprises and inhibiting the process of localization.

The Strategy of Scientific and Technological Development of the Russian Federation indicates the importance of the problem of inconsistency of priorities and tools to support the scientific and technological development of the Russian Federation which is a barrier to the formation of production chains to create added value of high-tech products and services.

The current situation with the formation of international alliances in the automotive industry and an increase in the level of production localization is a real example of the inclusion of Russian enterprises in the global value chains. As a result, such cooperation can to some extent contribute to the solution of the above problem and provide positive dynamics for the scientific and technological development of the country.

## References

1. President Decree on the Strategy of Scientific and Technological Development of the Russian Federation from 01.12.2016 [online], Available at: http://kremlin.ru/acts/bank/41449/print (2019)
2. T. Khasanov, Models in range: Why the localization of industry in Russia is stalling. [online], Available at: https://iz.ru/836469/timur-khasanov/modeli-v-riad-pochemu-lokalizatciia-avtoproma-v-rossii-buksuet (2019)
3. The order of the Government of the Russian Federation of July 17, 2015 N 719 "About confirmation of industrial products production on the territory of the Russian Federation" [online], Available at: http://docs.cntd.ru/document/420289297 (2015)
4. Suppliers of parts for LADA XRAY are named. Avtovesti, 17.12.2015 [online], Available at: https://auto.vesti.ru/news/show/news_id/657351/ (2015)
5. Erns & Young LLC Independent Auditor's Report on PJSC “AVTOVAZ” (2019)
6. Baker Tilly Rus Audit Report of the Independent Auditor PJSC "SOLLERS" (2019)
7. V.E. Dement'ev, E.S. Novikova, E.V. Ustyuzhanina, Russia's place in global value chains. *Nation Interests: Priorities and Security*, **1**(334), 17-30 (2015)
8. T.S. Naumova, E.I. Shakirzyanova, Russia's participation in global added value chains. *Azimuth of Scientific Research: Economics and Management*, **7**(23), 363-366 (2018)
9. E.D. Suslin, Russia in the global chains of added value. *Innovation and Investment*, **3**, 138-142 (2016)
10. M. Hannibal, G. Knight, Additive manufacturing and the global factory: Disruptive technologies and the location of international business. *International Business Review*, **27**(6), 1116-1127 (2018)
11. A.O. Laplume, B. Petersen, J.M. Pearce, Global value chains from a 3D printing perspective. *Journal of International Business Studies*, **47**(5), 595-609 (2016)
12. J.S. Arlbjørn, T. Lüthje, Global operations and their interaction with supply chain performance. *Industrial Management and Data Systems*, **112**(7), 1044-1064 (2012)

# US-China Trade War: Causes and Outcomes

*Larisa* Kapustina[1*], *Ľudmila* Lipková[2] , *Yakov* Silin[1] , and *Andrei* Drevalev[1]

[1]Ural State University of Economics, Marketing and International Management Department of.465, 62, 8 Marta St., Ekaterinburg, 620219, Russian Federation
[2]University of Economics in Bratislava, International Economic Relations and Economic Diplomacy Department, Dolnozemská cesta 1, 852 35 Bratislava 5, Slovak republic

**Abstract.** The article reviews the chronology and analyses the causes and possible consequences of the trade war between the US and China. The authors identify four main reasons that led to the greatest trade conflict between the two economies in history associated with intentions of the US: a) to reduce the deficit of bilateral trade and increase the number of jobs; b) to limit access of Chinese companies to American technologies and prevent digital modernization of the industry in the PRC; c) to prevent the growth of China's military strength; and d) to reduce the federal budget deficit. It is confirmed and concluded that trade wars have no winners. Taking into account the two countries' GDP scale and export volumes, the US-China trade war will come amid a slowdown in the global production and international trade. The article analyses the outcomes of the trade confrontation based on the scenario methodology. The authors identify 4 scenarios by making assumptions about the response pattern and conjectural variation in the bilateral economic relations. The US protectionist policy has its political dimension. The constraints of the bilateral trade is one of the tools for slowing China's military and economic power growth.
**Key words.** trade war, protectionism, trade policy, economic sanctions, import tariff, trade balance, bilateral trade, high-tech capacity, global dominance, global supply chains

## 1 Introduction

Increasing economic power of China during the recent decades alongside with declining share of the United States in the global production and international trade in the beginning of the 21$^{st}$ century has led to a change in the geopolitical landscape of the world and emergence of the "Group of Two", or simply G2.

China became the leader in commodity exports in 2015 and became a dominant player in international trade. China's absolute nominal GDP reaches USD 14 092 million (with a share of 16.1% on world GDP, second place behind the US (USD 20 412 million with 23.3% share of world GDP)). The absolute GDP by purchasing power parity is higher in China (USD 23 159 million) compared to the US (USD 19 390 million) [1]. The middle

* Corresponding author: lakapustina@bk.ru

class is growing steadily: in 2002 80 million people belonged to the middle class in China, and in 2020 the middle class is expected to reach 700 million citizens of China, which will be about half of the total population [2]. China is the world's largest exporter with an annual export volume of USD 2,263.33 million, compared to the US exports of USD 1,546.72 million in the second place [3]. According to evaluation by some of the Chinese scientists, China's national economic power had surpassed that of the US in 2014: China has been the world's low-cost manufacturing centre and is becoming an export-oriented global technology hub [4]. Empirical studies with the use of econometric models have shown that China's economic influence has indeed increased, however, the America still holds leading position in all stock, credit, energy and commodity markets, and the US has remained the dominant power in the global economy. Therefore, there is still no unified understanding of the balance in the scale of the two economies in scientific literature [5].

The vision of its absolute dominance in the global economy in the US has come into conflict with the growing imbalance in bilateral trade with China, increasing competitiveness of high-tech companies based in the PRC, and an increase in China's investments exports. The US officials state that China is pursuing unfair trade policies, exploiting the benefits of trade liberalization and WTO membership, while at the same time keeping its domestic market safeguarded against foreign competition by providing subsidies and facilitating export through currency devaluation. The US accuses China of stealing scientific and technical knowledge and technologies from the American companies, violating the intellectual property rights and neglecting environmental protection requirements.

At the same time, researchers point out that when trying to "make America great again" D. Trump led the US to a direct violation of international law and multilateral agreements, guided exceedingly only by national interests. The protectionist trade policy of the US has been referred to as the policy of national egoism within the framework of the economic patriotism concept [6] and even economic terrorism [7]. The economic interests of the country's partners in the economic and political organizations and informal unions are neglected. It has become obvious that the US abandons the idea and the concept of free trade in order to preserve its status of the global leader which can result in spreading the de-globalization processes in the global economy, forming the regional mega-unions – as no country in the World can withstand the American protectionism without economic allies. The US is not interested in China's modernization, but the profound interdependence continues to contribute to development of the bilateral relationship [4].

Trade wars have no winners, as the global economy history proves. Both sides suffer losses, but the US has had a long record of winning negotiations on resolving trade contradictions and forcing other countries to step back. China has also shown its willingness to compromise, reduce the imbalance in bilateral trade to USD 200 billion and liberalize its domestic market for US companies. However, the requirements and sanctions imposed by the US have strong impact on the industrial policy of China and threaten the "Made in China 2025" plan. The latter implies that the PRC should secure leadership in 10 high-tech industries (robotics, artificial intelligence, etc.). The US is convinced that China's first successes in the implementation of the plan result from the use of American technologies and will try to restrict access to them for the China's industries.

Rafi Sheikh studies the US trade war with China and asks whether it is possible to win this war [8]. No trade war has the winner, but every trade war recognizes three losers: both trade partners and the global decline in trade, leading to a slowdown in the global economic growth. Carnegie addresses the issue of the escalation of Trump's trade war [9], Grossman and Helpman turn their attention to the issue of trade war and trade negotiations [10]. The current US-China trade war also involves negotiations which have led to many concessions, especially from a relatively highly protected Chinese market. Chunding, Chuntian, and

Chungwei address the economic impact of a possible trade war [11]. In his article, Bouët asks whether trade wars are beneficial and whether they can be easily won [12].

The article aims to examine the causes, identify the scenarios of the trade war, outline the consequences for the US, China, and the global economy as a whole. An attempt is made to assess the development of the trade war between the American superpower and China as the potential global economic leader, taking into account the high degree of interdependence of the two economies, their mutual interest in maintaining presence in the largest markets. The topic is especially urgent and relevant as exporting and other international activities of the Group of Two as well as their direct confrontation influence the global GDP growth rates.

## 2 Methodology

In order to identify the causes and forecast the outcomes of the trade confrontation the article relies on the general scientific methods of analysis and synthesis. The authors analyse the statistics of the bilateral trade as well as the timeline of the trade war. The analysis highlights the milestones that reveal national interests of the two countries. Projecting the internal reasoning allows defining the aims and points of special importance for the both countries.

The outcomes are analysed based on the scenario methodology. The authors formulate four scenarios by drawing conclusions from the assumptions about the response pattern, conjectural variation in the bilateral economic relations and feedback intensity in terms of the systems theory: a) exacerbating feedback: conflict escalation in case of mirroring and matching responses with the tendency of intensification; the trade war will escalate into Cold War II, the sanctions will be applied to all the bilateral trade and will continue to intensify; b) balancing feedback: finding compromise solutions in case the countries reach consensus to avert the trade war; China will make considerable concessions; a significant part of the sanctions will be cancelled; c) zero response: new protectionist measures are not taken; the trade conflict is frozen on the already implemented bilateral measures, and the business makes extensive use of commercial schemes to circumvent sanctions; d) asymmetric response: trade war can escalate, involve military confrontation and lead to the Hot World War III.

## 3 Chronology and causes of the trade war

It is considered that March 23, 2018 was the formal date when the trade war began with D. Trump signing the "Presidential Memorandum Targeting China's Economic Aggression" and introducing tariffs on steel and aluminium [13]. However, the tensions in the economic relations of the US and China had appeared and were discussed earlier. The WTO granted China the status of a market economy in 2017, which aroused criticism from the US, because the decision limited opportunities for protectionism against companies from China. The US refused to recognize China as a market economy [14], which was the first step towards the confrontation within the "Group of Two". Trump's confrontational policy was reflected in the National Security Strategy, adopted in December 2017. It introduced restrictions on China's investments in American technology, tightened exports control and expanded the list of dual-use products that could not be shipped to China. The Entity List was introduced: US companies were banned from doing business with enlisted companies, including the ZTE Corporation which was accused of violating US sanctions against Iran.

**Table 1.** Trade war milestones.

| Time period, general context | US actions | China's actions |
|---|---|---|
| April – May 2017<br>Top-level negotiations to resolve trade disbalances;<br>100-day plan for trade talks | Investigation on steel and aluminium imports is initiated.<br>The US allows China to sell cooked poultry to the US | US companies get greater access to China's agriculture, energy, and financial markets |
| February – April 2018<br>Investigation into China's acts, policies and practices relating to technology transfer, intellectual property and innovation.<br>The US initiates a WTO case against China for discriminatory licensing.<br>The US releases the official statement (May 2018) | Global safeguard tariffs:<br>30% on solar panels;<br>20% on washing machines;<br>25% on steel imports;<br>10% on aluminium imports.<br>Measures targeting China: restricting investment in key technology sectors; imposing import tariffs on aerospace, IT, communication and machinery; including ZTE in the Entity List | 15-25% tariffs on 128 product categories including fruit, wine, seamless steel pipes, pork and recycled aluminium.<br>178.6% antidumping duties on sorghum imports from the US |
| July 2018<br>Ongoing negotiations; internal discussion of the new lists of restrictions | 25% tariff on 818 products (imports worth $34 billion) | 25% tariff on 545 products (imports worth $34 billion), including agricultural products, autos and aquatic products |
| August 2018<br>The parties exchange preliminary lists.<br>China files WTO claim and complain against the US | 25% tariff on 279 goods (imports worth $16 billion) including: semiconductors, chemicals, plastics, motorbikes and electric scooters | 25% tariffs on 333 goods (imports worth $16 billion) including: coal, copper scrap, fuel, buses and medical equipment |
| September 2018<br>China cancels the trade negotiations and releases the White Paper stating the official position | 10% tariff (announced subject to further increase up to 25% in 2019) on $200 billion worth imports from China | 5% and 10% tariffs on $60 billion worth imports |
| December 2018<br>G20 summit in Buenos Aires.<br>The US and China agree not to increase tariffs for 90 days | The US announces that the new list of tariffs will be delayed | China increases import of agricultural and energy products, and lowers tariffs on cars and auto products from 25% to standard 15% |
| May – June 2019<br>Ongoing trade negotiations before the G20 summit | 25% tariff (increase from 10%) on $200 billion worth imports.<br>Huawei and five other companies of China are added to the Entity List | 25%-20%-10% tariffs introduced for $60 billion worth of imports (increased from 10%-10%-5% correspondingly) |
| June 2019<br>G20 summit in Osaka.<br>The parties agree to avoid increasing tariffs | The ban on deals with Huawei is reconsidered. 110 products are excluded from the 25%-tariffs | China announces its plans to increase import of agricultural products |

Source: Authors' own elaboration based on [4].

Despite reaching the agreement at the G20 summit in Osaka, the parties very soon started exchanging threats to increase the tariffs. Chinese companies were reported to stop buying agricultural products from the US. The US accused China of currency manipulations aimed to gain competitive advantage and partly neutralize the effects of tariffs. China, in turn, initiated the third WTO case against the US questioning the reasons for imposing the tariffs. Neglecting the results of the G20 summit The US introduced a new list of tariffs on $125 billion worth of imports from China. China then imposed 5% tariff on

US crude oil and other goods worth $75 billion. The both parties later excluded some goods as it was becoming more and more evident that businesses of the both countries were paying too much for the actions of the governments.

Unilateral sanctions against ZTE and Huawei have significantly reduced their competitiveness and impeded the development of China's high-tech industries. China is unlikely to make significant concessions and abandon the development of the digital economy. On the contrary, it is likely to pursue the development of its own technologies, as well as put under boycott American goods and services.

In particular, one of the sensitive topics of the trade negotiations in agriculture is China's import of the US-produced soybeans. China had stopped importing soybeans from the US in July 2018 but resumed buying US soybeans in December 2018. Agricultural exports of the US have later been in the centre of negotiations. As a result, China excluded soybeans, as well as other farm products from additional tariffs in September 2019.

The US wants to increase the number of jobs by returning the US capital to its home country and conducting reindustrialization. And at the same the country seeks ways to reduce the costs of serving as a global leader. The Americans blame China for many of their problems [4]. By developing the Belt and Road Initiative, China also reduces American influence in the APR [15].

The causes of the US protectionist actions, and not only against China, can be identified as internal and external – from the perspective of the US own interests. Supporting domestic producers by limiting foreign competition can reduce overall consumption in the US but that will also increase the production volume of mainly steel and aluminium products to which increased tariffs apply. In addition to supporting domestic producers, the chronically passive balance of the US current account can also be considered to be the internal factor. Mutual trade with the PRC is a significant factor influencing the increasing current account deficit of the US.

We can identify four main causes, or incentives for the trade war between the US and China.

1. The trade war is supposed to reduce the deficit of bilateral trade and bring American jobs back home [4]. Out of the $796 billion worth US trade deficit in 2017, China accounted for $376 billion, or 47%, almost a half [13]. The US acknowledges several problems in the trade with the PRC, the trade balance deficit being the most important one. The issue has been emerging for decades and still has an increasing trend (although the US trade deficit with China reached a historic low in May 2019). The US does not consider trading with China "fair" (figure 1).

Analysis of the commodity structure of exports and imports shows that China imports mainly American-interdependent products, while the US imports China's final products: "the US mechanical and electrical products from China are as high as 50% of the total of the top ten commodities in Sino-US trade. It can also be seen that the technological differences between China and the US in this field are not very large" [17].

Out of the total Chinese exports, 19% goes to the US. But only 8.3% of the US exports go to China [18]. In 2018, exports from the US to the PRC dropped by 21% as a result of the introduction of protectionist measures, but reduction was only by 12% in the trade flow from China to the US [19].

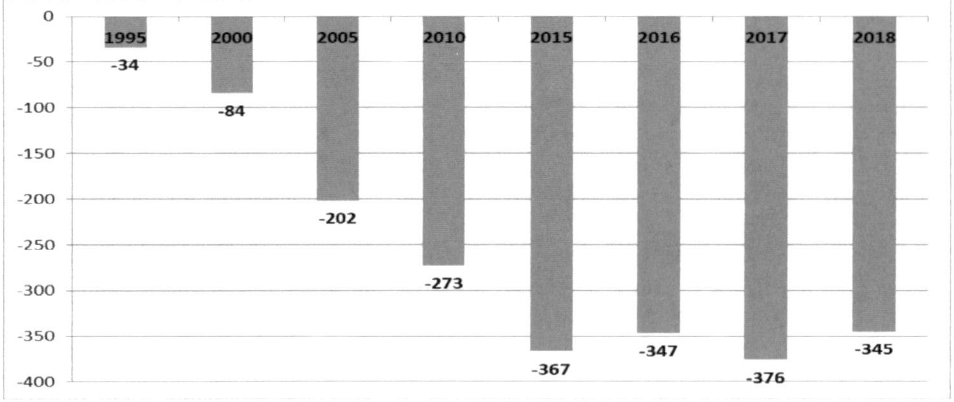


**Fig. 1.** US-China Trade Balance

Note: All values in the graph are in billions of USD.

Source: Authors' own elaboration based on [16].

2. The trade war is supposed to reduce high-tech capacity of China [4]. The US is not satisfied with China's requirements on creating joint ventures for technology transfer as a contribution to authorised share capital of local companies. Another sensitive topic is Chinese public investment creating unfair competition in global markets [14]. The US has been alerted by China's success in implementing a strategic plan for production modernization, increase in production of robots, lithium batteries, network equipment, etc. The US has increased import tariffs up to 25% on electronic products from China, including telecommunication and network equipment [13].

3. The trade war is supposed to prevent the growth of China's military strength. Markov believes that it is absolutely unacceptable for the US to let China achieve superiority in the military sector, even in the long run [20]. Consequently, the US is taking measures to ensure its competitive advantage in the national security sector and to prevent China from using American dual-use technologies.

4. The trade war is supposed to cut the federal budget deficit. According to Dongsheng Di, Gal Luft, & Dian Zhong, "the US will need additional sources of income like tariffs in order to balance its budget, and tariffs on Chinese products is viewed as a main source of such income" [21]. The budget deficit of the US federal government grew and exceeded USD 21 trillion, which was attributed partly to tax cuts in December 2017. The Chinese government also has a much healthier fiscal position and is free to compensate any industries harmed by a trade war. By contrast, the US government is facing a large budget deficit of some 4 percent of GDP that is set to rise in the next few years [22].

## 4 Outcomes and possible scenarios of the trade war

The tensions in US-China trade relations threaten the entire global economy, not just the relations inside the Group of Two. As a result of the US-China trade cut, global growth can be expected to decline by 0.5% in 2020 [23].

The World Bank analysts have used a Computable General Equilibrium (CGE) Model to make an economic forecast: introduction of a 25% tariff on all goods of Chinese imports to the US will reduce world exports by 3% and the global product by 1.7%. "The analysis shows that a US-China tariff escalation could reduce global exports by up to 3 percent (674 billion dollars) and global income by up to 1.7 percent (1.4 trillion dollars) with losses across regions". "The biggest declines in incomes are recorded by China and the US, up to 3.5% (USD 426 billion) and 1.6% (USD 313 billion), respectively. The sectors most affected include: agriculture, chemicals and transport equipment in the US; and electronic equipment, machinery and other manufacturing in China" [24].

The OECD reckons that around a third of the content of the US imports from China is actually of foreign origin. The value added of China's exports to the US is perhaps USD 329 billion – some 2.7 percent of China's USD 12 trillion economy. So even if new tariffs slashed China's exports to the US by 25 percent, the direct hit to the China's GDP would be 0.7 percent. It would be a noticeable impact, but it would still leave the Chinese economy growing at 6.1 percent a year [22].

If exports from China to the US will be halved as a result of high import tariffs, then the direct loss of China's GDP will be at least 0.43%. Taking indirect effects into account, the trade war will result in a 1.12% reduction in China's GDP. In case of imposing increased tariffs on all Chinese imports to the US, GDP losses will reach 4% [25].

According to some experts, the response of China in the trade war with the US may be more painful for the American economy than for China [26]. American companies will partly lose their consumers in China because of nationalistic appeals to boycott products of American companies, including Apple, McDonald's, KFC. Alongside with that, the internal China's demand for the products restricted by the US (e.g. ZTE and Huawei) is increasing. Sanctions against ZTE have put the company in a very difficult position, because most of the chips for the telecommunication equipment production were imported from the US, and suppliers from China and Korea could not replace that amount [26].

The impact on consumers will depend on an individual's consumption choices. For example, those Americans who enjoy tech products, usually assembled in China, are likely to feel short-changed sooner than those relying on domestically (American) produced goods, as the imported goods are made more expensive by the increased tariffs applied. A recent study shows that imposing the tariffs on $200 billion worth of the imports would cost each American household, on average USD 127 a year [27].

According to Bloomberg estimates, if all Sino-US mutual imports become subject to import tariffs, the global GDP will fall by USD 600 billion. And even if the trade war stops, previously implemented restrictions will result in 2 years' time in a slow-down of GDP growth by 0.2% in the US and by 0.5% in China [28].

Experts from China estimate the overall damage from the increase in US import tariffs from 0.3% to 0.7% of the PRC's GDP growth, which is a relatively "modest" estimation [13]. In particular, Terence Tai-Leung Chong & Xiaoyang Li forecast that the trade war will reduce employment in China by 1.1% and GDP by 1.0%, which may be considered not that sensitive for China [29].

C. Fred Bergsten has identified three scenarios for the development of a trade war: "a G0 in which the US is no longer willing to lead but China is not yet able or willing, and whether such (a likely) regime will be stable or unstable; a new G1, sooner or later, led by China; and a cooperative G2 in which the US and China agree to share leadership" [30]. We can foresee four scenarios of the further US-China trade war development.

### 4.1 Scenario 1. Trade war will escalate into Cold War II.

Cultural traditions and social values in the US and China are different. The countries will not give up their political and economic systems and will not transform them according to the practices imposed of suggested by the opposite side. China will have to reduce export-oriented production. In contrast to the Cold War I, China has a chance to win since it was able to build a competitive economy and form TNCs that made their way to the Global 500. The US will surround China "with powerful American military capabilities, creating

NATO-like adversarial alliances, isolating it economically and imposing costs when it did things the US did not like" [4].

China, in its turn, will seek to strive to push the US out from the APR. The US will explain its unilateral sanctions with the willingness to strike better deals, to reach an "honest" deal with China, but serving the interests of the US economy. The country will be achieving its goals by presenting ultimatums to its trade partner, if their national interests or security are considered harmed. As a result, more and more global problems will be addressed by unions of countries not involving the US (climate agreements, nuclear agreement with Iran, TPP – Trans-Pacific Partnership, China – EU summit in July 2018, etc.). The US will be gradually becoming isolated in the international community. The role of the US in global supply chains and international trade will decrease due to the deterioration of the country's authority. The US may be excluded from a number of regional trade and economic associations of countries. When some US companies are winning, others suffer losses. There is growing discontent among American business community and their overseas partners (with recent examples including Ford and Boeing, equipment manufacturing and chemical industry, etc.).

The US government bond market will suffer, inflation rate in the domestic consumer market will increase. The risk of unilateral economic sanctions imposed by the US will push other countries to search for alternative suppliers and markets.

China has become a leader in the Regional Comprehensive Economic Partnership (RCEP) of 16 countries, the scale of which expanded beyond the Comprehensive and Progressive Agreement for Trans-Pacific Partnership (CPTPP) without the US: 30% of global GDP, 50% of the world population [20]. Exporting to RCEP countries from the PRC accounts for 26% of the total China's exports. The development of mega-union will strengthen the leading role of China in the region, but on the other hand, may also cause a decrease in international trade standards, since it does not cover issues of labor, human rights, e-commerce, public procurement [14].

At the same time, China's investment climate will suffer, the implementation of the "Made in China - 2025" plan is under the risk of failure, which may slow down technological progress in the country [14]. China has gradually increased its expenditure on R&D, but it still lags behind the US. R&D share in the GDP accounts for 2.1% and 2.8% respectively. According to Kerry Liu, in view of the importance of Made in China 2025 to economic future, it will be very difficult for China to make substantial concessions during the trade war [31].

### 4.2 Scenario 2. Consensus to avert the trade war will be reached.

A trade war will end with a compromise. It is possible if the US learns to respect the other parties' core interests and major concerns. China has already declared its readiness to open its car market, liberalize the banking sector, strengthen the protection of intellectual property rights, and limit the transfer of American technologies to Chinese business. China may agree to buy as much of the agricultural produce from American farmers as they can supply.

Edwin L.-C. Lai states that China is willing to seek a compromise and is ready to increase imports of goods and services from the US, cut down government subsidies to its businesses, and make the transfer of American technology more transparent [32]. However, it is not likely to cancel the increased import tariffs previously introduced by the US. The scenario does not exclude the possibility that the trade conflict can be frozen up for a long period of time: "Although deep and warm peace among great powers is rare in history, a cold peace or the US-China détente is a realistic goal" [4].

Both countries are deeply interested in exporting products and services on a bilateral basis and in developing global supply chains. The negotiated settlement agreement and the end of the war is possible, since both countries will lose more than they can gain as a result of the war. We can expect that China and the US will remain strategic partners despite the increased competition between them. Business-to-business ties will become closer, and mutual understanding of cultures will improve.

China will increase imports of LNG, agricultural raw materials and food, as well as high-tech products. The US administration will maintain a relationship of trust with major trading partners and will hold back from imposing unilateral sanctions, thus returning to respecting the WTO rules. The largest American companies will play a major role in this process by lobbying and promoting free trade and conveying the idea that a trade war against the whole world is not beneficial for the US.

### 4.3 Scenario 3. The trade conflict is frozen on the already implemented bilateral measures.

The trade war is becoming prolonged, high import tariffs are not cancelled. At the same time, companies in both countries are adapting to export-import operations under economic sanctions, developing commercial schemes to circumvent trade restrictions. It can be assumed inevitable to some extent because foreign economic relations are driven by companies who are main actors in the international economy, and not countries. Benefits of foreign activities are valuable incentives for companies to look for ways to circumvent sanctions and embargoes in order to preserve and increase profits. In this scenario the imbalance in trade between the US and China will be decreasing and - according to official statistics – will reach a target level of USD 200 billion, possibly even less than that. However, in fact, the goods of Chinese origin will be supplied to the American market from other countries, e.g., neighbouring countries in the APR.

As Vinogradov, Salitsky and Semenova note, Chinese companies have already begun to move production units to Vietnam, Bangladesh, Ethiopia and other countries with relatively cheap labor, given the fact that China largely exports products of TNCs that have placed final assembly production units in the PRC. Foreign firms account for 59% of Chinese exports to the US, and 12% are products of American companies located in China [13]. China's exports to the US are typically Chinese-assembled goods that contain many foreign parts and components – and are often American-branded. A further 37 percent of the US imports from China consist of parts and components on which US-based manufacturers rely.

TNCs have begun to restructure their global supply chains and may move the final assembly of goods from China to neighboring Asian countries. It will result in changes in the geographic structure of China's exports and imports. While maintaining the final assembly in China, the intermediate links of the global production chain can be made cheaper. For example, China supplies USD 2 billion/year worth iPhones to the US, but only a small part of the product's value is created in China, since the gadget is made in Taiwan of components supplied from many countries: Germany, Korea, Japan, the USA, The Netherlands, Singapore, etc. [6]

Trade wars lead to changes in export-import flows, orientation to other countries, emergence of regional free trade zones and economic alliances. Lawrence J. Lau notes the shift in China's foreign economic relations focus to such countries and regions as the EU, ASEAN, Japan and Russia [25]. Analysts have identified a trend towards intensifying China-EU trade and economic ties, and moreover, the Asian market is becoming more important for European business than that of the US [14]. Consumer spending has grown four times faster in China than in the US in the last decade. Badar Alam Iqbal & Nida

Rahman believe that the trade war will result in losses for both conflicting parties but it might come as a benefit for other countries: "Being the two largest economies of the world, giving in for disengaging trade with each other is bound to have a ripple effect on the global system of trade" [33]. In the view of the US protectionism, countries with large domestic markets will be the ones to gain a competitive advantage.

Since the beginning of the trade war, alongside with the US limiting imports from China, six countries in Southeast Asia and Taiwan have organized supplies of nearly 1,600 new categories of goods that they have never sold in the US before [34].

At the same time the Chinese market is crucial for many US companies, therefore restricting access to Chinese consumers causes great financial damage to corporations [7]. As the result, the companies seek possible ways to circumvent sanctions and to lobby and promote their business interests. Under this scenario, the US will achieve its goal of reducing the trade deficit in bilateral trade with China, but with poor economic efficiency.

### 4.4 Scenario 4. Trade war will escalate into Hot World War III

A trade war can escalate from Cold to Hot stage, which is especially unwanted scenario considering that the US allows limited use of nuclear weapon. According to Terence Tai-Leung Chong & Xiaoyang Li, the trade conflict between China and the US is of fundamental nature and it cannot be easily resolved [29]. The conflict is associated with the race for global economic domination. The US has a serious advantage in cyberspace and it seeks to "digitally colonize" the global economy. According to Ashmanov, almost all neural platforms in the world are US-based, including first of all those belonging to Google and Facebook [35]. "The US shows with its actions, that they are ready to easily throw any inconvenient contracts and obligations away into the trash can, and also to ignore the UN and other international organizations, granting itself the right for unilateral military actions, the US is moving towards the role associated with the British Empire after the World War I" [6]. This scenario seems unlikely now, but it is not excluded.

Trump has made matters worse by acting unilaterally against China in a way that would appear to breach World Trade Organization rules. Indeed, potential allies find Trump's "America First" rhetoric repulsive. All this has given China the political high ground – "China doesn't want a trade war, but we're not afraid to fight a trade war" has become Beijing's official line [22].

## 5 Conclusion

The greatest trade war in economic history can result in a change in the international trade architecture, slow-down of financial markets. The countries can be divided into two blocks supporting the US or China, and at the same time, forming the mega-alliances of economies, as well as regional currency zones. The Asia's role in globalization processes and the development of global supply chains is likely to strengthen. The US strives to weaken its main competitor and maintain dominance in the global arena: in the economy, politics and the national security.

The current foreign trade policy of the US aims to slow down the still rapid economic growth of the PRC and its growing importance in the world economy. The China's government in its turn has a goal to achieve leadership in robotics, biotechnology and artificial intelligence. It will provide financial support to high-tech industries, and will do everything possible not to let the US stop or slow down the modernization and digitalization of the China's economy.

The current situation in US-Chinese trade relations is referred to as the Cold Trade War. The common economic interests of both countries are more significant than disputes. The mutual interdependence of both economies has led to negotiations that will result in a new "fairer" trade agreement. The new US-China trade agreement should include the so-called a new trade architecture that will regulate not only the use of classical tools in mutual trade such as tariffs, quotas, and sanitary and phytosanitary certificates, but also other regulations as well as intellectual property protection, technology transfers, and Chinese state subsidies to manufacturers and exporters. The new agreement will also deal with questions beyond trade, such as cyber espionage and the expansion of US firms' access to the Chinese market.

Finally, we note that the US protectionist attack on their trading partners, especially on the People's Republic of China, has a political dimension in addition to its economic dimension. China is designated by the US leaders as the US' chief strategic rival in the future. Therefore, various constraints of trade with the US and other ways to slow down China's economic growth are also tools for slowing China's political power growth.

## References

1. Statistics Times, *List of Countries by GDP (PPP)* [online], Available at: http://statisticstimes.com/economy/countries-by-gdp-ppp.php (2018)
2. Statista, *Number of the middle class population in China in 2002 and 2020 (in millions)* [online], Available at: https://www.statista.com/statistics/875874/middle-class-population-in-china (2019)
3. Statista, *Top 20 export countries worldwide in 2017 (in billion U.S. dollars*) [online], Available at: https://www.statista.com/statistics/264623/leading-export-countries-worldwide/ (2019)
4. Z. Suisheng, D. Guo, A New Cold War? Causes and Future of the Emerging US-China Rivalry. *Vestnik RUDN. International Relations*, **19**(1), 9-21 (2019)
5. D. Zhang, L. Lei, Q. Ji, A.M. Kutan, Economic policy uncertainty in the US and China and their impact on the global markets. *Economic Modelling*, **79**, 47-56 (2019)
6. Y.A. Savinov, A.N. Zelenuk, E.V. Taranovskaja, G.A. Orlova, A.V. Skurova, Increased protectionism in US trade policy. *Russian Foreign Economic Bulletin*, (1), 36-51 (2019)
7. I. Sinitsyn, P. Kuimov, The economic terrorism of the USA with regard to Europe, Russia and other countries. *Vlast,* (9), 236-241 (2018)
8. S. Rafi Sheikh, *Les États-Unis ne pourront pas gagner la guerre commerciale contre la Chine* [online] Available at: https://www.mondialisation.ca/les-etats-unis-ne-pourront-pas-gagner-la-guerre-commerciale-contre-la-chine/5626497 (2018)
9. A. Carnegie, *Trump's Trade War Escalates* [online], Available at: https://www.foreignaffairs.com/articles/2018-06-25/trumps-trade-war-escalates (2018)
10. G.M. Grossman, E. Helpman, Trade wars and trade talks. *Journal of Political Economy,* **103**(4), 675-708 (1995)
11. L. Chunding, H. Chuantian, L. Chuangwei, Economic Impacts of the Possible China–US Trade War. *Emerging Markets Finance and Trade*, **54**(7), 1557-1577 (2018)
12. A. Bouët, *Les guerres commerciales sont-elles bonnes et faciles à gagner?* [online], Available at: http://parisinnovationreview.com/article/les-guerres-commerciales-sont-elles-bonnes-et-faciles-a-gagner (2018)
13. A.O. Vinogradov, A.I. Salitsky, N.K. Semenova, US-China Economic Confrontation: Ideology, Chronology, Meaning. *Vestnik RUDN. International Relations*, **19**(1), 35-46 (2019)

14. S.I. Dolgov, Y.A. Savinov, International trade: USA on the warpath. *Russian Foreign Economic Bulletin,* (9), 7- 20 (2018)
15. Y. Silin, L. Kapustina, I. Trevisan, A. Drevalev, China's economic interests in the "One Belt, One Road" initiative. *SHS Web of Conferences: Innovative Economic Symposium 2017 – Strategic Partnership in International Trade,* **39** (2017)
16. US Census Bureau, *Trade in Goods with China* [online], Available at: https://www.census.gov/foreign-trade/balance/c5700.html (2019)
17. Y. Deng, F. Pan, Dependence analysis of Sino-US trade. *Journal of Physics: Conference Series,* **1176**(4) (2019)
18. Statista, *Total value of U.S. trade in goods (export and import) worldwide from 2004 to 2018 (in billion U.S. dollars)* [online], Available at: https://www.statista.com/statistics/218255/total-value-of-us-trade-in-goods-worldwide-since-2004/ (2019)
19. US Census Bureau, *Trade in Goods with China* [online], Available at: https://www.census.gov/foreign-trade/balance/c5700.html (2019)
20. V.V. Markov, China and the USA: From economic rivalry in Asia-Pacific to trade war. *The Herald of the Diplomatic Academy of the MFA of Russia. Russia and the World,* **4**(18), 110-119 (2018)
21. D. Dongsheng, G. Luft, D. Zhong, Why did Trump launch a trade war? A political economy explanation from the perspective of financial constraints. *Economic and Political Studies*, **7**(2), 203-216 (2019)
22. P. Legrain, *Why China will win the trade war* [online], Available at: https://foreignpolicy.com/2018/04/13/why-china-will-win-the-trade-war (2018)
23. A.N. Da Costa, *The early victims of Trump's trade war* [online], Available at: https://www.bbc.com/news/business-45028014 (2018)
24. C. Freund, M. Ferrantino, M. Maliszewska, M. Ruta, Impacts on global trade and income of current trade disputes. *Macroeconomics, Trade Investment (MTI) Practice Notes*, **2**, 11 (2018)
25. L.J. Lau, The sky is not falling! *Economic and Political Studies*, 1-26 (2019)
26. Y. Wang, *Last warning: what US Hi-Tech companies lose in trade war with China* [online], Available at: https://www.forbes.ru/tehnologii/360641-poslednee-preduprezhdenie-chto-poteryayut-tehnologicheskie-kompanii-ssha-v (2018)
27. Q. Bui, N. Irwin, *How Much Will the Trade War Cost a Typical American Family? Around $60 (So Far)* [online], Available at: https://www.nytimes.com/interactive/2018/07/12/upshot/trade-war-cost-families.html (2018)
28. B. Holland, C. Sam, A $600 Billion Bill: Counting the Global Cost of the U.S.-China Trade War [online], Available at: https://www.bloomberg.com/graphics/2019-us-china-trade-war-economic-fallout/ (2019)
29. T.T.L. Chong, X. Li, Understanding the China–US trade war: causes, economic impact, and the worst-case scenario. *Economic and Political Studies*, **7**(2), 185-202 (2019)
30. C.F. Bergsten, China and the United States: The Contest for Global Economic Leadership. *China & World Economy*, **26**(5), 12-37 (2018)
31. K. Liu, Chinese manufacturing in the shadow of the China–US trade war. *Economic Affairs*, **38**(3), 307-324 (2018)

32. E.L.C. Lai, The US–China trade war, the American public opinions and its effects on China. *Economic and Political Studies*, **7**(2), 169-184 (2019)
33. B.A. Iqbal, N. Rahman, J. Elimimian, The future of global trade in the presence of the Sino-US trade war. *Economic and Political Studies*, **7**(2), 217-231 (2019)
34. N. Dembinskaya, *How Beijing bypasses American bans in trade war* [online], Available at: https://ria.ru/20190607/1555320044.html (2019)
35. D. Smirnov, *Can the West with one click of a computer turn off our factories, banks and power stations* [online], Available at: https://www.ural.kp.ru/daily/26992/4053140 (2019)

# Impact of digitalization on consumer behaviour and marketing activity in financial markets

*Yury* Korobov[1*]

[1]Saratov Socio-Economic Institute (branch of Plekhanov Russian Economic University), Department of Banking, Money & Credit, 410003 Saratov, Russia

**Abstract.** The article reveals the impact of digitalization of the economy on the behaviour of consumers of financial services: increasing the relevance of financial needs, increasing the consumer independence, changing the model of consumer behaviour. The directions for improving the marketing activities of banks were identified: a shift in focus from image advertising to product advertising, and increased attention on quality characteristics of financial products.
**Key words:** financial market, digital economy, consumer behaviour, financial needs, marketing of financial services, advertising of financial services, quality of financial services

## 1 Introduction

The development of the digital economy is nowadays one of the most important factors of economic development throughout the world.

Perhaps most of all digitalization affects financial markets, especially the banking market. Common problems of transformation in banking in modern conditions were explored by King [1], Wehinger [2], etc., as well as the author of this article [3, 4].

One of the consequences of the digitalization of financial markets is a change in the consumer behaviour. Among studies concerning consumer behaviour change caused by the development of digital economy should be mentioned publications by Lecinski [5], Muller, Damgaard et al. [6], Su [7], etc.

Consumer behaviour changes require increased attention to the quality of financial services. The study of fundamental and applied aspects of the quality of banking services was carried out by Crosby [8], Parasuraman, Zeithaml and Berry [9], Plotnikova [10], etc. Issues of quality assessment of banking services were studied by Novatorov [11], Ikhsanova [12], Mukhamedzyanova [13], etc.

The purpose of this article is to discuss the impact of digitalization on behaviour of consumers of financial services, as well as on marketing activities of financial institutions – specifically, on advertising policy and quality assurance.

* Corresponding author: yurikor1965@gmail.com

## 2 Data and methods

Main data sources for the study were analytical materials of the Bank of Russia, domestic and foreign consulting institutions.

Methodological basis of the study was the systematic approach. In the research process, general scientific methods and techniques were used: scientific abstraction, grouping, comparison, etc.

## 3 Results and discussion

### 3.1 Changes in behaviour of consumers of financial services

In my opinion, the most important changes in behaviour of consumers of financial services caused by the digitalization of economy include: 1) increasing the relevance of financial needs; 2) increasing the consumer independence; 3) changing the model of consumer behaviour.

#### *3.1.1 Increasing the relevance of financial needs*

The main factor in formation of consumer behaviour, of course, are needs. The peculiarity of the needs satisfied by financial services is their secondary nature and, as a result, low relevance for many consumers [4]. Material products are able to satisfy the primary needs that arise from the nature of a man as a biological organism and as a member of society – material, spiritual, and social needs. The importance of meeting these needs (for example, such as the need for food, clothing, housing) is obvious to any person, and awareness of them does not require any external stimuli. Producers of material products designed to meet such needs do not have to make additional efforts to explain their purpose.

Financial products cannot directly meet primary needs, their acquisition is for the consumers not an end in itself, but an intermediate step. Financial needs which they satisfy (financing, investing, cash flow streamlining) are of a secondary, derivative nature, their importance to consumers is often not obvious. For example, a person first has a need for housing, but a need for a loan to acquire it can arise only under certain conditions: awareness of the possibility of borrowing, lack of prejudice against using a loan, etc. Thus, realization of financial needs requires additional efforts as well as certain level of financial literacy. Financial needs of many people remain in a latent, dormant state, and need external stimuli to become relevant. The development of information environment of financial market caused by the digitalization of economy is precisely a powerful external stimulus for increasing the relevance of financial needs.

#### *3.1.2 Increasing the consumer independence*

In addition, with the development of remote banking services (Internet banking, mobile banking), the degree of independence of consumers of financial services significantly increases (up to their self-sufficiency). A consumer gets the opportunity to use financial services on his own, without any help of a bank employee. A financially literate customer in the digital economy can become a "bank for oneself".

Previously [3], the author of this article hypothesized that the person of the digital economy is no longer just Homo sapiens (an intelligent person who appeared hundreds of thousands years ago) or even Homo economicus (economically-minded person formed in the conditions of development of the market economy), but *Homo financial* (a person with financial literacy and financial thinking) (Fig. 1).

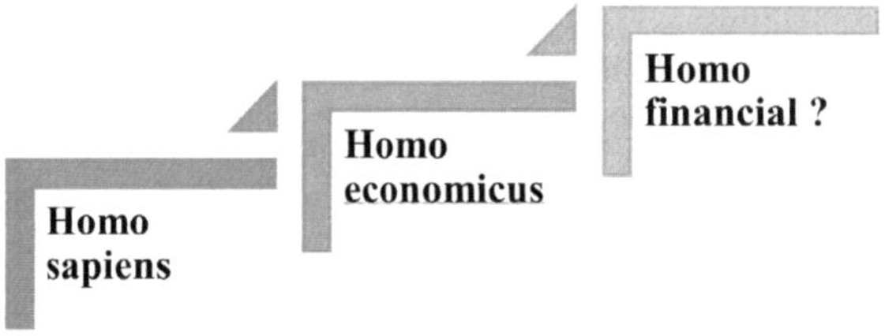


**Fig. 1.** A hypothesis of human evolution in the digital economy

Source: Own processing.

In addition to this, the development of Internet services providing information on financial markets dramatically increases the degree of independence and freedom of choice of consumers of financial services. Long gone is the time when, in order to compare the financial services of different banks, a consumer had to visit or ring up them one by one. Today, a consumer sometimes does not even need to explore their web sites – he just goes to the site that integrates information about financial services of different banks. And this leads to the change in behaviour of consumers in the financial market.

### *3.1.3 Changing the model of consumer behaviour*

The traditional model of consumer behaviour in the financial services market looked like this: a client identifies a financial need, then he chooses a bank for service, and then in this particular bank he chooses an optimal financial product for himself (Fig. 2).

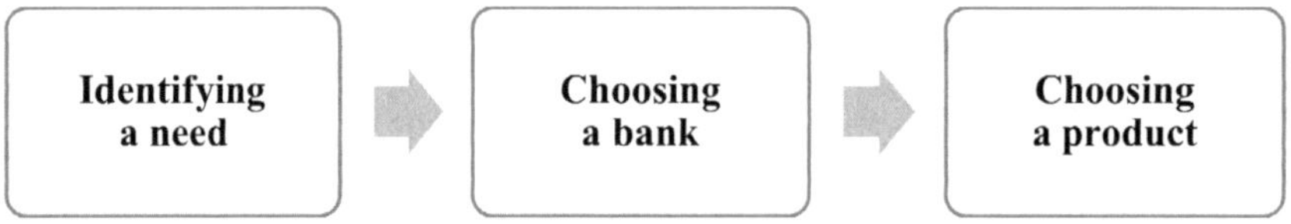


**Fig. 2.** Traditional model of consumer behaviour in financial markets

Source: Own processing.

In the era of digitalization of the economy, the second and third steps switch places: the client first selects the product with the help of Internet, and only after that turns to the institution whose financial product most satisfies the client's need (Fig. 3).

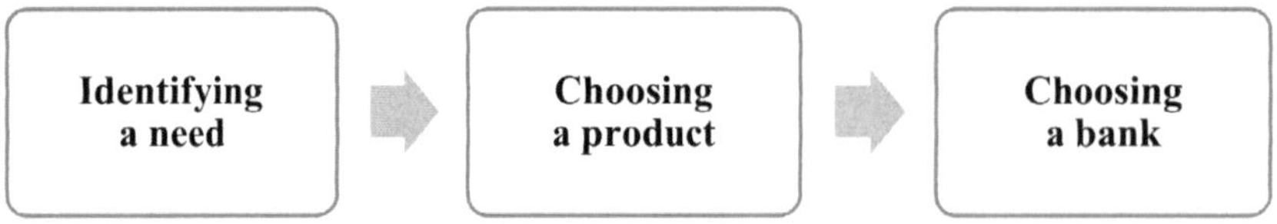


**Fig. 3.** New model of consumer behaviour in financial markets

Source: Own processing.

The fact that in modern conditions the purchase of a product is preceded by the online research conducted by the customer, was proved, in particular, in a study by Google. This online decision-making moment was called the Zero Moment of Truth [5].

According to ForresterAnalytics, already in 2017 56% of global B2C eCommerce occurred via marketplaces and this will increase to 67% by 2022 [14]. In 2018, the Bank of Russia launched the Marketplace project, the purpose of which is to organize a system of remote retail distribution of financial products and services.

### 3.2 Changes in marketing activity of banks

Changes in the behaviour of consumers of financial services require adequate changes in the marketing activity of banks – specifically, in advertising policy and in quality assurance.

#### *3.2.1 Returning to product advertising*

In the 20th century, the advertising policy of banks went through a certain evolution, which can be seen on the example of changing forms of banking advertising.

Depending on the advertised objects, there are three main forms of banking advertising:

- product advertising (carried out in order to attract the attention of customers to certain banking services or products);
- need advertising (carried out in order to awaken the needs of potential customers in banking services);
- image advertising (aimed not at the sale of a particular service, but at attracting the attention of consumers to a particular bank).

Until the mid-1970s, the dominant form of banking advertising was product advertising. The advantage of this form of advertising was its information saturation: a consumer receives extensive information about a specific financial product. At the same time, there were two significant disadvantages. Firstly, the attractiveness of such an advertisement is rather low: it may be interesting only to those consumers who have already identified specific financial need and are engaged in a targeted search for a product that will satisfy this need; for the rest of the consumer audience, this form of advertising will not be attractive. Secondly, due to the universalization of banking, all banks offer the same products, and in product advertising information about the service provider is secondary, not memorable, and by the time the client realizes the need for the product he has seen in advertising, he can simply forget about what bank published it.

In order to solve the first problem, banks began to use the second form of advertising – need advertising, which contains information not so much about the service itself as about the final need, which the customer can satisfy as a result of the acquisition of this service. This found expression, in particular, in giving a special name to services aimed at meeting certain final needs (for example, car loans, holiday savings, etc.). Such advertising attracted better attention, but the information about the publishing bank was still poorly remembered by consumers.

Therefore, at the end of the 20th century, under conditions of increasing competition in the banking market, image advertising became the main form of bank advertising. Such advertising does not contain specific information about a particular banking product, but is intended to draw the client's attention to the bank's brand and form its favourable image.

This approach worked well under the conditions of the old model of consumer behaviour, when a client first chose a bank, and then a product. Nowadays, when these stages of consumer choice are reversed, the advertising policy again has to focus not on a bank image, but on a banking product.

### *3.2.2 Focusing on quality*

In this regard, the quality of banking services becomes the main instrument of bank marketing, and the problem of determining its criteria is one of the most important problems of banking theory and practice.

The problem of the quality of banking services is still among the discussion. The presence of different points of view on this issue is not surprising: since banking services (like any other services) are by definition intangible, their quality is largely a subjective category.

For example, in American literature for many years, quality of banking services was associated with courtesy of service. Friendly and qualified service was the main criterion for choosing a bank for most consumers. However, with increasing attention to this problem, other approaches to the definition of quality began to appear. For example, Crosby [8] wrote that quality of banking services is determined by the compliance with requirements of consumers.

Hence, it is important to determine the criteria for the quality of banking services. The quality criteria in the real economy sector are extensively studied, and do not cause any particular discussion, since material products are tangible and its quality can be measured objectively with the help of special equipment. Financial services are intangible and it is quite difficult to form a single set of quality criteria for them.

I share the methodological approach proposed by German experts [15] that looks at the criteria for the quality of banking services from two perspectives: from the point of view of a bank and from the point of view of a client. According to this study, from the bank's point of view, the quality was determined by the speed of internal work processes, by the level of costs for error correction, by the efficiency of work processes, by the level of motivation of employees, by labour productivity, by the level of credit risk, etc. From the client's point of view, the criteria for the quality of banking services were: speed of service, urgency of operations, presence of errors and inaccuracies, office hours of the bank, quality of consulting (depth, active or passive), personal side of relations with the bank, etc.

Developing this approach and considering that for each type of banking services the set of criteria is specific, following quality criteria can be proposed [16] (Table 1).

**Table 1.** Quality criteria for the most common banking services

| Banking services | Quality criteria | |
|---|---|---|
| | from the bank's point of view | from the client's point of view |
| 1) credit services | - the state of the loan portfolio;<br>- the endowment of the loan process with financial resources;<br>- interest margin, etc. | - the initial parameters of the loan;<br>- the speed of consideration of the loan application;<br>- results of using the loan, etc. |
| 2) deposit services | - the cost of attracting financial resources;<br>- the competitiveness of the proposed conditions compared to competitors;<br>- the risk of early withdrawal of funds by the client, etc. | - the possibility of effective investment of free funds;<br>- the quality of consulting;<br>- guaranteeing the safety of funds, etc. |
| 3) cash and settlement services | - the number of errors;<br>- the cost of correcting them;<br>- the duration of operations, etc. | - the level of digitalization of services;<br>- the speed of payments, etc. |

Source: Own processing.

## 4 Conclusion

The most important changes in behaviour of consumers of financial services caused by the digitalization of economy are: increasing the relevance of financial needs; increasing the consumer independence; changing the model of consumer behaviour.

With the development of remote banking services, the degree of independence of consumers of financial services increases up to their self-sufficiency. A financially literate customer in the digital economy can become a "bank for oneself".

The development of web services providing information on financial markets increases freedom of choice of consumers of financial services which leads to the change in the model of consumer behaviour. Nowadays a customer first selects the product with the help of Internet, and only after that turns to the institution whose financial product most satisfies the client's need.

The directions for improving the marketing activities of banks are: a shift in focus from image advertising to product advertising, and increased attention on quality characteristics of financial products.

In determining the quality criteria, the bank's point of view and the client's point of view should be distinguished.

Directions for further research in this area may be: detailing the quality criteria for certain types of banking services, as well as determining the significance (weight) of each criterion.

## References

1. B. King, *Bank 3.0: Why banking is no longer somewhere you go, but something you do*. 1st ed. Wiley (2012)
2. G. Wehinger, Banking in a challenging environment: Business models, ethics and approaches towards risks. *OECD Journal*, **2** (2012)
3. Y. Korobov, Global banking: transformation, innovation & competition. *SHS Web of Conferences: Innovative Economic Symposium 2017 – Strategic Partnership in International Trade*, **39**, 01012 (2017)
4. Y. Korobov, Banking competition and its development in Russia. *SHS Web of Conferences: Innovative Economic Symposium 2018 – Milestones and Trends of World Economy*, **61**, 01009 (2019)
5. J. Lecinski, *Winning the zero moment of truth*. Vook Inc. (2011)
6. P. Muller, M. Damgaard, A. Litchfield, M. Lewis, J. Hörnle, *Consumer behaviour in a digital environment*. Brussels: European Parliament (2011)
7. B. Su, The Evolution of consumer behavior in the digital age [online], Available at: https://medium.com/analytics-for-humans/the-evolution-of-consumer-behavior-in-the-digital-age-917a93c15888 (2016)
8. P.B. Crosby, *Quality without tears*. New York: McGraw-Hill (1984)
9. A. Parasuraman, V.A. Zeithaml, L.L. Berry, A conceptual model of service quality and its implications for future research. *Journal of Marketing,* **49**(4), 41-50 (1985)
10. M.V. Plotnikova, Качество банковских кредитных услуг: аспекты теоретического анализа [Quality of bank credit services: aspects of theoretical analysis]. *Вестник СГСЭУ [Bulletin of SSSEU]*, **1** (2014)
11. E.V. Novatorov, Методы измерения качества банковских услуг [Methods of measuring the quality of banking services]. *Банковское дело [Banking]*, **7** (2010)

12. L.R. Ikhsanova, *Качество банковских услуг как фактор повышения эффективности деятельности региональных банков [The quality of banking services as a factor in improving the efficiency of regional banks]*. Dissertation thesis (2011)
13. Y.S. Mukhamedzyanova, Оценка качества банковского обслуживания: выбор методики [Assessment of the quality of banking services: choice of methodology]. *Деньги и кредит [Money & Credit]*, **2** (2014)
14. ForresterAnalytics, Online Marketplaces Tracker [online], Available at: https://www.forrester.com/report/Forrester+Analytics+Online+Marketplaces+Tracker+2015+To+2018+Global/-/E-RES143794 (2018)
15. W. Drewes, *Qualitätsmanagement bei Banken*. (1992)
16. Y.I. Korobov, M.V. Plotnikova, К вопросу о качестве банковских услуг [To the question of quality of banking services]. *Экономическая безопасность и качество [Economic Security and Quality]*, **2** (2019)

# Cross-national analysis of the older workers' employment rate

*Oleh* Kratt[1], *Inna* Kirnos[1*]

[1]Kremenchuk Mykhailo Ostrohradskyi National University, Pershotravneva Street 20, 39600 Kremenchuk, Ukraine

**Abstract.** The share of older people in population structure is constantly rising in developed countries. This leads to a decrease of working age population. Labor market situation of Eastern European countries is aggravated by labor force migration. An effective response to challenges of ageing is productive ageing policy aimed at attracting older workers to labor market. The countries with remarkable success in older workers' employment are defined in the article. Research method – analysis of extreme values repeatability (5 leading countries and 5 outsider ones) of the statistical sample. An object of the study – is replenishment of human resources at the national labor market. Subject of the study is involving older workers to national labor market. There are three stages of the study. The first stage is ranking the countries according to the older workers' capability to implement their professional qualities at the labor market on a global scale. Capability domain of Global Age Watch Index is chosen as a criterion for ranking. Relative stability (the composition of countries) and relative volatility (annual change of country indicators' value) are the features of the leaders' group. The main feature of the outsider's group is absence of the state policy towards older workers' employment. The second stage of the study is ranking OECD countries by older workers' employment rate. PWC Golden Age Index. Is chosen as a criterion of ranking The values of the criterion is a result of the common policy in the framework of intergovernmental agreements. The proof is positive dynamics of the criterion in both groups. The third stage of study is ranking European countries by duration of the working life. State policy of productive ageing is directed towards prolongation of working life. So, direction of seniority duration characterizes policy effectiveness. Extreme values of seniority demonstrate mutual loyalty of the employers and employees. The decrease in the highest values of seniority and the suspension of their pace of dynamics suggests the existence of an upper limit of seniority. The growth rate of low values of seniority is increasing, which proves the mobility of the bottom line.

**Key words:** Global Age Watch Index, Golden Age Index, older people, ranking, productive ageing policy, labor market, duration of working life

* Corresponding author: innakirnos1976@gmail.com

## 1 Introduction

The share of older people in the population is increasing in the majority of countries. Decline in fertility rates and increased life expectancy are causing a significant shift in the global age structure. The ageing of the world's population introduces several major policy challenges, and its unprecedented nature means that we cannot look to earlier historical episodes for guidance on how this demographic upheaval will unfold or on how best to manage it [1]. Social structures developed in the times of growing population should be revisited in times of global ageing.

Declining and aging populations are often viewed as having negative effects on economic growth and employment. Decreasing share of younger people in a population is perceived as leading to a reduction in economic activity, whereas an increase in the proportion of older people is perceived as resulting in an economic burden due to higher requirement for pensions and health care. Most industrialized countries will experience a rapid shift towards increased elderly dependency ratios over the coming decades [2]. The growing number of retirees and shrinking labor force threatens the sustainability of the welfare systems, while the increase in life expectancy makes pensions expensive. Many countries have therefore implemented reforms that improve the financial sustainability of public pension programs by increasing the public pension age [3].

Europe is one of the world's oldest populations. Many European states have the lowest fertility rates and highest life expectancy rates in the world. Much of the focus of policy action has been on old-age pension reform but a more comprehensive set of reform may be necessary to encourage work at an older age. Policy response in EU to challenges of ageing became 'active ageing'. The new emphasis on employment in later life has been the main reason for the recent interest in active ageing and the fact that it has become a political priority in Europe [4]. Increasing the employment rate of older workers became a centerpiece in the employment strategy agreed upon by the Lisbon summit. The goal set in 2000 for the year 2010, to be achieved by all member states, was an employment rate of at least 50% among the population aged 55-64. In 2010 European Council declared the year 2012 as the European Year of Active Ageing and Solidarity between the Generations, with the goal of raising awareness of the contribution that older people make to society and encouraging policymakers and relevant stakeholders at all levels to take action to create better opportunities for active ageing. The 2012 Commission Communication 'Towards a job rich recovery' also emphasized the importance of introducing comprehensive measures to help extend working lives, such as tax-benefit incentives, access to lifelong learning through career guidance and training, flexible working arrangements for those who need them, and safe and healthy workplaces [5].

The main directions of pension reforms were as follows: rising retirement ages, increasing contribution years for entitlement, Eliminating or restricting early retirement options, introducing incentives for later retirement, changing indexation rules from wages to prices, extending the working period for the calculation of benefits to the entire working life, adjusting benefits to changes in life expectancy, adjusting eligibility conditions to changes in life expectancy [6]. Policy action towards prolonging work life include changes in benefits parameters for the older unemployed, reduced unemployment benefits for older persons, reduction in the duration of unemployment benefits, abolition of the job search exemption for the older unemployed, more activation during unemployment. All these measures strengthened incentives to continue working at an older age [7].

The governments implement policy measures not only towards older workers, but also towards the employers, providing them benefits for employing and training employees aged 50+. The German government has, for example, implemented a new policy package called 'Campaign 50 Plus', which provides wage subsidies to employers who employ people over

50 and assists in the retraining of such workers. In the Netherlands, employment policy promotes part-time jobs as a means of expanding job opportunities, including opportunities for older people. It is apparent from such examples that European countries have begun moving towards the promotion of employment for older people. On the other hand, the US government has had a policy of promoting the employment of older people for several decades. It introduced the Age Discrimination in Employment Act. This law was first enacted in 1967 and then revised in 1978 and 1986. It first aimed at eliminating age discrimination against people up to 65. The age limit was raised to 70 in the first revision, and then lifted altogether in the second revision. This has created an employment environment where there is no mandatory retirement or age limit for hiring, with a few exceptions [8].

Although active ageing is a common policy paradigm in the EU countries, some of them have significantly higher rates of older workers employment.

The main stages of the research are aimed at revealing the leading countries in attracting older workers to the labor market:

- ranking the countries according to the older workers' capability to implement their professional qualities at the labor market on a global scale,
- ranking OECD countries by employment level of older workers,
- ranking countries by duration of working life at European level.

## 2 Data and methods

The research is based on the information sources of different scales: the global, supranational and national ones. The global level includes the Global Age Watch Index reports provided by Help Age International [9-11] and the Golden Age Index Report provided by Price Waterhouse Coopers [12, 13]. Eurostat data regarding duration of working life is used for supranational level analysis [14]. A country is used as a taxonomic unit of analysis. Ranking the leading and outsider countries is carried out by means of extreme value's repeatability analysis. An object of the study is resources' replenishment at the national labor market. The subject of the study is involving older workers to the national labor market.

The global level is represented by 91-96 countries for the period of 2013-2015. The countries are divided into two groups – leaders and outsiders. The leaders are chosen by the maximum value of the "overall index rank". The outsiders are the countries with the minimal value of the indicator. Countries of each group are presented by "overall index value", "capability rank" and "capability value". In addition, the "average values for a group of countries" are calculated. The set of indicators allows ranking countries by the efficiency of the state policy towards attracting older workers to the labor market.

The supranational level is represented by the two categories of countries. The countries of the first category are OECD developed economies. 35 OECD countries are ranked to two groups for period 2014-2016. The first group is represented by five counties with the maximal values of the Golden Age Index. The second one includes five countries with the minimal values of the Golden Age Index. The values of the indicator "average values for a group of countries" are also calculated for these countries. The EU countries represent the second category of the supranational. 35 EU countries are ranked to two groups: the leaders and the outsiders by the maximum and minimal values of the indicator "duration of the working life" in 2016-2018. In addition, the indicator «average chain growth rate over the period» is calculated.

All the indicators presented and calculated are aimed at ranking countries according to the state policy efficiency in the domain of prolonging older workers' stay at labour market.

Choosing the leaders provides an opportunity for further research in investigation of their good practices in older workers policy.

# 3 Results

## 3.1 Ranking the countries according to the older workers' capability to implement their professional qualities at the labor market on a global scale

The Global Age Watch Index (GAWI) characterizes state policy regarding older people at the global level. It is calculated from 2013 and now the data for 2013, 2014 and 2015 years is available. The popularity of *GAWI* is increasing. In 2013 the Index covered 91 countries, while in 2014-2015 their number increased to 96. The aim of *GAWI* is to access the quality of life and wellbeing of older people. It highlights the successes and fails in the state policy and responses to the global ageing challenges. Also, it stimulates demand for and supply of sufficient age- and sex-disaggregated data as necessary to study policy-relevant topics on ageing [5].

*GAWI* is characterized by the following indicators: $OIR_i^{cod}$ – overall index rank; $OIV_i^{cod}$ – overall index value. The upper index of the indicator represents ISO country code [11], the bottom index marks the last two figures of the year.

*GAWI* includes four domains, representing country ranks in different spheres. Capability domain (*IC*) highlights the opportunities of older workers to access to the labor market - their employment levels, and educational attainment of older people. Employment level of people aged 55-64 shows their opportunities at labor market. Educational attainment shows their human and social capital potential. Capabilities' domain is characterized by the indicators: $CR_i^{cod}$ – capability rank; $CV_i^{cod}$ – capability value – *CV*). $\overline{OIV}_i^{cod}$ и $\overline{CV}_i^{cod}$ – average values for a group of countries. Comparing $CR_i^{cod}$ and $OIR_i^{cod}$, $CV_i^{cod}$ and $OIV_i^{cod}$ allows to assess the influence of the domain to *GAWI* country rank. Table 1 shows the dynamics of the above said indicators for the period of 2013-2015 [3, 4].

**Table 1.** Dynamics of $CR_i^{cod}$ and $CV_i^{cod}$, $OIR_i^{cod}$ and $OIV_i^{cod}$ during 2013-2015.

| 2013 | | | | | 2014 | | | | | 2015 | | | | |
|---|---|---|---|---|---|---|---|---|---|---|---|---|---|---|
| The leading countries according to $CR_i^{cod}$ value | | | | | | | | | | | | | | |
| Country | $CR_{13}^{кр}$ | $CV_{13}^{кр}$ | $OIR_{13}^{кр}$ | $OIV_{13}^{кр}$ | Country | $CR_{14}^{кр}$ | $CV_{14}^{кр}$ | $OIR_{14}^{кр}$ | $OIV_{14}^{кр}$ | Country | $CR_{15}^{кр}$ | $CV_{15}^{кр}$ | $OIR_{15}^{кр}$ | $OIV_{15}^{кр}$ |
| NO | 1 | 85.4 | 2 | 89.8 | NO | 1 | 76.2 | 1 | 93.4 | NO | 1 | 76.3 | 2 | 89.3 |
| US | 2 | 76.6 | 8 | 83.8 | AU | 2 | 70.2 | 13 | 76.1 | CH | 2 | 75.0 | 1 | 90.1 |
| AM | 3 | 76.5 | 51 | 50.5 | SE | 3 | 65.7 | 2 | 88.3 | DE | 3 | 68.4 | 4 | 84.3 |
| AU | 4 | 76.3 | 11 | 79.8 | US | 4 | 65.0 | 8 | 83.5 | US | 4 | 65.7 | 9 | 79.3 |
| SE | 5 | 74.3 | 1 | 89.9 | DE | 5 | 62.8 | 5 | 86.3 | SE | 5 | 65.6 | 3 | 84.4 |
| $\overline{CV}_{2013}^{cod}$ = 77.82 | | $\overline{OIV}_{2013}^{cod}$ =78.76 | | | $\overline{CV}_{2014}^{cod}$ = 67.98 | | $\overline{OIV}_{2014}^{cod}$ = 85.52 | | | $\overline{CV}_{2015}^{cod}$ = 70.2 | | $\overline{OIV}_{2015}^{cod}$ = 85.48 | | |
| Outsider countries according to $CR_i^{cod}$ value | | | | | | | | | | | | | | |
| JO | 91 | 1.6 | 88 | 11.4 | PS | 96 | 1.3 | 94 | 4.5 | PS | 96 | 1.8 | 93 | 12.3 |
| RW | 90 | 5.3 | 87 | 16.6 | JO | 95 | 1.6 | 90 | 17.0 | JO | 95 | 4.4 | 85 | 28.6 |
| MZ | 89 | 6.7 | 83 | 25.5 | MZ | 94 | 4.2 | 95 | 4.1 | MZ | 94 | 4.5 | 94 | 4.5 |
| TZ | 88 | 7.3 | 90 | 4.6 | TR | 93 | 6.0 | 77 | 36.1 | TR | 93 | 7.0 | 75 | 36.3 |
| AF | 87 | 9.4 | 91 | 3.3 | RS | 92 | 9.9 | 78 | 35.4 | IQ | 92 | 11.9 | 87 | 23.2 |
| $\overline{CV}_{2013}^{cod}$ = 6.06 | | $\overline{OIV}_{2013}^{cod}$ = 12.28 | | | $\overline{CV}_{2014}^{cod}$ = 4.6 | | $\overline{OIV}_{2014}^{cod}$ = 19.42 | | | $\overline{CV}_{2015}^{cod}$ = 5.92 | | $\overline{OIV}_{2015}^{cod}$ = 20.98 | | |
| Other countries | | | | | | | | | | | | | | |
| UA | 35 | 48.7 | 66 | 40.2 | UA | 85 | 15.2 | 82 | 31.0 | UA | 44 | 34.8 | 73 | 37.0 |
| CZ | 22 | 54.2 | 25 | 62.5 | CZ | 28 | 44.5 | 25 | 64.8 | CZ | 17 | 56.2 | 22 | 65.6 |

Source: Own processing.

Change in the leaders' group composition shown in the Table1 demonstrates the activity and discretion in the IC domain development. Leaving the leaders' group by Armenia (AM) in 2014 and Australia (AU) is explained by $CR_{2013}^{AM} > OIR_{2013}^{AM}$ (significant difference) and $CR_{2014}^{AU} > OIR_{2014}^{AU}$ respectively. *IC* domain is a part of *GAWI*, which explains the arrival of Germany (DE) in the group in 2014 with $CR_{2014}^{DE} = OIR_{2014}^{DE}$. In 2015 Switzerland entered the leaders' group. It has $CR_{2015}^{CH}$ 2, a $OIR_{2015}^{CH} = 1$. The domains of GAWI are interrelated and mutually conditioned. So, the difference between $CV_i^{cod}$ and $OIV_i^{cod}$ should be rather small. Presence of Norway (NO), USA (US) and Switzerland (SE) in the group during three years proves: 1) the relative stability of *IC*, while the value of $CV_i^{cod}$ doesn't form simultaneously but during long-term period; 2) the relative volatility of *IC*, while the values of countries' values change annually. For example, Switzerland changed the values of $CR_i^{cod}$ and $OIR_i^{cod}$ each year, wherein stayed in the group. Two circumstances characterize the leaders' group in general. Firstly, the European presence is rising. In 2013 2 countries, in 2014 – 3 countries and 4 countries in 2015 were the European ones among the 5 leaders. Secondly, IC domain decreases the value of overall GAWI rank: in 2013 the values of $\overline{CV}_{2013}^{cod}$ and $\overline{OIV}_{2013}^{cod}$ differed insignificantly, in 2014 and 2015 their difference was considerable. The synthesis of circumstances determines the regional aspect of the possibility of working longer for older workers.

The composition of outsiders' group was changing each year. Jordan (JO) and Mozambique (MZ) were present in the group during three years. Rwanda (RW), Tanzania (TZ) and Afghanistan (AF) left the group in 2014 and were replaced by Palestine (PS), Turkey (TR) and Serbia (RS). In 2015 Serbia was replaced by Iraq (IQ). A distinguishing feature of the outsiders' group is absence of the state policy towards older workers' employment. Thus, Tanzania's rank was 86, Ukraine´s (UA) was 85. Ukraine differs from other countries in the group by significant volatility of $CR_i^{cod}$ : in 2013 its rank was 35, in 2015 – 44. This fluctuation points to the absence of consistent state policy towards older workers. The opposite situation is in Czech Republic (CZ), where fluctuations of $CR_i^{cod}$ and $CV_i^{cod}$ are insignificant, which testifies the existence of consistent policy towards older people's employment.

### 3.2 Ranking OECD countries by older workers' employment rate

Golden Age Index is calculated annually by PWC for OECD countries ( $GAI_i^{cod}$ ). It is calculated on the basis of labor market indicators and weighted by their significance: employment rate of aged 55-64 (40%); employment rate of aged 65-69 (20%); gender gap in employment (ration women/men) (10%); incidence of part-time work for 55-64 (10%); full-time earnings, 55-64 relative to 25-54 (ratio) (5%); average effective labor force exit age (years) (5%); participation in training of 55-64 age group (ratio, relative to employed persons aged 25-54) (5%) [6]. The maximal index score is 100. Table 2 shows the dynamics of $GAI_i^{cod}$ indicator for the leading and outsider countries in 2014-2016 [6, 7].

**Table 2**. Dynamics of leading and outsider countries by $GAI_i^{cod}$

| 2014 | | 2015 | | 2016 | |
|---|---|---|---|---|---|
| The leading countries according to $GAI_i^{cod}$ value | | | | | |
| Country | $GAI_i^{cod}$ | Country | $GAI_i^{cod}$ | Country | $GAI_i^{cod}$ |
| IS | 97.2 | IS | 99.5 | IS | 98.7 |
| NZ | 82.4 | NZ | 84.7 | NZ | 85.3 |
| IL | 78.3 | IL | 80.6 | IL | 82.3 |
| SE | 78.2 | SE | 80.1 | ES | 81.8 |
| ES | 76.5 | ES | 78.8 | SE | 81.2 |
| Outsider countries according to $GAI_i^{cod}$ value | | | | | |
| TR | 37.8 | TR | 36.5 | TR | 38.4 |
| LU | 43.2 | LU | 41.3 | LU | 38.8 |
| SI | 41.9 | SI | 44.7 | SI | 48.2 |
| GR | 42.0 | GR | 46.4 | GR | 48.9 |
| BE | 45.4 | BE | 47.7 | BE | 49.6 |

Source: Own processing.

Statistical sample $GAI_i^{cod}$ shown in the Table 2 is quantitatively and qualitatively different from *GAWI* sample. The quantity of countries is three times as less. They are more homogeneous while the majority represents developed countries with high per capita income and high ranks of Human Development Index. So, the composition of the leaders' and outsiders' groups was unchanging as also their ranks. The majority of leaders' group is represented by the European countries Iceland (IS), Sweden (SE) and Spain (ES). New

Zealand (NZ) and Israel (IL) thought to be the benchmark countries regarding attitude towards older people. The values of $GAI_i^{cod}$ for all countries of the group were rising in 2015 and 2016, which implies the existence of a targeted policy.

The outsiders' group includes only the European countries. Turkey (TR) (outsider by $CR_i^{cod}$ in 2014 and 2015), Slovenia (SI) and Greece (GR) represent Balkan Peninsula. The values of $GAI_i^{cod}$ for Balkan countries have risen. Presence of Belgium (BE) and Luxemburg (LU) in the group allows supposing, that there could be several reasons for absence of the state productive ageing policy. Some countries have no resources to tackle the problem of ageing. The other countries do not have a problem and therefore do not need special state programs. The values of $GAI_i^{LU}$ in 2015 and 2016 were decreasing, the values of $GAI_i^{BE}$ were rising, which proves the existence of different initial conditions in developed countries.

### 3.3 Ranking countries by duration of the working life

The EU countries carry out common policy of labor market. Indicator $DWL_i^{cod}$ – «duration of the working life» characterizes labor relations [8]. Eurostat data of $DWL_i^{cod}$ for the period of 2008-2018 allows to reveal dynamics' tendencies of $DWL_i^{cod}$ by means of indicators «average chain growth rate over the period» $DWL_i^{cod}$ ($\overline{T}_{пр}^{л} DWL_i^{cod}$) and «basic growth rate for the period» $DWL_i^{cod}$ ($T_{пр}^{б} DWL_i^{cod}$). Data in Table 3 shows dynamics of five countries with minimal and maximal values of $DWL_i^{cod}$ for the period 2016-2018, and also $\overline{T}_{пр}^{л} DWL_i^{cod}$ and $T_{пр}^{б} DWL_i^{cod}$ for countries with extremal values of $DWL_i^{cod}$ in 2018.

**Table 3.** Dynamics of working life duration

| 2016 | | 2017 | | 2018 | | For the period | |
|---|---|---|---|---|---|---|---|
| Country | $DWL_i^{cod}$ | Country | $DWL_i^{cod}$ | Country | $DWL_i^{cod}$ | $\overline{T}_{пр}^{л} DWL_i^{cod}$ | $T_{пр}^{б} DWL_i^{cod}$ |
| Countries with the longest duration of the working life | | | | | | | |
| IS | 47.4 | IS | 47.0 | IS | 46.3 | 0.50 | 4.51 |
| CH | 42.4 | CH | 42.5 | CH | 42.7 | 0.35 | 3.14 |
| SE | 41.3 | SE | 41.7 | SE | 41.9 | 0.57 | 5.28 |
| DK | 40.3 | NL | 40.1 | NL | 40.5 | 0.03 | 0.25 |
| NL | 39.9 | DK | 39.7 | DK | 39.9 | 0.42 | 3.85 |
| $\overline{DWL}_{2016}^{cod} = 42.26$ | | $\overline{DWL}_{2017}^{cod} = 42.20$ | | $\overline{DWL}_{2018}^{cod} = 42.26$ | | | |
| Countries with the shortest duration of the working life | | | | | | | |
| TR | 28.5 | TR | 29.0 | TR | 29.4 | 1.78 | 17.13 |
| MK | 31.0 | ME | 31.2 | MK | 31.5 | 0.22 | 1.94 |
| ME | 31.2 | MK | 31.3 | IT | 31.8 | 2.04 | 15.11 |
| IT | 31.3 | IT | 31.7 | ME | 32.0 | 0.77 | 7.07 |
| BG | 31.7 | GR | 32.5 | GR | 32.4 | 0.22 | 1.89 |
| $\overline{DWL}_{2016}^{cod} = 30.74$ | | $\overline{DWL}_{2017}^{cod} = 31.14$ | | $\overline{DWL}_{2018}^{cod} = 31.42$ | | | |

Source: Own processing.

The essence of $DWL_i^{cod}$ excludes the opportunity of significant changes in its values, which explains stability in group composition in Table 3. The majority of countries with long working life comes from Northern Europe (northern group). All countries with short seniority come from South (southern group). Geographic orientation of the extreme values of $DWL_i^{cod}$ is probably explained by the official retirement age fixed by law. At the same time, it is more difficult for people of pre-retirement age to find job than for the younger ones. Therefore, retirement age doesn't determine the durability of working life directly. A methodology of seniority calculation is also different in countries, which also explains how countries enter the table groups.

Seniority is a result of the desires and opportunities of the parties to build employment relations. Productive ageing policy, conducted by the government, contains mechanisms and tools aimed to convince the parties to reach a compromise for their interests' sake. It is possible that in civil society the employers and the older workers can realize the necessity of compromise in labor relations on their own. This may explain the absence of Switzerland (CH), Denmark (DK) and Netherlands (NL) in the table 2 and presence in the northern group.

Decrease in values of $DWL_{2017}^{IS}$ and $DWL_{2018}^{IS}$ implies the existence of the upper limit of working life duration in Iceland (*IS*). The values of $\overline{T}_{np}^{л}DWL_i^{cod}$ and $T_{np}^{б}DWL_i^{cod}$ for northern group countries demonstrate a slowdown in the pace of dynamics, which also argues for the existence of an upper limit of seniority. Not slowdown but suspension of growth of $DWL_i^{cod}$ in the northern group justify the equality $\overline{D}\overline{WL}_{2016}^{cod} \approx \overline{D}\overline{WL}_{2017}^{cod}$, $\overline{D}\overline{WL}_{2016}^{cod} = \overline{D}\overline{WL}_{2018}^{cod}$. Growth rates of the southern group are increasing ($\overline{D}\overline{WL}_{2016}^{cod} < \overline{D}\overline{WL}_{2017}^{cod} < \overline{D}\overline{WL}_{2018}^{cod}$), which proves the mobility of the bottom line of $DWL_i^{cod}$.

## 4 Discussion and Conclusion

1. Use of *IC* (capability) domain of the *GAWI* on a global scale allowed to reveal the extreme values (5 maximum and 5 minimal ones) of $CV_i^{cod}$. Dynamics of extrema of $CV_i^{cod}$ for 2013-2015 demonstrated individual stability (composition of countries in a group) and collective stability ($\overline{C}\overline{V}_i^{cod}$). Maximal values of $\overline{C}\overline{V}_i^{cod}$ were as follows: 2013. – 77.82, 2014 – 67.98, 2015. – 70.20. Minimal values of $\overline{C}\overline{V}_i^{cod}$ were equal to: 2013 – 6.06, 2014 – 4.60, 2015 -5.92. The average maximal value of $\overline{C}\overline{V}_i^{cod}$ in 2013 was 12.84 times as higher as the average minimal value of $\overline{C}\overline{V}_i^{cod}$, in 2014 – 14.77 times and 2015 – 11.85 times (Table 1). A significant difference can be explained by the essence of the IC domain – the opportunity for older people to find job in the labor market. Demand for older workers arises due to the unfavorable demographic situation in the country and the inability (unwillingness) to import labor force. In case of favorable demographic situation, the state pursues a policy unfavorable for the work of retirees. The policy of "giving way to the young" is implemented in case of chronic unemployment caused by low level of economic development. For example, many retirees resigned in Ukraine in 2012 and 2013 when an income tax was imposed on retirement benefits for working pensioners. Therefore,

the domain does not have any common sense in several countries while there is no need to involve older persons into work.

2. Indices and rankings are a worldwide recognition of the ability of public authorities to convince labor market actors in the necessity for cooperation, which enables successful implementation of policy aimed at tackling aging. Calculation of $GAI_i^{cod}$ for OECD countries is important while this organization is a worldwide recognized platform for discussions on the most pressing issues, including older workers to the labor market. Values of $GAI_i^{cod}$ are the part of the analytical foundation for discussions and formal agreements between OECD countries. For example, a report by Price Waterhouse Coopers on the UK Economic Outlook. Section 4.2 discusses the updated results for Youth Employment and Golden Age indices [12]. At the same time, the values of $GAI_i^{cod}$ are the result of a common policy implemented by national institutions in the framework of formal intergovernmental agreements. An example is a dynamics of $\overline{GAI}_i^{cod}$ in both groups of countries: the values of $\overline{GAI}_i^{cod}$ was increasing annually at the same pace in each group (Table 2). The collection and analysis of information, collective discussions of ways and forms of development, the adoption of joint agreements make the OECD policy effective. Governments' mutual learning, multilateral surveillance and appropriate pressure are the most effective OECD tools. However, despite the OECD common policy, each country chooses its own methods and forms of its implementation. The government of a particular country takes into account demographic and economic situation, national traditions and perceptions in society. Assignment of countries in Table 2 to one or another group in no way divides them into the "positive" or "negative" ones.

3. An employee's seniority is the duration of his relationship with the employer. The duration of the relationship demonstrates, on the one hand, the employee's interest in realizing his professional qualifications, and demand for his qualities in the labor market on the other hand. The average duration of working life in country ( $DWL_i^{cod}$ ) characterizes the extent of mutual trust in employee-employer relationship at the national labor market. State policy of productive ageing is directed at an increase of the value of $DWL_i^{cod}$. So, dynamics direction of $DWL_i^{cod}$ shows the effectiveness of productive ageing policy. Extreme values of $DWL_i^{cod}$ shows the extreme displays of mutual loyalty of employers and employees. Countries with maximum loyalty (northern group) have high value of $OIR_i^{cod}$. For example, in 2015. Iceland took the 7th rank, Switzerland – 1st rank, Sweden – 3rd, Netherlands – 6th, Denmark – 11th. This means that countries also had high ranks in other domains: «Income security» (Measured by pension coverage, poverty rate in old age, the relative welfare of older people, and living standards using GNI per capita as a proxy), «Health status» (Measured by three indicators: life expectancy at 60, healthy life expectancy at 60, and psychological wellbeing. Good physical and mental health is critical to social and economic engagement of older people), «Enabling environment» (Measured by access to public transport, physical safety, social connections and civic freedom). Therefore, duration of the working life is a complex result of older people wellbeing. While average value of $\overline{T}_{np}^{л} DWL_i^{cod}$ for northern group of countries was 0.37, its value for countries of the southern group was 1.006. This means that growth rate of the southern group is 2.72 times higher than the rate of the northern one. Average value of $T_{np}^{б} DWL_i^{cod}$

in northern group was equal to 3.406, southern group has 8.63. Relatively low growth rates of $DWL_i^{cod}$ in northern group and relatively high ones in the southern group prove shrinking the gap between groups. The difference between $\overline{D}\overline{WL}_{2016}^{cod}$ for the southern and northern group was 37.48%, for $\overline{D}\overline{WL}_{2017}^{cod}$ – 35.52%, for $\overline{D}\overline{WL}_{2018}^{cod}$ – 34.50%. Reducing difference between the groups is a sign of the common policy in the EU countries. In no way does, duration of working life determines the country level of socio-economic development.

## References

1. D. Bloom, D. Canning, G. Fink, Implications of Population Ageing for Economic Growth. *Oxford Review of Economic Policy*, **26**(4), 583-612 (2010)
2. S. Harper, Economic and Social Implications of Aging Societies. *Science*, **346**(6209), 587-591 (2014)
3. OECD, *Pensions at a glance 2017: OECD and G20 indicators*, Paris: OECD Publishing (2017)
4. A. Walker, T. Maltby, Active ageing: A strategic policy solution to demographic ageing in the European Union. *International Journal of Social Welfare*, **21**(1) (2012)
5. T. Feifs, T. Weber, O. Vargas Llave, K. Fric, M. Curtarelli, I. Mandl, Role of Governments and Social Partners in Keeping Older Workers in the Labour Market. Eurofound, Dublin (2013)
6. M. Kohli, C. Arza, The Political Economy of Pension Reform in Europe. In: *R. H. Binstock, & L. K. George (Eds.), Handbook of aging and the social sciences*, 251-264 (2011)
7. A. Sonnet, H. Olsen, T. Manfredi, Towards More Inclusive Ageing and Employment Policies: The Lessons from France. *De Economist*, **162**, 315-339 (2014)
8. A. Seike, Pensions and Labour Market Reforms for the Ageing Society. In: *Eds. H. Conad, V. Heindorf, F. Waldenberger, Human Resource Management in Ageing Society*, 29-42 (2008)
9. HelpAge International, Global AgeWatch Index 2013: Insight report (2013)
10. HelpAge International, Global AgeWatch Index 2014: Insight report (2014)
11. HelpAge International, Global AgeWatch Index 2015: Insight report (2015)
12. PWC. PWC Golden Age Index. The potential $2 trillion prize from longer working lives [online], Available at: https://www.pwc.co.uk/economic-services/golden-age/pwc-golden-age-index-june2017-final-v2.pdf (2017)
13. PWC. PWC Golden Age Index. Unlocking a potential $3.5 trillion prize from longer working lives [online], Available at: https://www.pwc.co.uk/economic-services/golden-age/golden-age-index-2018-final-sanitised.pdf (2018)
14. Eurostat. Duration of working life, Annual data [online], Available at: https://appsso.eurostat.ec.europa.eu/nui/show.do?dataset=lfsi_dwl_a&lang=en (2019)

# Comparison of neural networks and regression time series in predicting export from Czech Republic into People´s Republic of China

*Tomáš* Krulický[1]*, *Tomáš* Brabenec[2]

[1]University of Žilina, The Faculty of Operation and Economics of Transport and Communications, Univerzitná 8215/1, 01026 Žilina, Slovakia
[2]University of Economics, Faculty of Finance and Accounting, nám. W. Churchilla 1938/4, 13067 Prague, Czech Republic

**Abstract.** The People´s Republic of China is one of the largest, but also the most demanding markets in the world. The trade is limited by a number of barriers, strong competition and unusual environment for trades from other parts of the world. Despite those limitations, Czech exporters are able to establish themselves in the Chinese market, exporting mainly machines and vehicles. To predict future export trends is very difficult; however, these predictions can be crucial not only for individual exporters but also for the whole national economy. For predictions, economists use causal, intuitive or statistical methods. The objective of the contribution is to compare the accuracy of equalizing time series by means of regression analysis and artificial neural networks for a possible prediction of future export trends on the example of the Czech Republic export to the People´s Republic of China. For the purposes of analysis by means of statistical methods, the data obtained from monthly statements from the period starting from the year 2000 and ending in July 2018. First, a linear regression is carried out and subsequently, neural networks are used for regression. Finally, the results are compared. It appeared that in practice, mainly all retained neural networks are applicable. However, the first of them showed significant deviations within a very short period of time.
**Key words:** export, artificial neural networks, regression analysis, time series, prediction

## 1 Introduction

Since the first computers were created, the best programmes have been striving for creating an algorithm that would be able to imitate the activity of human brain. The result of these efforts is later referred to as artificial intelligence, or artificial neural networks [1].

According to [2] artificial neural networks are inspired by biological neural network, where a fundamental building block is a neural cell, or neuron. [3] adds individual neurons

* Corresponding author: krulicky@mail.vstecb.cz

are connected by weighted links. Such links and the ability to adapt the weights (to learn) on the basis of training algorithms in the data provide a wide range of possible applications. [4] states that compared to routine methods, neural networks have a number of advantages. [5] mentions e.g. their easy use, clarity, high accuracy when analyzing complex patterns, ability to work with a large amount of data, etc. On the other hand, their disadvantages consist in the method of creating individual artificial neural networks models and requirement for extensive sample data [6].

There are a number of possibilities for the application of neural networks [7]. According to [8] artificial neural networks can be used e.g. for regression or classification of time series. [9] state that time series is a form of chronological sequence of data. Simply said, time series is a sequence of values of an indicator (economic, physiological, technological, etc.). [10] adds that time series can be found e.g. in physics, medicine, economics, but also in meteorology. In various fields of human activities, there are efforts to define the development of the given indicator in the past, to identify the factors behind the variety and make predictions for the future [11]. If a company wants to prosper and be competitive in market environment, it should perform a regular financial analysis of its activities, assess its achievements and failures and use the results for predicting the future development of the company [12].

And prediction, namely the prediction of the development of the Czech export to the PRC is what this contribution deals with. [13]. sees export as a macroeconomic term describing the overall volume of products and services purchased by foreign entities expressed in financial terms. [14] state that both the Czech Republic and the People´s Republic of China rank among the countries with open economy. According to [15] the Czech Republic is the PRC 18th largest market. [16] add that the CR exports to China show a long-term upward trend. The largest increase was recorded in 2007 (by almost 60%) and in 2010 (by 53%). There was also a large increase in 2011 (by more than 30%). In 2012 and 2013, the growth was between 8-11%. It shows that the export volumes have tripled since 2007. In the last five years they have increased roughly by 70% [17]. In 2017, Czech export to China amounted to CZK 56 billion, which represents a year-to-year increase by 20% and historic record export from the CR to China. China is considered one of the largest, but also the most demanding world markets. Business is characterized by countless barriers, strong competition and unusual environment for Europe. Despite of this, Czech exporters are steadily successful in establishing themselves in Chinese market. In 2014 Czech export to China amounted to CZK 42.5 billion [15]. According to [18] Czech export to China consists especially of machines and vehicles, accounting for more than a half of direct export. Furthermore, the Czech Republic also exports industrial and consumer goods, mainly motor vehicle components, pumps, switchboards, telephones, circuits, toys, prams, and tires.

The objective of the contribution is to compare the accuracy of balanced time series by means of regression analysis and neural networks using the example of the Czech Republic (CR) export to the People´s Republic of China.

## 2 Data and methods

Data used for analysis are available on websites of the World Bank, etc. For the purposes of the analysis, the information / data on the Czech Republic export to the People´s Republic of China. The period for which the data are available is monthly statement for the period from January 2000 to July 2018, that is, 223 input data. Currency is Euro.

Descriptive characteristics of the data are given in Table 1.

**Table 1.** Data set characteristics

| Samples | Month (Input variable) | Export (output – target ) |
|---|---|---|
| Minimum (Training) | 36526.00 | 2454273 |
| Maximum (Training) | 43252.00 | 201110297 |
| Mean (Training) | 39916.52 | 72336830 |
| Standard deviation (Training) | 1949.82 | 57137498 |
| Minimum (Testing) | 36586.00 | 2941297 |
| Maximum (Testing) | 43282.00 | 189894269 |
| Mean (Testing) | 39702.30 | 68339319 |
| Standard deviation (Testing) | 2174.13 | 61536318 |
| Minimum (Validation) | 36951.00 | 6079350 |
| Maximum (Validation) | 43040.00 | 204067342 |
| Mean (Validation) | 40047.88 | 78077699 |
| Standard deviation (Validation) | 3096.16 | 71615096 |
| Minimum (Overall) | 36526.00 | 2454273 |
| Maximum (Overall) | 43282.00 | 204067342 |
| Mean (Overall) | 39904.26 | 72594815 |
| Standard deviation (Overall) | 1963.77 | 57933645 |

Source: Own processing.

An interesting phenomenon is the development of the export over time. Figure 1 shows graphs representing statistical characteristics including an input data histogram.

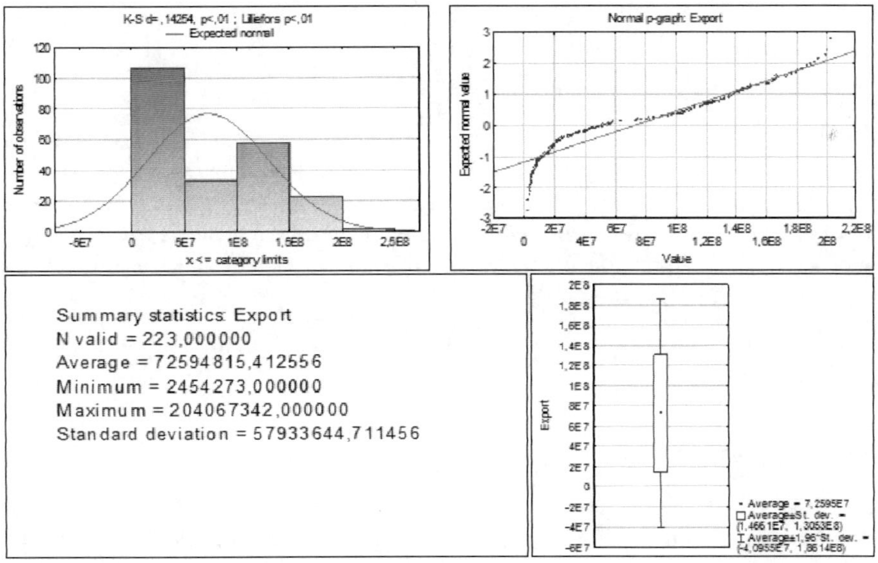


**Fig. 1.** Graph of basic statistical characteristics

Source: Own processing

The histogram shows that the major changes were in the lowest quintile. For data processing, DELL´s Statistica software, version 12 will be used. Firstly, linear regression will be carried out. Subsequently, neural networks will be used for regression.

Linear regression will be performed on the examined sample for the following functions:

- Linear,
- Polynomial,
- Logarithmic,
- Exponential,
- Distance weighting polynomial,
- Negative-exponential smoothing polynomial.

Firstly, correlation coefficient, that is, the dependence of CR export to PRC on time will be calculated. We will work with the significance level 0.95.

Subsequently, regression using neural networks will be carried out. Multilayer perceptron network and radial basis function network will be generated. Time will be an independent variable. A depended variable will be CR export to PRC. The time series will be divided into three sets – training, testing and validation. The first set will contain 70% of input data. Based on the training data set, neural structures will be generated. The two remaining data sets will contain 15% of input data each. Both data sets will be used for verification of the generated neural structure, or model reliability. The delay of the time series will be 1. 10,000 neural networks will be generated, out of which five with the best characteristics* will be retained. The hidden layer will contain no less than two and no more than 50 neurons. For the radial basis function network, the number of neurons in the hidden layers will be no less than 21 and no more than 30. For the multilayer perceptron network, the following distribution function in the hidden and output layer will be considered:

- Linear,
- Logistic,
- Atanh,
- Exponential,
- Sinus.

Other settings will be default (according to the ANS tool – automated neural networks).

Finally, the results of linear regression and regression by means of neural networks will be compared. The comparison will not be through residual analysis (minimum and maximum values, the dispersion of residuals, etc.) but as expert view and through the experience of the evaluator (economist).

## 3 Results

### 3.1 Linear regression

The correlation coefficient is 0.9605, which indicates a statistically significant direct dependence of export on the development over time. The coefficient of determination achieves the value of 0.9226.

A scatter plot (see Figure 2) was drawn up, where the individual points were fitted with a regression curve, linear in this case. The curve parameters are shown in the figure.

* The method of least squares will be used. Generating of networks will be finished when there is no improvement, i.e. when the sum of the squares is not lower. We will retain the neural structures whose sum of the residual square compared to the actual development of CR export to PRC is as low as possible (zero in ideal case).

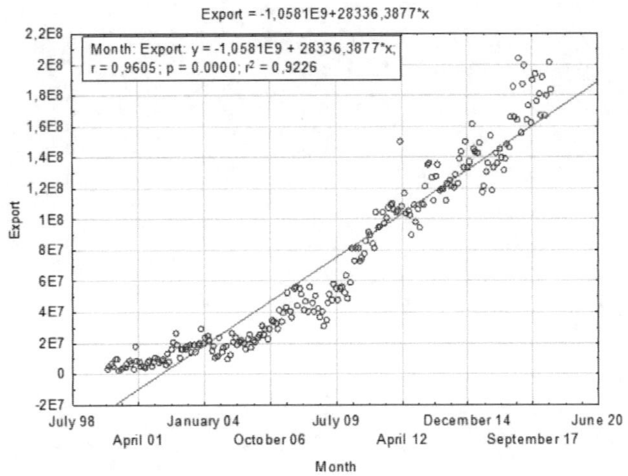


**Fig. 2.** Scatter plot representing the CR export to PRC, fitted with regression curve – linear function

Source: Own processing.

The red line represents the regression function. Linear function quite corresponds with the development of CR export to PRC.

Figure 3 represents fitting the scatter plot with a polynomial function.

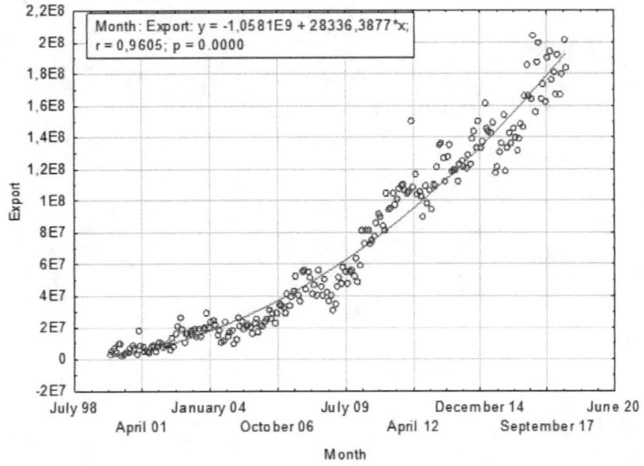


**Fig. 3.** Scatter plot of CR export to PRC fitted with a regression curve –polynomial function

Source: Own processing.

In this case, the solid red line also represents the regression curve. Its shape corresponds with the development trend of the CR export to PRC, optically even better than the curve in the Figure 2.

Figure 4 shows a scatter plot fitted with a logarithmic function.

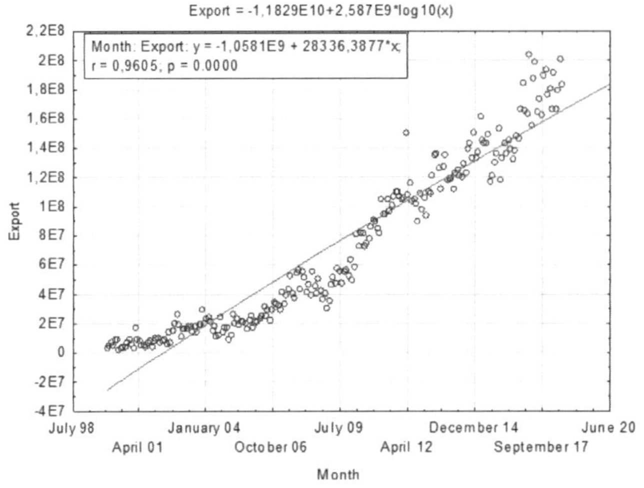


**Fig. 4.** Scatter plot of CR export to PRC fitted with a regression curve – logarithmic function

Source: Own processing.

The logarithmic function curve is similar to the linear function. However, its position in the graph is a bit different. It results from the position of the individual points in the graph that logarithmic function is not convenient for regression.

Figure 5 shows a scatter plot of the CR export to PRC fitted with a LOWESS function.

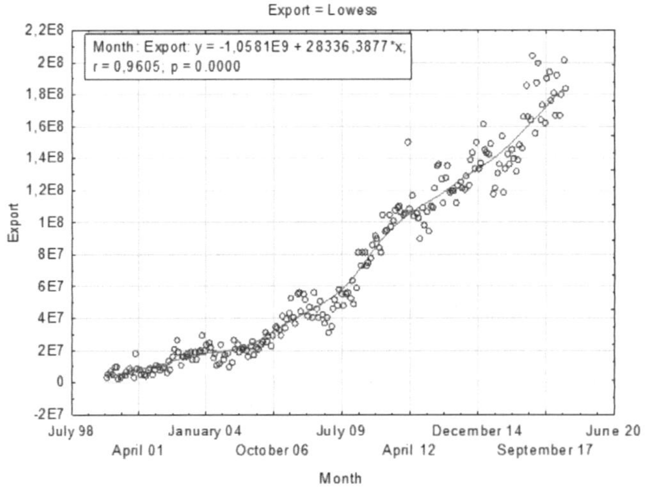


**Fig. 5.** Scatter plot of CR export to PRC fitted with regression curve – LOWESS

Source: Own processing.

LOWESS (locally weighted scatterplot smoothing) function, or also LOES (locally estimated scatterplot smoothing) function is calculated as a regression function in partial intervals. The LOWESS function curve balances the time series and follows its trend. The graph fitted with the LOWESS curve is very interesting. It enables to reveal the development trend of the monitored variable.

Figure 6 shows a scatter plot of the CR export to PRC development fitted with the function obtained by means of the method of the least squares of weighted distance.

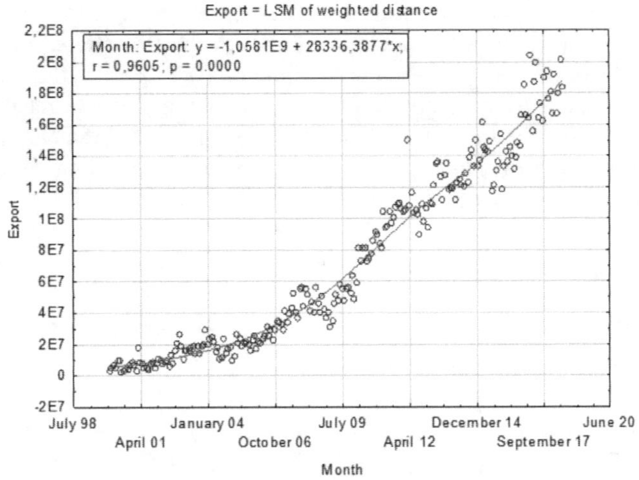


**Fig. 6.** Scatter plot of CR export to PRC fitted with regression curve – function LSM of weighted distance

Source: Own processing.

The curve follows the development of export in the entire interval much better than the linear, polynomial and logarithmic function.

Figure 7 shows fitting the graph with the function obtained by means of the least squares method negatively – by exponential smoothing.

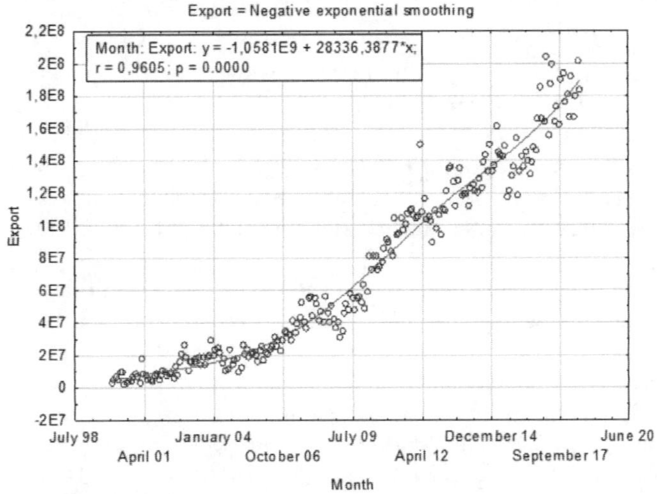


**Fig. 7.** Scatter plot of CR export to PRC fitted with regression curve – the LSM function – negatively exponential smoothing

Source: Own processing.

Even this curve appears to be interesting and convenient for a possible forecast. It also has better results than the linear, polynomial and logarithmic function.

The correlation coefficient indicates a significant statistical direct dependence of the target variable on the development over time [19]. Its value achieves 0.9605. The coefficient of determination is 0.9226. If assessing the results only by visual comparison of the CR export to PRC and the regression line shape, considering also the simple linear regression, it could be stated that the development trend is best illustrated by the LOWESS function. Other convenient ones are the curve obtained by means of the method of least squares, more specifically by negative-exponential smoothing and weighted distance. All the three functions follow the trend of the CR export to PRC development trend. It should be also stated that the other functions (linear, polynomial and logarithmic functions) are not convenient for balancing the time series, as they simplify its course.

### 3.2 Neural structures

Based on the established procedure, 10,000 neural networks were generated, out of which 5 with the best characteristics were retained. The generated structures are given in Table 2.

**Table 2.** Retained neural networks

| Network | Train. perform. | Test. perform. | Valid. perform. | Train. error | Test. error | Validation error | Train. algorithm | Error function | Activation of hidden layer | Output activation function |
|---|---|---|---|---|---|---|---|---|---|---|
| RBF 1-25-1 | 0.985012 | 0.980579 | 0.987450 | 4.771255E+13 | 7.444074E+13 | 5.083755E+13 | RBFT | Sum.quart. | Gauss | Identity |
| RBF 1-24-1 | 0.986758 | 0.988762 | 0.988485 | 4.217975E+13 | 4.178269E+13 | 4.393104E+13 | RBFT | Sum.quart. | Gauss | Identity |
| RBF 1-22-1 | 0.984784 | 0.984158 | 0.987483 | 4.843365E+13 | 5.948545E+13 | 4.857253E+13 | RBFT | Sum.quart. | Gauss | Identity |
| RBF 1-29-1 | 0.987128 | 0.985992 | 0.987617 | 4.100611E+13 | 5.522412E+13 | 4.713247E+13 | RBFT | Sum.quart. | Gauss | Identity |
| RBF 1-28-1 | 0.988870 | 0.986798 | 0.987984 | 3.547655E+13 | 4.972306E+13 | 4.766382E+13 | RBFT | Sum.quart. | Gauss | Identity |

Source: Own processing.

The networks are only the radial basis function networks. There is only one variable (time) in the input layer. The neural networks in the hidden layer contain 22-28 neurons. Logically, the output layer contains only one neuron and one output variable – CR export to PRC. In all networks, training algorithm RBFT was applied. In addition, for activating the hidden layer all neural networks used the same function, Gauss curve. Similarly, for activating the output layer the same function (identity function) is used (for more details, see Table 2).

Another interesting fact is a training, testing and validation performance. In general, we are looking for a network with ideally the same performance in all data sets (as stated before, the division of data into data sets was random). The error should be as small as possible. As for the error, it should be noted that export in EUR, that is, a detailed export is concerned. This will be reflected in the calculation of error.

The performance of the individual data sets is expressed in the form of the correlation coefficient. The values of the individual data sets by specific neural networks are given in Table 3.

**Table 3.** Individual data sets correlation coefficients

| | Exports (Training) | Exports (Testing) | Exports (Validation) |
|---|---|---|---|
| 1.RBF 1-25-1 | 0.985012 | 0.980579 | 0.987450 |
| 2.RBF 1-24-1 | 0.986758 | 0.988762 | 0.988485 |
| 3.RBF 1-22-1 | 0.984784 | 0.984158 | 0.987483 |
| 4.RBF 1-29-1 | 0.987128 | 0.985992 | 0.987617 |
| 5.RBF 1-28-1 | 0.988870 | 0.986798 | 0.987984 |

Source: Own processing.

It results from the table that the performance of all retained neural structures is very high. The correlation coefficient value of all training data sets is in the interval of 0.984 to almost 0.989. The correlation coefficient of the validation data sets for all neural networks is above 0.987. To choose the most suitable neural structure, a detailed analysis of the results obtained must be carried out. Table 4 shows the basic statistical characteristics of the individual data sets for all neural structures.

**Table 4.** Statistics of individual data sets by retained neural structures

| Statistics | 1.RBF 1-25-1 | 2.RBF 1-24-1 | 3.RBF 1-22-1 | 4.RBF 1-29-1 | 5.RBF 1-28-1 |
|---|---|---|---|---|---|
| Minimum forecast (Training) | -8418893 | 3472318 | 9180892 | 1251742 | 4918553 |
| Maximum forecast (Training) | 190822276 | 186070372 | 181203377 | 187395786 | 180517887 |
| Minimum forecast (Testing) | -7161402 | 4078977 | 9180892 | 68287 | 5140141 |
| Maximum forecast (Testing) | 188177859 | 185241939 | 176690372 | 188231841 | 179655155 |
| Minimum forecast (Validation) | 3894649 | 3794002 | 9180893 | 5962466 | 6206731 |
| Maximum forecast (Validation) | 190126526 | 185920027 | 180856649 | 188088008 | 180308749 |
| Minimum residuals (Training) | -34967735 | -27425467 | -27924255 | -26036132 | -22551514 |
| Maximum residuals (Training) | 33790350 | 41984963 | 53868193 | 35793399 | 44478206 |
| Minimum residuals (Testing) | -26450342 | -22970109 | -14708615 | -25960011 | -26960987 |
| Maximum residuals (Testing) | 19614324 | 29080375 | 40809726 | 23105018 | 31063549 |
| Minimum residuals (Validation) | -16056606 | -13641111 | -14345343 | -14204405 | -13700966 |
| Maximum residuals (Validation) | 25744050 | 22330673 | 23210693 | 24535173 | 26859519 |
| Minimum standard residua (Training) | -5 | -4 | -4 | -4 | -4 |
| Maximum standard residuals (Training) | 5 | 6 | 8 | 6 | 7 |
| Minimum standard residuals (Testing) | -3 | -4 | -2 | -3 | -4 |
| Maximum standard residuals (Testing) | 2 | 4 | 5 | 3 | 4 |
| Minimum standard residuals (Valid.) | -2 | -2 | -2 | -2 | -2 |
| Maximum standard residuals (Valid.) | 4 | 3 | 3 | 4 | 4 |

Source: Own processing.

Ideally, the individual statistics of neural network are identical across all the data sets (minimum, maximum, residuals etc.). In the case of balanced time series differences can be seen both between the individual data sets of one neural network and between the individual neural networks. Bigger differences can be seen in the residuals characteristics. However, it is not possible to clearly identify which of the retained neural networks has the most suitable results.

Figure 8 represents a line graph predicting the actual development of the CR export to PRC and the development of forecasts using the individual generated and retained neural networks.

It follows from the graph that all neural networks predict different development of export in the individual intervals. The important thing, however, is not the similarity of the individual networks forecasts but the similarity (or the degree of commonality) with the actual export development.

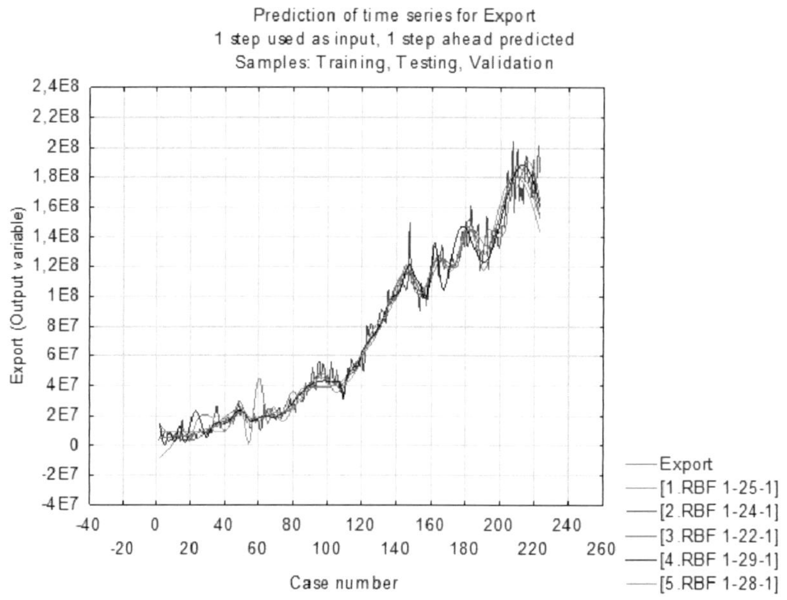


**Fig. 8.** Line graph – development of CR export to PRC predicted using neural network compared to the actual development in the monitored period

Source: Own processing.

Examining the course of the individual curves and their conformity with the actual CR export to PRC development it can be concluded that all neural networks follow the development of the variable relatively well. Only the 1st RBF 1-25-1 network in the interval between the 4th and 80th monitored case show significant differences from the actual development.

## 4 Conclusion

The objective of the contribution was to compare the accuracy of balancing the time series using regression analysis and neural networks on the example of Czech Republic export to People´s Republic of China.

In general, each forecast is determined by a certain fulfilment probability degree. When predicting the future development of any variable, the forecast of its development based on the data from the previous periods is sought to be made. Although most factors influencing the target variable can be considered, certain simplification of reality always occurs and thus there is always a certain degree of probability that the forecast will come true [20]. Both in the case of linear regression and regression by means of neural networks, quite a significant simplification occurs. We work with two variables only – input (time) and output (CR export to PRC) variable. Other input variables, often significantly influencing the level of CR export to PRC (international political situation, taxation in both countries, cost of production factors, government export subsidies, size of both countries, raw material wealth of both countries, standards of living, consumption behaviour of the inhabitants, etc.) are thus not considered. The crucial factor is the purpose of the calculation. Such simplification will be sufficient for predicting the future development of the individual states economy development based on the monitored variable and the calculation will have adequate information value. However, when planning the transportation capacity of a specific company, the simplification in the form of balancing time series will not be sufficient, as it holds true that the aggregated variables can be predicted more easily than the partial variables.

It can also be stated that due to the simplification it is not possible to predict emergency situations and their impact on the CR export to PRC (at least not in the long-term run). It would thus be ideal to make a short-term (days) forecast; however, it is not possible to obtain data for the purpose of a short-term forecast.

The CR export to PRC can be predicted based on the statistical, causal and intuitive methods. In this contribution, the focus was on comparing the statistical methods. However, this only helped us to create a framework of the monitored variable development. What is important is to process the information about the possible future development of economic, politic or legal environment. If we are able to predict the development, it could be considered when monitoring the selected variable. At the same time, however, the evaluator should be considered, an economist who is able to correlate with the resulting variable determined based on the framework of the statistical methods and specified based on the causal links on the basis of their knowledge and experience.

Optically, the LOWESS curve appeared to be the most convenient in terms of the linear regression. The other convenient ones were the curve obtained by means of the method of least squares by negative exponential smoothing and by means of the methods of least squares of weighted distance. All neural networks turned out to be applicable. Only the 1st RBF 1-25-1 neural network showed more significant deviations in a very short time interval.

The objective of the contribution was achieved.

## References

1. D. Chicea, S.M. Rei, A fast artificial neural network approach for dynamic light scattering time series processing. *Measurement Science and Technology*, **29**(10) (2018)
2. M.S. Hossain, Z.C. Ong, Z. Ismail, S. Noroozi, S.Y. Khoo, Artificial neural networks for vibration based inverse parametric identifications: A review. *Applied Soft Computing*, **52**, 203-219 (2017)
3. M.V. Cho. Learning Process in a Neural Network Model. *Journal of the Korean Physical Society*, **74**(1), 63-72 (2019)
4. M. Vochozka, Formation of complex company evaluation method through neural networks based on the example of construction companies´ collection. *AD ALTA – Journal of Interdisciplinary Research*, **7**(2), 232-23 (2017)
5. Q.J. Zhu, L.C. Tian, X.H. Yang, L.F. Gan, N. Zhao, Y.Y. Ma, Advantages of Artificial Neural Network in Neutron Spectra Unfolding. *Chinese Physics Letters*, **31**(7) (2014)
6. Z. Rowland, J. Vrbka, Using artificial neural networks for prediction of key indicators of a company in global world. *16th International Scientific Conference on Globalization and its Socio-Economic Consequences*, pp. 1896-1903. ISBN 978-80-8154-191-9 (2016)
7. D. Sánchez, P. Melin, Modular Neural Networks for Time Series Prediction Using Type-1 Fuzzy Logic Integration. *Studies in Computational Intelligence: Design of intelligent systems based on fuzzy logic, neural networks and nature-inspired optimization*, **601**, 141-145 (2015)
8. M. Deng, W.T. Yang, G.L. Liu, R. Jin, F. Xu, Y. Zhang, Heterogeneous Space-Time Artificial Neural Networks for Space-Time Series Prediction. *Transactions in GIS*, **22**(1), 183-201 (2018)

9. M.K. Rafsanjani, M. Samareh, Chaotic time series prediction by artificial neural network. *Journal of Computational Methods in Sciences and Engineering,* **16**(3), 599-615 (2016)
10. B. Wang, S.H. Xu, X.H. Yu, P.C. Li, Time Series Forecasting Based on Cloud Process Neural Network. *International Journal of Computational Intelligence Systems*, **8**(5) 992-1003 (2015)
11. F. Fernandez-Navarro, M.A. de la Cruz, P.A. Gutierrez, A. Castano, C. Hervas-Martinez, Time series forecasting by recurrent product unit neural network. *Neural Computing & Applications*, **29**(3), 779-791 (2017)
12. T. Klieštik, J. Vrbka, Z. Rowland, Bankruptcy prediction in Visegrad group countries using multiple discriminant analysis. *Equilibrium – Quarterly Journal of Economics and Economic Policy*, **13**(3), 569-593 (2018)
13. Ministry of Industry and Trade – MIT, *Export v ekonomice [Export in economy]* [online], Available at: https://www.mpo.cz/ (2018)
14. J. Gourdon, S. Monjon, S. Poncet, Trade policy and industrial policy in China: What motivates public authorities to apply restrictions on exports? *China Economic Review,* **40**, 105-120 (2016)
15. Business info.cz., China: Trade and Economic Cooperation with the Czech Republic, [online], Available at: https://www.businessinfo.cz/cs/clanky/cina-obchodni-a-ekonomicka-spoluprace-s-cr-19054.html (2018)
16. V. Stehel, P. Šuleř, Foreign trade between China and the Czech Republic. *Littera Scripta*, **9**(3), 84-95 (2016)
17. V. Humlerová, Czech-Chinese Business Cooperation Case Study. *31st International-Business-Information-Management-Association Conference Innovation management and education excellence through vision 2020*, pp. 3185-3191 (2018)
18. P. Higgins, T. Tha, W. Zhong, Forecasting China's economic growth and inflation. *China Economic Review,* **41**, 46-61 (2016)
19. T. Klieštik, M. Mišánková, K. Valášková, L. Švábová, Bankruptcy prevention: new effort to reflect on legal and social changes. *Science and Engineering Ethics*, **24**(2), 791-803 (2018)
20. Ch. Bolton, V. Machová, M. Kováčová, K. Valášková, The power of human-machine collaboration: Artificial intelligence, business automation, and the smart economy. *Economics, Management, and Financial Markets*, **13**(4), 51-56 (2018)

# Base erosion and profit shifting by Google and platform market competition in Korea

*Taehee* Lee[1*]

[1]Kookmin University, College of Business Administration, 77 Jeongnueung-ro, Seoul, Korea

**Abstract.** The adoption of Google Taxes is often mentioned with respect to corporate taxes avoided for a MNE that does not have a permanent establishment in the domestic market. None has looked at the implications of avoided corporate taxes on platform market competition. Using Google as a representative MNE that employs the DIDS framework, the present study, using publicly disclosed financial statements information, documents that if a MNE with a permanent establishment in Korean market had generated profits of 520.9b KRW in 2017.
**Key words:** profit, market, google, erosion

## 1 Introduction

Multinational enterprises (MNE's) often exercise transfer pricing strategies among her subsidiaries to shift profits across different jurisdictions for various purposes. Such practices have long been subject to regulatory scrutiny, mainly by national taxing agencies, for profit shifting by MNE's inevitably entails avoided corporate income taxes on the part of the national tax agencies. OECD first coined the acronym BEPS (Base Erosion and Profit Shifting) to reflect the phenomenon and accordingly launched BEPS Action Plan in 2014 [1].

At the age of digital economy, the issue of BEPS has developed into a more heated controversy as MNE's, notably Google, Apple, Facebook, Amazon (GAFA), heavily engage in R&D activities, create intellectual properties (IP's), and use transfer pricing of IP's as a tool for transnational tax planning. MNE's lease (licence to use) IP's to their affiliates in low tax jurisdictions and these affiliates sublease IP's to another affiliates with customer base in high tax jurisdictions.

Such practices of GAFA-like MNE's (hereafter tech MNE's) are more controversial than those of traditional MNE's because tech MNE's do not establish a permanent establishment (PE) in their respective customer base countries. Most of the time, tech MNE's engage in business activities via intangible assets (IP's). An "avoided PE's (APE)" is then defined as a foreign company (usually a MNE) who makes substantial sales through activity in the customer base country while avoiding the creation of a taxable presence in that jurisdiction.

Like many other countries in EU, a heated debate has arisen recently regarding BEPS issue in Korean platform market by tech MNE's. This study adds some evidence to the

* Corresponding author: thlee@kookmin.ac.kr

debate by estimating Google's sales, operating expenses, income taxes in Korea if Google were subject to Korean tax law. The study also draws some implications as to what impact tech MNE's BEPS activities has on competition in the Korean platform market. Prior to the current study, none has addressed the issue using accounting figures based on publicly disclosed financial information.

[2] review and critique two prominent theories in the international business and international economics literatures regarding the role of multinational enterprises (MNEs) in host country development: the "spillovers" perspective on the impact of MNE investment in host countries and the liabilities of foreignness (LOF) view that specifies the constraints MNEs must overcome to succeed in local, developing country markets.

According to [3, 4], MNEs are central agents, they argue, for promoting economic growth meanwhile they complement domestic savings, transfer technology and management skills, increase competition, and stimulate entrepreneurship, but another group of researchers opposes this view suggesting that MNEs are more likely to crowd out local firms, use technology that is inappropriate for local circumstances, actively constrain potential technology spillovers, and reduce the domestic capital stock and tax basis due to transfer price manipulation and excessive profit repatriation [5, 6].

## 2 Documentation of Profit Shifting in Korean Market

### 2.1 Estimation of sales revenue

The present paper estimates the sales revenue of Google in Korea by using publicly available information disclosed by Google Asia Pacific Ltd (hereafter, Google Singapore). From the financial statements filed to Accounting and Corporate Regulatory Authority (ACRA), a national regulatory body of Singapore, sales revenue of Google Singapre amounted to $15.7b in 2017 [7].

*The above statement of financial position should be read in conjunction with the accompanying notes.*

**Google Asia Pacific Pte. Ltd.**
**Statement of profit or loss and other comprehensive income**
**For the year ended 31 December 2017**

| | | Notes | **2017 US$'000** | 2016 US$'000 |
|---|---|---|---|---|
| Revenue | | 27,39 | 15,684,946 | 11,131,903 |
| Cost of revenue | 77.3% | 28 | (12,122,171) | (9,017,866) |
| **Gross profit** | | | 3,562,775 | 2,114,037 |
| Sales and marketing expenses | | 28,39 | (2,229,279) | (1,131,110) |
| General and administrative expenses | | 28 | (834,716) | (513,783) |
| **Operating expenses** | 96.8% | | (3,063,995) | (1,644,893) |
| Other income | | 29 | 135,338 | 50,726 |
| Other expenses | | 29 | (141,597) | (101,351) |
| **Net other expenses** | | | (6,259) | (50,625) |
| **Profit before income tax** | 3.2% | | 492,521 | 418,519 |
| Income tax expense | | 30 | (1,436) | (52,221) |
| **Profit for the year** | 3.1% | | 491,085 | 366,298 |
| **Other comprehensive income** | | | | |
| Cash flow hedges – effective portion of changes in fair value | | | (87,207) | 138,561 |
| Cash flow hedges – reclassified to profit or loss | | | (42,786) | (30,989) |
| **Other comprehensive income for the year, net of tax** | | | (129,993) | 107,572 |
| **Total comprehensive income for the year** | | | 361,092 | 473,870 |
| Profit is attributable to: | | | | |
| Owners of Google Asia Pacific Pte. Ltd. | | | 491,085 | 366,298 |
| Total comprehensive income for the year is attributable to: | | | | |
| Owners of Google Asia Pacific Pte. Ltd. | | | 361,092 | 473,870 |

**Fig. 1.** Revenues, costs and profits of Google Singapore in ratios

Source: Own processing.

Given that sales revenue of $15.7b by Google Asia Pacific Ltd. represents the major chunk of Google sales revenue in APAC markets, the portion of Korean market within the APAC markets is estimated based on two underpinnings.

First, footnote 10 of financial statements of Google Singapre discloses that the firm hedges foreign currency risk by engaging in foreign currency options and forward contracts of three currencies: Australian Dollar (AUD), Korean Won (KRW) and Japanese Yen (JPY). This implies that the Korean market is likely to be at least in the third with respect to Google sales in APAC markets.

Second, footnote 34 of the same document discloses that the firm has transactional currency exposures arising from sales and purchases that are denominated in currencies other that the functional currency (US$) of the firm: Japanese Yen (JPY), Singapore Dollar (SGD), Taiwanese Dollar (TWD), Australian Dollar (AUD), Hong Kong Dollar (HKD), Euro (EUR), Malaysian Ringgit (MYR), Korean Won (KRW), New Zealand Dollar (NZD) and Thai Baht (THB). The firm also discloses its exposure to foreign currency risk at the end of the fiscal year as follows.

Taken together, the disclosed financial statements information reveals that the volume of Google transactions in Korea accounts for a significant portion of Google Singapore's foreign exchange risk. Thus, by using the KRW share in the foreign currency risk exposure of Google Asia Pacific Ltd. as a proxy for the proportion of sales revenue in Korea relative to the total sales revenue of Google Singapore. Using the portion of KRW with respect to other currencies with respect to both the receivables and payables, 13.6% is used as a proxy for the portion of Google sales in Korea.

Multiplying 13.6% to sales revenue of Google Asia Pacific Ltd. of $15.7b gives an estimated sales revenue of $2.1b in 2017 from the Korean market.

## 2.2 Estimation of costs and profits

Several points are noteworthy from Google Singapore's comprehensive income statement captured in figure 1. First, cost of revenue and operating expenses are 77.3% and 96.8% of sales revenue, respectively. High cost ratios show that Google Singapore is at disadvantage with respect to the cost structure. Further inquiry of financial statements reveals that such cost structure is attributable to high service and license costs. Footnote 28 (Expenses) reveals that service and license costs comprise a large percentage (87.7%) of major operating costs.

Second, profitability ratios are extremely low thanks to high cost ratios. Profit before income tax ratio and net income ratio are only 3.2% and 3.1% of sales revenue, respectively. This is in sharp contrast to the profitability of its Korean competitor in the Korean platform market, NAVER, which displayed 25.6% and 16.5% for the commensurate ratios in 2017.

We next tried to estimate the profits that were shifted from the Korean market. Using the estimated revenue of $2.1b in Section 2.1 of this paper as a starting point, Google profit in 2017 from the Korean market is estimated as follows.

First, the cost of sales (41.1%) and R&D expenses (15.0%), in percentages of sales, of Alphabet Inc., were applied to the estimated sales revenue from the Korean market (₩2,139.5B). Had a hypothetical Google subsidiary in Korea with permanent establishment existed, it would have incurred cost of sales and R&D expenses similar to its ultimate parent company, Alphabet Inc. [8].

Second, sales & marketing expenses (14.2%) and general & administrative expenses (5.3%), in percentages of sales, of Google Singapore., were applied to the estimated sales revenue from the Korean market. This assumes that these costs of the hypothetical Google subsidiary in Korea with permanent establishment would resemble those of Google Singapore because they do not include royalty payments of subleasing IP's.

Third, after accounting for major operating expenses (75.7%), the hypothetical Google subsidiary in Korea with permanent establishment would generate ₩520.9b of operating profit in 2017 which is 24.3% of sales revenue. Given that the hypothetical Google subsidiary is not subject to tax jurisdiction in Korea, such profits are shifted from Korea, thus avoiding corporate taxes which would have been levied in Korea.

**Table 1.** Estimating Google sales and profits in Korea

| | Google Asia Pacific Ltd. | | Google Sales in Korea | |
|---|---|---|---|---|
| | Amount (US$ B) | % | Amount (KRW B) | % |
| Revenue | 15.685 | 100.0 | 2,139.5 | 100.0 |
| Cost of Sales | 12.122 | 77.3 | 879.7 | 41.1 |
| Gross Profit | 3.562 | 22.7 | 1,259.7 | 58.9 |
| R&D expense | | | 320.9 | 15.0 |
| Sales expense | 2.229 | 14.2 | 304.1 | 14.2 |
| Admin. expense | 0.835 | 5.3 | 113.9 | 5.3 |
| Operating exp. | 15.186 | 96.8 | 1,618.6 | 75.7 |
| Operating income | 0.499 | 3.2 | 520.9 | 24.3 |

Source: Own processing.

## 3 Conclusions

The adoption of Google Taxes is often mentioned with respect to corporate taxes avoided for the MNE does not have a permanent establishment in the domestic market. None has looked at the implications of avoided corporate taxes on platform market competition. This is particularly important in country like Korea in which domestic platform companies are competing with MNE like Google.

If a domestic firm or a MNE with a permanent establishment in Korean market had generated profit of ₩520.9b in 2017, it would have booked a tax expense of ₩126.1b, an equivalent of applying a nominal tax rate of 24.2% to the pre-tax profit. In 2017

Such profit shifting and exemption of tax expense may have a profound impact on platform market competition. Immensity of the impact on platform market competition can be described in several ways. First, the amount, ₩126.1b is equivalent of hiring 1,292 top-tier SW engineers in Korea. In 2017, an average top-tier SW engineer earned about 0.1b a year [9]. Alternatively, with the same amount of tax expense, if exempted, a company can hire 1,125 highly skilled SW engineers in U.S. [10]. These equivalence stories imply that profit shifting and tax avoidance by MNE's unleashed by the DIDS structure may have a profound impact on domestic platform markets at the detriment of local platform companies.

## References


1. OECD, "*2015 BEPS Action Plan 1: Report on addressing the tax challenges of the digital economy*" (2015)
2. J. Oetzel, J.P. Doh, MNEs and development: a review and reconceptualization. *Journal of World Business,* **44**(2), 108-120 (2009)
3. R.E. Caves, Multinational Firms, Competition, and Productivity in Host-Country Markets. *Economica*, **41**(162), 176-193 (1974)
4. N. Lowe, M. Kenney, Foreign investment and the global geography of production: Why the Mexican consumer electronics industry failed. *World Development,* **27**(8), 1427-1443 (1999)
5. M. Haddad, A. Harrison, Are there positive spillovers from direct foreign investment? Evidence from panel data for Morocco. *Journal of Development Economics,* **42**(1), 51-74 (1993)
6. B.J. Aitken, A.E. Harrison, Do domestic firms benefit from direct foreign investment? Evidence from Venezuela. *American Economic. Revue,* **89**(3), 605-618 (1999)
7. Google Asia Pacific PTE. LTD., *Financial statements year ended 31 December 2017* (2017)
8. Alphabet Inc., *Financial statements year ended 31 December 2017* (2017)
9. Korea Software Industry Association, ″Average Wage of S/W engineers in 2017″ [online], Available at : https://search.naver.com/search.naver?where=image&sm=tab_jum&query=SW%EA%B8%B0%EC%88%A0%EC%9E%90+%ED%8F%89%EA%B7%A0%EC%9E%84%EA%B8%88+2017#imgId=blog32837065%7C7%7C221087391588_2146533436&vType=rollout (2017)
10. U.S. News, ″How Much Does a Software Developer Make ?″ [online], Available at : https://money.usnews.com/careers/best-jobs/software-developer/salary (2017)

# Machine learning forecasting of CR import from PRC in context of mutual PRC and USA sanctions

*Veronika* Machová[1*], *Jan* Mareček[1]

[1]Institute of Technology and Business, School of Expertness and Valuation, Okružní 517/10, 37001 České Budějovice, Czech Republic

**Abstract.** Mutual trade restrictions between the USA and the PRC caused by the USA feeling of imbalance of trade between these two countries have significantly influenced not only the trade between these two states but also the overall atmosphere of the international trade in the last few years. The objective of the contribution is to find out whether machine learning forecasting is capable of equalizing time series so that the model effectively forecasts the future development of the time series even in the context of an extraordinary situation caused by such factors as the mutual sanctions of the USA and PRC. The dataset shows the course of the time series at monthly intervals starting from January 2000 to June 2019. There is regression carried out using neural structures. Three sets of artificial neural networks are generated. They are differ in the considered time series lag. 10,000 neural networks are generated, out of which 5 with the best characteristics are retained. The mutual USA and PRC sanctions did not affect the success rate of the machine learning forecasting of the CR import from the PRC. It is evident that the mutual sanctions shall affect the trade between the CR and the PRC.
**Key words:** machine learning, mutual sanctions, import, artificial neural networks, forecasting, time series

## 1 Introduction

Stehel and Šuleř [1] argue that China is for the Czech Republic the principal exporter and the second principal importer with ratio 12.4% and 10.3%; however, the Czech Republic has a substantial trade deficit with People's Republic of China. Chinese imports to the Czech Republic have substantially increased with the policy focused on the investment goods and products used for further production [2]. According to Higgins, Tha and Zhong [3] it is mostly transportation equipment and machines, industrial and market products that are the predominating negotiable groups. Liu [4] declares that both economies – the Czech and Chinese one – are focused rather on the export than import. What is typical for the

* Corresponding author: machova@mail.vstecb.cz

current situation is the decrease in the import from China to the Czech Republic, thus copying the global trend when China has reduced its export recently. In contrast to the change in the import from China to the Czech Republic, there is a steady increase in export of goods and services from the Czech Republic to China [5]. The drop of import and the growth of export significantly reduced the balance from minus 421 to 387 billion. All the same it cannot be currently expected that the Czech Republic would encourage the export to China more than the import from the country in the near future. Nevertheless, a lot of complaints have been made from Czech enterprises and end-to-end consumers about variable quality of the import from China. If these Czech enterprises keep on exercising a growing influence in China, a gradual improvement of the balance sheet could be expected in the following years [6]. Ministry of Foreign Affairs [7] publishes on its websites that the deficit of the Czech Republic in trading with China rose from 75.8 billion to 90.9 billion in the first two months of 2019. While the Czech export has fallen in the year-on-year balance roughly by 4% this year, the import has increased by 17%. Different states generally tend to impose economic sanctions nowadays more than before, but rarely do they achieve intended political or economic aims. Li; Crăciun et al. [8, 9] argues that what plays the key role in assessing the effectiveness of sanctions are cultural standards of target countries. The author's study reveals that in order for the effective and prosperous sanctions to be imposed as a means of foreign policy, the initiators of the sanctions must consider their cultural consequences and responsibility. The USA imposes a wide range of sanctions against China. Although the American sanction did not radically affect the trade between the countries, their impact on producers and consumers was rather profound in both countries [10].

In order to thoroughly assess the import development, it is convenient to consider employing methods such as artificial neural networks (ANNs). Kliestik [11] argues that these networks constitute computational models which reflect biological neural networks and behaviour of neurons. These networks can be applied in various areas. According to Pao [12] artificial neural networks are currently widely used for resolving possible future issues including without limitations value prediction. Sayadi et al. [13] declares that one of the big advantages of artificial neural networks is the ability to learn, generalize etc. On the other hand, Rowland and Vrbka [14] argue that disadvantages of ANNs lie in their requirements for extremely precise information and the risk of illogical behaviour of the network. Horák and Krulický [15] point out that ANN is one of the most conventional methods which is used as a prediction method that requires a more complex model, is inclined to non-linearity and employs more variables. For that reason, ANN is most effectively used in the financial sector. The authors' study reports that ANNs as a prognostic model could considerably exceed the minimum return. Cho, Kim and Bae [16] proclaim that ANN techniques achieve better results than traditional statistical methods. Comprehensive and exact results are also yielded by modern hybrid models which combine statistical and ANN methods. Li et al. [17] emphasize that traditional statistical methods are, though, very popular for their simplistic interpretation, understanding and a fine prognostic performance. Balcaen and Ooghe [18] received a surprising result having compared selected models that resulted from the discrimination analysis with models that were based on logistic regression. They found out that the accuracy of the model crucially depends on ratios included in the model itself rather than the method from which the model was derived.

The economic sector may also use other models than already mentioned neural networks. These are for example discrimination analysis, ARIMA model, cluster analysis and decision trees [19]. According to Majerčíková and Bartošová [20] multiple discrimination analysis is a diagnostic method which consists in assessing and interpreting future economic risks. The advantage of this analysis is its low time consumption and simplicity. Kliešťik, Vrbka and Rowland [21] see its drawbacks in its non-comprehensiveness. ARIMA model (Autoregressive Integrated Moving Average) is one of the most widely used linear models applied in predicting time series [22]. The models are mostly used for a short-term prognosis in situations when information on variables is not available or when the model does not have a good prognosis [23]. Junttila [24] sees the advantages of the model mostly in its ability to readily adapt to changes throughout the time series. On the other hand, he spots disadvantages in them being applicable only for time series ranging from no less than 50 observations; furthermore, final models are hard to interpret and their practical application is very difficult. Majerová and Kliešťik [25] argues that the main purpose of the cluster analysis is to generalize and define relations in the group of variables. Their clusters are created according to their similarity. The aim of the cluster analysis is to classify an exact number of objects into several relevant homogeneous clusters. What is required is that the objects inside one cluster should be extremely similar while the objects in different clusters should be as different as possible [26]. According to Šuleř [27] this method is convenient for its simplicity; it allows using statistical software etc. The drawback of the cluster analysis is its inclination to distorting the results. Chollet [28] in his study points out that decision trees are data structures that are able to predict output values for specific inputs or to classify input data points. Their advantage is an easy visualization and interpretation. On the other hand, their disadvantage is that they cannot be always used since they require a more difficult continuous data processing [29].

Rowland, Šuleř and Vochozka [30] aimed at comparing the accuracy of equalizing the time series through ANNs and regression analysis giving the trade balance between the Czech Republic and China as an example. At first they carried out linear regression followed by regression via ANNs. Then they professionally compared both methods. According to their study ANNs appear very useful for predicting time series. Weijin and Yuhui [31] suggest a model which is used for predicting import and export trades in one sector. They concluded that a non-linear prediction may focus not only on the combination of data and improving their accuracy, but it can also vividly reflect non-linear characteristics of the prognosis system. Horák, Šuleř and Vrbka [32] focused on comparing accuracies of equalizing time series through the regression analysis and ANNs giving the export from the USA to China as an example. The aim of the study was to illustrate a possible use and advantages of ANNs in practice. The study compares two statistical methods while artificial intelligence does not apply in either of these techniques. On the other hand, a lot of economic sectors showed better results. It was found out that ANNs can effectively learn dependences of time series. As applied by analogy, Vrbka, Rowland and Šuleř [33] received the same results in their study where they compared the accuracy of the equalization of time series through regression analysis and ANNs giving the trade balance between the EU and China as an example.

## 2 Data and methods

Mutual trade restrictions between the USA and the PRC caused by the USA feeling of imbalance of trade between these two countries have significantly influenced not only the trade between these two states but also the overall atmosphere of the international trade in the last few years. In a way the restrictions also affect the trade between the countries that are out of this trade war. It is therefore very interesting to see how the trade war is reflected in the relations that are not related to it at first sight. This contribution is focused on the machine learning forecasting of the CR import from the PRC. Its objective is to find out whether machine learning forecasting is capable of equalizing time series so that the model effectively forecasts the future development of the time series even in the context of an extraordinary situation caused by such factors as the mutual sanctions of the USA and PRC.

The dataset shows the course of the time series at monthly intervals starting from January 2000 to June 2019. The course of the time series is shown in Figure 1.

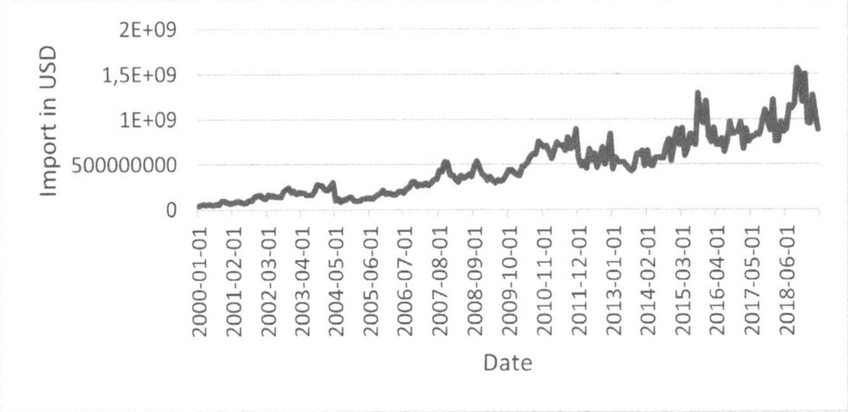


**Fig. 1.** Development of CR import from PRC between January 2000 to July 2019

Note: The values in the entire text are given in USD.

Source: World Bank [34], own processing.

The Figure clearly shows the overall increase of the CR import from the PRC. In the first years of the monitored period the increase was rather gradual. In May 2004, the import decreased significantly almost to the third of the previous month. It was almost as if the development was the same as back in 2004. On the contrary, the highest values were achieved in October 2018 and January 2019, when the import volume was approximately USD 1.5 bn. per month. There is also a question if it is possible to find seasonal trends in the development of the time series. We will work with the assumption when setting the methodology of work. The difference between the month with the highest volume of import achieved and the month with the smallest volume of the CR import from the PRC is more than 35 higher. Table 1 shows the basic statistical characteristics of the dataset.

**Table 1**. Basic statistical characteristics of examined dataset

| Samples | Date (Input variable) | Month (Input variable) | Year (Input variable) | Import (Output (target)) |
|---|---|---|---|---|
| Minimal (Training) | 36556.00 | 1.00000 | 2000.000 | 4.421794E+07 |
| Maximum (Training) | 43646.00 | 12.00000 | 2019.000 | 1.513269E+09 |
| Average (Training) | 40050.47 | 6.45455 | 2009.115 | 4.752065E+08 |
| Standard deviation (Training ) | 2001.16 | 3.50688 | 5.473 | 3.213289E+08 |
| Minimal (Testing) | 36585.00 | 1.00000 | 2000.000 | 4.983804E+07 |
| Maximum (Testing) | 43677.00 | 12.00000 | 2019.000 | 1.563688E+09 |
| Average (Testing) | 40130.09 | 6.00000 | 2009.371 | 4.666358E+08 |
| Standard deviation (Testing) | 2418.96 | 3.38683 | 6.691 | 3.892610E+08 |
| Minimal (Validation) | 36646.00 | 1.00000 | 2000.000 | 5.050822E+07 |
| Maximum (Validation) | 43373.00 | 12.00000 | 2018.000 | 1.204702E+09 |
| Average (Validation) | 40412.86 | 6.71429 | 2010.086 | 5.065008E+08 |
| Standard deviation (Validation) | 3403.28 | 3.44159 | 9.284 | 4.082950E+08 |
| Minimal (Overall) | 36556.00 | 1.00000 | 2000.000 | 4.421794E+07 |
| Maximum (Overall) | 43677.00 | 12.00000 | 2019.000 | 1.563688E+09 |
| Average (Overall) | 40116.30 | 6.42553 | 2009.298 | 4.786420E+08 |
| Standard deviation (Overall) | 2069.19 | 3.45140 | 5.669 | 3.305002E+08 |

Source: Authors.

There will be regression carried out using neural structures. We will generate multilayer perceptron networks (MLP). Three sets of artificial neural networks will be generated. They will differ in the considered time series lag:

1. 1-month lag,
2. 5-month lag,
3. 10-month lag.

Time series lag indicates the number of data used for the calculation of the following value (that is, based on the value of one previous month in the first case, five previous months in the second case, and ten previous months in the third case). Larger time series lag can result in averaging values. A small lag can result in extreme fluctuations in an equalized time series. Each time series lag entails greater demands on the complexity of the artificial neural structure, specifically the number of the neurons in the input layer will be increased. The calculation is then more demanding in terms of the IT equipment. In other settings, the experiments will match.

The independent continuous variable will be time. The seasonal fluctuation will be represented by categorical variable in the form of the month and the year in which the value was measured. We will thus work with the monthly seasonality of the time series. The variable in the form of the year will help to capture the development of the time series. The dependent continuous variable will be the CR import from the PRC.

The time series will be divided into three sets – training, testing and validation. The first group will contain 70% of the input data. Based on the training data set, neural structures will be generated. The remaining two datasets will contain 15% of the input information each. Both testing and validation data sets will serve to verify the reliability of the neural structure, or the model created. 10,000 neural networks will be generated, out of which 5 with the best characteristics[2]will be retained. The hidden layer will contain at least two neurons and at most 8 neurons for Experiment 1 and 9 neurons for Experiments 2 and 3. For the distribution functions in the hidden and output layers, the following functions will be considered: Linear, Logistic, Atanh, Exponential, Sinus.

Other settings will remain default (according to the ANN – automated neural networks tool). It is also possible to manually adjust the weights of the neural networks using the VNN (own neural networks). However, if there is a more appropriate weights distribution, it will most likely be a coincidence.

Finally, there will be a comparison of the three experiments carried out and it will be determined if the chosen methodology of machine learning forecasting is correct in terms of the direction and what time lag is the closest to the correct result.

## 3 Results

### 3.1 Experiment 1 (1-month lag of the time series)

Table 2 shows an overview of the neural networks retained from the Experiment 1.

[2] The least squares method will be used. Generating of networks will be terminated if there is no improvement, i. e. the reduction in the value of the sum of the squares. We will thus retain the neural structures whose sum of the residual squares to the actual CR import from the PRC is as low as possible (zero in ideal case).

**Table 2.** Retained neural networks from Experiment 1

| Network | Train. perform | Test. Perform. | Valid perform. | Train. error | Test. error | Valid. error | Train. Algorit. | Error function | Activation of hidden layer | Output activation function |
|---|---|---|---|---|---|---|---|---|---|---|
| MLP 3-10-1 | 0.963883 | 0.966223 | 0.947668 | 3.596561E +15 | 4.769768E +15 | 6.024952E +15 | BFGS (Quasi-Newton) 139 | Sum of squares | Tanh | Exponential |
| MLP 3-8-1 | 0.957052 | 0.963829 | 0.950740 | 4.274898E +15 | 5.077201E +15 | 5.136616E +15 | BFGS (Quasi-Newton) 101 | Sum of squares | Tanh | Tanh |
| MLP 3-9-1 | 0.947655 | 0.966741 | 0.947826 | 5.170708E +15 | 5.067522E +15 | 5.938017E +15 | BFGS (Quasi-Newton) 210 | Sum of squares | Exponential | Tanh |
| MLP 3-9-1 | 0.942154 | 0.958476 | 0.948380 | 5.739409E +15 | 5.941721E +15 | 6.174524E +15 | BFGS (Quasi-Newton) 157 | Sum of squares | Exponential | Logistic |
| MLP 3-9-1 | 0.944905 | 0.960153 | 0.947592 | 5.481413E +15 | 5.579881E +15 | 6.427599E +15 | BFGS (Quasi-Newton) 47 | Sum of squares | Tanh | Logistic |

Source: Authors.

It results from the table that there have been retained neural networks with 8, 9, and 10 neurons in the hidden layer of the neural network. For the activation of the hidden layer, the function of the hyperbolic tangent and exponential function are used. The output layer of the neurons is activated using the exponential, logistic, and hyperbolic tangent functions. What is important is the performance of the neural networks, expressed by the correlation coefficient. The value of the correlation coefficient is high in all data sets of all networks. It achieves the values higher than 0.94 (in the case of the first network, it is above 0.96 in two data sets), which indicates a high degree of direct dependence. Based on this parameter it is evident that the retained neural structures are a plausible model of the actual time series, and are thus able to forecast the future development of the time series. However, the level of error in all three data sets is relatively high. Figure 2 represents the comparison of the actual course of the time series and equalized time series.

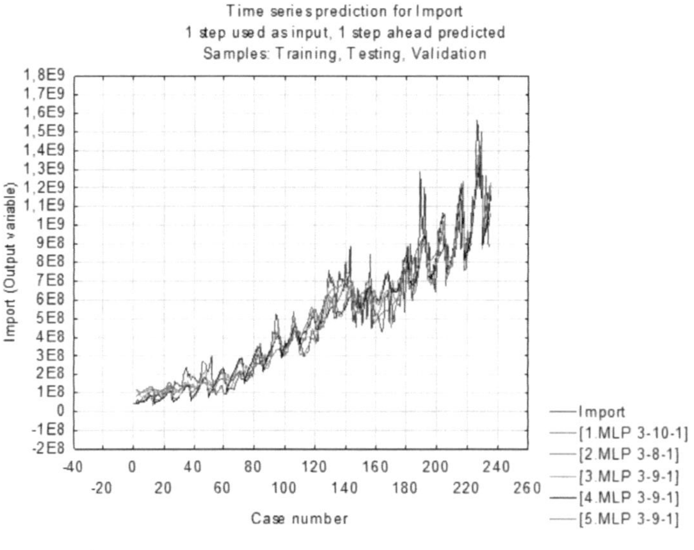


**Fig. 2.** Equalized time series of retained networks – Experiment 1

Source: Author.

Figure 2 shows that all equalized time series are able to follow the course of the actual development of the CR import from the PRC quite well. There have been identified deficiencies in the 40$^{th}$ case, 150$^{th}$ and 190$^{th}$ case, but not significant ones. In order to be able to determine the applicability of the networks, it is necessary to focus on their ability to forecast the future development. We will focus on estimating the CR development from the PRC for the period July 2019-December 2020, that is, for the period of 18 months. The concrete results are shown in Table 3.

**Table 3.** Development of forecasts for period of August 2019-January 2020 by networks retained from Experiment 1

| Date | MLP 3-10-1 | MLP 3-8-1 | MLP 3-9-1 | MLP 3-9-1 | MLP 3-9-1 |
|---|---|---|---|---|---|
| 31 July 2019 | 1.065574E+09 | 1.231325E+09 | 1.231142E+09 | 1.157255E+09 | 1.182213E+09 |
| 31 August 2019 | 1.181902E+09 | 1.292336E+09 | 1.294608E+09 | 1.228682E+09 | 1.248790E+09 |
| 30 September 2019 | 1.372477E+09 | 1.415371E+09 | -1.424833E+09 | 4.432479E+07 | 1.348762E+09 |
| 31 October 2019 | 1.569194E+09 | 1.488292E+09 | -1.424833E+09 | 4.421795E+07 | 1.412637E+09 |
| 30 November 2019 | 1.730247E+09 | 1.510818E+09 | -1.424833E+09 | 4.421794E+07 | 1.448757E+09 |
| 31 December 2019 | 1.824783E+09 | 1.513196E+09 | -1.424833E+09 | 4.421794E+07 | 1.466828E+09 |
| 31 January 2020 | 1.854371E+09 | 1.513268E+09 | -1.424833E+09 | 4.421794E+07 | 1.475707E+09 |
| 29 February 2020 | 1.287440E+09 | 1.420006E+09 | -1.424833E+09 | 4.421794E+07 | 1.375046E+09 |
| 31 March 2020 | 1.228724E+09 | 1.470567E+09 | -1.424833E+09 | 4.421794E+07 | 1.410437E+09 |
| 30 April 2020 | 1.189816E+09 | 1.502848E+09 | -1.424833E+09 | 4.421794E+07 | 1.436289E+09 |
| 31 May 2020 | 1.212808E+09 | 1.512284E+09 | -1.424833E+09 | 4.421794E+07 | 1.454421E+09 |
| 30 June 2020 | 1.330846E+09 | 1.513244E+09 | -1.424833E+09 | 4.421794E+07 | 1.468460E+09 |
| 31 July 2020 | 1.556136E+09 | 1.513268E+09 | -1.424833E+09 | 4.421794E+07 | 1.479600E+09 |
| 31 August 2020 | 1.875536E+09 | 1.513269E+09 | -1.424833E+09 | 4.421794E+07 | 1.488921E+09 |
| 30 September 2020 | 2.230543E+09 | 1.513269E+09 | -1.424833E+09 | 4.421794E+07 | 1.496195E+09 |
| 31 October 2020 | 2.537059E+09 | 1.513269E+09 | -1.424833E+09 | 4.421794E+07 | 1.501247E+09 |
| 30 November 2020 | 2.738303E+09 | 1.513269E+09 | -1.424833E+09 | 4.421794E+07 | 1.504351E+09 |
| 31 December 2020 | 2.813416E+09 | 1.513269E+09 | -1.424833E+09 | 4.421794E+07 | 1.505892E+09 |

Source: Authors.

It is evident that the 3. MLP 3-9-1 network´s forecast of the future development of the monitored variable is nonsensical. The results are negative. However, we obtained an axially symmetrical cure of the possible development of the CR import from the PRC. Other basically nonsensical results are provided by the network 4. MLP 3-9-1. The initial real values pass to the import zone which is by two orders lower. The forecast of the first retained neural network 1.MLP 3-10-1 also appears to be unrealistic. First, it shows an interesting development, but from August 2020, it starts to develop not in accordance with the previous values, as there is an extreme month-on-month increase in the CR import from the PRC. For selecting the most successful neural network, Figure 3 (development of time series and forecast) will be used.

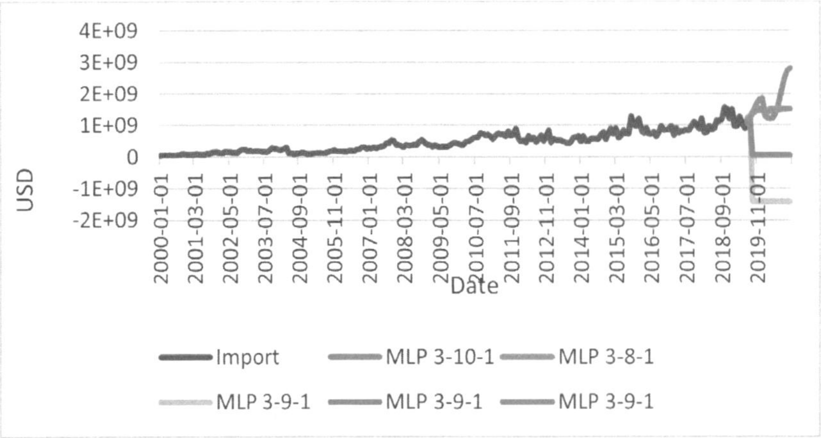


**Fig. 3.** Development of time series and forecasts by networks retained at Experiment 1

Source: Authors.

Due to possible distortion, there were omitted the neural networks where unrealistic forecasts of the future development of the CR import from the PRC were identified. The Figure thus captures the actual development of the time series along with the development forecast by 2.MLP 3-8-1 and 5.MLP 3-9-1. Both networks present a similar development of the time series, which ends in the monitored period, that is, in December 2020, using the similar value of the variable. For an overall comparison of the results of the partial experiments, the network 2.MLP 3-8-1 will be considered.

### 3.2 Experiment 2 (5-month lag of time series)

Table 4 shows the overview of the retained neural networks from Experiment 2, where a 5-month lag is assumed.

**Table 4.** Retained neural networks from Experiment 2

| Network | Train. perform. | Test. perform. | Valid. perform. | Train. error | Test. error | Valid. error | Train. algorit. | Error function | Activation of hidden layer | Output activation function |
|---|---|---|---|---|---|---|---|---|---|---|
| MLP 15-5-1 | 0.944810 | 0.931109 | 0.940005 | 5.380788E +15 | 8.664971E +15 | 6.583494E +15 | BFGS (Quasi-Newton) 59 | Sum of squares | Tanh | Tanh |
| MLP 15-7-1 | 0.950087 | 0.941681 | 0.941613 | 4.880342E +15 | 7.504740E +15 | 6.307628E +15 | BFGS (Quasi-Newton) 59 | Sum of squares | Tanh | Identity |
| MLP 15-3-1 | 0.951657 | 0.951374 | 0.941369 | 4.734524E +15 | 6.118437E +15 | 6.399299E +15 | BFGS (Quasi-Newton) 86 | Sum of squares | Tanh | Exponential |
| MLP 15-4-1 | 0.948516 | 0.949282 | 0.940584 | 5.025787E +15 | 6.288411E +15 | 5.796231E +15 | BFGS (Quasi-Newton) 73 | Sum of squares | Logistic | Identity |
| MLP 15-10-1 | 0.970359 | 0.953558 | 0.942601 | 2.920810E +15 | 5.761694E +15 | 5.662860E +15 | BFGS (Quasi-Newton) 182 | Sum of squares | Tanh | Exponential |

Source: Authors.

Due to the time series lag, the input layer contains 15 neurons. The table shows that the retained neural networks contain from 4 to 10 neurons in the hidden layer. The neural networks use the function of hyperbolic tangent and logistic function for the activation of the hidden layer. The output layer is activated using the function of hyperbolic tangent, exponential function and identity. The correlation coefficient in all data sets of all networks is very high, from more than 0.93 to more than 0.97. This indicates a high level of direct dependence. The retained neural networks shall thus be able to predict the future development of the CR import from the PRC thus precisely. The level of the error was determined by means of the sum of the least squares. The error is approximately the same as the error calculated in the first experiment. It is relatively high but regarding the high values of the variable it is still acceptable (for all networks and all data sets).

Figure 4 shows the comparison of the actual course of the time series and equalized time series.

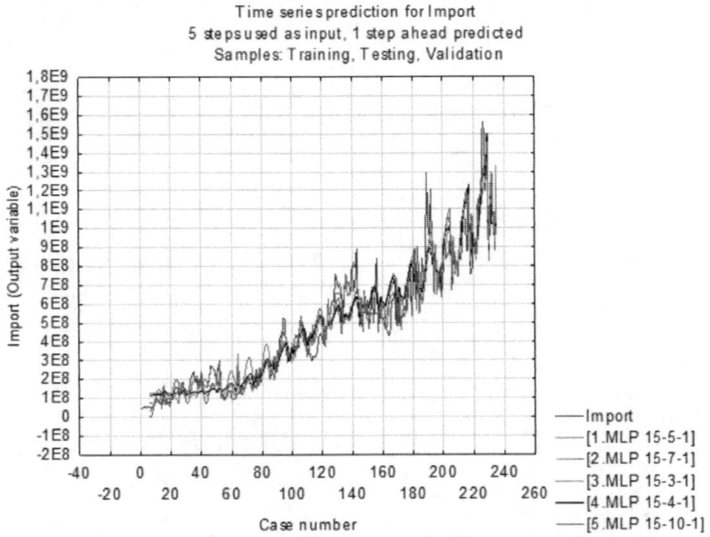


**Fig. 4.** Equalized time series of retained networks from Experiment 2

Source: Authors.

When comparing with Figure 2 (Experiment 1), better results are evident. The neural networks retained from the Experiment 2 are more accurate in equalizing time series. The only significant deviation is around the 190$^{th}$ case, where there is a sharp increase, a slight decline, repeated increase in the import volume, and then a significant decrease in the CR import (between September 2015 and December 2015. The values level off in January 2016).

In order to determine whether the networks are applicable, we will focus on their use for forecasting. We will forecast the development of the CR import from the PRC for the period of July 2019-December 2020. Table 5 shows specific data.

**Table 5.** Development of forecasts for the period of August 2019-January 2020 by the networks retained at Experiment 2

| Date | MLP 15-5-1 | MLP 15-7-1 | MLP 15-3-1 | MLP 15-4-1 | MLP 15-10-1 |
|---|---|---|---|---|---|
| 31 July 2019 | 1.028362E+09 | 1.123861E+09 | 1.329162E+09 | 1.092625E+09 | 1.221804E+09 |
| 31 August 2019 | 1.118063E+09 | 1.202537E+09 | 1.326695E+09 | 1.227136E+09 | 1.340655E+09 |
| 30 September 2019 | 1.196789E+09 | 1.301656E+09 | 2.406627E+09 | 1.357579E+09 | 1.449726E+09 |
| 31 October 2019 | 1.261490E+09 | 1.415406E+09 | 1.466920E+09 | 1.463281E+09 | 1.621533E+09 |
| 30 November 2019 | 1.312878E+09 | 1.524919E+09 | 1.239474E+08 | 1.568247E+09 | 1.929219E+09 |
| 31 December 2019 | 1.351882E+09 | 1.611339E+09 | 1.239474E+08 | 1.761729E+09 | 2.516401E+09 |
| 31 January 2020 | 1.380977E+09 | 1.659723E+09 | 1.239474E+08 | 2.089505E+09 | 3.692948E+09 |
| 29 February 2020 | 1.000428E+09 | 1.330634E+09 | 1.239474E+08 | 1.189211E+09 | 1.070819E+09 |
| 31 March 2020 | 1.332371E+09 | 1.577977E+09 | 1.239474E+08 | 1.731316E+09 | 1.178964E+09 |
| 30 April 2020 | 1.263561E+09 | 1.470854E+09 | 1.239474E+08 | 1.641207E+09 | 8.136747E+09 |
| 31 May 2020 | 1.300011E+09 | 1.500345E+09 | 1.239474E+08 | 1.752477E+09 | 1.246006E+09 |
| 30 June 2020 | 1.195967E+09 | 1.523270E+09 | 1.239474E+08 | 1.709404E+09 | 5.236437E+09 |
| 31 July 2020 | 1.289576E+09 | 1.626052E+09 | 1.239474E+08 | 2.023368E+09 | 8.511753E+09 |
| 31 August 2020 | 1.357345E+09 | 1.737773E+09 | 1.239474E+08 | 2.277100E+09 | 1.187036E+10 |
| 30 September 2020 | 1.403060E+09 | 1.861599E+09 | 1.239474E+08 | 2.459162E+09 | 1.497943E+10 |
| 31 October 2020 | 1.432786E+09 | 1.995227E+09 | 1.239474E+08 | 2.647926E+09 | 1.737552E+10 |
| 30 November 2020 | 1.451778E+09 | 2.127699E+09 | 1.239474E+08 | 2.882422E+09 | 1.899166E+10 |
| 31 December 2020 | 1.463856E+09 | 2.241309E+09 | 1.239474E+08 | 3.123546E+09 | 1.999224E+10 |

Source: Authors.

The results listed in the table represent a possible development of the time series. At first glance we could see that it will be necessary to omit the forecasts of the 3. MLP 15-3-1 network, which achieve very low values during the monitored period, and the 5. MLP 15-10-1 network, which achieve extremely high values. The detailed course of the predictions is shown in Figure 5.

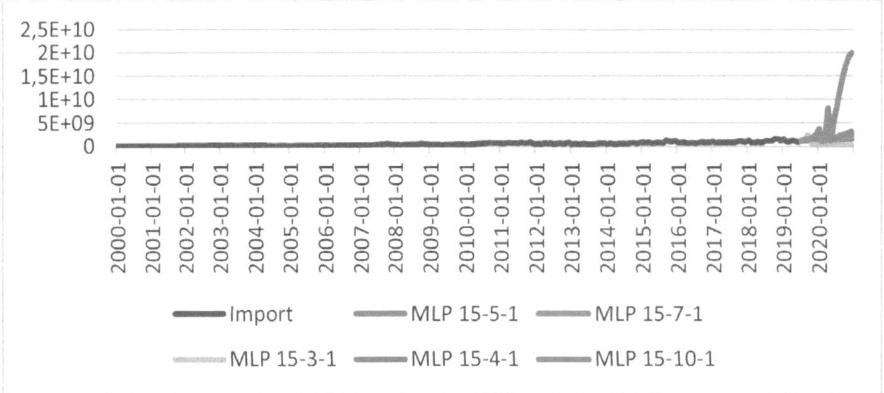


**Fig. 5.** Development of time series and forecasts by networks retained from Experiment 2

Source: Authors.

Comparing the actual development of the CR import from the PRC and the forecast development of the variable over time forecast using the artificial neural networks retained at Experiment 2, it can be concluded (especially considering the statistics given in Annex 2) that the 1. MLP 15-5-1 network is the best rated artificial neural structure with a potential to predict the further development of the CR import from the PRC. It is also able to capture the seasonal fluctuations as well as the overall trend of the time series.

### 3.3 Experiment 3 (10-month lag of time series)

For Experiment 3, 10000 artificial neural networks were generated, out of which 5 with the best characteristics were retained. For further details, see Table 6.

**Table 6.** Retained neural networks from Experiment 3

| Network | Train. perform. | Test. perform. | Valid. perform. | Train. error | Test. error | Valid. error | Train. algorit. | Error function | Activation of hidden layer | Output activation function |
|---|---|---|---|---|---|---|---|---|---|---|
| MLP 30-6-1 | 0.939665 | 0.946490 | 0.947035 | 5.447619E +15 | 6.944809E +15 | 5.530532E +15 | BFGS (Quasi-Newton) 67 | Sum of squares | Identity | Logistic |
| MLP 30-3-1 | 0.938466 | 0.946155 | 0.949107 | 5.538963E +15 | 6.894571E +15 | 5.356557E +15 | BFGS (Quasi-Newton) 43 | Sum of squares | Identity | Logistic |
| MLP 30-5-1 | 0.920542 | 0.933353 | 0.947078 | 7.205668E +15 | 8.467160E +15 | 4.892436E +15 | BFGS (Quasi-Newton) 11 | Sum of squares | Tanh | Logistic |
| MLP 30-9-1 | 0.938393 | 0.946123 | 0.952634 | 5.537254E +15 | 6.820576E +15 | 4.652061E +15 | BFGS (Quasi-Newton) 83 | Sum of squares | Identity | Logistic |
| MLP 30-7-1 | 0.922534 | 0.936408 | 0.948838 | 6.936388E +15 | 7.931314E +15 | 4.862199E +15 | BFGS (Quasi-Newton) 13 | Sum of squares | Tanh | Logistic |

Source: Authors.

The input variable and time series lag are represented by 30 neurons in the input layer. It results from the table that the neural networks with 3 – 9 neurons in the hidden layer were retained. For the activation of the hidden layer, the neural networks use the identity function and the function of hyperbolic tangent. The output layer of neurons is activated using the logistic function. The correlation coefficient of all networks is from more than 0.92 to almost 0.95, which indicates the high level of direct dependence. It appears that the retained neural structures are a very plausible model of the actual time series and are able to accurately forecast the further development of the time series. Even in this case, the method of the least squares was used for the calculation of error. Its value corresponds to the values from the previous experiments. Figure 6 shows the comparison of the actual course of the time series and equalized time series.

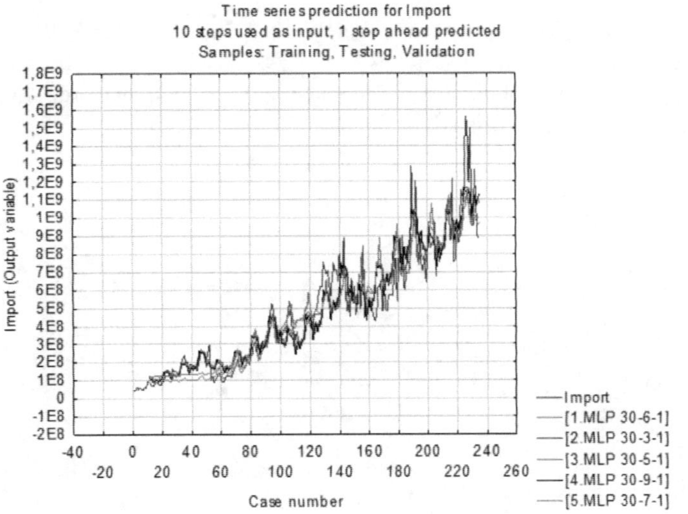


**Fig. 6.** Equalized time series of retained neural networks from Experiment 3

Source: Authors.

From Figure 6 it is evident that all equalized time series are able to copy the actual development of the CR import from the PRC. The course of all equalized time series is very similar. There can be some deficiencies identified around the 190th case and 225th case, where the equalized time series were not able to react adequately to extreme fluctuations

In the case of 10-month lag, we will also deal with forecasting the future development of the variable monitored. There will be forecast the development of the CR import from the PRC for the period of July 2019-December 2020. Concrete data are given in Table 6.

**Table 6.** Development of forecasts for the period of August 2019-January 2020 by the networks retained at Experiment 3

| Date | MLP 30-6-1 | MLP 30-3-1 | MLP 30-5-1 | MLP 30-9-1 | MLP 30-7-1 |
|---|---|---|---|---|---|
| 31 July 2019 | 1.084830E+09 | 1.086903E+09 | 9.676657E+08 | 1.126727E+09 | 1.004472E+09 |
| 31 August 2019 | 1.169410E+09 | 1.148619E+09 | 1.024224E+09 | 1.150604E+09 | 1.067836E+09 |
| 30 September 2019 | 4.421794E+07 | 4.421794E+07 | 1.091852E+09 | 4.421794E+07 | 1.133239E+09 |
| 31 October 2019 | 4.421794E+07 | 4.421794E+07 | 1.107948E+09 | 4.421794E+07 | 1.129745E+09 |
| 30 November 2019 | 4.421794E+07 | 4.421794E+07 | 1.150242E+09 | 4.421794E+07 | 1.180174E+09 |
| 31 December 2019 | 4.421794E+07 | 4.421794E+07 | 1.116857E+09 | 4.421794E+07 | 1.149804E+09 |
| 31 January 2020 | 4.421794E+07 | 4.421794E+07 | 1.058804E+09 | 4.421794E+07 | 1.100857E+09 |
| 29 February 2020 | 4.421794E+07 | 4.421794E+07 | 1.051114E+09 | 4.421794E+07 | 1.077613E+09 |
| 31 March 2020 | 4.421794E+07 | 4.421794E+07 | 1.093519E+09 | 4.421794E+07 | 1.124890E+09 |
| 30 April 2020 | 4.421794E+07 | 4.421794E+07 | 1.101035E+09 | 4.421794E+07 | 1.129029E+09 |
| 31 May 2020 | 4.421794E+07 | 4.421794E+07 | 1.101716E+09 | 4.421794E+07 | 1.134081E+09 |
| 30 June 2020 | 4.421794E+07 | 4.421794E+07 | 1.106164E+09 | 4.421794E+07 | 1.146630E+09 |
| 31 July 2020 | 4.421794E+07 | 4.421794E+07 | 1.118359E+09 | 4.421794E+07 | 1.159389E+09 |
| 31 August 2020 | 4.421794E+07 | 4.421794E+07 | 1.146699E+09 | 4.421794E+07 | 1.190990E+09 |
| 30 September 2020 | 4.421794E+07 | 4.421794E+07 | 1.169352E+09 | 4.421794E+07 | 1.214999E+09 |
| 31 October 2020 | 4.421794E+07 | 4.421794E+07 | 1.175210E+09 | 4.421794E+07 | 1.216847E+09 |
| 30 November 2020 | 4.421794E+07 | 4.421794E+07 | 1.184311E+09 | 4.421794E+07 | 1.229361E+09 |
| 31 December 2020 | 4.421794E+07 | 4.421794E+07 | 1.179678E+09 | 4.421794E+07 | 1.226067E+09 |

Source: Authors.

It results from the table that the networks 1. MLP 30-6-1, 2. MLP 30-3-1 and 4. MLP 30-9-1 forecast very low values. For this reason, they will not be considered in further attempts to identify the applicability of the neural networks for forecasting. This is to certain extent confirmed by the graph in Figure 7.

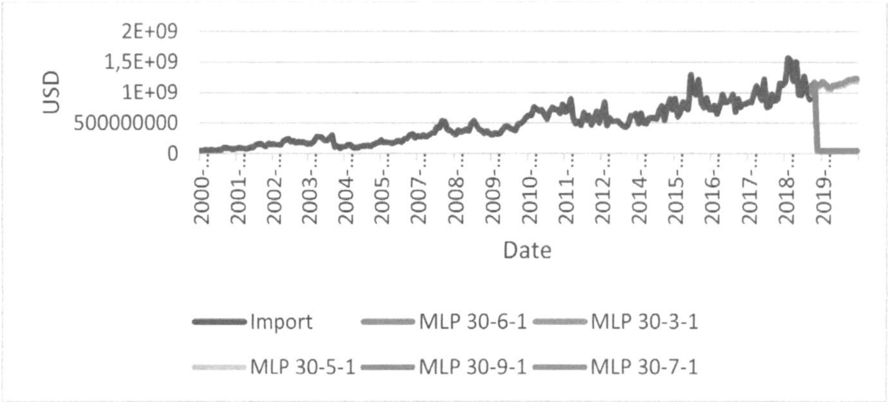


**Fig. 7.** Development of time series and predictions by networks retained at Experiment 3

Source: Authors.

Comparing the actual development of the CR import from the PRC and the expected development of the variable over time forecast by artificial neural networks retained at Experiment 3, it can be concluded that (especially considering the table of statistics in Annex 2) the network 5. MLP 30-7-1 shows the best predictions. However, it shall be added that the network 3. MLP 30-5-1 also shows very good characteristics.

## 4 Discussion

Within the research, three experiments were carried out, where the development of the time series represented by the CR import from the PRC was examined. Firstly, it was observed how the multilayer perceptron neural networks are able to equalize time series. Based on the research, other types of artificial networks were ignored, as it is considered to be proved that MLP show significantly better results in this type of tasks than other applicable neural networks. (Moreover, it is assumed that it is not suitable to use deep learning neural networks for this type of tasks). In each experiment, a total of 10,000 artificial neural networks were generated, out of which five best structures were retained. From the results described above it is evident that some of the retained neural structures were affected by overfitting. They were able to equalize the time series very well; however, their predictions do not correspond to the achieved characteristics. Other neural networks are applicable for forecasting the future development of the CR import from the PRC. The performance of the networks was always high, and the level of the error was acceptable. From each experiment, the most successful network was identified (although in two experiments, the difference between the most successful and the second most successful network was negligible). This network was subjected to further comparison (for more details, see Table 7).

**Table 7.** Comparison of the most successful neural structures from all experiments

| Date | Experiment 1 | Experiment 2 | Experiment 3 |
|---|---|---|---|
| | 2. MLP 3-8-1 | 1. MLP 15-5-1 | 5. MLP 30-7-1 |
| 31 July 2019 | 1.231325E+09 | 1.028362E+09 | 1.004472E+09 |
| 31 August 2019 | 1.292336E+09 | 1.118063E+09 | 1.067836E+09 |
| 30 September 2019 | 1.415371E+09 | 1.196789E+09 | 1.133239E+09 |
| 31 October2019 | 1.488292E+09 | 1.261490E+09 | 1.129745E+09 |
| 30 November 2019 | 1.510818E+09 | 1.312878E+09 | 1.180174E+09 |
| 31 December 2019 | 1.513196E+09 | 1.351882E+09 | 1.149804E+09 |
| 31 January 2020 | 1.513268E+09 | 1.380977E+09 | 1.100857E+09 |
| 29 February 2020 | 1.420006E+09 | 1.000428E+09 | 1.077613E+09 |
| 31 March 2020 | 1.470567E+09 | 1.332371E+09 | 1.124890E+09 |
| 30 April 2020 | 1.502848E+09 | 1.263561E+09 | 1.129029E+09 |
| 31 May 2020 | 1.512284E+09 | 1.300011E+09 | 1.134081E+09 |
| 30 June 2020 | 1.513244E+09 | 1.195967E+09 | 1.146630E+09 |
| 31July 2020 | 1.513268E+09 | 1.289576E+09 | 1.159389E+09 |
| 31 August2020 | 1.513269E+09 | 1.357345E+09 | 1.190990E+09 |
| 30 September 2020 | 1.513269E+09 | 1.403060E+09 | 1.214999E+09 |
| 31 October 2020 | 1.513269E+09 | 1.432786E+09 | 1.216847E+09 |
| 30 November 2020 | 1.513269E+09 | 1.451778E+09 | 1.229361E+09 |
| 31 December 2020 | 1.513269E+09 | 1.463856E+09 | 1.226067E+09 |

Source: Authors.

Table 7 shows significant differences between the neural networks. This is even better illustrated by Figure 8.

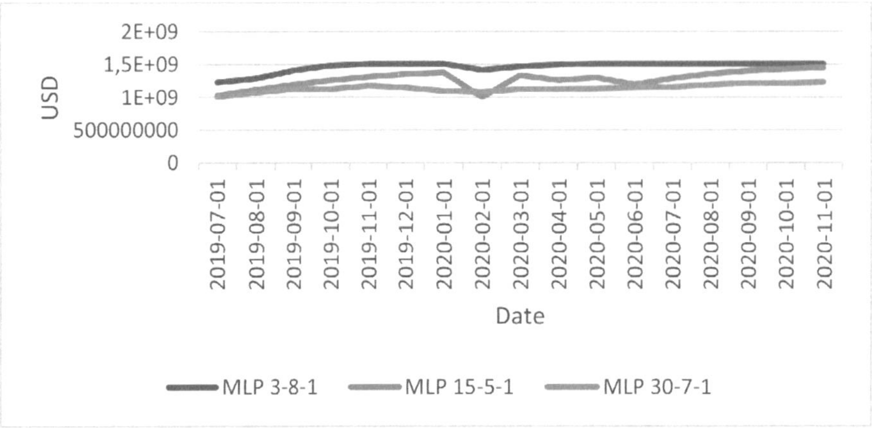


**Fig. 8.** Comparison of most successful neural networks from all experiments

Source: Authors.

The figure shows very similar courses of all selected neural structures. The retained network from the first experiment (with a 1-month lag of time series) shows the values calculated and derived from the fluctuation of the time series around the case No. 225. However, this is not optimal. The network from the second experiment assumes a positive development of the time series. It assumes seasonal fluctuations (the most significant one in March 2020), and accepts the overall increase in the CR import from China. The network retained in the third experiment (with a 10-month lag) equalizes the time series more than it is required. It could therefore be concluded that the best forecasting tool is the best network from the second experiment, 1. MLP 15-5-1, which is able to capture most precisely the development of the time series (with the exception of the case 190), and is able to forecast the development with its seasonal fluctuations.

## 5 Conclusion

The objective of the contribution was machine learning forecasting of the Czech Republic (CR) import from the People´s Republic of China (PRC).

The conclusion of the contribution can be as follows:

1. Multilayer perceptron networks are a suitable tool for forecasting the CR import from the PRC (which was proved both by research and experiments).
2. MLP networks are able to capture the trend of the whole time series as well as the seasonal fluctuations. Nevertheless, it is necessary to determine the right time series lag (that is, the time interval of the data used for the calculation of the forecast data).
3. It results from above that it is necessary to use a five-month lag.

The mutual USA and PRC sanctions did not affect the success rate of the machine learning forecasting of the CR import from the PRC. However, the question is whether the mutual USA and PRC sanctions affect significantly the trade between the CR and the PRC. In general, the PRC could be expected to look for new markets for its goods, which is mostly exposed to a duty. In such a case, the Czech Republic would also be considered. It is evident that the mutual sanctions shall affect the trade between the CR and the PRC. Nevertheless, the sanctions did not significantly influence the successfulness of the machine learning forecasting of the monitored variable development. The generated and retained neural networks (especially 1. MLP 15-5-1) have been very successful.

## References

1. V. Stehel, P. Šuleř, Foreign trade between China and the Czech Republic. *Littera Scripta*, **9**(3), 84-95 (2016)
2. T. De Castro, Z. Stuchlíková, China-V4 trade relations - a Czech Perspective. *International Scientific Conference on Current Trends and Perspectives in Development of China - V4 Trade and Investment*, pp. 40-60 (2014)
3. P. Higgins, T. Tha, W. Zhong, Forecasting China's economic growth and inflation. *China Economic Review,* **41**, 46-61 (2016)
4. Y. Liu, X. Li, H. Li, R&D subsidies and business R&D: Evidence from high-tech manufacturing firms in Jiangsu. *China Economic Review*, **41**, 1-22 (2016)
5. M. Vochozka, J. Vrbka, Estimation of the development of the Euro to Chinese Yuan exchange rate using artificial neural networks. *SHS Web of Conferences: Innovative Economic Symposium 2018 – Milestones and Trends of World Economy*, **61** (2019)
6. Czech Statistical Office, *Když se řekne zahraniční obchod [When you say foreign trade]* [online], Available at: https://www.czso.cz (2018)
7. Ministry of Foreign Affairs, China: Trade and Economic Cooperation with the Czech Republic. *BusinessInfo: Official portal for business and export* [online], Available at: https://www.businessinfo.cz/cs/clanky/cina-obchodni-a-ekonomicka-spoluprace-s-cr-19054.html (2019)
8. Y. Li, US Economic Sanctions Against China: A Cultural Explanation of Sanction Effectiveness. *Asian Perspective*, **38**(2), 311-335 (2014)
9. M. Crăciun, C. Raţiu, D. Bucerzan, A. Manolescu, Actuality of Bankruptcy Prediction Models used in Decision Support System. *International Journal of Computers Communications*, **8**(3), 375-383 (2013)
10. J. Yang, H. Askari, J. Forrer, H. Teegen, US Economic Sanctions Against China: Who Gets Hurt? *World Economy*, **27**(7), 1047-1081 (2004)
11. T. Klieštik, Models of autoregression conditional heteroskedasticity garch and arch as a tool for modeling the volatility of financial time series. *Ekonomicko-manažerské spektrum*, **7**(1), 2-10 (2013)
12. H. Pao, A comparison of neural network and multiple regression analysis in modeling capital structure. *Expert Systems with Applications*, **35**(3), 720-727 (2008)
13. A.R. Sayadi, S.M.M. Tavassoli, M. Monjezi, M. Rezaei, Application of neural networks to predict net present value in mining projects. *Arabian Journal of Geosciences*, **7**(3), 1067-1072 (2012)
14. Z. Rowland, J. Vrbka, Using artificial neural networks for prediction of key indicators of a company in global world. *Proceedings of the 16th International Scientific Conference Globalization and Its Socio-Economic Consequences*, pp. 1896-1903 (2016)
15. J. Horák, T. Krulický, Comparison of exponential time series alignment and time series alignment using artificial neural networks by example of prediction of future development of stock prices of a specific company. *SHS Web of Conferences: Innovative Economic Symposium 2018 – Milestones and Trends of World Economy*, **61** (2019)
16. S. Cho, J. Kim, J. K. Bae, An integrative model with subject weight based on neural network learning for bankruptcy prediction. *Expert Systems with Applications*, **36**(1), 403-410 (2009)

17. H. Li, L.Y. Hong, Q. Zhou, H.J. Yu, The assisted prediction modelling frame with hybridisation and ensemble for business risk forecasting and an implementation. *International Journal of Systems Science*, **46**(11), 2072-2086 (2013)
18. S. Balcaen, H. Ooghe, 35 years of studies on business failure: an overview of the classic statistical methodologies and their related problems. *The British Accounting Review*, **38**(1), 63-93 (2006)
19. M. Kováčová, T. Kliestik, P. Kubala, K. Valášková, M. Radisic, J. Borocki, Bankruptcy models: Verifying their validity as a predictor of corporate failure. *Polish Journal of Management Studies*, **18**(1), 167-179 (2018)
20. V. Majerčíková, V. Bartošová, The methodology for determining the value of a company. *Ekonomicko-manažerské spektrum*, **6**(1), 94-99 (2012)
21. T. Kliestik, J. Vrbka, Z. Rowland, Bankruptcy prediction in Visegrad group countries using multiple discriminant analysis. *Equilibrium-Quarterly Journal of Economics and Economic Policy*, **13**(3), 569-593 (2018)
22. H. Pao, A comparison of neural network and multiple regression analysis in modeling capital structure. *Expert Systems with Applications*, **35**(3), 720-727 (2008)
23. G. Mélard, J.M. Pasteels, Automatic ARIMA modeling including interventions, using time series expert software. *International Journal of Forecasting*, **16**(4), 497-508 (2000)
24. J. Junttila, Structural breaks, ARIMA model and Finnish inflation forecasts. *International Journal of Forecasting*, **17**(2), 203-230 (2001)
25. J. Majerová, T. Kliestik, Towards the European insolvency law harmonisation in the context of cluster analysis. *10th International Days of Statistics and Economics*, pp. 1141-1150 (2016)
26. M. Vochozka, Analysis of enterprises operating in the construction industry based on Kohonen networks. *12th International Days of Statistics and Economics*, pp. 1874-1883 (2018)
27. P. Šuleř, Using Kohonen´s neural networks to identify the bankruptcy of enterprises: Case study based on construction companies in South Bohemian region. *Proceedings of the 5th International conference Innovation Management, Entrepreneurship and Sustainability*, pp. 985-995 (2017)
28. F. Chollet, *Deep learning v jazyku Python: knihovny Keras, Tensorflow [Python deep learning: Keras, Tensorflow libraries]*. Prague: Grada Publishing (2019)
29. K. Valášková, T. Kliestik, M. Mišánková, The role of fuzzy logic in decision making process. *2014 2nd International Conference on Management Innovation and Business Innovation*, **44**, pp. 143-148 (2014)
30. Z. Rowland, P. Šuleř, M. Vochozka, Comparison of neural networks and regression time series in estimating the Czech Republic and China trade balance. *SHS Web of Conferences: Innovative Economic Symposium 2018 – Milestones and Trends of World Economy*, **61** (2019)
31. J. Weijin, X. Yuhui, A novel method for nonlinear time series forecasting of time-delay neural network. *Wuhan University Journal of Natural Sciences*, **11**(5), 1357-1361 (2006)
32. J. Horák, P. Šuleř, J. Vrbka, Comparison of neural networks and regression time series when predicting the export development from the USA to PRC. *Proceedings of 6th International Scientific Conference Contemporary Issues in Business, Management and Economics Engineering* (2019)

33. J. Vrbka, Z. Rowland, P. Šuleř, Comparison of neural networks and regression time series in estimating the development of the EU and the PRC trade balance. *SHS Web of Conferences: Innovative Economic Symposium 2018 – Milestones and Trends of World Economy*, **61** (2019)
34. World Bank [online], Available at: https://data.worldbank.org/indicator/TM.TAX.MRCH.WM.AR.ZS?end=2017&locations=CZ-CN&start=2014 (2019)

### *Annex 1*

### Basic statistics of equalized time series – 1-month lag

| Statistics | 1.MLP 3-10-1 | 2.MLP 3-8-1 | 3.MLP 3-9-1 | 4.MLP 3-9-1 | 5.MLP 3-9-1 |
|---|---|---|---|---|---|
| Minimal predictions (Training) | 1.019626E+08 | 8.243890E+07 | 2.664638E+07 | 4.594127E+07 | 5.604704E+07 |
| Maximal predictions (Training) | 1.397260E+09 | 1.445876E+09 | 1.407840E+09 | 1.329468E+09 | 1.333489E+09 |
| Minimal predictions (Testing) | 1.031296E+08 | 7.868087E+07 | 8.458917E+07 | 4.494537E+07 | 5.645213E+07 |
| Maximal predictions (Testing) | 1.289120E+09 | 1.218565E+09 | 1.284406E+09 | 1.236164E+09 | 1.264197E+09 |
| Minimal predictions (Validation) | 1.058761E+08 | 9.279675E+07 | 8.606640E+07 | 4.742774E+07 | 5.687259E+07 |
| Maximal predictions (Validation) | 1.218385E+09 | 1.140678E+09 | 1.224814E+09 | 1.199455E+09 | 1.210029E+09 |
| Minimal residuals (Training) | -2.212713E+08 | -2.934693E+08 | -2.625207E+08 | -2.799729E+08 | -2.703439E+08 |
| Maximal residuals (Training) | 4.849909E+08 | 3.854199E+08 | 4.829077E+08 | 4.761820E+08 | 4.753483E+08 |
| Minimal residuals (Testing) | -1.861218E+08 | -1.972519E+08 | -1.383202E+08 | -1.762328E+08 | -1.943952E+08 |
| Maximal residuals (Testing) | 2.767607E+08 | 3.451225E+08 | 2.908815E+08 | 3.275236E+08 | 2.994905E+08 |
| Minimal residuals (Validation) | -3.425319E+08 | -2.648245E+08 | -3.489605E+08 | -3.236024E+08 | -3.341759E+08 |
| Maximal residuals (Validation) | 3.042138E+08 | 2.507207E+08 | 2.246107E+08 | 2.583186E+08 | 2.893341E+08 |
| Minimal standard residuals (Training) | -3.689617E+00 | -4.488485E+00 | -3.650804E+00 | -3.695578E+00 | -3.651492E+00 |
| Maximal standard residuals (Training) | 8.087046E+00 | 5.894829E+00 | 6.715667E+00 | 6.285494E+00 | 6.420454E+00 |
| Minimal standard residuals (Testing) | -2.694937E+00 | -2.768274E+00 | -1.943066E+00 | -2.286286E+00 | -2.602394E+00 |
| Maximal standard residuals (Testing) | 4.007337E+00 | 4.843521E+00 | 4.086187E+00 | 4.248997E+00 | 4.009319E+00 |
| Minimal standard residuals (Validation) | -4.412902E+00 | -3.695045E+00 | -4.528513E+00 | -4.118224E+00 | -4.168222E+00 |
| Maximal standard residuals (Validation) | 3.919242E+00 | 3.498256E+00 | 2.914806E+00 | 3.287410E+00 | 3.608904E+00 |

### *Annex 2*

### Basic statistics of equalized time series – 5-month lag

| Statistics | 1.MLP 15-5-1 | 2.MLP 15-7-1 | 3.MLP 15-3-1 | 4.MLP 15-4-1 | 5.MLP 15-10-1 |
|---|---|---|---|---|---|
| Minimal predictions (Training) | 2.870538E+07 | 2.084195E+06 | 1.249895E+08 | 1.134611E+08 | 7.142744E+07 |
| Maximal predictions (Training) | 1.279052E+09 | 1.354799E+09 | 1.386515E+09 | 1.482664E+09 | 1.459338E+09 |
| Minimal predictions (Testing) | 9.253505E+07 | 9.400378E+07 | 1.278659E+08 | 1.229188E+08 | 1.065232E+08 |
| Maximal predictions (Testing) | 1.179155E+09 | 1.251712E+09 | 1.249326E+09 | 1.252984E+09 | 1.302836E+09 |
| Minimal predictions (Validation) | 1.019749E+08 | 6.793485E+07 | 1.272657E+08 | 1.183130E+08 | 9.380888E+07 |
| Maximal predictions (Validation) | 1.181279E+09 | 1.201416E+09 | 1.259741E+09 | 1.181576E+09 | 1.198775E+09 |
| Minimal residuals (Training) | -3.033300E+08 | -2.885160E+08 | -2.702194E+08 | -2.716622E+08 | -2.305341E+08 |
| Maximal residuals (Training) | 4.443217E+08 | 4.862127E+08 | 4.117377E+08 | 4.551818E+08 | 4.482359E+08 |
| Minimal residuals (Testing) | -2.821787E+08 | -3.106774E+08 | -2.805951E+08 | -2.616854E+08 | -2.627082E+08 |
| Maximal residua (Testing) | 3.845328E+08 | 3.119759E+08 | 3.143621E+08 | 3.107034E+08 | 3.160098E+08 |
| Minimal residuals (Validation) | -3.054264E+08 | -3.255626E+08 | -2.925264E+08 | -2.696388E+08 | -2.710663E+08 |
| Maximal residuals (Validation) | 2.292086E+08 | 2.769787E+08 | 2.457431E+08 | 3.414572E+08 | 3.138268E+08 |
| Minimal standard residuals (Training) | -4.135161E+00 | -4.129949E+00 | -3.927158E+00 | -3.832015E+00 | -4.265633E+00 |
| Maximal standard residuals (Training) | 6.057237E+00 | 6.959870E+00 | 5.983875E+00 | 6.420707E+00 | 8.293826E+00 |
| Minimal standard residuals (Testing) | -3.031382E+00 | -3.586261E+00 | -3.587235E+00 | -3.299963E+00 | -3.460976E+00 |
| Maximal standard residuals (Testing) | 4.130949E+00 | 3.601250E+00 | 4.018926E+00 | 3.918101E+00 | 4.163183E+00 |
| Minimal standard residuals (Validation) | -3.764249E+00 | -4.099222E+00 | -3.656780E+00 | -3.541682E+00 | -3.602115E+00 |
| Maximal standard residuals (Validation) | 2.824898E+00 | 3.487492E+00 | 3.071957E+00 | 4.485010E+00 | 4.170346E+00 |

### *Annex 3*

## Basic statistics of equalized time series – 10-month lag

| Statistics | 1.MLP 30-6-1 | 2.MLP 30-3-1 | 3.MLP 30-5-1 | 4.MLP 30-9-1 | 5.MLP 30-7-1 |
|---|---|---|---|---|---|
| Minimal predictions (Training) | 9.581047E+07 | 8.978954E+07 | 9.523406E+07 | 8.620605E+07 | 1.226069E+08 |
| Maximal predictions (Training) | 1.143141E+09 | 1.169006E+09 | 1.128150E+09 | 1.156245E+09 | 1.152624E+09 |
| Minimal predictions (Testing) | 1.275041E+08 | 1.225674E+08 | 9.857046E+07 | 1.188153E+08 | 1.261200E+08 |
| Maximal predictions (Testing) | 1.142398E+09 | 1.163636E+09 | 1.069460E+09 | 1.144003E+09 | 1.085287E+09 |
| Minimal predictions (Validation) | 1.274159E+08 | 1.139746E+08 | 9.620851E+07 | 1.102632E+08 | 1.244502E+08 |
| Maximal predictions (Validation) | 1.124260E+09 | 1.155792E+09 | 1.050281E+09 | 1.166400E+09 | 1.092989E+09 |
| Minimal residuals (Training) | -3.298498E+08 | -3.663393E+08 | -4.178500E+08 | -3.515643E+08 | -4.260971E+08 |
| Maximal residuals (Training) | 3.812593E+08 | 3.643545E+08 | 5.210952E+08 | 3.813481E+08 | 4.795802E+08 |
| Minimal residuals (Testing) | -2.326371E+08 | -2.291130E+08 | -2.144975E+08 | -2.476463E+08 | -2.300545E+08 |
| Maximal residua (Testing) | 4.212896E+08 | 4.000518E+08 | 4.942276E+08 | 4.196847E+08 | 4.784008E+08 |
| Minimal residuals (Validation) | -2.435260E+08 | -2.420521E+08 | -1.957826E+08 | -2.076164E+08 | -2.164618E+08 |
| Maximal residuals (Validation) | 1.888672E+08 | 1.813321E+08 | 3.057997E+08 | 1.845939E+08 | 2.816919E+08 |
| Minimal standard residuals (Training) | -4.469026E+00 | -4.922313E+00 | -4.922472E+00 | -4.724518E+00 | -5.116134E+00 |
| Maximal standard residuals (Training) | 5.165556E+00 | 4.895644E+00 | 6.138750E+00 | 5.124770E+00 | 5.758303E+00 |
| Minimal standard residuals (Testing) | -2.791572E+00 | -2.759282E+00 | -2.331060E+00 | -2.998620E+00 | -2.583201E+00 |
| Maximal standard residuals (Testing) | 5.055342E+00 | 4.817954E+00 | 5.371037E+00 | 5.081741E+00 | 5.371794E+00 |
| Minimal standard residuals (Validation) | -3.274627E+00 | -3.307242E+00 | -2.799055E+00 | -3.043960E+00 | -3.104309E+00 |
| Maximal standard residuals (Validation) | 2.539645E+00 | 2.477603E+00 | 4.371944E+00 | 2.706417E+00 | 4.039783E+00 |

# Anti-crisis approach in the industrial enterprise management: methodological tools of preventive regulation

*Olga* Maslak[1*], *Petro* Sokurenko[2], *Natalya* Grishko[1], *Ievgen* Buriak[1], *Mariya* Maslak[3]

[1]Kremenchuk Mykhailo Ostrohradskyi National University, Pershotravneva Street 20, 39600 Kremenchuk, Ukraine
[2]Kremenchuk institute of Higher Educational Establishment «Alfred Nobel University», Naberezhna Sicheslavska str. 18, 49000 Dnipro, Ukraine
[3]National Technical University «Kharkiv Polytechnic Institute», Kyrpychova str. 2, 61002 Kharkiv, Ukraine

**Abstract.** The paper covers the theoretical and methodological framework for assessment of the crisis management effectiveness and the enterprise bankruptcy probability. At the present stage of the Ukrainian economic development, the main weaknesses in the activities of enterprises include: loss of sales markets, imperfect production and sales policy, lack of working capital, inefficient financial management, significant production costs, which produce signs of crisis processes and determine the feasibility of preventive approach anti-crisis management use. At the same time, the choice of the topic is due to insufficient research of this scientific problem as a whole and its relevance for the economy of Ukraine, especially in the context of the search for external investment resources, cooperation on this basis with the European Union.
It was determined that one of the ways to solve the problem of predicting the probability of bankruptcy is the use of insolvency prediction models developed on the basis of the discriminant analysis method – Z-coefficients. The authors systematized the most well-known approaches to predicting the likelihood of bankruptcy of industrial enterprises, illustrated their advantages and disadvantages, the mutual consistency of use.
It has been established that additional opportunities for monitoring the effectiveness of anti-crisis measures are created by a system of indicators characterizing the level of effectiveness of the anti-crisis management system. Evaluating the effectiveness of the anti-crisis program only on the basis of an analysis of relative indicators is not enough: during the global financial crisis, sometimes convincing indicators of efficiency are absolute indicators and a projected trend of their change.
As part of this study, attention is paid to the analytical levers of the crisis management mechanism, which are aimed at preventive regulation and localization of crisis processes in the enterprise, restoring the effectiveness of production and marketing activities.
**Keywords**: crisis, crisis management, model, preventive regulation, bankruptcy, profitability, loss ratio, solvency.

* Corresponding author: 2nata.grishko@gmail.com

## 1 Introduction

The variability of the external environment, the need to apply new approaches to the stabilization of activities and the development of enterprises actualizes the issue of the anti-crisis approach for managing business entities. The effectiveness of anti-crisis measures depends on the speed of the enterprise response to the appearance of crisis processes' signs, which makes it advisable to apply a preventive approach to management. The aggravation of crisis processes in the production sphere, their systemic nature and the absence of an effective diagnostic apparatus for determining the depth of crisis phenomena, the scale of the crisis and effective mechanisms of crisis management determine the feasibility of research in this field.

The search for effective models of crisis management, theoretical and methodological apparatus for studying the impact of crisis processes on the activities of enterprises are in the interests of both foreign and domestic scientists: Azarenkova, Piskunov and Goikhman [1], Aref'eva [2], Bradley and Cowdery [3], Bryan, Fernando and Tripathy [4], Dikan and Nazarenko [5], Elexa, Hvolkova and Knapkova [6], Tobback [7]. The achievements of scientists and practitioners in predicting the likelihood of bankruptcy of business entities are really deep and significant, at the same time, the features of anti-crisis management technologies in domestic enterprises, the applied aspects of the evaluation model application in the realities of the national economy are not sufficiently studied.

Thus, the study of existing approaches to assessing the effectiveness of crisis management of enterprises leads to the conclusion that among modern concepts [2, 4-7] there is no objective method that would allow to give an unambiguous quantitative assessment of it. This indicates the need to develop an integrated approach to assess the degree of crisis based on the systematization of its main components and the establishment of relationships between them.

## 2 Data and Methods

The aim of the study is to formulate methodological approaches to assessing the effectiveness of crisis management in the operation of an enterprise based on preventive regulation.

The solution for the assigned tasks was carried out using the following methods: *theoretical generalization* – to study and systematize theoretical approaches to the definition and justification of the problem studied; *analysis and synthesis* – to determine the object and subject of research; *graphic; mathematical statistics* – when establishing the dependence of the integral indicator of assessing the effectiveness of the crisis management system.

Information base selected materials of official publications of the State Statistics Committee, data of domestic and foreign scientific literature, data of primary reporting documentation of enterprises, annual financial and statistical reporting.

## 3 Results

### 3.1 Current state and prerequisites of crisis management of Ukrainian industrial enterprises

All of the above indicates the feasibility of conducting research on applied aspects and theoretical and methodological approaches to crisis management of an enterprise. For industrial enterprises, these issues are priorities given the leading role of industry in the structure of the national economy and the dual nature of crisis processes in industrial development, in particular. On the one hand, crises are destructive in nature, they bring destabilization to the life processes of economic entities, and on the other hand, progressive changes that are revolutionary in nature take the beginning of abrupt growth at the point of bifurcation against the background of the crisis process. Thus, the occurrence of crises is a natural process and a starting point for constructive changes in business processes. A number of scientists note the cyclical nature of the emergence and, along with the subjective, objective prerequisites of crisis processes, a certain nature of socio-economic systems.

The emergence of a market economy in Ukraine is associated with the deployment of crisis processes in the activities of many enterprises. The dynamics of industrial production indices indicate a negative trend of a gradual decrease in production volumes by enterprises in Ukraine in 2010-2015. During 2016-2018 production stabilization was observed as a result of the increased profitability of the operating activities of Ukrainian enterprises (Fig. 1).

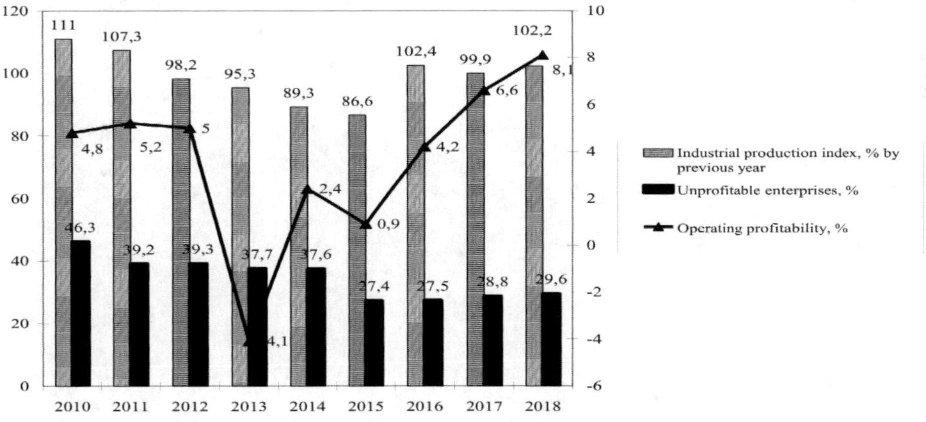


**Fig. 1.** Dynamics of indices of industrial production, unprofitable enterprises and profitability / unprofitability of operating activities of enterprises in Ukraine, %

Source: Formed by the authors on the basis of [8].

The worst values of the profitability index (namely, a loss ratio of 4.1%) were observed during 2013. At the end of the study period (January-August 2018), the profitability of operating activities increases to 8.1%, which is the best value of this indicator during 2010-2018.

The study of unprofitability and bankruptcy of enterprises of Ukraine during 2010-2018 (January-August). Indicates a significant number of unprofitable enterprises – 27.4-46.3% of the total. Despite a slight decrease (in 2014 – up to 37.6%, and in 2015 – up to 27.4%), the share of unprofitable enterprises in 2016 is still quite large. Since 2016, there has been a negative growth trend in the number of unprofitable enterprises, among the leading causes of this should be the influence of an unstable environment and inadequate actions of the enterprises themselves. Almost every 30 enterprises turned out to be unprofitable, with the load on the production capacities of industrial enterprises of Ukraine during 2014-2018. This indicates not a full measure of the use of the industrial sector potential.

Ukrainian large and medium-sized enterprises during the study period are characterized by non-utilization of production capacities, as a result, first of all, we get the lack of orders and a reduction in the demand for maintaining an excessive number of personnel, and, under these conditions, an increase in costs and a decrease in production profitability. At the same time, there was a negative irreversible upward trend in prices for raw materials and energy carriers, which also leads to an increase in expenditure aspects of the enterprises' s activities. The decrease in sales volumes and, consequently, profits, limits the financial development opportunities of enterprises, provokes financial insolvency and leads to a crisis state of enterprises.

The factors constraining production, based on materials from statistical studies, include: insufficient demand, lack of labor, lack of raw materials, materials, equipment, financial constraints. Negative financial results characterize, on the one hand, the crisis processes of the socio-economic system of an enterprise, on the other hand, the symptoms of a crisis are the appearance of signs of insolvency. All this limit the possibilities of self-financing of investment activities, as well as the development of the enterprise through research [9], of attracting external investment resources (Fig. 2).

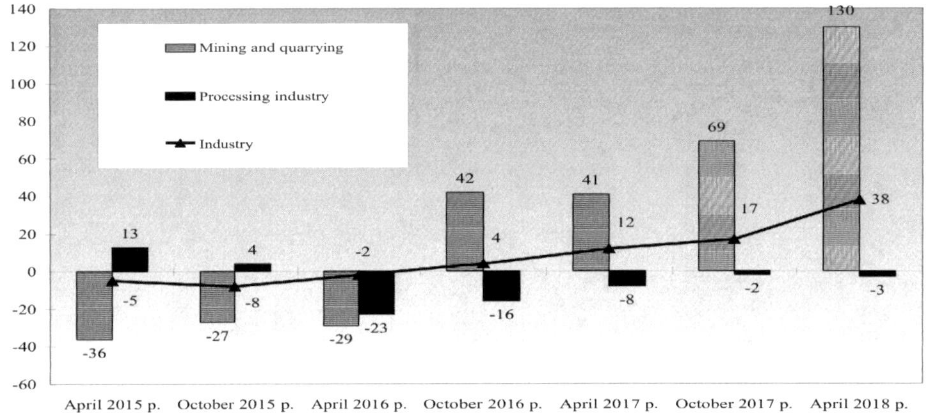


**Fig. 2.** Expected changes in the volume of capital investments of industrial enterprises of Ukraine in the current year compared with the previous year, %

Source: Formed by the authors on the basis of [8].

The competitive position of industrial enterprises in foreign markets within the European Union in a certain way actualizes the issue of finding reserves to strengthen their position in the market [10-15] and the introduction of progressive changes in domestic enterprises, aims to increase their investment attractiveness (tab. 1).

**Table 1.** Evaluation of the competitive position of industrial enterprises in foreign markets within the European Union

| Year / Quarter | 2018 | | | | | | | | | | | | | | |
|---|---|---|---|---|---|---|---|---|---|---|---|---|---|---|---|
| | I quarter | | | | | II quarter | | | | | III quarter | | | | |
| | improved | not changed | worsened | no export | balance | improved | not changed | worsened | no export | balance | improved | not changed | worsened | no export | balance |
| **Industry** | **3** | **39** | **4** | **54** | **-1** | **3** | **38** | **4** | **55** | **-1** | **3** | **39** | **4** | **54** | **-1** |
| Mining | 2 | 36 | 2 | 60 | 0 | 2 | 31 | 2 | 65 | 0 | 0 | 33 | 1 | 66 | -1 |
| Processing industry | 3 | 45 | 5 | 47 | -2 | 4 | 44 | 5 | 47 | -1 | 3 | 46 | 5 | 46 | -2 |

Source: Formed by the authors on the basis of [8].

The research results indicate potential opportunities for the indicator of business confidence of Ukrainian enterprises (Fig. 3), illustrate the positive dynamics, but at the same time, unfavorable development forecasts in the processing industry and in the industrial sector as a whole, actualize the issues of improving the crisis management system management for domestic enterprises.

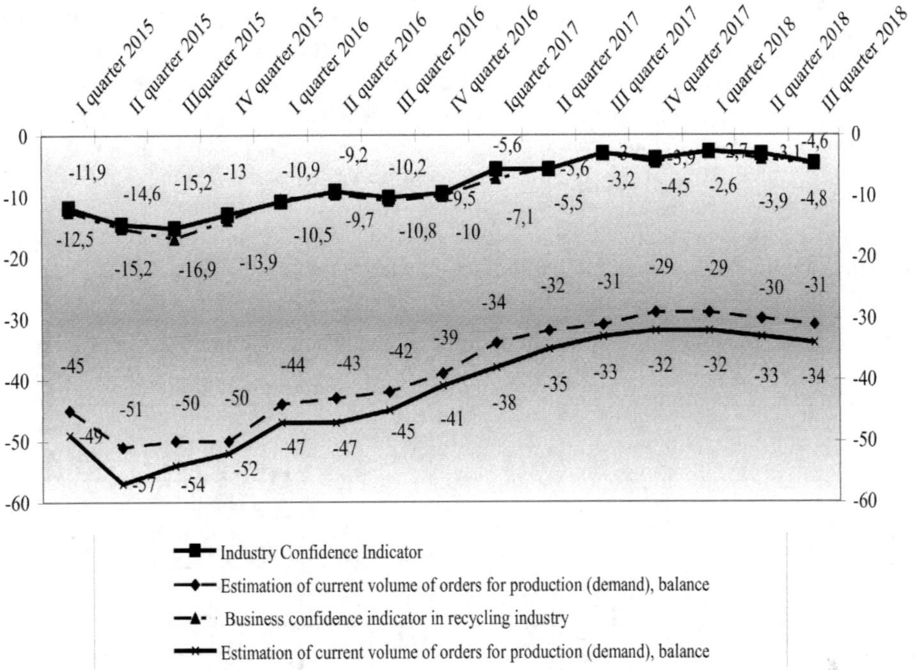


**Fig. 3.** Business confidence indicator in the industry of Ukraine (percent), %

Source: Formed by the authors on the basis of [8].

So, the anti-crisis management of an industrial enterprise must be preventive in nature and, given the specificity of a particular object, be complex.

## 3.2 Approaches to assessing the effectiveness of crisis management of the enterprise

The company's activity in market conditions is associated with the risk of capital loss from inefficient management: as a result of large losses, an enterprise usually fails to settle its debt obligations, and, as a result, its crisis state and bankruptcy. To prevent the loss of invested funds, an analysis of the probability of bankruptcy is intended, the purpose of which is to provide interested persons with accurate and timely information about the possibility of occurrence, the scale of crisis phenomena and their consequences for developing measures to prevent or minimize their negative impact.

Research approaches to predicting the probability of bankruptcy, formed by foreign and domestic scientists [1-5], have, of course, a number of advantages. These include a relatively small number of estimated indicators, a complex character based on the formation, the possibility of obtaining a quantitative assessment. At the same time, the formation of models based on a statistical sample of foreign enterprises limits the possibility of obtaining the correct result when used to form a forecast of the development of domestic enterprises.

The study of the dynamics of the company's profits and on this basis, obtaining a forecast for the retrospective assessment of this indicator for the future, allows us to determine the probability of a crisis in the profitability of the enterprise. In the case when the enterprise is unprofitable, it is possible to establish the presence or absence of sources of replenishment of own funds of the enterprise for its normal functioning and development prospects. The main method used at this stage is the calculation of the trend parameters of the company's net profit and the modeling of the main trend of its change. The most accurate forecast can be formed in conditions of a sufficient amount of retrospective sampling. To study the predicted development trend of the enterprise, forecast models were used based on determining the possible dynamics of indicators in the future according to the retrospective trend of actual indicators of the object of study. It should be noted that the period of more than 10 years should be considered necessary and sufficient statistical base (input information). Draw up a calculation table 2 on the performance indicators of the activities of Public Joint-Stock Company «Kremenchug Wheel Plant» (Kremenchug Wheel Plant- machine-building enterprise specialized in the production of wheels) for the period 2002-2017.

**Table 2.** Calculation of the parameters of the trend equation of the net profit of PJSC «KWP» for the period 2002-2017

| Year | Net profit, thousand UAH. (Actually) | Four quarter moving average | The sum of two adjacent values | Centered average, trend, thousand UAH | **Costs / Trend x 100** |
|---|---|---|---|---|---|
| 2002 | 8,223.5 | x | x | x | x |
| 2003 | 9,084.9 | x | x | x | x |
| 2004 | 20,812.1 | 11,487.1 | 26,663.5 | 13,331.8 | 156.1 |
| 2005 | 7,827.8 | 15,176.4 | 34,358.1 | 17,179.0 | 45.6 |
| 2006 | 22,980.9 | 19,181.6 | 39,682.2 | 19,841.1 | 115.8 |
| 2007 | 25,105.7 | 20,500.6 | 39,313.8 | 19,656.9 | 127.7 |
| 2008 | 26,088 | 18,813.2 | 33,367.1 | 16,683.5 | 156.4 |
| 2009 | 1,078 | 14,553.9 | 27,304.7 | 13,652.3 | 7.9 |
| 2010 | 5,944 | 12,750.8 | 20,908.5 | 10,454.3 | 56.9 |
| 2011 | 17,893 | 8,157.8 | 17,532.0 | 8,766.0 | 204.1 |
| 2012 | 7,716 | 9,374.3 | 23,477.3 | 11,738.6 | 65.7 |
| 2013 | 5,944 | 14,103.0 | 28,026.8 | 14,013.4 | 42.4 |
| 2014 | 24,859 | 13,923.8 | 32,536.8 | 16,268.4 | 152.8 |
| 2015 | 17,176 | 18,613.0 | 41,227.8 | 20,613.9 | 83.3 |
| 2016 | 26,473 | 22,614.8 | x | x | x |
| 2017 | 21,951 | x | x | x | x |
| **Variation** | 107.2 | 136.7 | 124.5 | 31.6 | |
| **Year** | **Quarter 1** | **Quarter 2** | **Quarter 3** | **Quarter 4** | |
| **2019** | | | 156.1 | 45.6 | x |
| **2020** | 115.8 | 127.7 | 156.4 | 7.9 | x |
| **2021** | 56.9 | 204.1 | 65.7 | 42.4 | x |
| **Uncorrected average** | 108.5 | 138.4 | 126.1 | 32.0 | 404.9 |
| **Adjusted average** | 107.2 | 136.7 | 124.5 | 31.6 | 400.0 |

Source: Formed by authors based on [16].

Thus, to study the projected development trend of PJSC «KWP», the data for the period 2002-2017 was used. Additional opportunities for the analyst are generally provided with the assessment of the quality of profits, if for a long period of time stable rates of profit growth are observed, and the share of profits from the main activity prevails in its structure (as is observed within the enterprise PJSC «KWP»), its quality is high.

## 4 Discussion

Integral assessment of the level of the crisis of the enterprise, as it is proposed for use, should be formed on the basis of a set of financial indicators that allow to fully characterize the state of the enterprise according to the probability of a crisis state for the period of the study and in the future as well. The methodology used combines two components of the assessment - analysis of solvency and its level of profitability.

On the basis of a statistical sample of indicators of industrial enterprises' financial analysis, a linguistic assessment of enterprises' state corresponding to various levels of the development of crisis processes was obtained. According to different levels of the crisis state, the limit values of the indices $K_{ij}$ та $Z_j$ are established (Table 3).

**Table 3.** Table 3. Limit values for integral indices Z

| Financial status | Limit value |
|---|---|
| Deep crisis | >2.5 |
| Light crisis | 0.26-2.5 |
| Sustainable works | <0.26 |

Source: Formed by authors based on [1-3].

A visual representation of the obtained calculation result is an absolute advantage of this technique, while the conditions for establishing the position of the enterprise in crisis zones are made, the choice of priorities of the anti-crisis program of PJSC «KWP» are being made. We will calculate the integral indicator of the assessment of the crisis state of the enterprise using table 4.

**Table 4.** The calculation of the integral indicator of assessing the crisis state of PJSC «KWP»

| Indicator | year | | | | | Calculation procedure Actual / conditional opt indicator | | | | |
|---|---|---|---|---|---|---|---|---|---|---|
| | **2014** | **2015** | **2016** | **2017** | **2018** | **2014** | **2015** | **2016** | **2017** | **2018** |
| Current rate of coverage, coeff. **(>2)** | 1.68 | 2.07 | 1.18 | 1.61 | 1.51 | 0.812 | 1.000 | 0.570 | 0.778 | 0.729 |
| Autonomy ratio **(>0.5)** | 0.64 | 0.66 | 0.58 | 0.57 | 0.51 | 0.970 | 1.000 | 0.879 | 0.864 | 0.773 |
| Absolute liquidity ratio **(>0.2)** | 0.02 | 0.07 | 0.17 | 0.099 | 0.11 | 0.118 | 0.412 | 1.000 | 0.582 | 0.647 |
| Profitability of operating activities, the fate of units **(max)** | 6.76 | 15.97 | 10.02 | 11.87 | 8.96 | 0.423 | 1.000 | 0.627 | 0.743 | 0.561 |
| Return on equity, the share of units **(max)** | 4.4 | 15.05 | 9.55 | 11.84 | 9.8 | 0.292 | 1.000 | 0.635 | 0.787 | 0.651 |
| Net profitability of sales of products (goods works, services), the fate of units **(max)** | 1.35 | 5.47 | 3.13 | 5.03 | 3.19 | 0.247 | 1.000 | 0.572 | 0.920 | 0.583 |
| **Integral indicator of assessing the crisis state of the enterprise $\sqrt{(1-X_i)^2}$ (min)** | | | | | | **1,488** | **0,588** | **0,809** | **0,600** | **0,859** |

Source: Formed by the authors on the basis of [16].

One of the ways to solve the problem of predicting the possibility of bankruptcy is also the use of insolvency prediction models developed on the basis of the discriminant analysis method – Z-coefficients. Some of the most well-known approaches to predicting the probability of bankruptcy are systematized in Table 5.

Summing up the data of an analytical assessment of the probability of bankruptcy, we can conclude that during 2013-2016. The probability of bankruptcy of PJSC «KWP» is low.

**Table 5.** Estimation of the probability of bankruptcy in the context of assessing the effectiveness of the crisis management of PJSC «KWP»

| Model | Individual factors included in the model / criterion | Calculation Results / Linguistic Assessment | | | | |
|---|---|---|---|---|---|---|
| | | **2014** | **2015** | **2016** | **2017** | **2018** |
| Beaver coefficient | **Kb** = *Cash flow / Total debts* | **-0.044** | **0.123** | **0.034** | **0.080** | **0.051** |
| | | considerable | exists | considerable | considerable | considerable |
| Altman model | **$z=0.717x_1+0.847x_2+3.107x_3+0.42x_4+0.998x_5$**<br>$x_1$ - (working capital / assets);<br>$x_2$ - profitability (net income / assets);<br>$x_3$ - profitability (profit before taxes and interest / assets); $x_4$ - financial stability (equity / raised capital);<br>$x_5$ - turnover (revenue / assets) / criterion - the minimum acceptable value of **z – 1.23** | **1.935** | **23.957** | **8.196** | **6.439** | **3.546** |
| | | the probability of bankruptcy exists | the probability of bankruptcy is low | the probability of bankruptcy is very low | the probability of bankruptcy is very low | the probability of bankruptcy exists |
| Lis model | **$z=0.063x_1+0.092x_2+0.057x_3+0.001x_4$**<br>$x_1$ - liquidity (working capital / assets);<br>$x_2$ - profitability (profit before taxes and interest / operating assets); x3 - retained earnings (reserves / operating assets);<br>$x_4$ - financial leverage (equity / debt)<br>Criterion - the minimum allowable value of **z – 0.37** | **2.343** | **4.382** | **2.009** | **1.393** | **1.007** |
| | | the probability of bankruptcy is low | the probability of bankruptcy is low | the probability of bankruptcy is negligible | the probability of bankruptcy is negligible | the probability of bankruptcy is considerable |
| Springate model | **$z=1.03x_1+3.07x_2+0.66x_3+0.4x_4$**<br>$x_1$ - liquidity (working capital / assets);<br>$x_2$ - profit before taxes and interest / assets;<br>$x_3$ - profit before taxes and interest / short-term debt; x4 - revenue / assets<br>Criterion - the minimum acceptable value of **z – 0.862** | **23.761** | **10.300** | **40.327** | **32.328** | **21.769** |
| | | the probability of bankruptcy is low | the probability of bankruptcy is low | the probability of bankruptcy is low | the probability of bankruptcy is low | the probability of bankruptcy is low |
| Model of the Ministry of Finance of Ukraine | **$z=1.04x_1+0.75x_2+0.15x_3+0.42x_4+1.8x_5-0.06x_6-2.16$**<br>$x_1$ - coverage ratio;<br>$x_2$ - the ratio of financial autonomy;<br>$x_3$ - the turnover ratio of the invested capital;<br>$x_4$ - the operating margin profitability ratio;<br>$x_5$ - asset profitability ratio;<br>$x_6$ - the turnover ratio of borrowed capital;<br>Criterion - the minimum allowable value of **z – 0.55** | **9.421** | **21.991** | **14.843** | **1.388** | **0.578** |
| | | the financial condition of the company is satisfactory | the financial condition of the company is satisfactory | the financial condition of the company is satisfactory | the financial condition of the company is satisfactory | additional expert research is needed |

Source: Formed by the authors on the basis of [1-5] according to [16].

At the same time, one should take into account the emergence of negative dynamics of indicators for the majority of evaluation models at the end of the study period – 2018. At the same time, according to Beaver's diagnostic model, the probability of bankruptcy of an enterprise should be recognized as possible and significant.

Although some authors argue about a sufficient level of predictability of the prospects for the enterprise using Z-coefficients – indicators of the probability of bankruptcy, common in Western countries, it should be pointed out that their use in Ukraine is limited as a result of calculating their weight values of other economic conditions than domestic ones.

Therefore, we supplemented the analysis with the model of the Ministry of Finance of Ukraine, according to calculations for which, at the end of the study period, the state of the enterprise is determined to be doubtful. We will carry out the main trends forecasting in the local indicators that are included in the integral indicator, which we use to calculate the complex indicator for assessing the crisis state of PJSC «KWP». It should be noted that the indicators for assessing the effectiveness of the financial management of the enterprise and the performance of the enterprise, which are internal and relevant, have been selected for consideration.

Trend models allow you to set the forecast values of financial ratios, assess the direction of developments and the priority of financial management measures.

**Table 6.** Forecasting models of local components of a complex integral indicator for evaluating the effectiveness of a crisis management system

| Local indicators | Model view | Coefficient of determination |
|---|---|---|
| Current rate of coverage, coeff. **(>2)** | $y = -0.1854x^4 + 2.2875x^3 - 9.7296x^2 + 16.348x - 7.04$ | $R^2 = 0.95$ |
| Autonomy ratio **(>0.5)** | $y = -0.0121x^4 + 0.1492x^3 - 0.6429x^2 + 1.0858x + 0.06$ | $R^2 = 0.86$ |
| Absolute liquidity ratio **(>0.2)** | $y = 0.0198x^4 - 0.2343x^3 + 0.9373x^2 - 1.4177x + 0.715$ | $R^2 = 0.78$ |
| Profitability of operating activities, the fate of units **(max)** | $y = -1.48x^4 + 18.627x^3 - 82.34x^2 + 148.04x - 76.09$ | $R^2 = 0.65$ |
| Return on equity, the share of units **(max)** | $y = -1.5025x^4 + 19.015x^3 - 84.603x^2 + 153.89x - 82.4$ | $R^2 = 0.88$ |
| Net profitability of sales of products (goods, works, services), the fate of units **(max)** | $y = -0.7783x^4 + 9.5667x^3 - 41.172x^2 + 72.343x - 38.61$ | $R^2 = 0.65$ |

Source: Formed by authors based on [16].

The lack of communication and the adequacy of the model to the experimental data reflect the coefficient of determination. The approximation of this indicator to the unit indicates the accuracy of the model and confirms the possibility of its use for the forecast. At the third stage, a list of targeted anti-crisis measures is developed, which correspond to the peculiarities of the financial condition of the company. We will carry out the ranking of indicators on the basis of the degree of development of negative dynamics and on this basis we will establish the phase of the development of the crisis process. The main requirements for input information in our proposed model are: accessibility, compliance with an integrated assessment approach, completeness.

According to the results of forecasting the dynamics of the coverage ratio for the next period, one should expect its decline, which is negative in the context of strengthening the financial state.

Forecast on extrapolation of the trend line of the current coverage indicator of PJSC «KWP» (with a forecast of a factor increase of 0,5 points) is presented in the fig. 4.

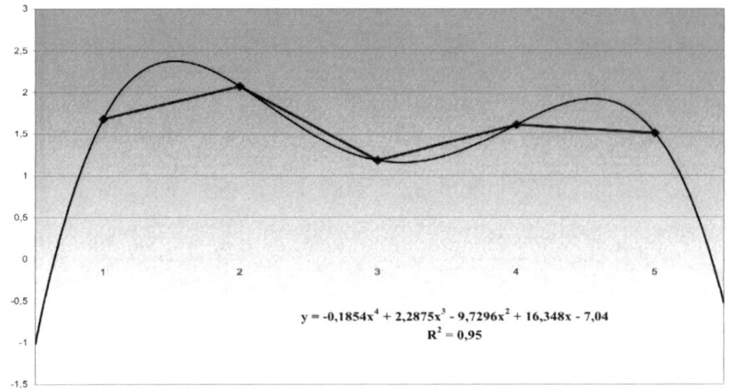


**Fig. 4.** Forecast on extrapolation of the trend line of the current coverage indicator of PJSC «KWP» (with a forecast of a factor increase of 0,5 points)

Source: Formed by the authors on the basis of [16].

Forecast on extrapolation of the trend line of the autonomy coefficient of PJSC «KWP» (with a forecast of a factor increase of 0,5 points) is presented in the fig. 5.

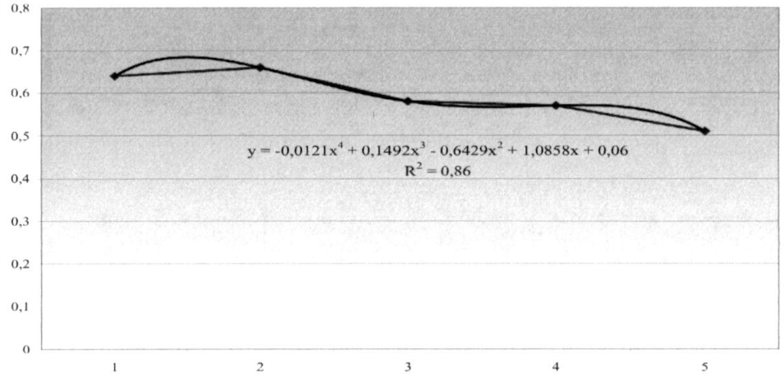


**Fig. 5.** Forecast on extrapolation of the trend line of the autonomy coefficient of PJSC «KWP» (with a forecast of a factor increase of 0,5 points)

Source: Formed by the authors on the basis of [16].

The forecast for the extrapolation of the trend line of return on equity of PJSC «KWP» (with a forecast of an increase in the factor of 0,5 points) is presented in the fig. 6.

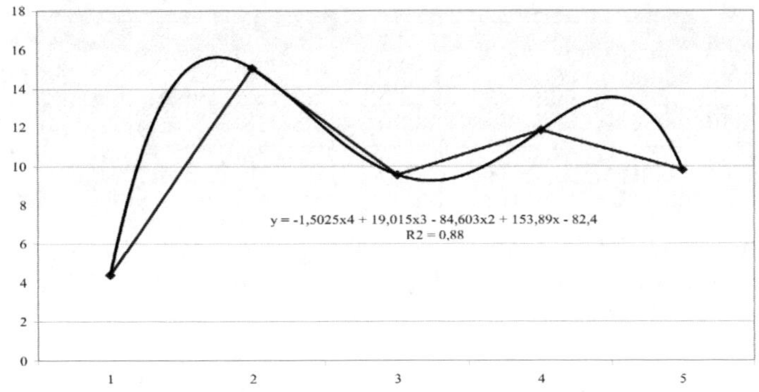


**Fig. 6.** The forecast for the extrapolation of the trend line of return on equity of PJSC «KWP» (with a forecast of an increase in the factor of 0,5 points)

Source: Formed by the authors on the basis of [16].

Forecast on extrapolation of the trend line of the absolute liquidity ratio of PJSC «KWP» (with a forecast of a factor increase of 0,5 points) is presented in the fig. 7.

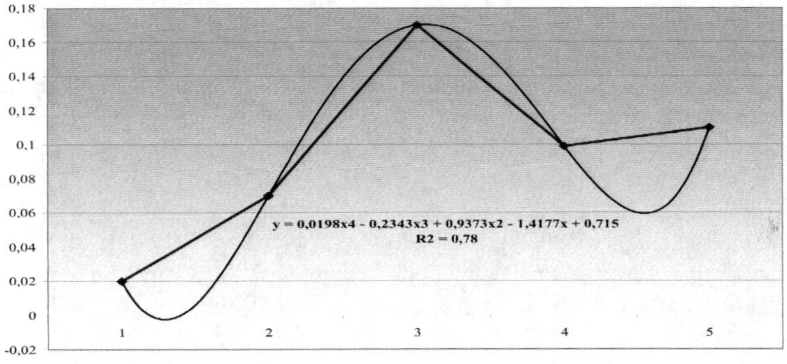


**Fig. 7.** Forecast on extrapolation of the trend line of the absolute liquidity ratio of PJSC «KWP» (with a forecast of a factor increase of 0,5 points)

Source: Formed by the authors on the basis of [16].

In particular, the last actual value of the indicator is 1.51, which is 32% lower than the standard (ordered). The forecast on the extrapolation of the trend line of the local components of the integrated integral indicator for evaluating the effectiveness of the crisis management system of PJSC «KWP» (with a forecast of an increase in the factor of 0,5 points) is illustrated in Fig. 4-7.

The polynomial trend forecast is chosen as the model form, which, compared to the linear dependence, more fully and correctly describes the dynamic changes and the forecast of the local indicator, the results of which are summarized in Table 7.

**Table 7.** Forecast values of local indicators for PJSC «KWP»

| Period | Current rate of coverage, coeff. | Autonomy ratio | Absolute liquidity ratio | Profitability of operating activities, the fate of units | Net profitability of sales of products, the fate of units |
|---|---|---|---|---|---|
| **01.06.2019** | 1.42 | 0.48 | 0.12 | 8.77 | 3.01 |
| **01.12.2020** | 1.38 | 0.46 | 0.14 | 6.98 | 3.00 |

Source: Formed by authors based on [16].

The predictive model of the general integral index is constructed as follows:

$$y = 0{,}102x^4 - 1{,}2789x^3 + 5{,}683x^2 - 10{,}527x + 7{,}509 \quad (1)$$

The adequacy of the constructed model is confirmed by the values of the correlation and determination coefficients, equal, respectively, 0.98 and 0.96. Based on the calculations, it has been determined that for PJSC «KWP», actions to improve it by strengthening the financial condition and increasing the profitability of production and sales will be successful.

## 5 Conclusion

Thus, the proposed toolkit of methodological support for the action of the anti-crisis program of an industrial enterprise is based on the developed forecast models that will allow evaluating the effectiveness of anti-crisis management both as a whole and individual measures, and is an element of information support for decision-making by lenders to finance the enterprise.

The results show that PJSC «KWP» is in a state of survival or inert existence, and according to the model for assessing the probability of Beaver's bankruptcy, the state is unstable. The increase in the level of competition, the curtailment of the scale of activities of the main consumers of wheeled products led to problems with the sale of the company's products. Ineffective financial management in the context of the formation of a non-optimal structure liability of the enterprise created the conditions for the emergence of the pre-crisis state of the enterprise.

## References

1. G.M. Azarenkova, R.O. Piskunov, M.I. Goikhman, *Anticrisis regulation of economic entities cash flow*. Kyiv: University of Banking of National Bank of Ukraine, pp. 125-127 (2014)
2. O.V. Aref'eva, *Anticrisis financial management of enterprise*. Lviv: Ukrainian Academy of Printing, pp. 75-79 (2011)
3. D. Bradley, Ch. Cowdery, Small Business: Causes of Bankruptcy [online], Available at: http://citeseerx.ist.psu.edu/viewdoc/download?doi=10.1.1.367.4484&rep=rep1&type=pdf (2002)
4. D. Bryan, G. Fernando, A. Tripathy, Bankruptcy risk, productivity and firm strategy. *Review of Accounting and Finance*, **12**(4), 309-326 (2013)
5. V.L. Dikan, I.L. Nazarenko, *Integrated method for determining the economic security level, risk assessment and probability of enterprises bankruptcy*. Kharkiv, Ukrainian State Academy of Railway Transport, pp. 54-80 (2011)

6. L. Elexa, L. Hvolkova, M. Knapkova, Anticrisis Management: Warning Signals Before Failure. *Marketing and Management of Innovations*, **3**, 98-111 (2019)
7. E. Tobback et al., Bankruptcy prediction for SMEs using relational data. *Decision Support System*, **102**, 69-81 (2017)
8. The State Statistics Service of Ukraine [online], Available at: www.ukrstat.gov.ua/
9. O. Maslak, N. Grishko, K. Vorobiova, O. Hlazunova, The offshorization of economy: the present realities. *SHS Web of Conferences: Innovative Economic Symposium 2017 – Strategic Partnership in International Trade*, **39** (2017)
10. M. Khomenko, K. Pryakhina, K. Latyshev, Prospects for development of Ukraine and EU in the field of renewable energy sources. *SHS Web of Conferences: Innovative Economic Symposium 2018 – Milestones and Trends of World Economy,* **61** (2019)
11. The Global Competitiveness Report 2017-2018 [Online], Available at: http://reports.weforum.org/global-competitiveness-index-2017-2018/at-a-glance-global-competitiveness-index-2017-2018-rankings/ (2018)
12. R. Bacik, J. Kloudova, J. Gonos, V. Ivankova, Management of Competitiveness and Economic Performance Based in the V4 Countries. *Marketing and Management of Innovations*, **3**, 73-88 (2019)
13. J. Sebestova, R. Sperka, S. Cemerkova, Determinants of regional business environment in Czech Republic: an expert dilemma. *Forum Scientiae Oeconomia*, **4**(4), 65-74 (2016)
14. E. Bartalosova, Rating of Innovation Performance and Competitiveness of the Visegrad Four in Comparison with the EU Member States. *17th International Scientific Conference on Finance and Risk 2015* (2015)
15. A. Kotaskova, Z. Rozsa, The Impact of Selected Factors on the Quality of Business Environment Assessment in the Czech Republic and the Slovak Republic. *International Journal of Entrepreneurial Knowledge*, **6**(2), 71-80 (2018)
16. Annual financial statements of enterprises, *Stock market infrastructure development agency of Ukraine (SMIDA)* [online], Available at: http:// www.smida.gov.ua/ (2019)

# The Institutional Regulatory Environment of the Digital Ecosystem: Theoretical Approach and Russian Experience

*Natalya* Masyuk[1*], *Marina* Bushueva[2], *Zinaida* Bragina[3]

[1]Vladivostok State University of Economics and Service, Gogolya st. 41, 690014 Vladivostok, Russia

[2]Russian University of Economics by GV Plekhanov (Ivanovo Branch), Ivanovo, Russia,

[3]Kostroma State University, Ulitsa Dzerzhinskogo 17, 156005 Kostroma, Russia

**Abstract.** Digitalization is an integral part of global change in the world economy. The successful creation and functioning of a digital economy are possible only in an adequate digital ecosystem. The main criterion for the formation of a digital ecosystem is the maturity and stability of its institutional environment, including the totality of new institutions. The creation of digital institutions is a new phenomenon of institutionalization and accompanies the introduction of digital technologies in the digital ecosystem. The article defines the concept of institutionalization in connection with the analysis of the economic interests of business entities operating in the regions of Russia. The definition of digital institutions as new social institutions in the digital age is formulated. The structure and basic elements of the institutional regulatory environment of the digital ecosystem that is necessary for regulating digital transformations in the digital ecosystem are determined.
**Keywords:** institutionalization, digital economy, digitalization, digital ecosystem, digital transformation, digital institutions, institutional regulatory environment.

## 1 Introduction

Institutionalization is the process of determining and consolidating social norms, rules, statutes, and roles, turning them into a system that can satisfy certain social and economic needs, which is largely determined by global changes in the global economy [1]. Globalization is accompanied by a deep institutional transformation that characterizes the current stage of development of the digital economy [2, 3].

Currently, the categorical apparatus of a new scientific direction – digital economy – is actively discussed in the scientific and methodological literature [4]. In the "Strategy for the Development of the Information Society of the Russian Federation for 2017-2030", the

* Corresponding author: masyukn@gmail.com

digital economy is understood as follows: "The digital economy is an economic activity in which the key factor in production is digital data, the processing of large volumes and the use of analysis results which are compared with traditional forms of management can significantly increase the efficiency of various types of production, technology, equipment, storage, sale, delivery of goods and services "[5]. The above definition is quite general, not reflecting the semantic meaning of the term; therefore, it is necessary to reveal the main features of the digital economy. These include:

- the creation of cyber-physical systems in which a person and a machine represent a single, harmoniously working organism
- the use of intelligent platforms in all areas of socio-economic activity;
- the introduction of the Internet of people, the Internet of things, the Internet of services (along with the already well-developed Internet of ideas) [6];
- application of Big Data technologies (processing of "big data");
- the use of modern information technologies, such as block-chain, ensuring transparency of transmission, reliability of data storage; nadge-technologies that promote the marketing of products (goods and services), etc. [7 ,8].

These signs do not have clearly defined areas of action. Their effects intersect, creating a positive synergistic interaction effect. For example, the cyber-physical system of the regional economy is based on the introduction of digital platforms using Big Data, block-chain, and nadge-technologies. In this system, based on the Internet of people, professional expert communities are formed, and Internet services allow enterprises in the region and its population to improve the quality of the services provided (at work, in private life). The Internet of things "aligns" the role of man and machine in the organization of production and business activities.

The relevance of the study is determined by the fact that the institutionalization of digital transformations for modern Russia is a special economic resource that allows you to adapt the institutional foundations of economic growth to Russian conditions at the present stage of development. With the development of digitalization, the institutional economy receives a new version of life - a new field of problems generated by a change in the medium of the economic entity, its reflection, and measurement, represented by indicators, coefficients, indices, etc. An innovative ecosystem of the digital economy is being formed [4, 9, 10, 11].

In the Strategy for the Development of the Information Society in the Russian Federation for 2017-2030. The ecosystem of the digital economy is defined as "a partnership of organizations that ensures the constant interaction of their technological platforms, applied Internet services, analytical systems, information systems of state authorities of the Russian Federation, organizations and citizens" [5]. However, there are other definitions of a digital ecosystem, for example: a "digital ecosystem" is a multi-structural relationship between the main actors of the digitalization of the economy (population, state, business structures) and the basic conditions for their functioning [12].

Growth points of the digital economy of Russia can be both sectors of the national economy and the subjects of the Federation [13]. Currently, the government's digitalization program for the Russian economy focuses on large industry holdings and state corporations. Indeed, within these associations, which have sufficient resources (financial, labor, intellectual), already have a high starting position, and, most importantly, relying on concentrated management, the solution to the problems of digitalization of the economy seems to be the most promising [14]. Currently, the introduction of elements of the digital economy, primarily digital platforms, is limited mainly by the development of social and communication services (Internet, mobile communications, utility bills, banking, online ticketing, etc.) [15]. Often the introduction of digital tools is accompanied by a conflict between the old, familiar methods of doing business and innovative ones, the introduction of innovations must be carried out consistently, resolving contradictions using local compromises [16].

## 2 Method and methodology

The theoretical and methodological basis of this study is based on a set of fundamental premises of the classical and modern economic theory about the objective conditionality of the development process, its sources, forms, orientation and limitations, based on the provisions of classical economic theory, the provisions of institutional theory using the methodological tools of genesis, evolution and development socio-economic systems [17, 18, 19]. This study was conducted on the basis of a system-institutional approach, which is a synergistic combination of the "concept of a system economy" and institutional theory. The concept of a systemic economy focuses on "complex" socio-economic formations in which certain combinations of internal forces and external conditions provide relative stability in time and space. At the same time, the methodology of a systemic economy expands the boundaries of systemic vision by introducing into the scope of interdisciplinary consideration the fields of scientific knowledge related to the areas of knowledge bordering on the economy in a given space: psychology, mathematics, sociology, computer science, regional studies, etc. Under the system (in the framework of the theory of economic systems) we understand the relatively isolated and relatively stable in space and time part of the world (considered as a system-containing space in), characterized by external integrity and internal manifold [20]. At the same time, the continuous process of the network economy should be taken into account [21]. Institutional economics uses the concept of "institute," the focus is currently on social institutions. Social institutions are value-normative complexes (values, rules, norms, attitudes, patterns, standards of behavior in certain situations), as well as bodies and organizations that ensure their implementation and approval in society. The signs of a social institution include: a set of institutions, social groups, the purpose of which is to satisfy one or another of the needs of society; system of cultural patterns, norms, values, symbols; a system of behavior in accordance with these standards and patterns; material and human resources necessary for solving problems; socially recognized mission, goal, ideology. All these signs will be used by us in the future to define the concept of "digital institutions".

## 3 Results

As a result of the study, the authors propose to introduce the concept of "digital institute" and "institutional regulatory environment of the digital ecosystem" into business circulation. According to the definition of a social institution given above, it can be argued that digital institutions are a form of social institutions. Moreover, by a digital institution, we mean the regulatory tools of the digital ecosystem, i.e. a set of various tools - electronic products and services that contribute to the development and implementation of digital technologies. According to the authors, the digital ecosystem for successful development requires the creation of certain conditions or a special institutional environment that will regulate digital transformation.

Based on this postulate, as well as on the basis of the above definition of "digital institution", a structural diagram of the institutional (regulatory) environment of the digital ecosystem was developed (Fig. 1) and its definition is given. It is proposed to understand the institutional regulatory environment as a system of electronic products and services that accompany the creation and development of digital institutions, as well as regulatory decision-making procedures and entities involved in regulatory decisions.

Digital institutions, in turn, consist of hard regulation tools (laws, other regulations, etc.), soft regulation tools (recommendations, clarifications, instructions, standards of self-regulatory organizations, etc.), as well as "regulatory sandboxes" under which special legal regimes are understood. Regulatory sandboxes are a separate type of special regime, which is temporarily facilitated conditions for the implementation of innovative projects. Their main advantage is a reduction in regulatory pressure on innovative business with minimal risks due to the limitation of the scope of experimental activity.

## 4 Discussion

In the process of digitalization, the most important aspect is its legislative support. In Russia, there are still many weaknesses in this direction. But in fairness, it should be noted that in solving this issue the Russian authorities are not marking time. So, the first package of amendments is already being prepared for introduction to the State Duma, regulating issues of personal data circulation, protection, and ease of use of digital signatures, the introduction of electronic workbooks, etc. [22]. It is assumed that in 2021 universities in Russia will start issuing new-type diplomas that are protected by the block-chain, etc. Despite the technological and financial difficulties, the penetration of digital technologies into innovative businesses cannot be stopped.

In Russia, among the areas whose representatives have already relied on digital transformation, are transport and logistics. In these industries, freight and passenger autopilot vehicles are already being tested, based on the Internet of things, big data, neural networks, and other technologies. Intelligent transport systems are also actively developing, including at the initiative of the state. Among the examples of the latter are the Plato system and automated weight and dimensional control systems (ASVGK), the number of which is growing year by year.

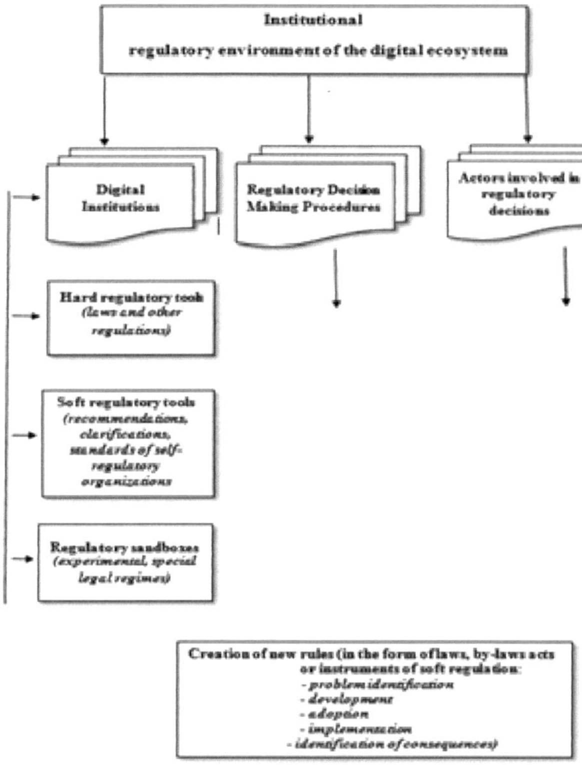


**Fig. 1.** Institutional regulatory environment of the digital ecosystem

Source: Own processing.

Car sharing also demonstrates the effectiveness of the synergistic approach. Participants in this market represent one of the most promising trends for the industry – mobility as a service (MaaS). In the future, this approach can completely change the urban transport environment. Car sharing companies are constantly introducing new tools – from chatbots to recognizing a user by face and voice. This facilitates interaction with customers and ensures the safety of the fleet.

Logistics also goes to a new level. Fleet and cargo control services are already quite routine. New possibilities appear: for example, special sensors monitor the position and movement of cargo in space, including the tilt axis, the temperature regime of transportation (for perishable goods or goods with special transport conditions), monitor the condition of the driver by eye movement, and so on.

One of the most revolutionary trends for Russian logistics in the future should be the development of platform solutions for the search and selection of freight carriers (an analog of Uber for freight transport), the organization of multimodal transportation [6]. Similar platforms are based on cloud technologies, Big Data and other tools. It is thanks to the synergy that, according to experts, they are able to completely transform the market, as has already happened in the field of taxi transportation, which has radically changed due to the so-called Uberization. Obviously, similar changes in the freight market are a matter of time.

## 5 Conclusion

The development of the digital economy is currently one of the most significant global trends, the consequences of which are felt in various spheres of life. Under these conditions, in many countries, strategies and plans for the formation of a digital economy have been developed and are being implemented. Digital institutions are emerging as a new kind of social institutions in society. In Russia in 2017, documents were also adopted that defined the prospects for this area, including the Strategy for the Development of the Information Society and the Digital Economy in the Russian Federation Program. It is obvious that digitalization in the near future will be the most important factor in the development of each region and the country as a whole. No wonder the digital economy is often called the chance for a breakthrough in many areas: in the economy, industry, infrastructure, the quality of life of a particular person, etc. Moreover, the regulatory institutional environment in which digital transformations take place will be of decisive importance.

## References

1. N. Baranov, Institutionalization in Russia: features of national model. *Political Expertise: PolitEx,* **3**(4), 69-87 (2007)
2. N.S. Revenko, The US digital economy in the era of information globalization: current trends. *The USA and Canada: Economics, Politics, Culture,* **8**(572), 78-100 (2017)
3. N.N. Masyuk, M.A. Bushueva, L.K. Vasyukova, A.E. Kiryanov, Platforms of digital experience and digital transformation in an innovative economy. *Proceedings of the III International Scientific and Practical Conference "Resonances Science"* (2018)
4. E. Avramakis, J. Anchen, A. Raverkar, Digital ecosystems: extending the boundaries of value creation in insurance. *Swiss Re Institute* [online], Available at: https://www.swissre.com/institute/research/topics-and-risk-dialogues/digital-and-technology/Digital-ecosystems.html (2019)
5. *The Strategy for the Information Society Development in the Russian Federation for 2017–2030*, Presidential Decree # 203 of 9 May 2017 was approved
6. V. Mesropyan, Digital platforms - a new market power [online], Available at: https://www.econ.msu.ru/sys/raw.php?o=46781&p=attachment (2018)
7. L.L. Avdeeva, Analysis of the prospects for the development of the digital economy in Russia and abroad. *Digital Economy and "4.0 Industry ": problems and prospects: works of the Scientific and Practical Conference with International Participation*, pp. 19-25 (2017)
8. N.D. Bublik, I.I. Lukina, D.V. Chuvilin, T.A. Shafikov, R.F. Janusova, Development of the digital economy in the regions of Russia: problems and opportunities (for example, the Republic of Bashkortostan). *Regional Economics and Management* [online], Available at: https://eee-region.ru/article/5313/ (2018)
9. A. Corallo, G. Passiante, A. Prertcipe, The Digital Business Ecosystem. *Edward Elgar Publishing Limited,* 41-42 (2007)
10. N.V. Smorodinskaja, Network Innovative Ecosystems and Their Role in the Dynamization of Economic Growth. *Innovations,* **7**(189), 27-33 (2014)
11. D. Jackson, What is an Innovation Ecosystem? [online], Available at: http://erc-assoc.org/sites/default/files/topics/policy_studies/DJackson_Innovation%20Ecosystem_03-15-11.pdf (2011)

12. V.V. Stepanova, A.V. Ukhanova, A.V. Grigorishchin, D.V. Yakhyaev, Evaluating digital ecosystems in Russia's regions. *Economic and Social Changes: Facts, Trends, Forecast,* **12**(2), 73-90 (2019)
13. O. Sukharev, *Institutional Change, Efficiency and Structure of Economy.* Lambert Academic Publishing (2011)
14. *Methodology of development of the economy, industry and services in the conditions of digitalization* / Ed. A.V. Babkin. SPb: Polytech-Press. DOI.10.18720 / IEP / 2018.6 (2018)
15. N.N. Masyuk, Z. Chen, L.K. Vasyukova, Change management and digital transformations in the era of business digitalization. *In the collection: The economy of the regions of Russia: current state and forecast prospects. A collection of articles based on the materials of the All-Russian Scientific and Practical Conference of teachers, graduate students, undergraduates of the Ivanovo branch of the Russian Economic University named after G.V. Plekhanov,* pp. 253-258 (2019)
16. N.N. Masyuk, M.A. Bushueva, L.K. Vasyukova, N.A. Mosolova, Innovative Managerial Decisions: Towards a Conflict-Compromise Approach. *Precedings in 32nd IBIMA Conference* (2018)
17. J.A. Shumpeter, *The Theory of Economic Development.* Russia, Moscow: Progress, (1982)
18. J.F. Moore, The Death of Competition: Leadership and Strategy in the Age of Business Ecosystems. *N.Y.: Harper Business* (1997)
19. C.W. Wessner, Entrepreneurship and the Innovation Ecosystem. Policy Lessons from the United States. *The Papers on Entrepreneurship, Growth and Public Policy* (2004)
20. A.B. Kleiner, System economy as a platform for the development of modern economic theory. *Economic Issues,* **6**, 7, (2013)
21. Z.V. Bragina, M.L. Irodov, A.V. Maslova, *From network interactions to a network economy.* Russia, Yaroslavl: MUBiNT (2016)
22. A.A. Berezutsky, Informal aspects of the institutionalization of the Russian political elite. *Management consulting,* **5**, 171-177 (2017)

# The Ukraine and EU trade balance: prediction via various models of time series

*Valerii* Matskul[1*], *Diana* Okara[2], *Nataliia* Podvalna[1]

[1]Odessa National Economic University, 65082 Odessa, Ukraine
[2]Odessa State Academy of Civil Engineering and Architecture, 65029 Odessa, Ukraine

**Abstract.** This article is the first to study, simulate and forecast the monthly dynamics of the trade balance between Ukraine and the European Union for the period from 2005 to 2019. In the presented work, three types of models were used for modeling and forecasting: Automated Neural Networks, additive models ARIMA *ARIMAS (Autoregressive integrated moving average with season component) and Holts model with a damped trend. When modeling using the Automated Neural Networks module, various ensembles of networks and neuron activation functions in hidden layers were used. It turned out that Automated Neural Networks have poor prognostic ability (as in the case considered by us, when modeling insufficiently long series of dynamics). Therefore, when modeling and forecasting the dynamics of the Ukraine-EU trade balance, classical (so-called Box-Jenkins) time series models were used. In this case, the time series is divided into several components (in our case, terms): the main trend is the trend, the seasonal component and the random component (the so-called white noise). By smoothing the initial series, a trend was found, and an analysis of the autocorrelation functions revealed a one-year seasonality. Eliminating the trend and the seasonal component, we obtained a random component, which has a Gaussian distribution. This made it possible to apply first the ARIMA* ARIMAS additive model, and then the Holt model of exponential smoothing with a damped trend. Adequate models of Ukraine-EU trade balance dynamics have been obtained, according to which the forecast has been made. A comparative analysis of the models used. The best model was chosen for forecasting, which allowed to get a good forecast (in comparison with actual data).
**Key words:** trade balance, export-import, neural networks, Box-Jenkins models

## 1 Introduction

In recent years, international trade has begun to attract the close attention of many scientists. Among recent works, we note publication [1], which explores the dynamics of the trade balance between the EU and China. At the same time, Ukrainian scientists mainly investigate the international trade of Ukraine as a whole, not highlighting its export-import operations

* Corresponding author: valerii.matskul@gmail.com

with the EU. Note article [2], in which an attempt is made on the basis of annual statistical data for the period 2010-2017 years to conduct a correlation and regression analysis of the export of goods and services of Ukraine. A factor was included as one of the predictors in the model - time, and it turned out to be the most significant. Economists around the world have always paid close attention to the study, modeling and forecasting of the dynamics of economic processes. A particularly rapid development of modeling and forecasting using time series models began after the release of the classic Box-Jenkins book [3]. Note that these methods have not lost relevance at present. So in the works [4, 5], the Box-Jenkins models were used to model and predict the dynamics of unemployment. And in the work [6], Box-Jenkins models were compared with Automated Neural Networks when modeling the dynamics of Czech GDP. Examples of the application of various models of time series are described in the book [7]. The subject of this work is the study of the dynamics of trade relations between Ukraine and the EU countries. The relevance of this study is indisputable, since the EU has become the largest trading partner of Ukraine in recent years.

## 2 Materials and methods

Data for analysis is available on the website of the State Statistics Service of Ukraine [8]. The time period for which the data will be available is the monthly balance (as the difference between total export and imports between the Ukraine and the EU) starting from January 2005 and ending in March 2019. This is about 171 input data. The unit is the $ million. Preliminary data processing was carried out in MS Excel spreadsheets. When modeling and computing was used DELL STATISTICA software, version 12.

### 2.1 Neural networks

When searching for the best model of the neural network, the "Automated Neural Networks" program module of the DELL STATISTICA software is used.

### 2.2 Box-Jenkins models

We use the following Box-Jenkins additive models of time series:

$$y_t = f(t) + S_t + e_t \tag{1}$$

where $y_t$ is a level of time series at time $t = 1, 2, \ldots$; $f(t)$ - trend is traced the long-term and evolution is deterministic in time; $S_t$ - seasonal fluctuations; $e_t$ - random fluctuations (white noise).

## 3 Results and Discussion

The descriptive characteristics of the data are given in Table 1.

**Table 1.** Characteristics of the data file

| Variable | Average | Minimum | Maximum | Standard deviation |
|---|---|---|---|---|
| Export | 1269.819 | 648.6 | 2222.1 | 302.87 |
| Import | 1816.365 | 561.1 | 3302.7 | 526.95 |
| Trade balance (TB) | -546.546 | -1356.3 | 370.8 | 334.38 |

Source: Authors.

Analyzing the main characteristics of the dynamic range of the trade balance, we conclude that there are no abnormal levels (the so-called "emissions"). Figure 1 therefore shows the dynamics export-import and trade balance.

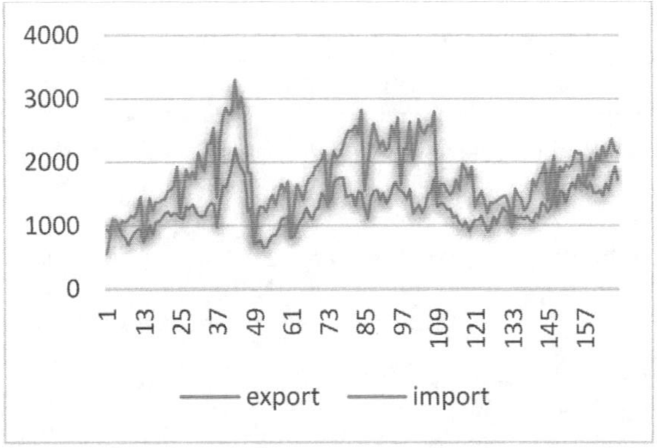


**Fig. 1.** The dynamics of the export-import between the Ukraine and EU

Source: Authors.

Figure 2 shows the dynamics of the trade balance between Ukraine and EU

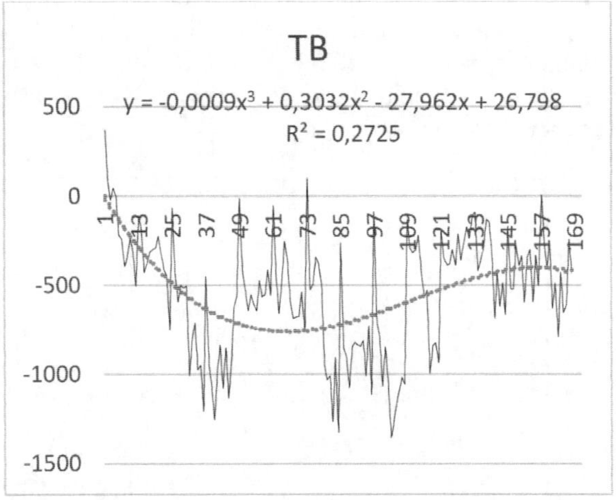


**Fig. 2.** The dynamics (with trend) of the trade balance (TB) between Ukraine and EU

Source: Authors.

Everywhere in the future, a dynamic series of the trade balance (variable TB) is used for modeling and forecasting. First, identify the trend. Note that some researchers are trying (see, for example, [1, 6]) to define a trend as a regression of the levels of the time series by time, requiring a high level of determination of R2> 0,95. Numerous examples of modeling time series (see, for example, [7]) suggest that in the case of an unobvious main trend, the trend should be selected by smoothing the initial dynamic series. Figure 3 below shows the dynamic range of the trade balance after 25-fold exponential smoothing.

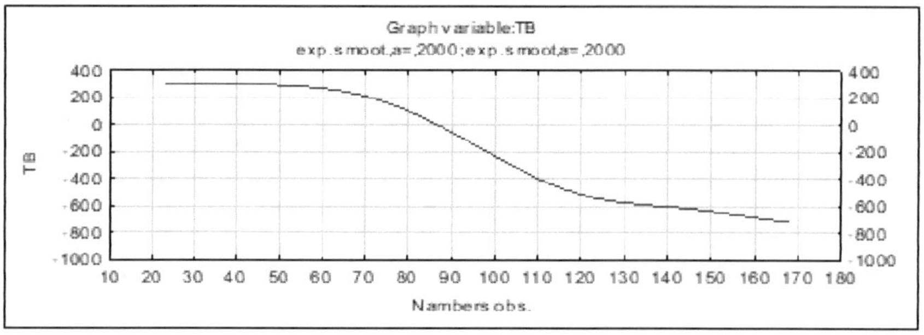


**Fig. 3.** Dynamic range of the trade balance after 25-fold exponential smoothing

Source: Authors.

Analyzing the smoothed series, we conclude that the trend has the form of a third-order parabola (see Figure 2).

### 3.1 Neural networks

The module "Automated neural networks" DELL STATISTICA software was used. We will generate multi-layer perceptron networks, and basic radial function neural networks. SALDO will be determined as the dependent variable. Time series will be divided into training sets, testing and control sets, in the following ratio: 70%, 15% and 15%. A delay in time series will be equal to 1 and 1,000 neural networks will be generated, out of which 5, proving the best results, will be preserved. As the activation functions of neurons of the hidden layer, a set of all functions available in the module was used. Since the number of observations is small, we used the multiple subsample method. As a result, the best neural network with 30 neurons in the hidden layer and radial basis activation functions was selected. Figure 4 presents the results of modeling and forecasting.

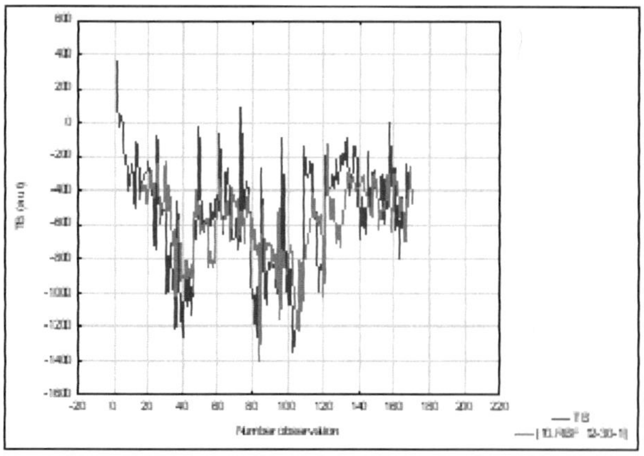

**Fig. 4.** Neural network modeling and prediction results

Source: Authors.

Note that the neural network showed poor prediction results compared to the actual data for 2019. Therefore, we do not provide numerous intermediate results obtained by modeling neural networks.

### 3.2 Box-Jenkins models

First, we will identify the time series. Analyzing the graph of the autocorrelation (Figure 5) and partial autocorrelation functions presented below, we conclude that there is a second-order autocorrelation at lag 2 and annual seasonality with a lag of 12.

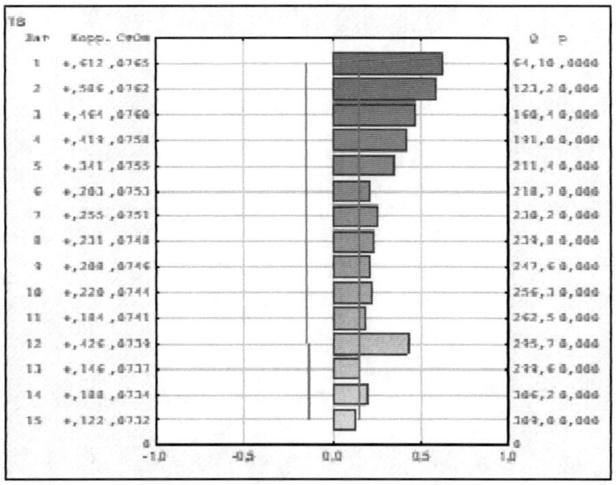


**Fig. 5.** The graphs of the autocorrelation function

Source: Authors.

Partial autocorrelation function is presented below (Figure 6).

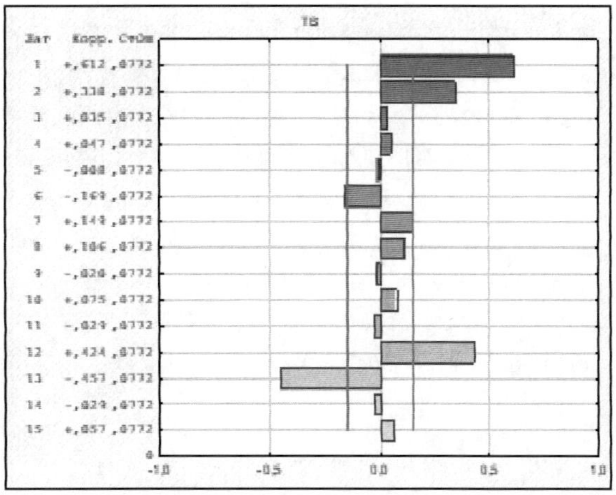


**Fig. 6.** The graphs of the partial autocorrelation function

Source: Authors.

As it is known, for correct application of the ARIMA*ARIMAS model it is necessary that the time series be stationary. Therefore, we eliminate the trend by taking the third-order difference on lag 3. After this operation, we really get a stationary series, which is clearly shown by the following graph (Figure 7).

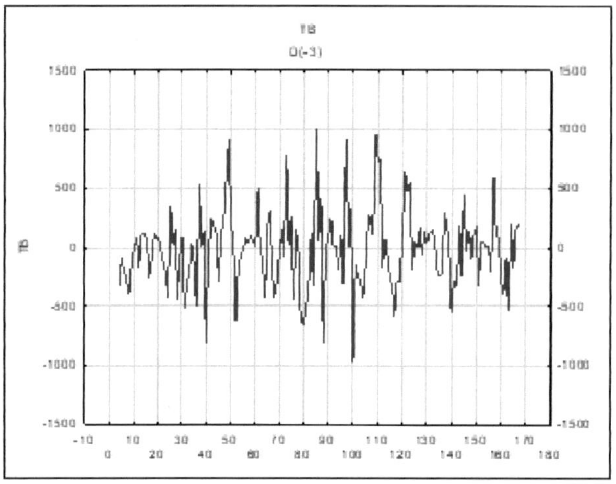


**Fig. 7.** The graph of the variable TB after elimination the trend

Source: Authors.

After the analysis of the dynamics, we use an additive model *ARIMA * ARIMAS* with the trend and seasonal lag 12:

$$y_t = p_{t-2}y_{t-2} + p_{t-1}y_{t-1} + f(t) - q_{t-1}e_{t-1} + S_t \quad (2)$$

where $y_t, t = \overline{1,168};$ is the levels of time series;

$$f(t) = -0,0009x^2 + 0,3032x^2 - 27,962x + 26,798 \quad (3)$$

is a trend component; $p_{t-2}$ and $p_{t-1}$ is the auto-regression coefficients of the second and of the first order; $q_{t-1}$ is a coefficient the moving average model; $e_{t-1}$ is an irregular component (random deviations or so-called white noise). We apply the Smoothed Moving Average of first order $S_t = c + Q_{t-1}S_{t-1}$ to find seasonal coefficients. Parameters of the model one should estimate at 95% confidence level (or 5% risk) with the condition of minimizing the MSE (mean square error):

$$MSE = \sum_{i=1}^{n} \frac{(y_i - \tilde{y}_i)^2}{n} \quad (4)$$

where $y_i$ is the actual value and $\tilde{y}_i$ estimates of indicator $y$ at the period $i$. As a result, we obtained the following parameter estimates (Table 2).

**Table 2.** Parameter estimates of the model ARIMA*ARIMAS

| Parameter | Estimate | Standard deviation |
|---|---|---|
| $p_{t-2}$ | 0.35976 | 0.07908 |
| $p_{t-1}$ | -0.6137 | 0.08725 |
| $q_{t-1}$ | -0.8768 | 0.05805 |
| $Q_{t-1}$ | -0.4498 | 0.06147 |

Source: Authors.

We are convinced of the adequacy of the obtained ARIMA*ARIMAS model by analyzing the histogram of residuals (Figure 8).

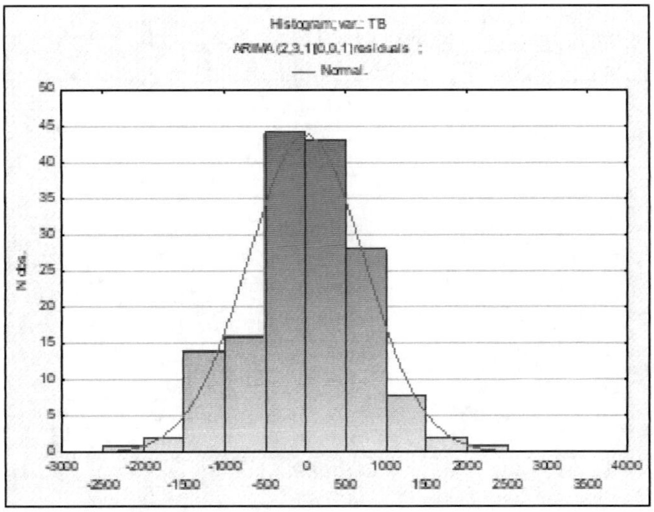


**Fig. 8.** The histogram of the residuals

Source: Authors.

Let us compare the results of forecasting according to the ARIMA*ARIMAS model with the actual data (for the first quarter of 2019) of the trade balance between Ukraine and the EU (Table 3).

**Table 3.** Forecasts for ARIMA*ARIMAS model and actual data

| Forecasts (ARIMA) | Actual |
|---|---|
| 195.1039 | 239.4 |
| -301.5741 | -409.3 |
| -210.9955 | -125.7 |

Source: Authors.

As we can see, the relative error of ARIMA forecasting is big enough. Therefore, an additive Holt model of exponential smoothing is applicable for prediction. The advantage of this model is that the selection of smoothing parameters can be achieved so that the most recent levels of the time series are taken into account when predicting. For example, the value of the smoothing parameter 0.1 means that 10% of the last values of the time series determine 90% of the forecast. To simulate again, we will use the "Time Series and Prediction" module of the STATISTICS program. Before building the Holt model, in order to find the optimal values of the smoothing parameters, we will conduct the so-called "Search on the grid". As a result, the following values were found: the smoothing parameter of the main process alpha = 0.4; smoothing seasonal component parameter delta = 0.1: trend damping parameter phi = 0.1. Table 4 shows the results of modeling and forecasting.

**Table 4**. Results of the modeling and forecasting (fragment) using the Holt model of exponential smoothing

| Exp.smooth.: Addit.season (12) Demph.trend,ad.seas.: Alpha=,400 Delta=,100 Phi=,100 TB | | | | |
|---|---|---|---|---|
| Observ. | TB | Smoothing TB | Resid. | Season comp. |
| 1 | 370.80 | 413.82 | -43.02 | 455.323 |
| 2 | 85.40 | 16.64 | 68.762 | 88.092 |
| 3 | -23.40 | -21.16 | -2.241 | 2.954 |
| 4 | 44.40 | -4.31 | 48.705 | 21.025 |
| 5 | -3.60 | 99.95 | -103.5 | 91.557 |
| 6 | -218.60 | 2.24 | -220.8 | 65.250 |
| 7 | -244.20 | -342.60 | 98.396 | -126.23 |
| 8 | -393.90 | -243.84 | -150.0 | -94.386 |
| 9 | -346.00 | -362.35 | 16.350 | -109.47 |
| 10 | -238.70 | -399.94 | 161.29 | -157.6 |
| 11 | -311.10 | -165.05 | -146.0 | -34.308 |
| 12 | -506.20 | -433.17 | -73.0 | -202.19 |
| 13 | -111.20 | 171.59 | -282.7 | |
| 14 | -195.70 | -387.75 | 192.05 | |
| 15 | -429.20 | -344.06 | -85.13 | |
| 166 | -612.10 | -667.12 | 55.022 | |
| 167 | -241.50 | -301.99 | -60.47 | |
| 168 | -432.00 | -489.31 | 57.307 | |
| 169 | | 233.62 | | |
| 170 | | -407.11 | | |
| 171 | | -131.08 | | |

Source: Authors.

We are convinced of the adequacy of the obtained the Holt model of exponential smoothing by analyzing the histogram of residuals and Normal curve (Figure 9).

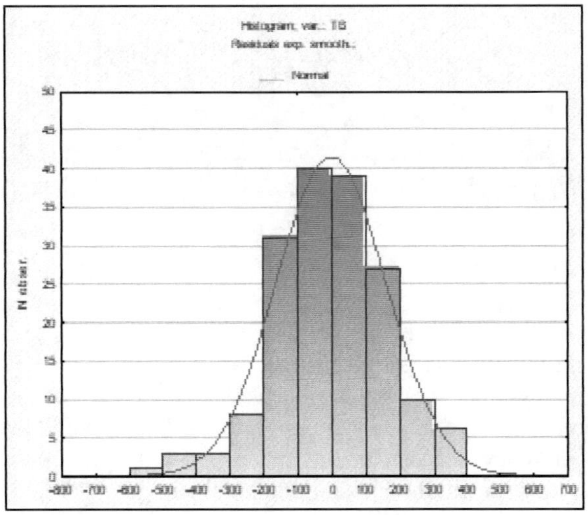


**Fig. 9.** The histogram of the residuals

Source: Authors.

The following Figure 10, which shows the original and simulated time series, clearly demonstrates the quality of the constructed model.

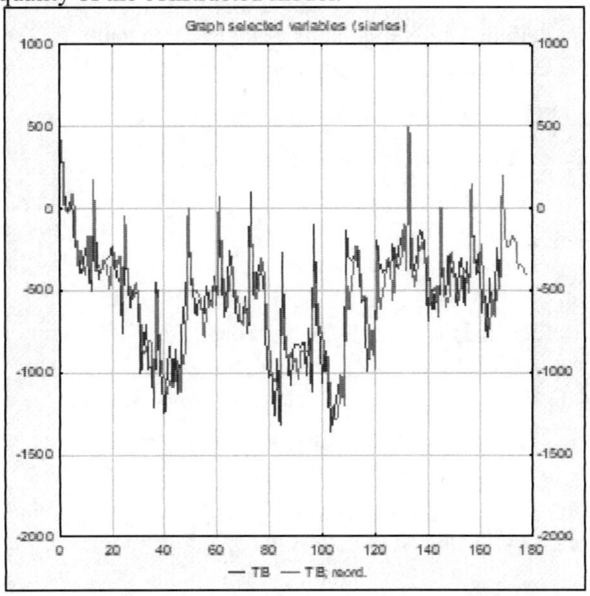


**Fig. 10.** Visualization of source data and results obtained using the Holt model of exponential smoothing

Source: Authors.

Let's compare the forecast results with the actual data of the trade balance between Ukraine and the EU for the first quarter of 2019. To do this, we calculate the relative prediction error by the formula:

$$\delta_{\max} = \max_{t} \left\{ \frac{\left| y_t^{predict} - y_t^{actual} \right|}{\left| y_t^{actual} \right|} \right\} = 0,0428 \tag{5}$$

So, the last model showed good prediction results, since the relative error does not exceed 0.0428 or not more than 4.28%.

## 4 Conclusion

The main objective of this work was the ability to predict the dynamics of the trade balance between Ukraine and the EU using various adequate mathematical models of time series. And as a result of the research, this goal has been achieved. Note that the Ukraine-EU trade balance is consistently negative. Since January 1, 2016, Ukraine and the EU have entered into an agreement on an expanded free trade zone (Deep and Comprehensive Free Trade Area). Thanks to this agreement, the EU has become the largest trading partner for Ukraine. At the same time, quotas and the requirement to comply with European standards significantly limit the export of Ukrainian goods to the EU. This is clearly seen from the studies: only in the first month of the year is the Ukraine-EU trade balance positive. That is, Ukraine very quickly "chooses" its quotas for the supply of export goods. The Ukrainian government intends to revise the agreement Deep and Comprehensive Free Trade Area. Once again, as a result of the research conducted in this work, it can be concluded that it is possible to build models of time series that are quite acceptable for predicting the trade balance. Therefore – the goals of the work are achieved.

## References

1. J. Vrbka, Z. Rowland, P. Šuleř, Comparison of neural networks and regression time series in estimating the development of the EU and the PRC trade balance. *SHS Web of Conferences*: *Innovative Economic Symposium 2018 – Milestones and Trends of World Economy*, **61**, 01031 (2019)
2. G.V. Golubova, The Statistical Analysis and Forecasting of Ukraine's Export. *Statistics of the Ukraine*, **2**, 6-12 (2018)
3. G.E.P. Box, G.M. Jenkins, Time Series Analysis. *Forecasting and Control* (1976)
4. C. Dzitsaki, Forecast of SARIMA models: An application to unemployment rates of Greece. *American Journal of Applied Mathematics and Statistics*, **4**, 5, 136-148 (2016)
5. N. Dzitsakis, P. Klareglon, Forecasting Unemployment Rates in USA Using Box-Jenkins Methodology. *Internathional Journal of Economics and Financial Issues*, **8(**1), 9-20 (2018)
6. L. Dvořáková, CZ GDP Prediction via neural networks and Box-Jenkins Method. *SHS Web of Conferences: Innovative Economic Symposium 2017 – Strategic Partnership in International Trade*, **39**, 01005 (2017)
7. Y. Kozak, V. Matskul, T. Shengelia, Mathematical methods and models for master of economics. *Practical applications* (2017)
8. The State Statistics Service of Ukraine [online], Available at: http://www.ukrstat.gov.ua (2019)

# Modern Aspects of the Motivation of the Staff of a Commercial Enterprise Through the Social Menu System

*Evelina P.* Pecherskaya[1*], *Tatiana M.* Tarasova [1]

[1]Samara State University of Economics, Soviet Army Str., 141, 443090 Samara, Russia

**Abstract.** At present, a very difficult situation is developing in the economy, which should be characterized as "personnel shortage". Every year this situation is complicated by the deteriorating demographic situation. Moreover, at present, in most companies, the incentive process is imperfect, since in most cases it does not depend on the results of the employee's work.
Well-thought-out stimulation helps to best manage the business behavior of staff, ensuring the welfare of the organization. Most domestic enterprises strive to create conditions to meet the needs of their employees and are very successful in this. Therefore, a high wage level and the offer of a full social package, training and career planning for employees, organizing corporate events, diagnosing the situation in a company with the involvement of external consultants and conducting surveys among staff gradually become familiar working conditions for employees of commercial organizations.
The article conducts a study of the incentive system for the labor activity of the "Fix Price" sales department, which was carried out on the basis of an analysis of the company's internal regulatory documents and by questioning the employees. This study revealed the main problems in the functioning of the current system of motivation, on the basis of which measures were proposed to improve the personnel incentive system and a model of external motivation of staff was proposed.
**Keywords:** staff motivation, staff incentive system, social package, staff hunger, wages.

## 1 Introduction

The relevance of the research lies in the fact that in a market economy, professional skills and personal qualities of employees help organizations gain an advantage over competitors [1]. Employees provide an enterprise profit, which can be significantly increased by paying sufficient attention to staff development, building loyalty to the organization in it, creating an atmosphere within the team suitable for successful work. Thus, staff incentives are the foundation for ensuring efficient use of resources [2].

* Corresponding author: pecherskaya@sseu.ru

Stimulation is the indicator that is able to objectively demonstrate the initiative of staff, the development of its potential. If employees have not enough incentives to work better, more efficiently, there is no clear focus on the results of the activity of the whole enterprise and each individual employee, then there will be no interest in work and the desire to work for the benefit of the company's owners under any circumstances [3].

Difficulties in the development and implementation of effective staff incentives are associated with constantly changing market conditions that have a significant impact on the company and its individual employees [4].

## 2 Problem Statement

At present, a very difficult situation is developing in the economy, which should be characterized as a "personnel shortage". There is an acute shortage of highly qualified specialists in most industries. Every year this situation is complicated by the deteriorating demographic situation. Moreover, at present, in most companies, the incentive process is imperfect, since in most cases it does not depend on the results of the employee's work [5]. Often, employees are paid a fixed salary; bonuses are issued on significant dates (excluding the employee's contribution to the work of the enterprise). In these criteria, a well-designed incentive process is the main element of the personnel management system, especially for fast-growing and dynamically developing companies [6].

## 3 Research Questions

Competently thought-out incentives help to best guide the business behavior of staff, ensuring the welfare of the organization. Most domestic enterprises strive to create conditions to meet the needs of their employees and are very successful in this. Therefore, a high wage level and the offer of a full social package, training and career planning for employees, organizing corporate events, diagnosing the situation in the company with the involvement of external consultants and conducting surveys among staff gradually become familiar working conditions for employees of commercial organizations [7].

The purpose of the research is to study the system of personnel incentives in the enterprise, conduct analysis and develop recommendations for its improvement.

The function of motivation is that it influences the labor collective of an enterprise in the form of incentives for effective work, social influence, collective individual incentive measures. These forms of influence intensify the work of the subjects of management; increase the efficiency of the entire system of management of an enterprise and organization.

## 4 Methods

The analysis of the incentive system for the organization's personnel will use the example of the Best Price LLC limited liability company, the FixPrice department of the city district of Chapayevsk.

The FixPrice chain of stores operates in the "fixed price store" format, also known as variety store, pound shop, dollar store and is a universal retail store selling a wide range of both food and non-food items at the same price.

Consumers of a store are citizens aged from 7 years to 70 years, with different levels of up-coming. On average, the FixPrice store serves up to 1,000 thousand consumers per day.

Traditionally it is believed that the main potential of any company is in personnel, since no organization can achieve its goals and survive in the competition without the skilled workers it needs.

Figure 1 shows the personnel management system, LLC "Best Price". As can be seen from the figure, the personnel management system consists of the following elements: personnel selection, personnel assessment, personnel placement, personnel adaptation, personnel training.

**Fig. 1.** Personnel management system of Best Price LLC

Source: Authors.

The study of the incentive system for the labor activity of the "FixPrice" sales department was carried out in two ways.

Personnel incentives were studied on the basis of an analysis of the company's internal regulatory documents: orders and directives of the general director, internal labor regulations of employees of the sales department, job descriptions and other local instructions. Analyzing the documents of the enterprise, attention was drawn to the following factors: wages; material benefits; intangible benefits; negative incentives (punishments); staff turnover.

The second way to study the incentive system for employees of the sales department "FixPrice", used in the work, is the questionnaire survey of employees using a specially designed questionnaire. The questionnaire included 29 questions. With the help of this questionnaire 20 people were interviewed, employees of the sales department. The questionnaire was anonymous and was not conducted in the presence of management. When drafting the questionnaire, it was taken into account that there is a set of external conditions in the context of work: remuneration (salary); job security; working conditions; status; the quality of leadership and intra-company relations between colleagues, with superiors, subordinates. The presence of these conditions, satisfying the individual, does not necessarily stimulate his work activity. However, their absence affects the satisfaction of employees, as they need, at least, to maintain a level that can be called a lack of dissatisfaction. Therefore, the external conditions in this theory are called dissatisfaction factors, or hygienic factors.

When selecting the questionnaire questions, it was also taken into account that there is a system of internal conditions related to the very content of the labor process: achievement; confession; a responsibility; advancement; proper work; growth opportunity. The absence of these conditions is not an indication of a large degree of dissatisfaction. However, if they are present in the labor process, they give it strong motivational incentives, resulting in a good job performance.

Next, we present a qualitative and quantitative analysis of the results obtained on the motivation system in the sales department of FixPrice.

Questioning of employees of the sales department «FixPrice». To the question "Is your work interesting for you?" 60% of employees answered: "Yes", 10% of employees answered: "Rather yes", 30% of employees of the sales department answered "Rather no". Among the frequent reasons for the lack of interest in the work were the following: "tired", "boring, everything has been known for a long time", "I do not expect anything good". Other employees are interested in work, because "communicate with interesting people", "constantly learn something new", "up-to-date on new products of consumer equipment". In our opinion, there is a negative trend of loss of interest in the work of employees of the sales department.

Answering the question "Do you perceive your work as something important, valuable and worthwhile?" The answers were divided as follows. "Yes" answered 45% of employees in the sales department. "More likely yes" was chosen by 25% of workers. "Rather not," responded 25% of workers. One worker does not know how to perceive his work as something important and valuable or worthless, useless. Not a single worker perceives their work as something worthless and useless. Most workers consider their work effective and advisory. These are 65% of employees. "Rather not," said 35% of the workers.

70% of workers believe that work consists of several frequently repeated operations, it is monotonous. And only 30% of workers were able to say that work consists of many rarely repeated operations, it is diverse; you can independently plan the work and determine its mode. It turns out that monotony is a factor that can cause (already causes) a growth of dissatisfaction among workers.

To the question "Do you often have to work under conditions of increased workload, with overwork, with over fulfilment of the norms?" Workers answered as follows: "often" – 40%, "sometimes" – 45%, "seldom" – 10%, never "- 5%. It can be concluded that the work of staff is stressful.

45% of respondents answered that they are satisfied with the work schedule and 55% of respondents answered that they are dissatisfied with their work schedule. In order to increase job satisfaction, a change in work schedule is necessary.

25% of workers believe that their working conditions are comfortable, the rest of the respondents answered "no" and "rather no" (75% of employees). A sufficiently large percentage of workers are dissatisfied with working conditions.

Most workers do not see the connection between the results of their work and their salaries. And this is despite the fact that in the sales department there is a bonus for over fulfilling the sales norms. According to the theory of expectation, the worker should have a stable idea that the results of his work depend on his efforts, that certain results for him result from his work, and also that the results he obtained ultimately have value for him. This is not happening in FixPrice LLC. In the absence of these conditions, or even one of them, the incentive process will be extremely complex or even unworkable.

We list the most important labor incentives for the employees of the "FixPrice" sales department (in descending order): remuneration, working conditions, relations with management, promotion, relationships with colleagues, professional development and self-actualization, satisfaction with the work done.

Employees of a commercial enterprise were asked to list other methods of additional incentives used in this enterprise, but no one took advantage of this opportunity.

Most of the employees of the FixPrice sales department noted that they were completely unhappy with the amount of wages – 65%. The other part of the workers (25%) wage "doesn't satisfy". And only 10% of employees chose the answer "rather satisfied". There were no workers, who were completely satisfied with the amount of wages, at this trade enterprise.

To the question "Are employees always given objective and deserved thanks?" The employees answered as follows: yes, rather – 40%, I don't know – 20%, probably not – 25%, no – 15%. Considering an unfair or unequal assessment of his work, a person loses the motivation for active creative, from the point of view of the organization, actions that can lead to many negative consequences.

20% of the employees of a commercial enterprise believe that the management quite objectively evaluates the results of their work. 20% of workers argue that the assessment of their work is more objective than subjective. 15% of employees have not yet understood the essence of the assessment of their work by management and responded to the question: "I do not know." 35% of workers are confident that the assessment of their work is more biased than objective. 10% of workers noted that the assessment of their work is completely biased. There is a rather negative trend. According to the theory of equality, one of the constant aspirations of people is the desire to receive a fair assessment of their actions. If justice is violated, if individual members of the organization receive an undeservedly high rating and reward, then the person feels offended, and this leads to frustration and dissatisfaction. Considering an unfair and unequal assessment of his work, a person loses his motivation for active creative, in terms of the goals of the organization, actions, which leads to many negative consequences. For example, a person may decide for himself that it is necessary to reduce costs, that he does not need to work much, to expend great efforts; criminal attempts to compensate for underpayment are possible, a person can leave the organization, etc.

75% of employees said that the management feel maximum supervision, control over their work. 25% of employees expressed their opinion on the average level of supervision and control over their work by management. And 0% of employees oversight, management control identified as minimal.

To the question "The results of your work depend on the degree of your efforts?" Half of the workers answered that rather yes – 50% of employees. The remaining half of the workers are confident that the results of their work do not depend on the degree of their efforts.

To the question "Is the staff" punished with a ruble "fairly, depriving bonuses, making cash deductions from wages, fines, collecting shortages, etc.?" The following answers were received: 5% of employees answered: "I completely agree", 10% workers – "rather agree", 15% – "I do not know", 60% – "rather disagree", 10% – "completely disagree". It turns out that 70% of workers do not agree with the fairness of material penalties.

The opinions listed above allow us to conclude that the majority of workers are not convinced either of fair remuneration or fair disciplinary action, which does not best affect their motivation.

## 5 Findings

An analysis of internal documents made it possible to draw the following conclusions about the incentive system for employees of the sales department of FixPrice.

In the Best Price LLC in order to improve the staff incentive system, the following activities were proposed and carried out:

**Stage 1.** At the first stage – determination of tariff rates for various categories of personnel, pre-payments and allowances, determination of the procedure for calculating and the size of collective bonuses and payments from profits.

**Stage 2.** At the second stage, a collective agreement was concluded between the administration of the company and employees.

The collective agreement enshrines: the size of the minimum tariff rate; tariff coefficients for all categories of employees; the procedure for compensation of wages in terms of inflation; forms of participation of employees in the incomes of the enterprise (collective bonuses, payments from profit); forms and sizes of social benefits and the procedure for their provision.

**Stage 3.** It was proposed to expand the functional responsibilities of the director:

It was proposed to include in the duties of the deputy director of the sales department the following functions: participation in the development of incentives; overall management of the employee incentive process; motivational management of managers; participation in the development and implementation of working standards (service standards, standards of relationships in a team); collection and processing of information about the results of shifts.

The company's accountant, in our opinion, should act as a functional unit or technical support unit for the personnel incentive process. The functions of the accountant of the companionable company "FixPrice" we proposed to include the following:

- participation in the development of a system of incentives for labor activity of personnel;
- planning costs associated with staff incentives;
- implementation of material remuneration of employees;
- collection and analysis of information on financial performance enterprises, its services, individual employees;
- determining the effectiveness of the incentive effect on employees' work activity.

**Stage 4.** Changing the wage system. It was proposed to change the wage system in the department of FixPrice. For line personnel (except for the category of workers), it took the following form (Figure 2).

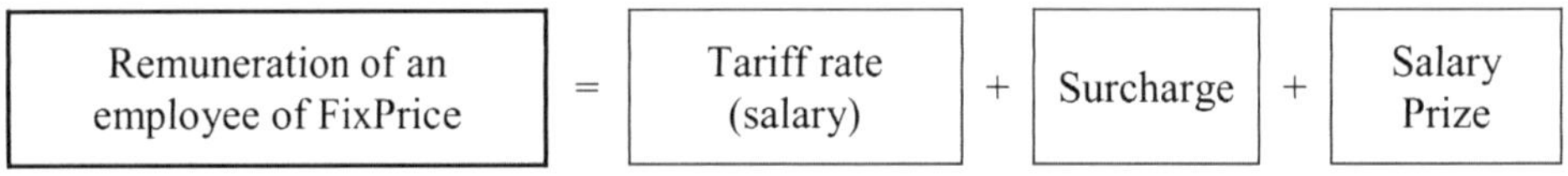


**Fig. 2.** The changed wage structure of workers of the companionable company "FixPrice"

Source: Authors.

The employee is actually paid time on the basis of the tariff rate.

Allowance – a stimulating payment to the tariff part, bearing a permanent or temporary character for irregular working hours, work experience, for high professional skills, high achievements in work, for intensity and intensity of work, for doing particularly important work, etc. Table 1 presents the sizes of the proposed surcharges for line staff.

**Table 1.** The size of the proposed allowances to the tariff salary of employees of the "FixPrice"

| Offered surcharges | Size of surcharge | Condition of payment of surcharge |
|---|---|---|
| 1. Premium for work experience | 10% of the salary | 2 years of uninterrupted work experience inthe "FixPrice" |
| | 20% of the salary | 3 years of uninterrupted work experience inthe "FixPrice" |
| | 30% of the salary | 4 years of uninterrupted work experience inthe "FixPrice" |
| | 40% of the salary | 5 years of uninterrupted work experience inthe "FixPrice" |
| 2. Surcharge for the level of education | 20% of the salary | The presence of higher professional education, confirmed by assigning the person who successfully passed the final certification, qualification "Master" or "Chartered Specialist" in the financial and economic field |
| | 10% of the salary | The presence of unfinished higher professional education in the financial and economic field is not less than 3 courses |

Source: Authors.

In FixPrice for line personnel (except for workers), this rate of reward is used as the growth in sales. Bonus conditions – the excess of the base sales of the month. The figure of 600 thousand rubles was taken as the base sales volume of the month. When this number was exceeded, the sales consultants and cashiers received a bonus of 50% of their salary (tariff rate).

The author proposed to divide the basic sales volume of the month into 3 shifts of workers, and take into account the labor contribution not of the entire sales department, but the labor contribution of each work shift. This introduces certain difficulties in fixing financial results, but this is a healthy element of competition in the work of staff.

In the new bonus system, the figure of 200 thousand rubles was taken as the base sales of the month. If this number is exceeded, the working shift of sales assistants is completed, and the cashier receives a premium of 60% of his salary (tariff rate). That is, the value of the premium for the implementation of sales volume was increased by 10%. But since there are seasonal sales fluctuations at trade enterprises, in certain months it is allowed to reduce the base output to 170 thousand. The data on the output, by the author, is set somewhat higher than the employee's existing capabilities to encourage them to move and develop.

The payment of management personnel was proposed to be built in the same way as the line staff. Table 2 presents the size of the proposed overhead for management personnel.

**Table 2**. The size of the proposed allowances to the tariff salary of the managerial staff of LLC "FixPrice"

| Allowances | Amount of Allowance | Condition of Allowance Payment |
|---|---|---|
| 1. Premium for work experience | 20% of the salary | 2 years of uninterrupted work experience in the "FixPrice" |
| | 30% of the salary | 3 years of uninterrupted work experience in the "FixPrice" |
| | 40% of the salary | 4 years of uninterrupted work experience in the "FixPrice" |
| | 50% of the salary | 5 years of uninterrupted work experience in the "FixPrice" |
| 2. Premium for management level | 50% of the salary | level 1 - leader |
| | 40% of the salary | level 2 - deputy heads |
| | 20% of the salary | Level 3 - Head of Direction |

Source: Authors.

The premium for the level of education for management personnel was not introduced, so initially people who have higher professional education are employed

For management personnel, the bonus rate is used – the complexity and intensity of work. Bonus conditions – the excess of the base sales of the month. When base sales are exceeded, management personnel FixPrice receives a bonus of 70% of their salary. The basic sales volume of the month is set at 500 thousand rubles.

All proposed changes will be valid for a year, after which the results of the effectiveness of the proposed measures will be summarized. Most employees of the sales department immediately noticed the difference between the old and the new staff incentive system.

**Stage 5.** At this stage, the question of the distribution of benefits among employees was addressed. Along with the employee benefits and guarantees established by the legislation of the Russian Federation, it is necessary to use benefits and compensations that improve the social security of employees, such as:

Benefits related to working time (flexible working hours, the provision of additional days of rest, etc.);

- various additional payments by employees in connection with anniversaries, holidays, vacations, childbirth, etc.;
- organizing preferential or fully paid meals for employees;

- providing workers with transport or compensation for the cost of travel to work;
- the provision of vouchers to employees and their families in health and recreation facilities; organization of recreation for children of the company's employees; pay for training employees and their children;
- payment of children's institutions to employees of the company; additional non-state provision of employees;
- insurance of employees.

In addition, social benefits include various payments to employees for personal celebrations, say, on the occasion of 10-, 20-, 30-year-olds, etc. anniversaries of official activity at the enterprise in combination with special leave. Moreover, the size of payments and the duration of additional leave may depend on the length of service in the enterprise.

Providing employees with an additional old-age pension also motivates employees. Staff turnover will be reduced, because employees will hold on to their jobs, knowing that after retirement they will receive additional support from the enterprise, which, in turn, will help them maintain their standard of living. Serious motivating role carries in itself the provision of employees the right to use social services. This kind of social services can be divided into four groups: catering staff; provision of housing and benefits for its acquisition; organization of medical care; social advisory assistance.

## 6 Conclusion

The proposed model of social policy in the organization can be called a cafeteria. Its essence lies in the fact that the organization provides employees with a certain basket of social benefits for a certain amount, from which they can independently choose certain social benefits and vary them within the existing budget, that is, make up for themselves the so-called "social menu".

To create high motivation among employees to achieve the company's goals is the main task of all managers, so the managers of Best Price LLC should be familiar with the basic theoretical concepts and practical approaches to motivating and encouraging staff to consciously apply them in their daily activities. At such a seminar it is necessary to consider the following questions:

- classical and modern theories of motivation;
- methods of diagnosis of labor motivation;
- types and forms of labor incentives; principles and methods for the formation of the fixed and variable parts of wages;
- development of a social package, etc.

The main goal of the seminar is to mobilize and set up the management team for active and fruitful work. Without the direct participation of all managers, the improvement of the incentive system will be ineffective.

Thus, the proposed measures will improve the staff incentive system in the sales department "FixPrice", and the proposed external motivation model will not only take into account the permanent part of wages, but also make the employee's income dependent on labor productivity and the final results of the sales department.

And in conclusion, we consider the possible risks that the management of the organization may face in the implementation of the author's recommendations.

Thus, incentive measures were proposed to retain staff in the future. The developed incentive methods will allow you to purposefully spend the funds allocated by the budget, to more clearly determine the items of expenditure on staff incentives. The active combination of material and non-material forms of incentives allows you to vary the size, frequency, variety and targeting of rewards.

The system of stimulating the work of workers has become personally oriented, so its foundation, above all, laid not only the results of the work of workers, but also their expectations in recognition of merit, etc., which confirms the presence of effective feedback in the organization.

## References

1. H. Aguinis, H. Joo, R.K. Gottfredson, What monetary rewards can and cannot do: How to show employees the money. *Business Horizons*, **56**, 241-249 (2013)
2. N. Commeiras, A. Loubes, A. Bories, Identification of organizational socialization tactics: The case of sales and marketing trainees in higher education. *European Management Journal*, **31**(2),164-178 (2013)
3. Y.K. Kuo, Organizational commitment in an intense competition environment. *Industrial Management and Data Systems*, **113**(1), 39-56 (2013)
4. L.C. Barton, V. Ambrosini, The moderating effect of organizational change cynicism on middle manager strategy commitment. *International Journal of Human Resource Management*, **24**(4), 721-746 (2013)
5. H. Yamamoto, The relationship between employees' perceptions of human resource management and their retention: From the viewpoint of attitudes toward job specialties. *International Journal of Human Resource Management*, **24**(4), 747-767 (2013)
6. Y. Cheng, I.S. Chen, C.J. Chen, H. Burr, H.M. Hasselhorn, The influence of age on the distribution of self-rated health, burnout and their associations with psychosocial work conditions. *Journal of Psychosomatic Research*, **74**(3), 213-220 (2013)
7. D.S. Schultz, V.M. Brabender, More challenges since Wikipedia: The effects of exposure to internet information about the Rorschach on selected comprehensive system variables. *Journal of Personality Assessment*, **95**(2), 149-158 (2013)

# Scientific instrumentarium for diagnostics of staffing at the Ukrainian machine-building enterprises: marketing approach

*Kateryna* Pryakhina[1*], *Andryi* Pochtovyuk[1]

[1]Kremenchuk Mykhailo Ostrohradskyi National University, Pershotravneva Street 20, 39600 Kremenchuk, Ukraine

**Abstract.** The article deals with scientific instrumentarium for diagnostics of staffing of Ukraininan machine-building enterprisesthanks to a set of methods and techniques of scientific knowledge (analysis to determine the conformity of production and economic objectives to staffing, Synthesis to investigate the quantitative dynamics of staff by typing posts, assessments to identify the inadequate level of professional and qualification correspondenceof the employee with the requirements of the employer), which allows to determine the causes of the problems of staffing.This will allow in the future to form a system of scientific and practical measures to bring the professional qualifications of the workers into line with the needs of employers in accordance with the portfolio of the orders of the enterprise.
**Key words:** staffing, marketing approach, machine-building enterprises, professional and qualification correspondence, conformity class

## 1 Introduction

The state of the development of machine-building enterprises is directly illustrated by the economic vector of the national economy and indirectly by the social basis in Ukrainian society.In order to meet the high mission of the society entrusted to machine building, enterprises of the industry should be guided by the external market environment, that is, in time to adapt to consumer preferences.Taking into account the constant nature of the changes in the market environment, the activities of machine-building enterprises should constantly adjust their own resources to the technological requirements of order execution.This refers to the component of activity as personnel management. Staffing is one of the most important aspects of human resources management, which needs new approaches to improve performance in changing business conditions. Marketing is one of the effective aspects of the formation of effective staffing in the modern conditions. The marketing approach to staffing means, that profession and qualification at the same timeare characteristics of the personnel needs of employers and proposals from the employees. Staffing of the enterprise is a complex of measures, which allows maximum achievement of professional-qualification balance of the needs of employers and the qualities of employees.

* Corresponding author: katerinapryakhina@gmail.com

## 2 Data and Methods

To solve the tasks, a wide range of theoretical and empirical methods of scientific knowledge was used: scientific abstraction – in determining the categorical apparatus of the staff of machine-building enterprises on the principles of marketing; diagnostic – in the building tools for personnel support; deductions and analogies – when combining the general (branch) and individual (enterprise subdivision) levels of research; combinatorics and grouping – in determining the qualification characteristics; synthesis – in the classification of types of positions on the basis of the vector and intensity; convergence – in determining the level of professional and qualification correspondence of the qualities of employees to the requirements of the employer, respectively, occupied position; graphic image – for a visual reflection of the results of the study.

The information and empirical base is the data of the State Statistics Service of Ukraine, reports of the leading machine-building enterprises of Ukraine, monographic research materials of domestic and foreign scientists, Internet resources. Sources of initial information are the materials of their own research. The legal basis for scientific research is the Laws of Ukraine, the Decree of the Cabinet of Ministers of Ukraine, the Economic Code of Ukraine.

## 3 Results

The choice of diagnostics is due to the fact that its object may be: the enterprise as a whole as a production-economic system, its elements; operation of separate units, production functions, activities. Diagnostic analysis determines the objective state of the selected parameters in the enterprise in accordance with the goals and objectives set, the study of deviations from the normal state of the object of diagnosis, the causes of these deviations, the duration of their course, their interpretation for the formation of conclusions and recommendations. The method of diagnosis – a system of theoretical and cognitive categories, scientific instruments and regulatory principles of research activities of subjects of management. The basis of the diagnostics of staffing is the triad of methods: analysis, synthesis and evaluation. In this case, the diagnosis involves an assessment of staffing. Consequently, the scientific instrumentarium for the diagnostics of human resources provision of enterprises consists of independent and simultaneously interconnected three stages, envisaging 24 procedures (Fig.1).

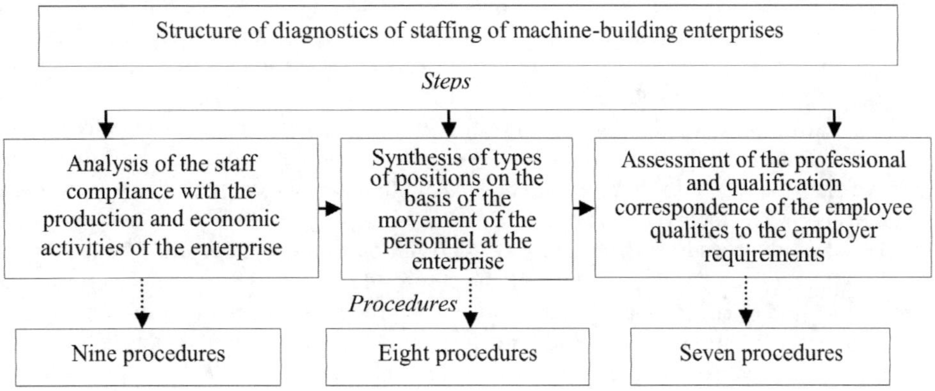


**Fig. 1.** Structure of diagnostics of staffing of machine-building enterprises

Source: Authors.

The first stage of the diagnostic assessment involves an analysis of the staff compliance with the production and economic activities of the enterprise and consists of nine procedures. The first procedure determines the investigated enterprises. Personnel support is unexplored in terms of sector specificity, because the work of personnel requires specific properties for the industry. The specialization in in machine building is determined by the profile of machine-building enterprises.

The second procedure is to determine the time limits of the study. If $i$ is a period of research (years), it is determined $\{2011-2018\}$.

The third procedure is to determine the list of indicators. The personnel policy of the enterprise is derived from its production activities and is aimed at increasing the profit of the enterprise. The study has selected absolute indicators that characterize the financial and economic activities of the companies – "net revenue from sales of products (goods and services)" and marked with a symbol $W_i$. Indicators that determine personnel policy are "payroll fund", "the average number of employees" and "share of the wage fund in the volume of sales". Payroll fund be symbolized $F_i$, the average number of employees – $S_i$, and share of the wage fund in the volume of sales – $d_i$.

The fourth procedure is the justification for evaluating the change in the results of economic activity of enterprises by means of statistical indicators: average level of interval series of dynamics with even intervals $\bar{Y}$, indicator of the rate of increase $T^{л}_{пр.}$ and the average indicator of the rate of increase $\bar{T}^{л}_{пр.}$. For analysis, such synthetic indicators are used: $T^{l}_{p}\,W$ – "rate of increase of net revenue from sales over the period", $T^{л}_{пр.}\,F_i$, – "rate of increase the payroll fund", $T^{л}_{пр.}S_i$ – "rate of increase of the average number of employees", $T^{л}_{пр.}d_i$ – "rate of increase of the share of the wage fund in the volume of sales"; $\bar{T}^{л}_{пр.}\,W_i$ – "average growth rate of net revenue from sales over the period", $\bar{T}^{л}_{пр.}\,S_i$ – "average growth rate of the average number of employees", $\bar{T}^{л}_{пр.}\,F_i$ – "average growth rate of the payroll fund"; $\bar{Y}(W_i)$ – "the average level of the interval series of dynamics of net revenue from sales of products", $\bar{Y}(F_i)$ – "the average level of the interval series of dynamics the payroll fund", $\bar{Y}(S_i)$ – "the average level of the interval series of dynamics the average number of employees", $\bar{Y}(d_i)$ – "the average level of the interval series of dynamics the wage fund in the volume of sales".

The fifth procedure is the definition of the compliance of human resources with the production and economic activity of the enterprise. With regard to economic expediency, the increase in sales proceeds leads to an increase in the payroll fund and the average number of employees and vice versa. As an example of three or more sub-sector enterprises, it is necessary to analyze the availability of this dependence. The analysis will prove the same tendency of the values: an increase in sales revenue will lead to an increase in the number of employees of the enterprise, which in turn will reflect such a trend towards the wage fund and the wage fund in the volume of sales, or it will detect a lack of correlation of indicators.

The sixth procedure is the selection of the structural unit of the enterprise for research. Taking into account the identity of the indicators of the sub-sector of machine-building complex and the marketing approach to personnel provision, which involves the personalized relations between employees and employer.

The seventh procedure provides for the description of the nomenclature of the products of the structural unit in the context of the needs of the personnel of the corresponding profession and qualification. Nomenclature of products reflects the full spectrum of technological operations, which respectively require personnel support. At the same time, taking into account sector specifics, each machine-building enterprise has its own portfolio of orders in accordance with its capacity and technological process. In today's management environment, the portfolio of orders of enterprise is constantly changing, the company can have both one order and $n$. Therefore, an enterprise portfolio can be defined:

$$P = \sum_{1}^{n} x_n, \tag{1}$$

where, P – portfolio of enterprise orders, $x_n$ – order of the enterprise, $n$ – serial number of the contract in the period. The portfolio of orders consists of predefined objects for the production of which the availability of resources and factors of production is required. In view of the marketing concept for efficient and balanced activity of the enterprise, this resource potential should be formed for each individual order.

The eighth procedure determines the calculation of the needs of the personnel of the relevant profession and qualification, as an example of two or more orders of the company that were in the portfolio of orders. This calculation was carried out with the help of a technological card naming products, which characterizes the interaction of technological processes. Each operation involves availability professionally qualified personnel and calculation of labor intensity.

Thus, the ninth, the last procedure of the first stage of diagnosis characterizes the generalization of the possibility of forming a staffing of the enterprise in accordance with contracts that make up a portfolio of orders.

The second stage of the diagnostic assessment is the synthesis of types of positions on the basis of the movement of the personnel at the enterprise, which consists of eight procedures. The tenth procedure determines the choice of a structural unit for further research, because in this structural unit main technological operations are executed (first stage). At the same time, the choice of the shop is justified by the use of methods of analogy and deduction.

There is a refinement from the general to the separate, that is, from the research of enterprises in the sub branch or industry to the structural unit, which ensures the implementation of major technological processes. The staffing of the enterprise is formed taking into account the specifics of its activity and the personified nature of labor relations. The method of analogy in the formation of staffing of the structural unit allows to generalize the staffing of enterprises in the sub-sector as a whole.

The eleventh procedure involves determining the range of indicators that demonstrate the movement of the staff in the structural unit [1]. Indicators are indicated by: $n_i^j$– "the number of staff to the end *i* period", $\hat{n}_i^j$ – "the number of accepted employees during *i* period", $\check{n}_i^j$ – "the number of dismissed workers within *i* period", $d_{n_i^j}$ – "part *j* categories of employees in accordance with the categories of the Ukrainian Classifier of professions in the total number of employees of the structural unit" [2], where $i$ – "the period of the studied period (calendar year)", the value is reduced to the last two digits of the year, so for example $i$ = {11–18}. Scope of definition *j* categories equals *j* = {4.1, 4.2, 4.3, 4.4, 4.7, 4.8, 4.9}, 4.1-4.9 – sections of the Ukrainian Classifier of professions.

The twelfth procedure determines the synthetic indicators in accordance with the absolute regarding the movement of the structural unit : $\text{Т}^{л}_{пр}N_i$ – "chain rate of increase of staff", $\text{Т}^{л}_{пр}\widehat{N}_i$ –"chain rate of increase the number of accepted employees", $\text{Т}^{л}_{пр}\check{N}_i$ – "chain rate of increase the number of dismissed workers", $\bar{T}^{л}_{пр.}N_i$ – "the average growth rate of staff", $\text{Т}^{л}_{пр}\widehat{N}_i$ – "the average growth rate of the number of accepted employees", $\text{Т}^{л}_{пр}\check{N}_i$ – "the average growth rate of the number of dismissed workers", $\bar{Y}_n$ – "the average level of the interval series of staff, $\bar{Y}_{\hat{n}}$ – "the average level of the interval series of the number of accepted employees", $\bar{Y}_{\check{n}}$ – "the average level of the interval series of the number of dismissed workers".

The thirteenth procedure is a characteristic of the overall dynamics of the number of staff of the structural unit. The total number of staff of the unit is determined by the formula: $N_i = \sum_{j=4.1}^{4.9} n_i^j$, total number of accepted employees $\widehat{N}_i = \sum_{j=4.1}^{4.9} \hat{n}_{i,}^j$ total number of dismissed workers $\check{N}_i = \sum_{j=4.1}^{4.9} \check{n}_i^j$. Consequently, the total number is calculated according to each post in terms of categories.

The fourteenth procedure determines the identification of trends in the movement of personnel of a structural unit in the context of positions in accordance with the categories of professions the Ukrainian Classifier of professions. For example, the number of posts in category 4.7 – "Electro-gas welder at the automobile and semi-automobile" at the end of the year 151 persons, while during the year 108 people were released, and 50 people were taken.

The fifteenth procedure is typing the positions of the unit in accordance with the signs of the movement of personnel: the vector and intensity. For the analysis of the personnel in the context of posts, an indicator is chosen – the movement of personnel, which involves two signs of vector and intensity. The vector is defined by the "change in the ratio of occupied and dismissed workers", the intensity – the rate of change in the number of adopted and dismissed personnel. The sixteenth procedure involves the allocation of four types of posts. It is synthesized four types of positions according to the indexes of the staff turnover (vector, intensity). The first type means no dynamics; the second – $\hat{n}$ (number of employed workers) = $\breve{n}$ (number of dismissed workers), the third – $\hat{n} < \breve{n}$, the fourth – $\hat{n} > \breve{n}$.

Consequently, the seventeenth, the last procedure of the second stage of the diagnosis is a synthesis of trends in the dynamics of staffing of the unit. This procedure involves determining the dynamics of the state by category in accordance with each post and the vector and the intensity of movement of staff in accordance with the type.

The third stage of the diagnostic is an assessment of the professional and qualification correspondence of the employee qualities to the employer requirements that consists of seven procedures. The need for an employer in the staff has two basic characteristics: "profession" and "qualification".

Eighteenth procedure – to determine the level of the professional and qualification correspondence of the employee professional qualities ($\beta_x$) to the needs of the employer ($\alpha_x$), and the qualifications of the employee ($\beta_y$) to the needs of the employer ($\alpha_y$). Symbol $\alpha$ means the position occupied by the worker, $\alpha_x$ – requirements of the employer on the qualifications of the employee in relation to the profession in accordance with the position occupied, $\alpha_y$ – requirements of the employer on the qualifications of the employee in terms of qualifications in accordance with the position occupied, $\beta_x$– professional qualities of an employee according to educational documents, $\beta_y$ – qualification of an employee according to the educational documents, $x$ – professional characteristics, and $y$ – qualification characteristic. There are three variants of the ratio $\alpha_x$ and $\beta_x$: «=» – full compliance; «≈» – partial compliance; «≠» – discrepancy. In the case of a qualification, then there are such correlation variants $\alpha_y$ and $\beta_y$: «>» – excess qualification; «<» – insufficient qualification; «=» – full compliance with qualifications [3].

To determine qualification characteristics, it is necessary to monitor the relationship between educational and educational qualifications levels and degrees of education, which are defined by normative documents, since the personnel who work at the enterprise received education at different times. For the first time, the educational and educational qualification levels are mentioned in the Law of Ukraine «About education» of 23.05.1991 # 1060-XII: educational levels – primary general education; basic general secondary education; complete secondary education; vocational and technical education; basic higher education; Higher Education. Educational and qualification levels: skilled worker; Junior Specialist; bachelor; specialist, master's degree [4]. This law has expired.

Resolution of the Cabinet of Ministers of Ukraine "On Approval of the Regulations on Educational Qualification Levels (Graduate Education)" of 20.01.1998 p. # 65 provides the following educational qualification levels (graduate education): skilled worker, junior specialist, bachelor, specialist and master [5]. It should be noted that the Regulation is in force, and the last changes were introduced in 2013. The letter of the Ministry of Education and Science of Ukraine of 25.04.2001 № 1 / 9-168 determined that the structure of higher education in Ukraine included the following educational qualification levels: junior specialist; bachelor; specialist; master [6]. Compared to previous documents there is no level – a skilled worker, there are also no mention of educational levels. The letter of the Ministry of Education and Science is also determined as not having expired.

The Law of Ukraine "On Higher Education" of 17.01.2002 # 2984 stipulated that the structure of higher education includes educational and educational-qualification levels. Educational levels: incomplete higher education; basic higher education; full higher education; educational qualification levels: junior specialist; bachelor; specialist, master's degree [7]. It should be noted that between the above normative documents is followed by partial continuity. Law of Ukraine "On Higher Education" of 01.07.2014 # 1556-VII It is radically different from previous documents, because it is oriented towards European standards and determines the level of higher education: the initial (short cycle), the first (bachelor), the second (master's), the third (educational-scientific), the scientific level. Higher education at each level of higher education involves the award of a corresponding degree of higher education: junior bachelor, bachelor, master, doctor of philosophy, doctor of sciences [8]. It should be noted that modern machine building requires the expansion of the spectrum of functions of junior specialists, which determines their significance in the staffing structure.

Consequently, on the basis of the analysis of normative documents, a combined system of characterization was proposed "*y*", which also uses educational levels and educational qualification levels (table 1).

**Table 1.** Combined system of characterization "*y*"

| Qualification | Definition |
|---|---|
| 1. Complete secondary education | A person who does not have an educational qualification level has completed a secondary school of three levels, that is, an unskilled worker. |
| 2. Vocational and technical education | A person who has an educational qualification level - a skilled worker, according to the Law "On Education" dated May 23, 1991, No. 1060-XII, graduated from a vocational and technical institution (college). |
| 3. Incomplete higher education | A person with an educational qualification level of a junior specialist is currently a junior bachelor in the Law on Higher Education and an initial level (in the Law "On Education" junior specialist and junior bachelor "). Training is provided by: technical colleges and colleges. |
| 4. Basic higher education | A person who has an educational qualification bachelor's degree. Provide colleges, other higher education institutions of the second-fourth accreditation level. In the current Law "On Higher Education", an educational degree obtained at the first level of higher education and awarded by a higher education institution. |
| 5. Higher Education | A person who has an educational or qualification level of a specialist or a masters degree. Preparations are provided by higher education institutions. |

Source: Authors.

Thus, according to Table. 1, five qualification characteristics are defined.

The profession is defined according to the Ukrainian Classifier of professions as the ability to perform such work, which requires a person of a certain qualification. In turn, the specialty is a complex of skills, skills and knowledge acquired during the learning process. So, for the title of the position the occupations used are given in the Classifier. At the same time, in the job description according to each post should indicate the specialty that the employee must possess. Therefore, to determine the characteristics "*x*" you need to use: CMU Resolution # 1117 of 1110.2007 "On Approval of the State List of Professions for the Training of Qualified Workers in Vocational Education Institutions" [9]; CMU Resolution # 839 of 20.07 2007 "On Approval of the List of Specialties on which the Training of

Specialists in Higher Educational Institutions under the Educational-Qualification Level of a Junior Specialist is carried out" [10]; CMU Resolution # 1719 of 13.12. 2006 "On the List of Directions for the Training of Specialists in Higher Educational Institutions for the Baccalaureate Degree" [11] and the CMU Resolution # 787 of 27.08.2010 "On Approval of the List of Specialties, according to which the training of specialists in higher educational establishments is carried out according to the educational and qualification levels of a specialist and a master's degree ", which have now become null and void [11].

The modern training of applicants is in accordance with the resolution of the Cabinet of Ministers of Ukraine of 29.04.2015, # 266 "On Approval of the List of Fields of Knowledge and Specialties for which Higher Education Institutions are Prepared" and which should be used from 2019 [12].

Consequently, professional characteristics are set according to the specialty (direction of training) specified in the education document. However, the demand for professionals is determined not by the level of training, but by how well they are prepared in the light of the requirements of employers.

The nineteenth procedure involves the allocation of 9 conformity classes: A, B, C, D, E, F, G, H, L, which can be divided into three groups. The first group of classes A, B, C is characterized by the fact that $\alpha_x = \beta_x$. Class A corresponds to the full compliance of the employer's professional qualification requirements with regard to the qualities of the worker $\alpha_x = \beta_x$ and $\alpha_y = \beta_y$. Class B is characterized $\alpha_x = \beta_x$ and $\alpha_y < \beta_y$, that is, the profession has full compliance, and the qualification of the employee exceeds the employer's requirements for the position occupied. Class C is characterized $\alpha_x = \beta_x$ and $\alpha_y > \beta_y$, means compliance with the professional requirements of the employer, but the employee has a lower qualification.

The second group of classes D, E, F is characterized by the fact that $\alpha_x \approx \beta_x$. Class D is characterized $\alpha_x \approx \beta_x$ and $\alpha_y = \beta_y$. Class E is characterized $\alpha_x \approx \beta_x$ and $\alpha_y < \beta_y$, the profession partially meets the requirements of the position, and the qualification of the employee is greater than necessary. Class F is characterized $\alpha_x \approx \beta_x$ and $\alpha_y > \beta_y$, the profession partially meets the requirements of the position, and the qualification of the employee is less than necessary.

Third group of classes G, H, L is characterized $\alpha_x \neq \beta_x$. Class G is characterized $\alpha_x \neq \beta_x$ and $\alpha_y = \beta_y$, which implies compliance with the requirements of the employer to the qualities of the work regarding qualifications and professional misconduct. Class H is characterized $\alpha_x \neq \beta_x$ and $\alpha_y < \beta_y$, provides for the lack of correspondence between the professional requirements of the employer and the qualities of the employee, and the qualifications of the employee more than required occupation. Class L is characterized $\alpha_x \neq \beta_x$ and $\alpha_y > \beta_y$, the lack of correspondence between the professional requirements of the employer and the qualities of the employee, and the qualification of the employee is less than the occupied position

The twentieth procedure – the definition of indicators to determine the level of professional qualifications compliance or inconsistency. These are indicators that are marked with symbols: $n_i^{\kappa/j}$– number of staff to the end period $i$, $k$ – conformity class {A, B, C, D, E, F, G, H, L} and $j$ = {4.1, 4.2, 4.3, 4.4, 4.7, 4.8, 4.9}, $n_i^k$ – the number of staff of the conformity class at the end period $i$, $d_{n_i^k}$ – staff share of the relevant class in the total number of employees of unit, $dn_k^j$– staff share $j$ – category in conformity class $k$. The total number of the unit staff according to the class is indicated: $\sum n_i^k$, and the total number of the staff according to the category – $\sum n_i^j$.

The twenty first procedure – the use of absolute indicators made it possible to form synthetic indicators for analysis: $T_{np.}^{л}(n_i^k)$, – "chain rate of increase the number of personnel of conformity class”, $\bar{T}^{\text{л}}_{\text{пр.}}k$ – “the average growth rate of the number of personnel of conformity class”, $\bar{Y}_k$ – “the average level of the interval series of staff of the classes”, and $k$ = {A, B, C, D, E, F, G, H, L }.

The twenty-second procedure is the implementation of the characteristics of the professional qualification of the staff of the structure unit in accordance with the classes in terms of the categories of Ukrainian Classifier of professions.

The twenty-third procedure is to find out the value of the coefficient of professional and qualification correspondence ($K_i^{pqc}$). Since staffing means one hundred percent correspondence of the interests of the employee and employer, the value $K_i^{pqc}$= 1. But the current state of staffing of Ukrainian machine-building enterprises is indicative of an imperfect HR policy. Therefore, it is proposed to introduce a scale of coefficients ($m^k$) for each conformity class (table. 2) [3].

**Table 2.** Coefficients according to the conformity class

| k | A | B | C | | D | E | F | G | H | L |
|---|---|---|---|---|---|---|---|---|---|---|
| $m^k$ | 1 | 0.7 | 0.6 | | 0.8 | 0.5 | 0.4 | 0.3 | 0.2 | 0.1 |

Source: Authors.

According to Table 2 indicator value $m^A$= 1, since this class characterizes the full compliance of the qualities of the employee with the requirements of the employer. Class D has a value of 0.8, because it describes the states in which the employee does not fully realize his skills and abilities. Class B has been assigned 0.7 because the class extends the professional qualities of the employee. Class C is characterized by $m^C$= 0,6, this class is typical for those employees who have experience in the company in accordance with the received profession, which compensates for the lack of qualifications. Class E has a coefficient of 0.5, is disadvantageous to both parties. A similar to the previous one – class F – has $m^F$= 0,4.

The most risky classes are unprofitable for the enterprise are G, H, L. Class G has a value of 0,3, because the knowledge and skills of the employee absolutely do not meet the requirements of the position, although the qualification is characterized as $\alpha_y = \beta_y$. Class H has $m^H$= 0,2. Class I has value of the coefficient 0,1. In general, the presence of employees of this class indicates problems in the staffing of the enterprise, because an employee can not perform work in accordance with his position.

Consequently, in accordance with the given coefficients, the calculation of the integral index of professional qualification correspondence in terms of expression:

$$K_i^{pqc} = \frac{\sum_k^9 m^k \cdot n_i^k}{N_i} \qquad (2)$$

The twenty-fourth procedure, the last procedure of the third stage of diagnosis, is the generalization of revealing the insufficient level of professional and qualification correspondence of the personnel by distributing the staff to the conformity class. Clarification of the value of the coefficient of professional and qualification correspondence will determine the degree of compliance of the qualities of personnel to the tasks of the enterprise.

## 4 Discussion and Conclusion

The choice of diagnostics as a tool for scientific knowledge is substantiated by the ability to identify problems, their causes and formulate recommendations for their elimination. The first stage of the diagnostic assessment involves an analysis of the staff compliance with the production and economic activities of the enterprise and consists of nine procedures. The second stage of the diagnostic assessment is the synthesis of types of positions on the basis of the movement of the personnel at the enterprise, which consists of eight procedures. The third stage of the diagnostic is an assessment of the professional and qualification correspondence of the employee qualities to the employer requirements that consists of eight procedures. The diagnostic process consists of three stages and includes 24 procedures, which are complementary and interferential.

Using deductive method and analogy has allowed to choose as the research plane the leading link in the technological process of the enterprise of the machine-building complex.

Instrumentarium of the diagnosis involves the use of a system of indicators that characterize financial and economic activity of the enterprise and personnel policy; the staff movement and the level of professional and qualification correspondence. According to the selected indicators, the synthetic ones are defined: the average level of the interval series of dynamics with equal intervals, the chain and average growth rate. Thus, the selected indicators will allow to carry out diagnostics of staffing of the domestic machine-building enterprises, to identify the problems and causes of their occurrence, which will form a way for further modernization of staffing.

## References

1. O. Zakharova, Features of estimation of movement of personnel of the enterprise. *Formation of a market economy: a collection of scientific works Special Issue in the 21st Century: Recent Trends, Social Dimension, Innovative Development,* pp. 439-452 (2012)
2. Classifier of professions DK 003: 2010, 746 (2010)
3. K. Pryakhina, Personnel: assessment of professional qualification correspondence. *Economic Sciences*, **2**(25), 10-14 (2017)
4. "On Education": Law of Ukraine dated May 23, 1991 No. 1060-XII (1991)
5. "On Approval of the Regulation on Educational-Qualification Levels (Graduate Education)": Resolution of the Cabinet of Ministers of Ukraine dated January 20, 1998 No. 65. Official Bulletin of Ukraine. 1998. No. 3. Art. 108 (1998)
6. "About educational qualification levels (degree education)": Letter of the Ministry of Education and Science of Ukraine of 25.04.2001 № 1 / 9-168. Guide. 2001. 08. № 30. (2001)
7. "On Higher Education" Law of Ukraine of January 17, 2002 No. 2984-III (2002)
8. "About higher education": Law of Ukraine dated 01.07.2014 № 1556-VII (2014)
9. "On Approval of the State List of Professions for the Training of Qualified Workers in Vocational Education Institutions": Decree of the Cabinet of Ministers of Ukraine dated September 11, 2007 No. 1117. Official Bulletin of Ukraine. No. 70. Art. 2656. p. 11 (2007)
10. "On approval of the list of specialties for which the training of specialists in higher educational institutions is carried out at the educational qualification level of a junior specialist": Resolution of the Cabinet of Ministers of Ukraine dated 20.06.2007 No. 839. Official Bulletin of Ukraine. No. 46. Art. 1877. p.23 (2007)
11. "On approval of the list of specialties for which the training of specialists in higher educational institutions is carried out at the educational and qualification levels of a specialist and master": Resolution of the Cabinet of Ministers of Ukraine dated August 27, 2010 No. 787. Government Courier. No. 175 (2010)

12. "On Approval of the List of Fields of Knowledge and Specialties under which Higher Education Institutions are Prepared": Resolution of the Cabinet of Ministers of Ukraine dated April 29, 2015 No. 266. Government Courier. No. 102 (2015)
13. O. Maslak, N. Grishko, K. Vorobiova, O. Hlazunova, M. Maslak, Developing the intra-firm technology transfer system at the industrial enterprise based on matrix approach. *Problems and Perspectives in Management*, **15**(3), 242-252 (2017)

# The Methodology of Simple Comparisons as a Replacement for Comprehensive Benchmarking in Controlling

*Pavel* Rousek[1*], *Simona* Hašková[1]

[1]Institute of Technology and Business, School of Expertness and Valuation, Okružní 517/10, 37001 České Budějovice, Czech Republic

**Abstract.** The paper aims to define an alternative to the traditional process of comprehensive benchmarking in controlling. Traditional benchmarking offers invaluable outputs, but the cost is a high process, time, and cost. The newly defined method of simple comparisons is less demanding from a process, time, and financial point of view, but the accuracy of the results is lower.
**Key words:** benchmarking, controlling, replacement

## 1 Introduction

The aim of this paper is a simplification of the traditional benchmarking process, which is complicated, lengthy, expensive, and, moreover, often difficult to interpret. This paper seeks for a method which is simpler, shorter, cheaper, and easier to interpret.

The character of contemporary economic systems can be described as adaptive depending on changes and effects of microeconomic and macroeconomic factors [1]. Growing global competition helps to liberalize national economies; therefore, companies need to set the quality of their products and services above the competition, to use better technologies than their competitors and to manage their costs below competition [2]. In other words, many companies have to try to be better, faster and cheaper than their competitors [3].

General strategic business practices are based primarily on vision, mission, values and management style [4]. What follows are the respective analyses of the internal and external environment of the organization [5]. The purpose of the analysis is to identify the internal potential of the organization and the market challenges arising from relationships with stakeholders and the trends of the market macro-environment. The strategic framework identifies the key factors to be analysed [6]. To collect the information needed for analysis, the organization uses a management information system, which consists of an internal information system and an external information system. A variety of methods and procedures are used for sorting and analysing information, with basic methods including gap analysis, SWOT analysis, benchmarking, Porter´s five market forces model, Ansoff market-products matrix, General Electric matrix, Boston Consulting Group matrix, etc. [7].

* Corresponding author: rousek@mail.vstecb.cz

This paper focuses on the benchmarking method, which is understood as a catalyst for improvement and innovation [8]. It is a tool to support the development of critical areas of business and management to achieve or exceed set goals based on learning from best practices and understanding the processes to achieve these goals.

A comparative literature survey showed that there are different types of benchmarking and many process tests models [9]. In some cases, the model has been uniquely developed to perform a particular type of benchmarking, which may present problems as confusion among users about whether they should use only a unique benchmark model that has been developed for a particular type or whether use any model for any type of benchmarking; from the user´s perspective, it may be difficult to choose the best model from the available models for the purpose given as each model differs in terms of the number of phases involved, the number of steps involved, the application, etc. [10]. Gurumurthy and Kodali [11] analysed the basic benchmarking classification schemes and thus the unique benchmarking models that are developed for each type. The conclusions of the analysis led the authors to design a universal benchmarking model that can be used for all types of benchmarking. The limitation of the model represents its general highly conceptual framework, which can lead to erroneous conclusions without previous verifying its usefulness in practice.

## 1.1 Benchmarking

According to Eschenbach and Siller [12] benchmarking is a continuous process of comparison of our products, services, processes, methods with the best products, services, processes, methods. The data for the benchmarking process comes from Activity-Based Costing and Activity-Based Management. Successful benchmarking leads to improved competitiveness or the competitive position of the organization.

The purpose is to introduce competition which should lead to improvement. This can reveal emerging differences (common and different features) and propose measures to increase efficiency and effectiveness. This method is based on the experience of others and promises a valuable explanation of the differences in performance, processes, and costs.

Types of benchmarking according to Bengt and Östblom [13]:

- Internal benchmarking is performed within a single enterprise between individual business processes, products or functions.
- External benchmarking seeks to compare and measure with another business, providing the opportunity to learn from the best in the same industry.
- Strategic benchmarking is used to improve the overall performance of an enterprise by examining a long-term strategy.
- Process benchmarking is used to make more effective use of a given process or activity within an enterprise.
- Performance benchmarking is used to determine the position of an enterprise by comparing the performance characteristics of services with other enterprises.
- Functional benchmarking is used by businesses to compare with its competitors in the industry.

## 1.2 Basic Benchmarking Parameters

The basic parameters of benchmarking should be thoroughly considered and analysed during the preparation and analysis:

- **Objects:** Products, services, processes, methods, structures, strategies, activities
- **Target values, parameters:** Performance, cost, quality, customer satisfaction, process time, storage time

- **Comparative entities, partners:** Other departments, other business segments, competitors, improved business

### 1.3 Advantages and Disadvantages of Benchmarking

Eschengach and Siller [12] present the basic advantages and profits of benchmarking:

- Use for analysis, planning, control, communication.
- Benefits of competition.
- New impulses form best practice.
- Comparing actual status and potential for improvement.
- More focus on market requirements.

Eschengach and Siller [12] present the basic disadvantages and risks of benchmarking:

- Ensuring comparable conditions to allow a proper comparison.
- Difficult to obtain detailed information about the competition.
- Time-consuming process.
- Bad results are unwelcome.

A prerequisite is that the entire process must be cost-effective. The benefits of benchmarking must exceed the cost of benchmarking. This paper aims to reduce the cost of benchmarking, reduce the consumption of time, and reduce the consumption of other resources.

### 1.4 Benchmarking Code of Ethics

The ethical context of benchmarking should be addressed in the Code of Ethics, as the benchmarking process will reveal a lot of sensitive information. APQC has developed a Benchmarking Code of Conduct defining the basic moral, ethical and legal bases of benchmarking [14]:

1. Principle of Legality.
2. Principle of Exchange.
3. Principle of Confidentiality.
4. Principle of Use.
5. Principle of Contact.
6. Principle of Preparation.
7. Principle of Completion.
8. Principle of Understanding and Agreement.

### 1.5 Cost Benchmarking

Cost benchmarking is a special form of benchmarking which aims to identify potential areas for improvement in overhead costs

Continuous efforts to compare the costs of the value-added chain with the best companies, identify the causes of the differences, and introduce corrective actions. It is one of the basic cost management tools applied in current business practice. This is a special form of benchmarking where costs are tracked along the entire value chain. In addition to cost items, the relevant process steps of the best-in-class enterprise are determined. It also aims to identify the potential for improvement in overheads.

## 2 Data and Methods

Practical experience with benchmarking methods, theoretical knowledge summarized in the above literary research, experience with teaching these methods to university students and feedback from these students were used to create this paper.

Interim conclusions were compared with real data from the Czech environment. These interim conclusions were verified based on financial statements (balance sheet, profit and loss account, cash flow) from the financial statements of selected companies.

## 3 Results

### 3.1 Simple Comparisons

A simplified process of the simple comparisons with omitted phases, which, while increasing the accuracy, complicate its execution:

**Phase 1**: Determining the comparison object

Basic comparison object is determined by what improvement should be made. Areas for improvement vary from products and services, through processes and methods to organizational structure etc.

**Phase 2:** Setting of target values

Quantitative parameter for the selected object and its target values needs to be set. Depending on the object, the quantitative parameter can be performance, cost, quantity, customer satisfaction, process time, storage time, etc.

**Phase 3:** Selection of comparative entities

Three basic examples of comparative entities are described below: market leader, market average and macroeconomic average.

**Phase 4:** Data analysis

The analysis of own data is performed as well as the analysis of data of comparative entities.

**Phase 5:** Identification of differences

Outputs of the previous phase (own results and comparative entity results) are compared, differences are identified, and their causes are found.

**Phase 6:** Implementation of measures

Negative differences are critically examined to eliminate these gaps. The proposed measures are implemented.

### 3.2 Comparisons to the Market Leader

If there is an indisputable leader in a specific market, it is advisable to learn from it. An example of such a leader is Apple for the smartphone industry, Google for internet search engine, Amazon for retail, Coca-Cola for soft drinks market, McDonald's for fast food restaurants, etc.

The advantage of this approach is the comparison to a specific entity, a specific company that is the best in its industry and learning from the best practice. The disadvantage is the potentially complicated acquisition of data on selected companies.

### 3.3 Comparisons to the Industry Average

If there is no clear pattern from which to learn, it is appropriate to compare the data to the industry average. This data is available through various applications, aggregation software, and aggregated databases.

The advantage of this solution is clear data availability. The disadvantage is the financial and time-consuming data acquisition. It is also necessary to change the interpretation of results. Instead of learning from the best enterprise, it is a comparison with an imaginary average enterprise. So, we cannot learn from the average, but we try to overcome the average.

### 3.4 Comparisons to the Macroeconomic Average

There are situations where a comparison with both the market leader and the market average is problematic. This is an example of monopoly markets where the analysed company is both a market leader and the only competitor. Then it is necessary to use the average data at the macroeconomic level.

The key advantage is the potential use in any situation. The main disadvantage is that the analysed company is being compared to an entity that is virtually created and far from the original company (different industry, different size, and different parameters).

## 4 Conclusions

Traditional benchmarking is a very beneficial discipline that uses best practice. Benchmarking outputs are therefore very valuable. On the other hand, the traditional benchmarking process is very complicated, lengthy, expensive, and often difficult to interpret. Simple comparisons seek to eliminate these disadvantages.

The simplification of the process creates an inseparable disadvantage, which is too much simplification of reality. This should be borne in mind when interpreting the results.

Therefore, the presented method of simple comparisons does not attempt to replace traditional benchmarking in its complexity. Rather, it seeks to complement a similar methodology for cases where we prefer simplicity, low-time, and low-budget instead of accuracy.

The goal of the paper was fulfilled. The methodology of simple comparisons as a replacement for comprehensive benchmarking was proposed. The alternative to the traditional process is defined. The newly defined method of simple comparisons is less demanding from a process, time, and financial point of view.

## References

1. M. Dudin, N. Lyasnikov, A. Kuznecov, I. Fedorova, Innovative transformation and transformational potential of socio-economic systems. *Middle East Journal of Scientific Research*, **17**(10), 1434-1437 (2013)
2. Z. Mohamed. *Effective management of benchmarking projects*. Routledge (2010)
3. A. McAfee, E. Brynjolfsson, T. H. Davenport, D. J. Patil, D. Barton, Big data: the management revolution. *Harvard business review,* **90**(10), 60-68 (2012)

4. P. Mirvis, B. Googins, S. Kinnicutt, Vision, mission, values. *Organizational Dynamics,* **39**(4), 316-324 (2010)
5. J. Gausemeier, A. Fink, O. Schlake, Scenario management: An approach to develop future potentials. *Technological Forecasting and Social Change*, **59**(2), 111-130 (1998)
6. G. Von Krogh, I. Nonaka, A. Manfred, Making the most of your company's knowledge: a strategic framework. *Long range planning,* **34**(4) 421-439 (2001)
7. M.M. Helms, J. Nixon. Exploring SWOT analysis–where are we now? A review of academic research from the last decade. *Journal of strategy and management*, **3**(3), 215-251 (2010)
8. S. Talluri, A benchmarking method for business-process reengineering and improvement. *International Journal of Flexible Manufacturing Systems,* **12**(4), 291-304 (2000)
9. D. Madsen, K. Slåtten, D. Johanson. The emergence and evolution of benchmarking: a management fashion perspective. *Benchmarking: An International Journal*, **24**(3) 775-805 (2017)
10. D.T. Zurell, J. Pagel, J.S. Cabral, T. Münkemüller, D. Gravel, S. Dullinger et al., Benchmarking novel approaches for modelling species range dynamics. *Global change biology*, **22**(8), 2651-2664 (2016)
11. A. Gurumurthy, R. Kodali, Benchmarking the benchmarking models. *Benchmarking: An International Journal*, **15**(3), 257-291 (2008)
12. R. Eschenbach, H. Siller. *Profesionální controlling: Koncepce a nástroje [Professional controlling: Concepts and tools].* 2nd edition. Prague: Wolters Kluwer (2012)
13. K. Bengt, S. Östblom. *Benchmarking: Jak napodobit úspěšné [Benchamarking: How to imitate successful].* 1st edition. Prague: Victoria Publishing (1995)
14. J. Nenadál, *Měření v systémech managementu jakosti [Measurement in quality management systems].* 2nd edition. Prague: Management Press (2004)

# Evaluation of the EU Policy Concerning the Basic Economic Functions of a Modern Government in a Mixed Economy

*Pavel* Rousek[1*]

[1]Institute of Technology and Business, School of Expertness and Valuation, Okružní 517/10, 37001 České Budějovice, Czechia

**Abstract.** The paper analyses the budget of the European Union in terms of its redistribution power and stabilization efficiency. The redistribution of finance from rich regions to poorer ones is an important function of public budgets. The EU budget only partially fulfils this function. Stabilizing economic development is another important function of public budgets. Also, the effectiveness of the EU budget in this area is not sufficient and could be significantly strengthened.
**Key words:** macro-economy, policy, government, mixed economy

## 1 Introduction

There are some basic economic functions of a government in a modern mixed economy. The evaluation of the correspondence of theory with the practice of the European Union is the subject of this paper.

### 1.1 Macroeconomic Government Functions

Samuelson and Nordhaus [1] define the modern mixed economy as "a combination of private enterprise working through the marketplace and government regulation, taxation, and programs." The importance of the public sector for the modern economy is also underlined by another Nobel Prize winner [2]. This shows that the importance of the public sector and large national economies are considerable. This applies to all the major economies of the globalized world, such as the United States, the European Union or China.

There are three basic macroeconomic functions of any government in a modern mixed economy [3]:

- Allocation function to achieve efficiency,
- Redistribution function to achieve equity,
- Stabilization function to achieve stability.

* Corresponding author: rousek@mail.vstecb.cz

#### *1.1.1 Allocation Function*

The allocation function means the amount and structure of funds directed to the production of public and private goods and services. Generally, it means public goods and services providing: national defence, government administration etc. These goods and services cannot be provided through the market mechanism. However, these goods are essential for consumers [4].

Broadway and Keen [5] deal with allocation efficiency and optimal direction of transfers. However, this is rather a microeconomic topic. From a macroeconomic perspective, redistribution and stabilization are more important.

#### *1.1.2 Redistribution Function*

The redistribution function means the degree of redistribution of public funds to reduce inequalities in the distribution of income and property. The tools to achieve these goals are taxes and subsidies.

Redistribution through public budgets is an important function of modern government. The influence of political rivalry on redistribution in society is investigated, for example, by [6]. In their article, they abstract from the influence of the voter-median and focus more on the rivalry of the two political parties. Other aspects of redistribution are dealt with by Wildasin, namely the results of redistribution in the labour market [7] or the interaction of redistribution and migration [8].

#### *1.1.3 Stabilization Function*

The stabilization function means the possibility of public finances to influence the main macroeconomic indicators such as stable economic growth, unemployment, inflation rate, external trade balance, etc.

The importance of stabilizing macroeconomic variables through public budgets shows the interest of experts in this area. The impact of fiscal rules on the stabilizing function of fiscal policy is examined by [9]. Their research is based on data from 21 OECD countries in almost thirty years. One of their most important findings is that „fiscal rules are unable to affect the inflation-destabilising nature of discretionary policy, if any, probably because of the higher importance of central banks in that respect“ [9].

## 2 European Union Policies

### 2.1 EU Policy of Redistribution

Following a policy of effective redistribution means allocating budget resources to the EU Member States with the low gross domestic product per capita. This allocation of funds enables the reduction of inequality in the product (or income) in various locations of the European Union.

**Table 1.** GDP per Capita in EU Member States (EUR per Capita in 2017).

| Country | GDP |
|---|---|
| Austria | 43,130.57 |
| Belgium | 38,833.46 |
| Bulgaria | 7,173.23 |
| Croatia | 11,594.40 |
| Cyprus | 23,106.04 |
| Czechia | 18,185.42 |
| Denmark | 51,059.28 |
| Estonia | 17,981.82 |
| Finland | 40,908.83 |
| France | 34,505.75 |
| Germany | 40,362.55 |
| Greece | 16,369.01 |
| Hungary | 12,586.79 |
| Ireland | 63,537.10 |
| Italy | 28,373.02 |
| Latvia | 13,611.12 |
| Lithuania | 14,442.73 |
| Luxembourg | 98,369.68 |
| Malta | 25,946.56 |
| Netherlands | 43,610.43 |
| Poland | 12,292.05 |
| Portugal | 18,758.30 |
| Romania | 9,458.17 |
| Slovakia | 15,651.27 |
| Slovenia | 20,844.71 |
| Spain | 25,109.36 |
| Sweden | 48,754.87 |
| United Kingdom | 35,947.30 |
| **European Union** | **30,190.86** |

Source: Author; Data: [10].

A high level of redistribution means a strong negative correlation between GDP per capita and EU budget expenditure. This would mean that the highest spending goes to countries with the lowest per capita GDP.

**Table 2.** Structure of the EU Budget per Capita by EU Member States (EUR per Capita in 2017).

| Country | Total | Agriculture | Regional policy | R&D[1] | Citizenship[2] | Outside the EU[3] | Admin[4] |
|---|---|---|---|---|---|---|---|
| Austria | **203.23** | 140.70 | 9.91 | 42.16 | 7.51 | 0.00 | 2.94 |
| Belgium | **653.54** | 46.86 | 27.23 | 128.28 | 22.72 | 0.00 | 428.45 |
| Bulgaria | **263.21** | 140.89 | 87.11 | 25.90 | 7.76 | 0.01 | 1.54 |
| Croatia | **156.94** | 74.11 | 64.47 | 10.25 | 2.76 | 3.23 | 2.12 |
| Cyprus | **243.98** | 87.72 | 75.25 | 56.52 | 9.69 | 0.00 | 7.04 |
| Czechia | **369.58** | 107.03 | 245.08 | 14.46 | 1.50 | 0.00 | 1.51 |
| Denmark | **257.76** | 183.05 | 13.71 | 47.77 | 2.77 | 0.00 | 10.45 |
| Estonia | **493.26** | 179.11 | 214.94 | 45.92 | 46.75 | 0.00 | 6.54 |
| Finland | **274.25** | 159.58 | 47.07 | 55.44 | 5.63 | 0.00 | 6.52 |
| France | **203.34** | 137.79 | 18.48 | 39.04 | 2.58 | 0.00 | 5.41 |
| Germany | **134.57** | 74.75 | 25.39 | 29.24 | 2.00 | 0.00 | 2.81 |
| Greece | **472.47** | 262.47 | 143.21 | 29.31 | 34.90 | 0.00 | 2.56 |
| Hungary | **410.85** | 153.91 | 217.19 | 34.48 | 3.65 | 0.00 | 1.62 |
| Ireland | **392.81** | 325.85 | 10.99 | 39.19 | 5.61 | 0.00 | 11.18 |
| Italy | **161.12** | 81.15 | 26.84 | 24.08 | 5.29 | 0.00 | 4.57 |
| Latvia | **370.93** | 198.53 | 147.19 | 13.75 | 7.28 | 0.00 | 4.18 |
| Lithuania | **539.04** | 244.78 | 234.64 | 45.41 | 10.59 | 0.00 | 3.63 |
| Luxembourg | **3,245.55** | 67.50 | 84.43 | 408.01 | 60.50 | 0.00 | 2,625.12 |
| Malta | **468.88** | 32.26 | 239.25 | 30.56 | 151.18 | 0.00 | 15.63 |
| Netherlands | **143.02** | 53.34 | 11.36 | 55.71 | 15.76 | 0.00 | 6.86 |
| Poland | **313.67** | 107.29 | 197.72 | 5.66 | 2.23 | 0.02 | 0.76 |
| Portugal | **383.25** | 123.02 | 226.17 | 24.76 | 5.15 | 0.00 | 3.67 |
| Romania | **238.63** | 167.69 | 63.23 | 5.09 | 1.61 | 0.00 | 1.01 |
| Slovakia | **303.47** | 113.80 | 150.30 | 35.22 | 2.07 | 0.00 | 2.08 |
| Slovenia | **230.67** | 112.98 | 62.03 | 45.12 | 5.32 | 0.04 | 5.16 |
| Spain | **208.06** | 126.89 | 47.23 | 27.57 | 3.83 | 0.00 | 2.53 |
| Sweden | **154.27** | 77.09 | 22.05 | 37.65 | 13.52 | 0.00 | 3.96 |
| UK | **97.52** | 56.66 | 10.01 | 25.31 | 2.57 | 0.00 | 2.03 |
| **EU** | **219.50** | **111.09** | **100.05** | **25.63** | **3.81** | **0.27** | **14.66** |

[1]Research and development / [2]Citizenship, freedom, security and justice / [3]Actions and programmes outside the EU / [4]Administration

Source: Author; Data: [11].

The hypothesis of a negative correlation between GDP per capita and budget spending per capita has not been confirmed. The correlation coefficient is even positive (0.616). One of the possible causes is a high administrative expenditure in rich countries (Luxembourg and Belgium) due to the headquarters of many European Institutions. By excluding these countries (Luxembourg and Belgium) from the analysis, a negative correlation appears, but it is still not significant (-0.354).

On closer analysis we can see that the only item that confirms the assumption is the regional policy, which is targeting redistribution:

- Agriculture (-0.099).
- Regional policy (-0.503).
- Research and development (0.741).
- Citizenship, freedom, security and justice (0.176).
- Actions and programmes outside the EU (-0.180).
- Administration (0.688).
- **Total (0.606).**

## 2.2 EU Policy of Stabilization

Following a macroeconomic stabilization policy means allocating budget resources to the EU Member States with a high macroeconomic multiplier. This allocation of funds enables better control of macroeconomic output using fewer budget resources.

**Table 3.** Macroeconomic Multiplier in EU Member States (2007 to 2017)

| Country | Marginal Propensity to Consume | Macroeconomic Multiplier |
|---|---|---|
| Austria | 0.492 | 1.968 |
| Belgium | 0.535 | 2.153 |
| Bulgaria | 0.381 | 1.616 |
| Croatia | 0.418 | 1.719 |
| Cyprus | 1.926 | N/A |
| Czechia | 0.521 | 2.087 |
| Denmark | 0.429 | 1.751 |
| Estonia | 0.389 | 1.636 |
| Finland | 1.091 | N/A |
| France | 0.502 | 2.006 |
| Germany | 0.410 | 1.694 |
| Greece | 0.487 | 1.950 |
| Hungary | 0.283 | 1.395 |
| Ireland | 0.075 | 1.081 |
| Italy | 0.902 | 10.242 |
| Latvia | 0.704 | 3.382 |
| Lithuania | 0.655 | 2.897 |
| Luxembourg | 0.214 | 1.272 |
| Malta | 0.192 | 1.238 |
| Netherlands | 0.270 | 1.370 |
| Poland | 0.547 | 2.209 |
| Portugal | 0.647 | 2.835 |
| Romania | 0.538 | 2.165 |
| Slovakia | 0.517 | 2.071 |
| Slovenia | 0.640 | 2.778 |
| Spain | 0.668 | 3.013 |
| Sweden | 0.417 | 1.716 |
| United Kingdom | 0.703 | 3.365 |
| **European Union** | **0.490** | **1.961** |

Source: [12-13].

Focusing on stabilization brings a strong positive correlation between the macroeconomic multiplier and EU budget expenditure. This would mean that the highest spending goes to countries with the greatest impact.

**Table 4.** Structure of the EU Budget per Capita by EU Member States (EUR per Capita in 2017).

| Country | Total | Agriculture | Regional policy | R&D[1] | Citizenship[2] | Outside the EU[3] | Admin[4] |
|---|---|---|---|---|---|---|---|
| Austria | **203.23** | 140.70 | 9.91 | 42.16 | 7.51 | 0.00 | 2.94 |
| Belgium | **653.54** | 46.86 | 27.23 | 128.28 | 22.72 | 0.00 | 428.45 |
| Bulgaria | **263.21** | 140.89 | 87.11 | 25.90 | 7.76 | 0.01 | 1.54 |
| Croatia | **156.94** | 74.11 | 64.47 | 10.25 | 2.76 | 3.23 | 2.12 |
| Cyprus | **243.98** | 87.72 | 75.25 | 56.52 | 9.69 | 0.00 | 7.04 |
| Czechia | **369.58** | 107.03 | 245.08 | 14.46 | 1.50 | 0.00 | 1.51 |
| Denmark | **257.76** | 183.05 | 13.71 | 47.77 | 2.77 | 0.00 | 10.45 |
| Estonia | **493.26** | 179.11 | 214.94 | 45.92 | 46.75 | 0.00 | 6.54 |
| Finland | **274.25** | 159.58 | 47.07 | 55.44 | 5.63 | 0.00 | 6.52 |
| France | **203.34** | 137.79 | 18.48 | 39.04 | 2.58 | 0.00 | 5.41 |
| Germany | **134.57** | 74.75 | 25.39 | 29.24 | 2.00 | 0.00 | 2.81 |
| Greece | **472.47** | 262.47 | 143.21 | 29.31 | 34.90 | 0.00 | 2.56 |
| Hungary | **410.85** | 153.91 | 217.19 | 34.48 | 3.65 | 0.00 | 1.62 |
| Ireland | **392.81** | 325.85 | 10.99 | 39.19 | 5.61 | 0.00 | 11.18 |
| Italy | **161.12** | 81.15 | 26.84 | 24.08 | 5.29 | 0.00 | 4.57 |
| Latvia | **370.93** | 198.53 | 147.19 | 13.75 | 7.28 | 0.00 | 4.18 |
| Lithuania | **539.04** | 244.78 | 234.64 | 45.41 | 10.59 | 0.00 | 3.63 |
| Luxembourg | **3,245.55** | 67.50 | 84.43 | 408.01 | 60.50 | 0.00 | 2,625.12 |
| Malta | **468.88** | 32.26 | 239.25 | 30.56 | 151.18 | 0.00 | 15.63 |
| Netherlands | **143.02** | 53.34 | 11.36 | 55.71 | 15.76 | 0.00 | 6.86 |
| Poland | **313.67** | 107.29 | 197.72 | 5.66 | 2.23 | 0.02 | 0.76 |
| Portugal | **383.25** | 123.02 | 226.17 | 24.76 | 5.15 | 0.00 | 3.67 |
| Romania | **238.63** | 167.69 | 63.23 | 5.09 | 1.61 | 0.00 | 1.01 |
| Slovakia | **303.47** | 113.80 | 150.30 | 35.22 | 2.07 | 0.00 | 2.08 |
| Slovenia | **230.67** | 112.98 | 62.03 | 45.12 | 5.32 | 0.04 | 5.16 |
| Spain | **208.06** | 126.89 | 47.23 | 27.57 | 3.83 | 0.00 | 2.53 |
| Sweden | **154.27** | 77.09 | 22.05 | 37.65 | 13.52 | 0.00 | 3.96 |
| UK | **97.52** | 56.66 | 10.01 | 25.31 | 2.57 | 0.00 | 2.03 |
| **EU** | **219.50** | **111.09** | **100.05** | **25.63** | **3.81** | **0.27** | **14.66** |

[1]Research and development / [2]Citizenship, freedom, security and justice / [3]Actions and programmes outside the EU / [4]Administration

Source: Author; Data: [11].

Nor has the second hypothesis assuming the implementation of an effective EU stabilization policy been confirmed (-0.172). In this case, most funds would have to be invested in countries with the highest macroeconomic multiplier (such as Italy), where public spending will be reflected in GDP ten times.

Similar results as for the overall budget also apply to each of its components, as can be seen in the overview of partial correlation coefficients:

- Agriculture (-0.117).
- Regional policy (-0.128).
- Research and development (-0.160).
- Citizenship, freedom, security and justice (-0.202).
- Actions and programmes outside the EU (-0.078).
- Administration (-0.131).
- **Total (-0.172).**

## 3 Conclusions

The analysis has shown that the overall EU budget does not target redistribution, mainly because of the reimbursement of the costs associated with the seat of major authorities in Belgium and Luxembourg. The regional policy sub-item is redistributive. This implies two proposals, either to relocate some European institutions to poorer regions or to strengthen the importance of the regional policy.

Furthermore, the analysis showed that neither the overall EU budget nor its components are targeting effective stabilization. This would be achieved if expenditure were made in countries with a high marginal consumption rate and therefore with a high macroeconomic multiplier that would ensure a significant impact of the funds spent. The proposal is, therefore, to change targeted budgetary expenditure to such countries.

## References

1. P. Samuelson, W. Nordhaus, *Economics.* 19th ed. New Delhi: Tata McGraw Hill (2010)
2. J. Stiglitz, *Economics of the Public Sector.* 3rd ed. New York: W. W. Norton and Company (2000)
3. S.K. Singh, D. Tripati Rao, Is monetary policy symmetrical in its effect on sectoral output? *Economics, Management, and Financial Markets*, **13**(1), 11-31 (2018)
4. P. Garcia-Duran, L.J. Eliasson, Squaring the circle: Assessing whether the European union´s pursuit of bilateral trade agreements is compatible with promoting multilateralism. *Journal of Self-Governance and Management Economics*, **6**(1), 7-32 (2018)
5. R. Boadway, M. Keen, Efficiency and the optimal direction of federal-state transfers. *International Tax and Public Finance*, **3**(137), 137-155 (1996)
6. A. Lindbeck, J. Weibull, Balanced-budget redistribution as the outcome of political competition. *Public Choice*, **52**(3), 273-297 (1987)
7. D. Wildasin, Income redistribution in a common labor market. *The American Economic Review*, **81**(4), 757-774 (1991)
8. D. Wildasin, Income redistribution and migration. *The Canadian Journal of Economics*, **27**(3), 637-656 (1994)
9. A. Sacchi, S. Salotti, The impact of national fiscal rules on the stabilisation function of fiscal policy. *European Journal of Political Economy*, **37**(C), 1-20, (2015)
10. Eurostat [online], Available at: https://ec.europa.eu/eurostat (2019)
11. European Parliament [online], Available at: https://www.europarl.europa.eu/portal/en (2019)
12. P. Rousek, M. Vochozka, M. Psárska, Analysis of the consumption on the goods and services market in the cybernetic model of the 15 older EU member states. *AD ALTA: Journal of Interdisciplinary Research*, **9**(1), 252-254 (2019)
13. P. Rousek, Analysis of consumption on the goods and services market in the cybernetic model of selected country. *SHS Web of Conferences: Innovative Economic Symposium 2018 – Milestones and Trends of World Economy* (2019)

# Machine learning forecasting of USA and PRC balance of trade in context of mutual sanctions

*Zuzana* Rowland[1], *Jaromír* Vrbka[1], *Marek* Vochozka[1*]

[1]Institute of Technology and Business, School of Expertness and Valuation, Okružní 517/10, 37001 České Budějovice, Czech Republic

**Abstract.** The USA decided to regulate the trade more by imposing tariffs on specific types of traded goods. It is therefore more interesting to find out whether the current technologies based on artificial intelligence with time series influenced by extraordinary factors such as the trade war between two powers are able to work. The objective of the contribution is to examine and subsequently equalize two time series – the USA import from the PRC and the USA export to the PRC. The dataset shows the course of the time series at monthly intervals between January 2000 and July 2019. 10,000 multilayer perceptron networks (MLP) are generated, out of which 5 with the best characteristics are retained. It has been proved that multilayer perceptron networks are a suitable tool for forecasting the development of the time series if there are no sudden fluctuations. Mutual sanctions of both states did not affect the result of machine learning forecasting.
**Key words:** forecasting, trade balance, machine learning, mutual sanctions, artificial neural networks

## 1 Introduction

In essence, machine learning is a sub-area of artificial intelligence. Machine learning focuses on many algorithms and techniques that allow a computer system to learn - that is, a change in the internal state of the system that will be able to adapt effectively to the environmental changes [1-3]. Ghoddusi, Creamer and Rafizadeh [4] state in their study that machine learning enables to create new opportunities for innovative research in the field of finance and energy economy. The authors focus on applications in areas such as demand forecasting, energy price forecasting (eg natural gas, energy and oil), business strategy, risk management, macro / energy trend analysis and data processing. The authors' analysis suggests that artificial neural networks (ANNs) and genetic algorithms are among the most popular techniques used in energy economy documents. Zhong and Enke [5] in their work, argue that large data analysis techniques associated with machine learning algorithms still play a very important role in many areas of application, including stock market investment. The authors state that today few studies are concerned with the prediction of daily returns from the stock markets, especially in the implementation of powerful machine learning techniques such as deep neural networks for analysis. Deep neural networks use a wide range of different deep

* Corresponding author: vochozka@mail.vstecb.cz

learning algorithms based on a combination of network structure, activation functions, and model parameters, and their performance depends on the level of data representation. Thanks to more complex algorithms of learning and other hidden layers, deep neural networks are recognized as advanced and significant technology in the field of artificial and computational intelligence [6]. Samanpour, Ruegenberg and Ahlers [7] report in their study that an important advantage of deep learning is the analysis and learning of large amounts of data unattended, making it a very valuable and useful tool for analysing huge amounts of data where raw data is largely unclassified and unmarked. Liu and Xie [8] contrast the performance of predicting machine learning methods and nine econometric methods, including a new hybrid method that combines model averaging and machine learning using social media and film industry data. The authors' results show that traditional econometric methods reflect the long-term trend better, while machine learning methods have the advantage of solving short-term noise. However, once the sample heterogeneity is rectified, this hybrid method tends to strike the right balance in solving both trend and noise, resulting in excellent forecast efficiency. According to Carbonneau, Laframboise and Vahidov [9], full cooperation in supply chains is ideal, which the participating companies should try to achieve. However, several factors prevent actual progress in this direction. Therefore, there is a need to predict the demand of participants if there is no complete information on the demand of other participants. The authors investigate the applicability of advanced machine learning techniques, including neural networks, recurrent neural networks, and support vector machines, to predict distorted demand at the end of the supply chain. The authors compare these methods to other traditional methods, including naive forecasting, trends, moving average and linear regression. The authors' results suggest that while repetitive neural networks and support vector machines perform best, their predictive accuracy was not statistically significantly better than the regression model.

Means such as artificial neural networks (ANNs) can be used to estimate the trade balance. Kliešik [10] claims that these networks are certain computational models that were inspired by biological neural networks. The application of these networks is applicable in many areas. Pao [11] states that artificial neural networks are nowadays used to solve possible future obstacles, especially for the forecasting of values. According to Sayadi et al. [12], the main advantages of ANNs are the ability to generalize and the ability to learn. The main disadvantages of ANNs, according to the Rowland and Vrbka [13], include the possibility of illogical network behaviour and that they demand very high-quality data. According to Vochozka and Horák [14], ANNs is one of the most popular methods used as a prognostic method that requires a more complex model, uses more variables and tends to be non-linear. Therefore, artificial neural networks are suitable for auxiliary adjustments of the financial sector. According to Cho, Kim and Bae [15], ANNs techniques perform better than traditional statistical methods. Accurate and complex results are also provided by modern hybrid models, which enable to combine statistical methods with artificial neural network techniques. Li et al. [16] state that traditional statistical methods are very popular due to their easy comprehensibility, interpretation and acceptable predictive performance Balcaen and Ooghe [17] achieved a surprising result, comparing selected methods based on discriminant analysis and methods based on logistic regression. The authors found that the accuracy of the model is much more influenced by the indicators that are included in the model than by the method by which the model was derived.

In the economic sphere it is possible to use other methods than just the above mentioned artificial neural networks. Thus, it is possible to use, for example, cluster analysis, decision trees, discriminant analysis and the known ARIMA model [18]. Multiple discrimination analysis is one of the traditional statistical models. According to Crăciun et al. [19], multiple discriminant analysis is a method of diagnosis consisting of measuring and interpreting future economic risks. According to Kliešik, Vrbka and Rowland [20], the advantage of discriminatory analysis is its simplicity and that it is not time consuming, the authors see a disadvantage in its incompleteness. Pai and Lin [21] state that the ARIMA model, or autoregressive integrated moving average, is one of the most widespread linear models in time series predictions. According to Mélard and Pasteels [22], this model is mainly used for short-term prediction in a situation where there is no data available to explain variables or whether the model has poor predictive properties. Junttila [23] sees the advantages of the model primarily in the ability to quickly adapt to changes during a time series, and the main disadvantages that practical applications of models are very demanding, resulting models are difficult to interpret and are only suitable for time series within at least 50 observations. The subject of cluster analysis is, according to Krulický [24], a generalization and a description of relations in a group of variables. Based on the similarity of variables, it creates their clusters. The purpose of this analysis is to classify a certain number of objects into several relevantly homogeneous clusters, where the objects within one cluster are required to be as similar as possible and the objects belonging to different clusters to be as different as possible. According to Šuleř [25], the main advantage of this method is its simplicity and the possibility of using statistical software. The disadvantage of the method is its susceptibility to distortion of results. According to Tuffell et al. [26], decision trees are data structures that allow predicting the output value for specific inputs or classifying input data points. Valášková, Klieštik and Mišánková [27] rank among their main advantages the easy visualization and interpretation, while their disadvantage is that they cannot be used at any time due to the more difficult processing of continuous data.

Rowland, Šuleř and Vochozka [28] set out to compare the accuracy of time series equalizing using regression analysis and artificial neural networks on the example of the trade balance of China and the Czech Republic. Initially, they performed linear regression followed by regression using artificial neural networks. They compared the two methods on a professional level. The results of the authors inform us that artificial neural networks seem to be very useful for the prognosis of time series. In their work, Weijin and Yuhui [29] present a model that is used to predict export and import transactions in one sector. The authors concluded that the nonlinear prognosis may not only address data combination and improving accuracy but may also vividly reflect the nonlinear characteristic of the prediction system. The aim of the authors Horák, Šuleř and Vrbka [30] was to compare the accuracy of time series equalizing using artificial neural networks and regression analysis on the example of US exports to China. The purpose of their study is to show the possible benefits and use of artificial neural networks in practice. The authors' results show that artificial neural networks can effectively learn time series dependencies. Similar findings were made by Vrbka, Rowland and Šuleř [31] in their work, where they compare the accuracy of time series equalizing using artificial neural networks and regression analysis on the example of trade balance between China and the EU.

## 2 Data and methods

In the last few years, mutual USA and PRC trade restrictions caused by the USA feeling of imbalance of the both countries trade significantly affect not the mutual trade of the two countries, but also the overall climate of the international trade. The USA decided to regulate the trade more by imposing tariffs on specific types of traded goods. It is therefore more interesting to find out whether the current technologies based on artificial intelligence with time series influenced by extraordinary factors such as the trade war between two powers are able to work. It is desirable that artificial neural networks are able to cope with such a task. For any businessman and economist, it is important to be able not only be aware of the current development but also to forecast the future development. The objective of the contribution is to examine and subsequently equalize two time series:

1. The USA import from the PRC.
2. The USA export to the PRC.

Based on the results it will be possible to forecast the future development of the balance of trade of the two trade war actors.

The dataset shows the course of the time series at monthly intervals between January 2000 and July 2019. The course of both time series is shown in Figure 1.

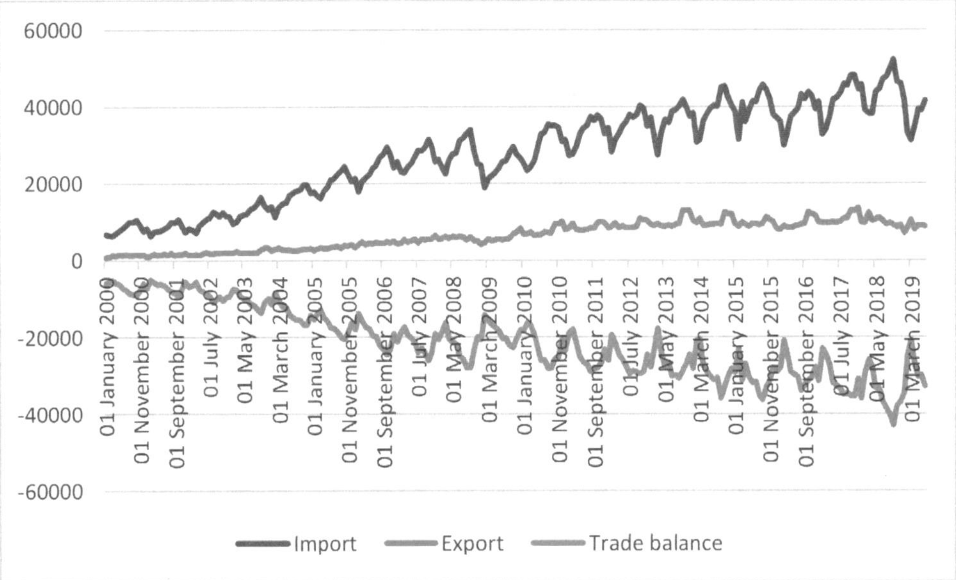


**Fig. 1.** Development of USA and PRC balance of trade between January 2000 and July 2019

Note: The values in the entire text are given in billions of USD.

Source: [32], own processing

The Figure shows seasonal fluctuations in the development of import within the individual years of the monitored period. There is a clear increase in the import from USD 6,902 bn in 2000 to USD 52,202 bn in October 2018. This means that the USA import from the PRC for the monitored period increased more than 7 times. On the contrary, in the case of export the seasonal effect is not apparent at first sight. The volume of the USA export to the PRC is significantly lower. There are many reasons, e.g. cheap labour in the PRC. However, this is not the subject of this contribution. On the contrary, what is important is that there are no seasonal fluctuations evident within one year at first sight. The balance of trade of both countries is negative from the USA perspective, which is clearly shown by the Figure. The shape of the curve is very similar to the inverse curve of imports. The development of export is less noticeable on the balance of trade curve. Table 1 shows the basic statistical characteristics of the dataset.

**Table 1.** Basic statistical characteristics of examined dataset

| Statistics | Date (Input variable) | Month (Input variable) | Year (Input variable) | Import (Output (target)) | Export (Output (target)) |
|---|---|---|---|---|---|
| Minimum (Training) | 36556.00 | 1.00000 | 2000.000 | 6375.60 | 863.10 |
| Maximum (Training) | 43646.00 | 12.00000 | 2019.000 | 48127.80 | 13147.80 |
| Average (Training) | 40050.47 | 6.45455 | 2009.115 | 27572.76 | 6415.19 |
| Standard deviation (Training) | 2001.16 | 3.50688 | 5.473 | 12039.78 | 3382.42 |
| Minimum (Testing) | 36585.00 | 1.00000 | 2000.000 | 6584.40 | 972.70 |
| Maximum (Testing) | 43677.00 | 12.00000 | 2019.000 | 52202.30 | 12382.10 |
| Average (Testing) | 40130.09 | 6.00000 | 2009.371 | 26146.36 | 5918.11 |
| Standard deviation (Testing) | 2418.96 | 3.38683 | 6.691 | 12282.23 | 3492.76 |
| Minimum (Validation) | 36646.00 | 1.00000 | 2000.000 | 7070.50 | 1227.50 |
| Maximum (Validation) | 43373.00 | 12.00000 | 2018.000 | 50015.00 | 13644.80 |
| Average (Validation) | 40412.86 | 6.71429 | 2010.086 | 29966.20 | 7025.04 |
| Standard deviation (Validation) | 3403.28 | 3.44159 | 9.284 | 21320.19 | 5528.71 |
| Minimum (Overall) | 36556.00 | 1.00000 | 2000.000 | 6375.60 | 863.10 |
| Maximum (Overall) | 43677.00 | 12.00000 | 2019.000 | 52202.30 | 13644.80 |
| Average (Overall) | 40116.30 | 6.42553 | 2009.298 | 27716.79 | 6431.99 |
| Standard deviation (Overall) | 2069.19 | 3.45140 | 5.669 | 12057.45 | 3429.19 |

Source: Authors.

There will be regression carried out by means of neural structures. The time series will be equalized separately for import and export. The balance of trade will be the difference of the two variables over the time. The procedure chosen corresponds with the situation where the seasonal fluctuations of both time series differ. Therefore, separated equalizing of each time series and the following calculation of the balance of trade will yield more accurate results.

Multilayer perceptron networks (MLP) will be generated. The considered time series lag will be 5 months. This means that the target (forecast) variable will be calculated using the data from the previous five months. This will enable to avoid any time series fluctuations caused by the latest data (in this case, this refers to the time series of the USA import from the PRC). A small lag of the time series can result I extreme fluctuations of the equalized time series. A larger time series lag can mean averaging values. However, each lag of the time series entails greater demands on the complexity of the artificial neural structure, specifically the neurons in the input layer (in the case of this calculation, it will be 15 neurons in the input layer of the artificial neural structure).

The independent continuous variable will be time. The seasonal fluctuations will be represented by the categorical variable in the form of the month and the year in which the value was measured. We will work with possible monthly seasonality of the time series. However, we should also capture the overall trend of the time series. They will be entered as continuous variables in the "Integer" regime. The dependent variable will be the USA import from the PRC or the USA export to the PRC.

The time series will be divided into three data sets: training, testing and validation. The first group will contain 70% of the input data. Based on the training data set, neural structures will be generated. The remaining two data sets will contain 15% of the input data each. Both groups will be used for the verification of the neural structure or the model created. 10,000 neural structures will be generated, out of which 5 with the best characteristics[1]will be retained. The hidden layer will contain at least two neurons and at most 9 neurons. The

[1] We will use the least squares method.

following distribution functions in the hidden and output layers will be considered: Linear, Logistic, Atanh, Exponential, Sinus.

Other settings will remain default. Finally, forecasting of both time series will be carried out till December 2020, and at the same time, by identifying the difference of export and import, the balance of trade will be calculated. However, this assumes that out of the retained neural structures the one capable of best forecasting of the development of the specific time series will be chosen.

# 3 Results

## 3.1 Import

Table 2 shows the overview of the neural networks retained in equalizing time series of the import, where the assumed lag is 5 months.

**Table 2.** Retained neural networks for equalizing import time series

| Network | Train. perform. | Test. perform. | Valid. perform. | Train. error | Test. error | Valid. error | Train. algorit. | Error function | Activation of hidden layer | Output activation function |
|---|---|---|---|---|---|---|---|---|---|---|
| MLP 15-3-1 | 0.984886 | 0.969392 | 0.985259 | 2049185 | 3621980 | 1610485 | BFGS (Quasi-Newton) 64 | Sum of squares | Logistic | Identity |
| MLP 15-7-1 | 0.991430 | 0.979645 | 0.987889 | 1152035 | 2003906 | 1290950 | BFGS (Quasi-Newton) 148 | Sum of squares | Tanh | Exponential |
| MLP 15-7-1 | 0.989698 | 0.977635 | 0.986896 | 1390544 | 2312432 | 1461439 | BFGS (Quasi-Newton) 114 | Sum of squares | Tanh | Exponential |
| MLP 15-10-1 | 0.989664 | 0.977584 | 0.985530 | 1397550 | 2337306 | 1638214 | BFGS (Quasi-Newton) 150 | Sum of squares | Tanh | Exponential |
| MLP 15-4-1 | 0.982478 | 0.965299 | 0.985481 | 2377804 | 4413279 | 1725236 | BFGS (Quasi-Newton) 34 | Sum of squares | Tanh | Exponential |

Source: Authors.

Due to the time series lag, the input layer contains 15 neurons. The Table shows that the networks with 4 – 10 neurons in the hidden layer of the neural network were retained. The neural networks use the functions of hyperbolic tangent and logistic function for the activation of the hidden layer. The output layer of neurons is activated using the exponential and identity functions. The correlation coefficient in all data sets achieves very high values in all networks. It is always above 0.96, which indicates the high degree of direct dependence. Based on this parameter it is evident that the retained neural structures represent a very plausible model of the actual time series and are therefore capable of forecasting the future development of the time series. The error was determined using the sum of the least squares, and its level is acceptable regarding the input data (for all networks and data sets).

Figure 2 shows the comparison of the actual course of the time series and equalized time series.

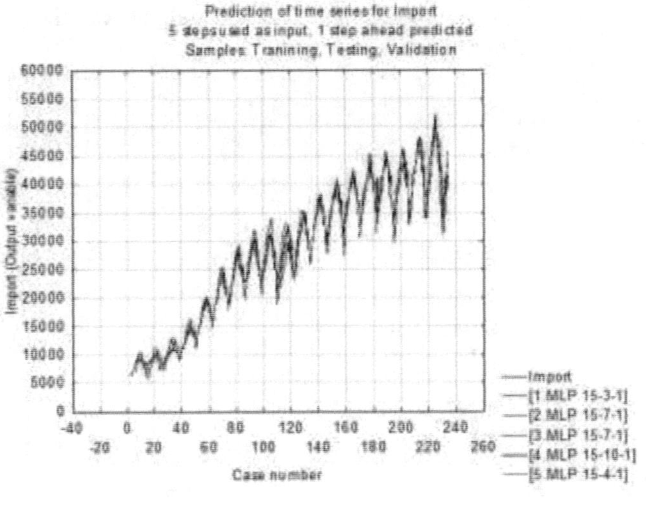


**Fig. 2.** Equalized time series of import

Source: Authors.

It is also evident from Figure 4 that all equalized time series are able to follow the course of the actual development of the USA import from the PRC. With some minor exceptions, all networks are able to capture not only the development trend, but also seasonal fluctuations.

In order to determine whether the networks are applicable or not, we will further deal with their use for forecasting. We will forecast the development of the USA import from the PRC for the period of August 2019 to December 2020. Concrete data are shown in Table 3.

**Table 3.** Development of forecasts for the period of August 2019-December 2020 by networks retained for equalizing import time series

| Date | MLP 15-3-1 | MLP 15-7-1 | MLP 15-7-1 | MLP 15-10-1 | MLP 15-4-1 |
|---|---|---|---|---|---|
| 31 August 2019 | 47515.57 | 47002.25 | 47040.21 | 48861.78 | 47797.37 |
| 30 September 2019 | 49604.57 | 49351.12 | 49124.91 | 51588.14 | 49334.92 |
| 31 October 2019 | 50764.12 | 51231.68 | 51143.95 | 52961.69 | 50106.41 |
| 30 November 2019 | 51112.63 | 50454.89 | 50574.42 | 50755.99 | 50462.89 |
| 31 December 2019 | 50848.47 | 47639.46 | 47206.94 | 46915.98 | 50694.26 |
| 31 January 2020 | 50248.03 | 48216.25 | 48999.36 | 46650.57 | 50964.55 |
| 29 February 2020 | 44783.80 | 35272.04 | 38268.77 | 34983.18 | 45836.01 |
| 31 March 2020 | 42825.99 | 35552.70 | 35592.07 | 37866.55 | 45078.61 |
| 30 April 2020 | 41978.92 | 41250.17 | 39894.06 | 42639.63 | 43639.14 |
| 31 May 2020 | 44659.23 | 43842.69 | 44274.29 | 43010.50 | 43976.81 |
| 30 June 2020 | 44337.75 | 47957.59 | 49797.58 | 45170.13 | 44301.28 |
| 31 July 2020 | 47481.72 | 51221.00 | 53123.88 | 52357.95 | 47687.89 |
| 31 August 2020 | 50947.31 | 54986.77 | 56840.84 | 59182.85 | 51304.98 |
| 30 September 2020 | 54119.11 | 59020.77 | 61001.23 | 65061.23 | 54268.33 |
| 31 October 2020 | 56519.55 | 62076.98 | 65109.56 | 67830.92 | 56228.17 |
| 30 November 2020 | 58001.95 | 60949.66 | 64774.65 | 64240.15 | 57365.47 |
| 31 December 2020 | 58675.62 | 57821.82 | 61761.25 | 59256.33 | 57994.40 |

Source: Authors.

The results shown in the Table represent a possible development of the time series. This is to certain extent confirmed by the graph in Figure 3.

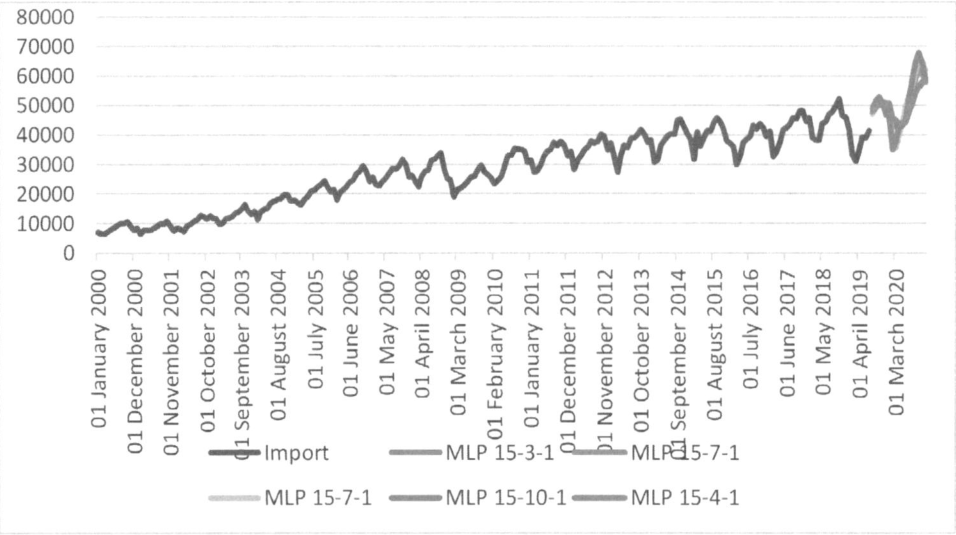


**Fig. 3**. Development of USA import from PRC and forecasts to December 2020

Source: Authors.

Comparing the actual development of the USA import from the PRC and assumed development of the variable over time forecast by means of neural networks with a 5-month lag, it can be concluded that (especially taking into account the Table of the statistics in the case of a 5-month lag in Annex 1), the 5. MLP 15-4-1 is the best rated artificial neural structure.

### 3.2 Export

Within equalizing the export time series, there also were generated 10,000 artificial neural networks, out of which 5 with the best characteristics were retained. For more details, see (Table 4).

**Table 4**. Retained neural networks of export time series

| Network | Train. perform. | Test. perform. | Valid. perform. | Train. error | Test. error | Valid. error | Train. algorit. | Error function | Activation of hidden layer | Output activation function |
|---|---|---|---|---|---|---|---|---|---|---|
| MLP 15-7-1 | 0.973572 | 0.966155 | 0.972225 | 286349.4 | 338708.9 | 333663.0 | BFGS (Quasi-Newton) 67 | Sum of squares | Exponential | Logistic |
| MLP 15-10-1 | 0.973730 | 0.963486 | 0.972162 | 284104.3 | 360204.8 | 317915.7 | BFGS (Quasi-Newton) 77 | Sum of squares | Exponential | Logistic |
| MLP 15-8-1 | 0.974078 | 0.962645 | 0.973662 | 280208.0 | 366654.9 | 303239.0 | BFGS (Quasi-Newton) 93 | Sum of squares | Exponential | Logistic |
| MLP 15-7-1 | 0.973724 | 0.963440 | 0.973038 | 283899.5 | 362903.6 | 308651.9 | BFGS (Quasi-Newton) 80 | Sum of squares | Exponential | Logistic |
| MLP 15-7-1 | 0.973842 | 0.961769 | 0.972307 | 282599.9 | 371616.4 | 323001.7 | BFGS (Quasi-Newton) 74 | Sum of squares | Exponential | Logistic |

Source: Authors.

In the case of export, there are also 15 neurons in the input layer. The Table shows that neural networks with 7-10 neurons in the hidden layer were retained. Neural networks use exponential function for the activation of the hidden layer. The output layer of neurons is activated by the logistic function. The performance of the artificial network's measures in terms of correlation coefficient achieves the level above 0.97 in the training data set. above 0.96 in the testing data set, and above 0.97 in the validation data set. This indicates high level of direct dependence. Based on this parameter it can be stated that the networks (if they do not suffer from overfitting) will be able to forecast the further development of the USA export to the PRC very accurately (as they are able to equalize the time series in the past). For the error calculation, the least squares method will be used. In the case of export, the error is rather small and thus a quality forecast of the time series can be assumed.

Figure 4 shows the comparison of the actual course of the time series and equalized tine series.

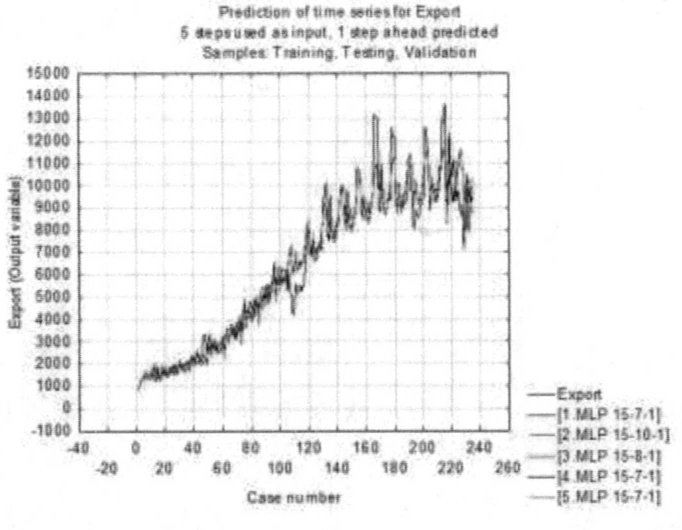


**Fig. 4.** Equalized time series of USA export to PRC

Source: Authors.

It is evident from Figure 4 that all equalized time series are able to copy the course of the actual development of the USA export to the PRC. The courses of all equalized time series are similar to each other. The neural networks are able to capture the development trend but also the seasonal fluctuations. However, some extreme fluctuations are not copied completely, especially in the cases around the case No. 115, specifically in the period between November 2008 and August 2009.

In the case of export, we will also deal with forecasting the further development of the monitored variable. We will forecast the development of the USA export to the PRC for the period between August 2019 and December 2020. Concrete data are shown in Table 5.

**Table 5**. Export forecast for the period August 2019-January 2020 by retained artificial neural networks

| Date | MLP 15-7-1 | MLP 15-10-1 | MLP 15-8-1 | MLP 15-7-1 | MLP 15-7-1 |
|---|---|---|---|---|---|
| 31 August 2019 | 10279.81 | 10739.16 | 10617.62 | 10166.11 | 10654.85 |
| 30 September 2019 | 10814.64 | 11132.83 | 11098.27 | 10731.32 | 11111.66 |
| 31 October 2019 | 11242.17 | 11337.84 | 11492.52 | 11230.89 | 11446.66 |
| 30 November 2019 | 11295.93 | 11009.44 | 11566.69 | 11264.85 | 11425.88 |
| 31 December 2019 | 10259.31 | 8417.86 | 10360.17 | 9751.11 | 10235.95 |
| 31 January 2020 | 5031.43 | 1788.88 | 3628.02 | 3083.52 | 4019.12 |
| 29 February 2020 | 7959.24 | 8028.68 | 6683.97 | 9776.34 | 6899.30 |
| 31 March 2020 | 9160.78 | 9694.63 | 9285.55 | 8493.03 | 8691.60 |
| 30 April 2020 | 8703.29 | 9826.61 | 10223.28 | 9339.56 | 9543.56 |
| 31 May 2020 | 7939.71 | 9665.11 | 9824.08 | 9601.72 | 9605.19 |
| 30 June 2020 | 8155.12 | 9539.10 | 9897.51 | 7796.75 | 9894.66 |
| 31 July 2020 | 8455.32 | 9508.48 | 10030.46 | 7752.58 | 10175.60 |
| 31 August 2020 | 8756.21 | 9206.95 | 10148.73 | 7700.18 | 10365.98 |
| 30 September 2020 | 8906.66 | 8228.94 | 10044.15 | 7316.84 | 10290.13 |
| 31 October 2020 | 8566.82 | 5725.59 | 9195.68 | 5868.67 | 9469.96 |
| 30 November 2020 | 6921.72 | 2075.81 | 6110.72 | 2740.77 | 6474.23 |
| 31 December 2020 | 3128.68 | 897.84 | 1536.10 | 956.15 | 1707.44 |

Source: Authors.

As already stated above, all predictions are similar to each other. This is also to certain extent confirmed by the graph in Figure 5.

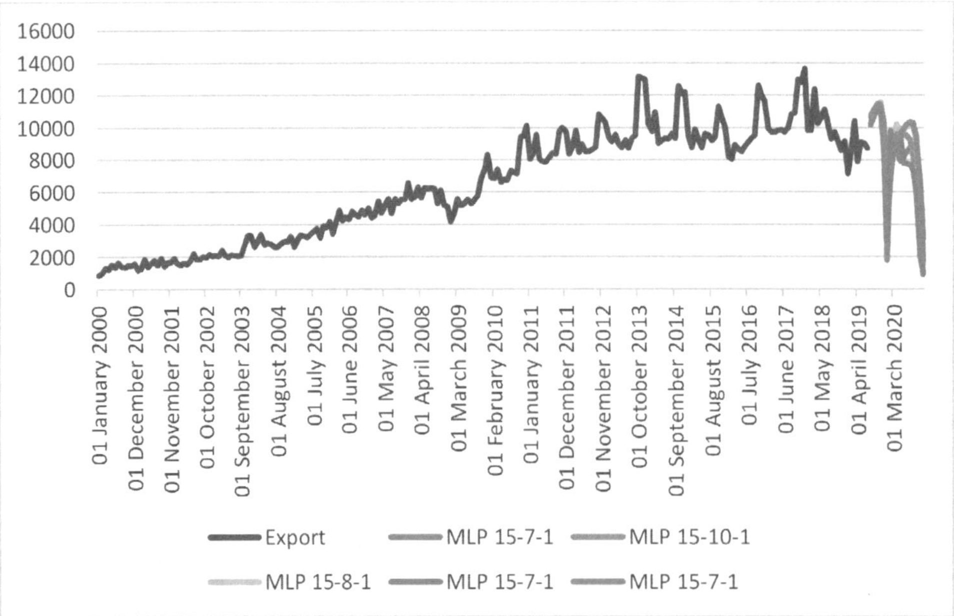


**Fig. 5.** Development of tine series and predictions of export by retained artificial neural networks

Source: Authors.

Comparing the actual development of the USA export to the PRC and the assumed development of the variable over time forecast using artificial neural networks retained at equalizing the export time series, it can be concluded (especially considering the table of statistics in Annex 2) that the network 1. MLP 15-7-1 achieves best results in forecasting the future development.

## 4 Results discussion

Within the research, equalizing of the time series of the USA import from the PRC and the USA export to the PRC was carried out. Time series lag was 5 months in both cases. This eliminated potential extraordinary fluctuations caused from the fluctuations of the time series in individual months. At first, we focused on how multilayer perceptron neural networks are able to equalize time series. In both cases, a total of 10,000 artificial neural networks were generated, and five of them with the best characteristics were retained. All retained artificial neural networks are more or less applicable for forecasting the future development of the USA import from the PRC, the USA export to the PRC, and for forecasting the balance of trade of both countries (from the perspective of the USA). The performance of the networks was always very high, while the error measured was rather small. For each time series, the best network was determined, and this was used for calculating the balance of trade (for more details, see Table 6).

**Table 6.** Comparison of the most sucessful structures from all experiments

| Date | USA import from PRC | USA export to PRC | Balance of trade |
|---|---|---|---|
| | 5. MLP 15-4-1 | 1. MLP 15-7-1 | |
| 31 August 2019 | 47797.37 | 10279.81 | -37517.56 |
| 30 September 2019 | 49334.92 | 10814.64 | -38520.27 |
| 31 October 2019 | 50106.41 | 11242.17 | -38864.24 |
| 30 November 2019 | 50462.89 | 11295.93 | -39166.96 |
| 31 December 2019 | 50694.26 | 10259.31 | -40434.95 |
| 31 January 2020 | 50964.55 | 5031.43 | -45933.12 |
| 29 February 2020 | 45836.01 | 7959.24 | -37876.77 |
| 31 March 2020 | 45078.61 | 9160.78 | -35917.83 |
| 30 April 2020 | 43639.14 | 8703.29 | -34935.86 |
| 31 May 2020 | 43976.81 | 7939.71 | -36037.09 |
| 30 June 2020 | 44301.28 | 8155.12 | -36146.15 |
| 31 July 2020 | 47687.89 | 8455.32 | -39232.56 |
| 31 August 2020 | 51304.98 | 8756.21 | -42548.76 |
| 30 September 2020 | 54268.33 | 8906.66 | -45361.67 |
| 31 October 2020 | 56228.17 | 8566.82 | -47661.35 |
| 30 November 2020 | 57365.47 | 6921.72 | -50443.75 |
| 31 December 2020 | 57994.40 | 3128.68 | -54865.72 |

Source: Authors.

Table 6 demonstrates the potential development of the monitored variables, especially the resulting variable – balance of trade. It forecasts the increase in the USA import from the PRC, decrease in the USA export to the PRC and thus also increasing negative USA-PRC balance of trade. This is even better illustrated by Figure 6.

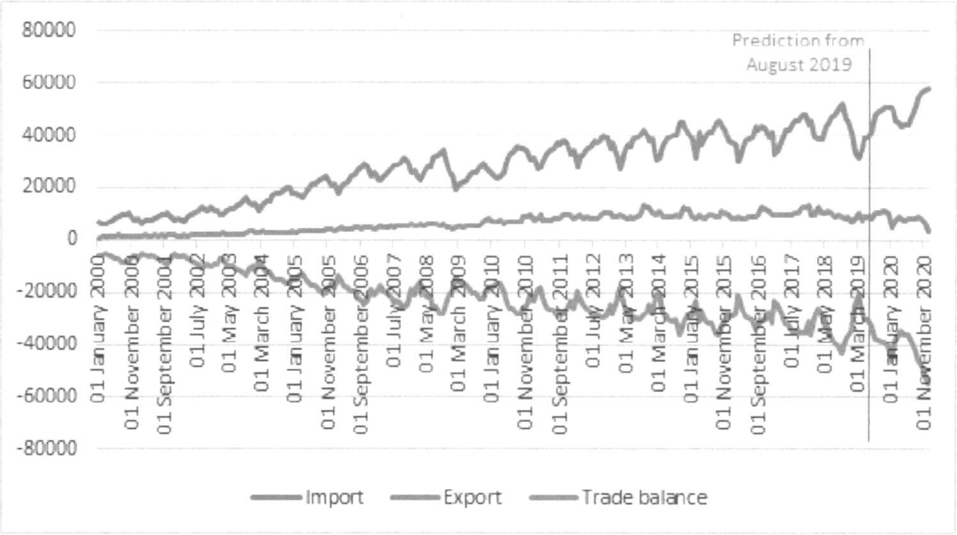


**Fig. 6.** Comparison of the most successful neural structures from all experiments

Source: Authors.

From the figure it is evident that the actual development of all three time series and the forecast development are very similar for all selected neural structures. It better illustrates what has been indicated by the table, that the difference between the USSA import to the PRC and the USA export to the PRC will grow. The balance of trade of both states will thus be even more negative than now.

## 5 Conclusion

The objective of the contribution was to forecast the United States of America (USA) and the People´s Republic of China (PRC) balance of trade development when using analysis and machine learning forecasting of the USA import from the PRC and the USA export to the PRC.

Both observed time series were equalized, and the most successful neural network was determined in both cases. On the basis of the difference between the USA export to the PRC and the USA import to the PRC, the balance of trade of both countries was calculated.

It could be stated that the objective of the contribution was achieved as:

1. It has been proved that multilayer perceptron networks are a suitable tool for forecasting the development of the time series if there are no sudden fluctuations.
2. MLP networks are able to capture the trend of the whole time series as well as its seasonal fluctuations.
3. It is necessary to consider the time series lag.
4. Mutual sanctions of both states did not affect the result of machine learning forecasting.
5. In the case of the USA export to the PRC and the USA import from the PRC, equalizing the time series had to (and will always have to be in the future) carried out separately. This way it is possible to improve the accuracy of forecasting the development of the balance of trade, as both time series show a completely different development.

Further research shall be focused on an experiment that would provide information on the most suitable time lag of the time series for this type of task and theses specific variables. It can be different for each variable and also different at sub-intervals of the monitored time series.

## References


1. N.K. Ahmed, A.F. Atiya, N.E. Gayar, H. El-Shishiny, An Empirical Comparison of Machine Learning Models for Time Series Forecasting. *Econometric Reviews*, **29**(5-6), 594-621 (2010)
2. P. Gogas, T. Papadimitriou, A. Agrapetidou, Forecasting bank failures and stress testing: A machine learning approach. *International Journal of Forecasting*, **34**(3), 440-455 (2018)
3. R. Milward, G.H. Popescu, K. Frajtová-Michalíková, Z. Musová, V. Machová, Sensing, smart, and sustainable technologies in Industry 4.0: Cyber-physical networks, machine data capturing systems, and digitized mass production. *Economics, Management, and Financial Markets*, **14**(3), 37-43 (2019)
4. H. Ghoddusi, G.G. Creamer, N. Rafizadeh, Machine learning in energy economics and finance: A review. *Energy Economics*, **81**, 709-727 (2019)
5. X. Zhong, D. Enke, Predicting the daily return direction of the stock market using hybrid machine learning algorithms. *Financial Innovation*, **5**(1), (2019)
6. J. Horák, T. Krulický, Comparison of exponential time series alignment and time series alignment using artificial neural networks by example of prediction of future development of stock prices of a specific company. *SHS Web of Conferences: Innovative Economic Symposium 2018 - Milestones and Trends of World Economy,* **61** *(*2019)
7. A.R. Samanpour, A. Ruegenberg, R. Ahlers, The Future of Machine Learning and Predictive Analytics. *Digital Marketplaces Unleashed*, Berlin, Heidelberg: Springer Berlin Heidelberg, 297-309 (2018)
8. Y. Liu, T. Xie, Machine learning versus econometrics: prediction of box office. *Applied Economics Letters*, **26**(2), 124-130 (2018)
9. R. Carbonneau, K. Laframboise, R. Vahidov, Application of machine learning techniques for supply chain demand forecasting. *European Journal of Operational Research*, **184**(3), 1140-1154 (2018)
10. T. Klieštik, Models of autoregression conditional heteroskedasticity garch and arch as a tool for modeling the volatility of financial time series. *Ekonomicko-manažerské spectrum*, **7**(1), 2-10 (2013)
11. H. Pao, A comparison of neural network and multiple regression analysis in modeling capital structure. *Expert Systems with Applications*, **35**(3), 720-727 (2008)
12. A.R. Sayadi, S.M.M. Tavassoli, M. Monjezi, M. Rezaei. Application of neural networks to predict net present value in mining projects. *Arabian Journal of Geosciences*, **7**(3), 1067-1072 (2012)
13. Z. Rowland, J. Vrbka, Using artificial neural networks for prediction of key indicators of a company in global world. *Proceedings of the 16th International Scientific Conference Globalization and Its Socio-Economic Consequences*, pp. 1896-1903 (2016)

14. M. Vochozka, J. Horák. Comparison of neural networks and regression time series in prediction of silver price development. *9th International Scientific Conference Company Diagnostics, Controlling and Logistics* (2018)
15. S. Cho, J. Kim, J.K. Bae. An integrative model with subject weight based on neural network learning for bankruptcy prediction. *Expert Systems with Applications*, **36**(1), 403-410 (2009)
16. H. Li, L.Y. Hong, Q. Zhou, H.J. Yu, The assisted prediction modelling frame with hybridisation and ensemble for business risk forecasting and an implementation. *International Journal of Systems Science*, **46**(11), 2072-2086 (2013)
17. S. Balcaen, H. Ooghe, 35 years of studies on business failure: an overview of the classic statistical methodologies and their related problems. *The British Accounting Review*, **38**(1), 63-93 (2006)
18. M. Kováčová, T. Kliestik, P. Kubala, K. Valášková, M. Radisic, J. Borocki, Bankruptcy models: Verifying their validity as a predictor of corporate failure. *Polish Journal of Management Studies*, **18**(1), 167-179 (2018)
19. M. Crăciun, C. Raţiu, D. Bucerzan, A. Manolescu, Actuality of Bankruptcy Prediction Models used in Decision Support System. *International Journal of Computers Communications*, **8**(3), 375-383 (2013)
20. T. Kliestik, J. Vrbka, Z. Rowland, Bankruptcy prediction in Visegrad group countries using multiple discriminant analysis. *Equilibrium-Quarterly Journal of Economics and Economic Policy*, **13**(3), 569-593 (2018)
21. P.F. Pai, C.S. Lin. A hybrid ARIMA and support vector machines model in stock price forecasting. *Omega*, **33**(6), 497-505 (2005)
22. G. Mélard, J.M. Pasteels. Automatic ARIMA modeling including interventions, using time series expert software. *International Journal of Forecasting*, **16**(4), 497-508 (2000)
23. J. Junttila, Structural breaks, ARIMA model and Finnish inflation forecasts. *International Journal of Forecasting*, **17**(2), 203-230 (2001)
24. T. Krulický, Using Kohonen networks in the analysis of transport companies in the Czech Republic. *SHS Web of Conferences*: *Innovative Economic Symposium 2018 – Milestones and Trends of World Economy*, **61** (2019)
25. P. Šuleř, Using Kohonen´s neural networks to identify the bankruptcy of enterprises: Case study based on construction companies in South Bohemian region. *Proceedings of the 5th International conference Innovation Management, Entrepreneurship and Sustainability*, pp. 985-995 (2017)
26. C. Tuffell, P. Král', A. Siekelová, J. Horák, Cyber-physical smart manufacturing systems: Sustainable industrial networks, cognitive automation, and data-centric business models. *Economics, Management, and Financial Markets*, **14**(2), 58-63 (2019)
27. K. Valášková, T. Kliestik, M. Mišánková, The role of fuzzy logic in decision making process. *2014 2nd International Conference on Management Innovation and Business Innovation*, **44**, pp. 143-148 (2014)
28. Z. Rowland, P. Šuleř, M. Vochozka, Comparison of neural networks and regression time series in estimating the Czech Republic and China trade balance. *SHS Web of Conferences: Innovative Economic Symposium 2018 – Milestones and Trends of World Economy*, **61** (2019)

29. J. Weijin, X. Yuhui. A novel method for nonlinear time series forecasting of time-delay neural network. *Wuhan University Journal of Natural Sciences*, **11**(5), 1357-1361 (2006)
30. J. Horák, P. Šuleř, J. Vrbka, Comparison of neural networks and regression time series when predicting the export development from the USA to PRC. *Proceedings of 6th International Scientific Conference Contemporary Issues in Business, Management and Economics Engineering* (2019)
31. J. Vrbka, J, Z. Rowland, P. Šuleř, Comparison of neural networks and regression time series in estimating the development of the EU and the PRC trade balance. *SHS Web of Conferences*: *Innovative Economic Symposium 2018 – Milestones and Trends of World Economy,* **61** (2019)
32. World Bank [online], Available at: https://data.worldbank.org/indicator/TM.TAX.MRCH.WM.AR.ZS?end=2017&locations=CN-US&start=2014 (2019)

## Annex 1

**Basic statistics of equalized import time series with 5-month lag**

| Statistics | 1.MLP 15-3-1 | 2.MLP 15-7-1 | 3.MLP 15-7-1 | 4.MLP 15-10-1 | 5.MLP 15-4-1 |
|---|---|---|---|---|---|
| Minimal prediction (Training) | 5752.93 | 7952.83 | 8335.37 | 8239.21 | 8144.24 |
| Maximal prediction (Training) | 48664.89 | 47750.67 | 47773.06 | 47580.68 | 48129.11 |
| Minimal prediction (Testing) | 9259.27 | 9226.17 | 9230.93 | 9160.76 | 9370.74 |
| Maximal prediction (Testing) | 48955.84 | 49144.91 | 48949.81 | 50189.47 | 48249.50 |
| Minimal prediction (Validation) | 11050.71 | 9440.66 | 9456.50 | 9598.51 | 9230.38 |
| Maximal prediction (Validation) | 48266.83 | 47781.93 | 47405.82 | 49350.75 | 47896.93 |
| Minimal residuals (Training) | -7132.39 | -4323.94 | -4579.37 | -4409.89 | -7308.98 |
| Maximal residuals (Training) | 7907.11 | 8082.84 | 7739.53 | 7788.84 | 8020.51 |
| Minimal residuals (Testing) | -8006.76 | -3966.38 | -4111.01 | -4183.43 | -9100.39 |
| Maximal residua (Testing) | 3246.46 | 3928.01 | 3951.86 | 4231.92 | 3952.80 |
| Minimal residuals (Validation) | -4409.00 | -2874.81 | -2937.65 | -4393.50 | -4758.93 |
| Maximal residuals (Validation) | 2948.14 | 4569.47 | 3537.80 | 4950.74 | 3961.35 |
| Minimal standard residua (Training) | -4.98 | -4.03 | -3.88 | -3.73 | -4.74 |
| Maximal standard residuals (Training) | 5.52 | 7.53 | 6.56 | 6.59 | 5.20 |
| Minimal standard residuals (Testing) | -4.21 | -2.80 | -2.70 | -2.74 | -4.33 |
| Maximal standard residuals (Testing) | 1.71 | 2.77 | 2.60 | 2.77 | 1.88 |
| Minimal standard residuals (Validation) | -3.47 | -2.53 | -2.43 | -3.43 | -3.62 |
| Maximal standard residuals (Validation) | 2.32 | 4.02 | 2.93 | 3.87 | 3.02 |

## Annex 2

**Basic characteristics of equalized export time series with 5-month lag**

| Statistics | 1.MLP 15-7-1 | 2.MLP 15-10-1 | 3.MLP 15-8-1 | 4.MLP 15-7-1 | 5.MLP 15-7-1 |
|---|---|---|---|---|---|
| Minimal prediction (Training) | 1300.71 | 1279.18 | 1309.90 | 1428.62 | 1383.41 |
| Maximal prediction (Training) | 11510.48 | 11556.53 | 11654.73 | 11562.20 | 11561.23 |
| Minimal prediction (Testing) | 1624.19 | 1619.56 | 1771.96 | 1693.04 | 1795.84 |
| Maximal prediction (Testing) | 11454.29 | 11498.79 | 11547.04 | 11525.11 | 11498.08 |
| Minimal prediction (Validation) | 1493.81 | 1586.39 | 1607.52 | 1592.13 | 1637.88 |
| Maximal prediction (Validation) | 11251.28 | 11367.60 | 11356.74 | 11297.36 | 11352.20 |
| Minimal residuals (Training) | -2904.28 | -2944.08 | -3048.53 | -2923.75 | -2955.03 |
| Maximal residuals (Training) | 2740.53 | 2762.95 | 2792.45 | 2678.17 | 2704.00 |
| Minimal residuals (Testing) | -2155.91 | -2307.25 | -2377.33 | -2179.79 | -2286.90 |
| Maximal residua (Testing) | 1982.08 | 1932.21 | 2135.10 | 2106.05 | 2284.37 |
| Minimal residuals (Validation) | -1093.76 | -1464.55 | -1270.61 | -1175.52 | -1305.33 |
| Maximal residuals (Validation) | 2463.54 | 2468.12 | 2324.76 | 2450.09 | 2456.53 |
| Minimal standard residua (Training) | -5.43 | -5.52 | -5.76 | -5.49 | -5.56 |
| Maximal standard residuals (Training) | 5.12 | 5.18 | 5.28 | 5.03 | 5.09 |
| Minimal standard residuals (Testing) | -3.70 | -3.84 | -3.93 | -3.62 | -3.75 |
| Maximal standard residuals (Testing) | 3.41 | 3.22 | 3.53 | 3.50 | 3.75 |
| Minimal standard residuals (Validation) | -1.89 | -2.60 | -2.31 | -2.12 | -2.30 |
| Maximal standard residuals (Validation) | 4.26 | 4.38 | 4.22 | 4.41 | 4.32 |

# Customer's View and Experience with Loyalty Programs in the South Bohemia

*Jaroslav* Stuchlý[1], *Tsolmon* Jambal[1], *Lenka* Ližbetinová[1*]

[1]Institute of Technology and Business, School of Expertness and Valuation, Okružní 517/10, 37001 České Budějovice, Czech Republic

**Abstract.** The aim of the article is to determine the attitudes of South Bohemian customers to the loyalty programs. The article is based on the partial output of research focusing on customer loyalty to loyalty programs. The research was conducted in 2018 by a questionnaire survey with 267 respondents. Respondents interviewed were customers who live in the South Bohemian Region. The results presented in the article are based on the part of the questionnaire, where the respondents answered how much they agree with the fourteen claims concerning loyalty programs. Attention is also focused on looking for dependencies of overall evaluation in relation to loyalty programs in terms of identification variables: gender, age, education and the size of the respondent's place of residence. Two-choice statistical tests and the Kruskal-Wallis test are used to determine dependencies. The statistical program R. was used to perform these tests.
**Key words:** loyalty program, customers, customer view, marketing

## 1 Introduction

Loyalty programs are among the means of sales support. Loyalty program can be defined as a marketing tool that serves to enhance the loyalty of existing customers and to gain confidence in future customers [1, 2]. The goal of loyalty programs is to motivate a customer who has already bought or used the service to become a permanent customer or stay longer with the product [3]. Furthermore, the loyalty program can be seen as an instrument of competitive advantage [4, 5]. However, it is not possible to use the "loyalty program". Given that most companies already have some type of loyalty program, the benefits and bonuses of loyalty programs have become commonplace. Many customers are a member of several competing loyalty programs at the same time, and their purchasing decisions are subject to those who offer better conditions at the moment. This is most often the case with loyalty programs of retail chains where chains offer almost identical goods and therefore it is not an obstacle for customers to get these goods in a different store [6, 7]. The Loyalty Program serves to maximize the company's turnover or for certain products. The customer should operate at all stages of the purchasing process [8, 9]. According to Brych [10], the loyalty program should be particularly comprehensible, playful, creative and should communicate regularly with clients. Zamazalová [11] sees their sense in creating a long-term relationship between the consumer and the company. The study by Waari, Bonuke and Kosgei [12]

* Corresponding author: lizbetinova@mail.vstecb.cz

confirms a direct relationship between loyalty programs benefits and customer loyalty. Their study was in the context of hotels with three to five stars. Based on their research, they concluded that the customer loyalty effect is enhanced or becomes stronger when customer satisfaction.

The main prerequisite for a good loyalty program is building customer loyalty, and the pay-out structure should be designed to make sure that the longer the customer stays on the program, the higher the return. This approach is particularly long-term, since keeping loyal customers is easier because they are more willing to pay higher prices [13]. Reichheld and Teal [14] claims that lack of loyalty reduces the company's performance by 25-50%. Relationship marketing has two positive effects, namely to strengthen the endurance of customers combined with the acquisition of new ones. The idea is that the longer customers stay with the brand, the better it will be. Revenue per customer will increase, operational costs will decline, the spread of positive messages between people will attract other customers, and loyal consumers are not so price-sensitive. Several factors influence a customer's satisfaction. These include, for example, satisfaction with goods, satisfaction with the operator, satisfaction with the sales environment or satisfaction with the whole business [15]. At present, an important factor is the use of IT resources to simplify the customization of loyalty programs [16-18]. The company can influence all the above mentioned factors, namely by purchasing marketing in the creation of a supply of goods, by ensuring affordable, quality goods in the required assortment, also by arranging on the sales area and the overall atmosphere in the shop [19], but above all by quality services based on staff, which is trained, willing and pleasant [20-23] and also by the company's overall approach to customers [24, 25].

In the Czech Republic, we can find a wide range of loyalty systems that have become one of the tools of customer engagement. Not only large large-scale retail chains use them, but also non-food stores or specialized shops and service providers including entities in transport and logistics chains [26-28]. They differ from each other by their parameters, offered bonuses and the way discounts are provided. However, they have one thing in common - to keep and attract customers and strengthen their loyalty [29, 30]. Tellis [13] sets out four important principles that can ensure that the bonus program achieves its goals and not just the extra cost of the company, namely: focusing on promising customers (loyal or big customers), building loyalty (reward for brand loyalty, the longer the customer stays on the program , the higher the yield), creating a value for the customer (creating a value the customer cannot just buy or buy on the market) and creating a more profitable than a cost centre (in the long run).

## 2 Data and methods

The aim of the article is to determine the attitudes of South Bohemian customers to forms and to set the conditions of the loyalty programs. The article draws on partial outcomes of research focusing on customer attitudes towards loyalty programs with a focus on the young generation. The young generation has a more positive attitude towards IT and loyalty programs and is more flexible in terms of accessing and sharing loyalty programs with each other. Research was conducted in 2018 by a questionnaire survey of 267 respondents. The interviewees were customers who have their permanent residence in the South Bohemian Region. The results presented in the article are based on the part of the questionnaire, where respondents could answer how much they identify with the 14 loyalty program submissions. They presented their attitudes on the Linkert scale from 1 to 5, where 1 is totally disagree, 2 disagree, 3 neutral, 4 agree and 5 totally agree.

The total number of respondents is 267 and the sample is composed of 43.5% of men and 56.5% of women. In the processing of the results, the basic statistical characteristics were used – relative and absolute frequency, weighted arithmetic mean. Attention is also focused on looking for dependencies of the overall evaluation of the relation to loyalty programs on the following identification variables: gender, age, education and the size of the size of respondent's municipality. For the overall evaluation variable (EV), we use the sum of the scores of all variants of the claim by individual respondents. In this calculation, it is necessary to negate the positive statement C10 - C11 and shifts 1 to 5 and 2 to 4. In particular, we will be interested in whether this evaluation variable depends significantly on the identification variables. We will use statistical two-choice tests and the Kruskal-Wallis test. The statistical program R was used to perform these tests.

## 3 Results and Discussion

The article presents partial research results that focus on the customers' view of the functioning and loyalty program settings. The dependence of their attitude is tested in terms of age, gender, education, and community size. Their attitudes are examined through identifying 14 claims on the Linkert scale from 1 to 5, where 1 is totally disagree, 2 disagree, 3 neutral, 4 agree and 5 totally agree. Table 1 and Figure 1 present the identification of respondents with claims C01 to C14.

**Table 1.** Customer´s view on the claims about working way of loyalty system

| Code | Claim | Percentage of respondents | | | | | | Mean |
|---|---|---|---|---|---|---|---|---|
| | | totally disagree | disagree | neutral | agree | totally agree | Sum | |
| C01 | Loyalty programs are many and sometimes customers do not have an overview of which programs are involved | 3.7 | 16.9 | 17.6 | 44.2 | 17.6 | 100 | 3.55 |
| C02 | Loyalty programs are so many that customers are no longer interested in registering | 4.1 | 13.9 | 25.1 | 43.4 | 13.5 | 100 | 3.48 |
| C03 | Mostly, there is not enough care about members of the loyalty program | 4.5 | 14.6 | 34.1 | 35.2 | 11.6 | 100 | 3.35 |
| C04 | Targeting offers to members of loyalty programs is often inaccurate | 3.7 | 22.1 | 33.3 | 34.5 | 6.4 | 100 | 3.18 |
| C05 | The awareness of members of the loyalty program about current and new benefits is not sufficient | 6.4 | 23.2 | 30.0 | 33.3 | 7.1 | 100 | 3.12 |
| C06 | Loyalty programs are often so complex that it is unclear under which conditions benefits are gained | 5.6 | 25.1 | 21.7 | 38.6 | 9 | 100 | 3.20 |
| C07 | Contacting loyalty program members with unintended offerings is common | 6.0 | 22.5 | 32.6 | 31.8 | 7.1 | 100 | 3.12 |
| C08 | Frequent problem is the time-limited possibility of applying the benefit | 3.4 | 15.7 | 27.0 | 36.7 | 17.2 | 100 | 3.49 |
| C09 | There are frequent technical problems with loyalty programs (for example, the card cannot identify the customer - it must be re-registered) | 20.2 | 36.0 | 23.2 | 14.2 | 6.4 | 100 | 2.51 |
| C10 | I'm not confronted with problems with loyalty programs | 6.4 | 24.3 | 26.2 | 31.8 | 11.2 | 100 | 3.17 |
| C11 | I like to use loyalty programs | 3.0 | 11.2 | 26.2 | 40.8 | 18.7 | 100 | 3.61 |
| C12 | A well-defined loyalty program motivates me to shop in specific stores | 5.2 | 12.0 | 24.7 | 42.3 | 15.7 | 100 | 3.51 |
| C13 | The benefits of membership motivate me to make more frequent purchases | 6.4 | 17.2 | 26.2 | 37.8 | 12.4 | 100 | 3.33 |
| C14 | I am not satisfied with the loyalty programs offered | 8.6 | 34.1 | 35.2 | 16.5 | 5.6 | 100 | 2.76 |
| Total | | 6,2 | 20.6 | 27.4 | 34.4 | 11.4 | 100 | 3.24 |

Source: Authors.

From the presented results (Table 1 and Figure 1), most of the customers use the loyalty programs in general (final mark 3.61 – the weighted arithmetic average of the respondents' rating) and the benefits of customer membership motivate for more frequent purchases (up to 60.2%, mark 3.51). On the other hand, 61.8% of customers are arguing that loyalty programs are many and customers sometimes have no insight into which of them are involved. Also, 56.9% of customers think that loyalty programs are so many that customers are no longer interested in registering for others. They also have a more negative view of the temporary nature of the use of benefits (53.9% it does not, and 27% to that is neutral). 46.8% customer agrees with the statement that is usually sufficient care for members of the loyalty program (34.1% to the set neutral), which creates an area for improvement. On the contrary, customers report that they do not have frequent technical problems (56.2%) in the operation of loyalty programs (eg card cannot identify the customer, etc.) and most of them deny that they would not be satisfied with the loyalty programs offered (22.1% are not satisfied and 35.2% state a neutral stance).

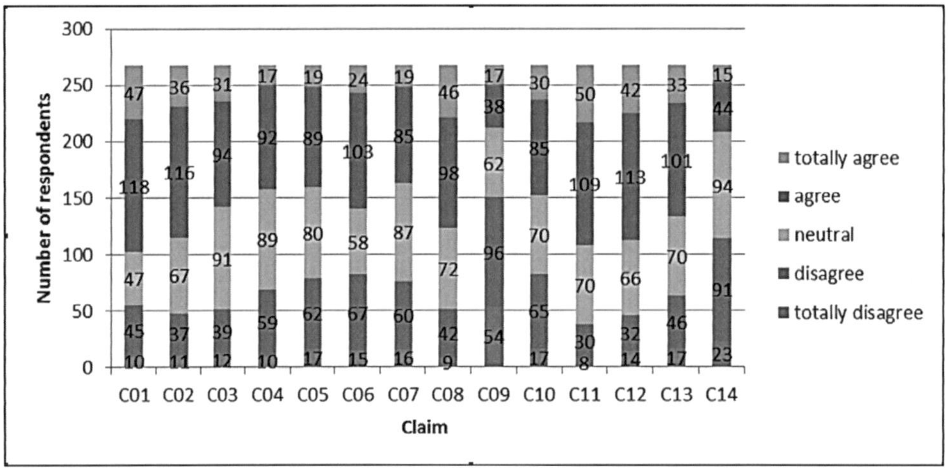


**Fig. 1.** Frequency distribution for the claims

Source: Authors.

In the next section, the dependence of the overall assessment of attitudes on loyalty programs is tested in relation to the identification variables: gender, age, education and the size of the respondent's residence. The overall evaluation variable (EV) is created as a sum of the scores of all variants of claims by individual respondents. Dependence is tested using statistical two-choice tests and the Kruskal-Wallis test. In Table 2, there are calculated basic numerical characteristics of the EV variable and its course is approximated by the box diagram and the absolute frequency histogram (see Figure 2).

**Table 2.** Numerical characteristics of EV evaluation

| mean | sd | IQR | coef.var. | skewness | kurtosis | min | 25% qu. | median | 75% qu. | max | n |
|---|---|---|---|---|---|---|---|---|---|---|---|
| 43.81 | 6.26 | 7 | 0.143 | -0.520 | 0.777 | 22 | 41 | 44 | 48 | 62 | 267 |

Source: Authors.

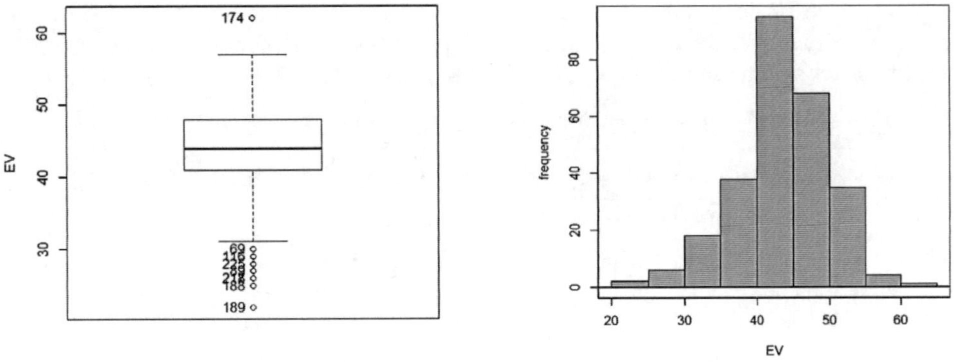


**Fig. 2.** Distribution of EV evaluation quantity - box diagram and histogram of absolute frequencies

Source: Authors.

The normality of data transformation is tested by Shapiro-Wilk's test, with outputs W = 0.9766, p-value = 0.000221. Based on the test result, normality was rejected at 5% significance level. It is probably caused by a number of outlying values and for this reason, there will be used non-parametric tests. Table 3 presents the conditional numerical characteristics of the EV variable by gender, and Figure 3 illustrates the conditional distributions.

**Table 3.** Conditional numerical characteristics of EV evaluating gender by gender

| sex | mean | sd | IQR | coef.var. | skewness | kurtosis | min | 25% qu. | median | 75% qu. | max | n |
|---|---|---|---|---|---|---|---|---|---|---|---|---|
| female | 43.91 | 5.78 | 7 | 0.132 | -0.326 | 0.287 | 27 | 41 | 44 | 48 | 57 | 151 |
| male | 43.67 | 6.86 | 8 | 0.157 | -0.651 | 0.953 | 22 | 40 | 44 | 48 | 62 | 116 |

Source: Authors.

Testing the independence of the EV variable on sex is performed using the Wilcoxon's two-choice test. The values obtained are W = 8726.5, p-value = 0.9604. The fact that EV is gender-neutral has already been deduced from the Figure 3 and the Table 3.

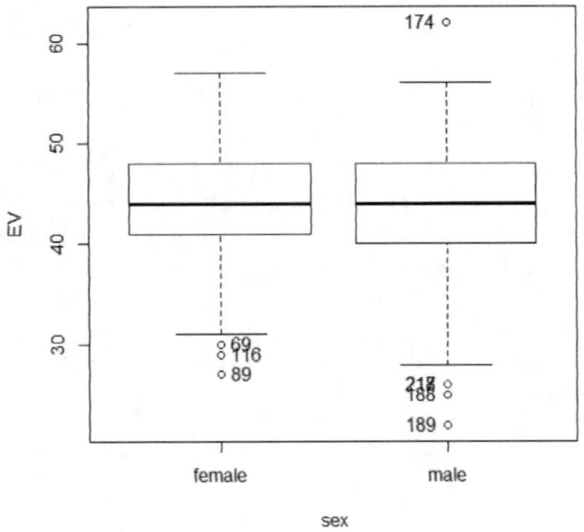


**Fig. 3.** Conditional distribution of the EV evaluation variable by gender

Source: Authors.

Also, the test failed to prove this dependence at a 5% level of significance. Consequently, the dependency test of the respondents is tested. For this purpose, the conditional numerical characteristics and the conditional distribution are presented in Table 4 and in Graph 4.

**Table 4.** Conditional numerical characteristics of EV evaluation by age

| age | mean | sd | IQR | coef.var. | skewness | kurtosis | min | 25% qu. | median | 75% qu. | max | n |
|---|---|---|---|---|---|---|---|---|---|---|---|---|
| 15-29 | 42.76 | 6.62 | 8 | 0.155 | -0.800 | 0.705 | 22 | 40 | 43 | 48 | 57 | 123 |
| 30-39 | 45.78 | 5.77 | 8 | 0.126 | 0.080 | 0.185 | 31 | 42 | 46 | 50 | 62 | 68 |
| 40-49 | 44.12 | 6.15 | 6.5 | 0.139 | -0.342 | 0.199 | 29 | 41.5 | 44 | 48 | 57 | 51 |
| 50-59 | 42.67 | 5.20 | 7.75 | 0.122 | 0.267 | -1.039 | 34 | 39 | 41.5 | 46.75 | 51 | 18 |
| 60 and more | 43.86 | 3.18 | 1.5 | 0.073 | -2.159 | 4.846 | 37 | 44 | 45 | 45.5 | 46 | 7 |

Source: Authors.

The conditional numerical characteristics and the conditional distribution are also presented in Graph 4.

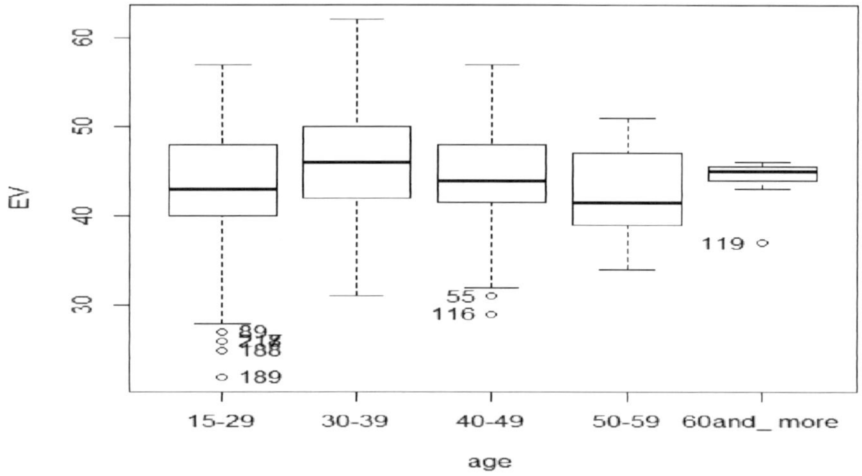


**Fig. 4.** Conditional distribution of the assessment variable EV by age

Source: Authors.

The Kruskal-Wallis test is used to test the independence of the EV variable for the ages and result is chi-squared = 9.1841, df = 4, p-value = 0.05666. Already from Chart 4 and Table 4 it is foreseeable that EV is age dependent. However, this dependence was not proven by the test at a 5% level of significance. However, there is confirmed a statistically significant EV dependence from age to 10% significance level. Another fact examined is dependence on the education of respondents. Conditional numerical characteristics are shown in Table 5.

**Table 5.** Conditional numerical characteristics of the EV evaluation variable by education

| education | mean | sd | IQR | coef.var. | skewness | kurtosis | min | 25% qu. | median | 75% qu. | max | n |
|---|---|---|---|---|---|---|---|---|---|---|---|---|
| university | 43.90 | 7.07 | 7 | 0.161 | -0.589 | 0.458 | 22 | 41 | 44 | 48 | 57 | 73 |
| upper secondary | 42.15 | 6.89 | 9 | 0.164 | -0.528 | 0.225 | 26 | 37 | 43 | 46 | 55 | 40 |
| lower secondary | 44.41 | 5.66 | 7.25 | 0.127 | -0.414 | 1.141 | 25 | 41 | 44 | 48.25 | 62 | 144 |
| primary | 41.10 | 4.15 | 1.75 | 0.101 | -1.069 | 1.995 | 32 | 41 | 41.5 | 42.75 | 47 | 10 |

Source: Authors.

Conditional distribution is shown on Figure 5.

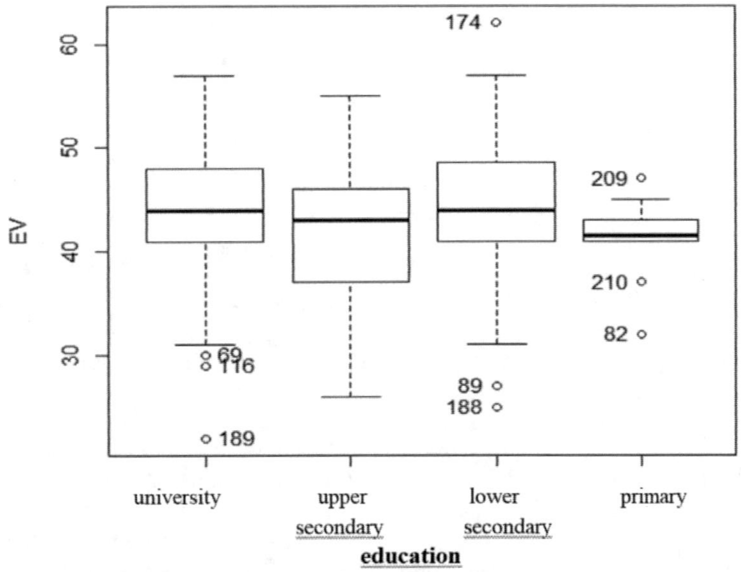


**Fig. 5.** Conditional distribution of EV evaluation value by education

Source: Authors.

The Kruskal-Wallis test is used to test the independence of the EV variable by education and the result of this test is: chi-squared = 6.1148, df = 3, p-value = 0.1062. There is evidence in Figure 5 and Table 5 that EV depends on education. However, the test did not confirm the significance of the statistical significance to the 5% significance level. Another examined dependency is the dependence of EV on the size of the respondents' residence (expressed in thousands of inhabitants). Conditional numerical characteristics are shown in Table 6.

**Table 6.** Conditional numerical characteristics evaluation values EV respondents by size of municipality

| municipality | mean | sd | IQR | coef.var. | skewness | kurtosis | min | 25% qu | median | 75% qu | max | n |
|---|---|---|---|---|---|---|---|---|---|---|---|---|
| (0;1] | 44.06 | 6.86 | 8 | 0.156 | -0.932 | 0.760 | 26 | 4 | 45 | 49 | 56 | 53 |
| (1;10] | 43.24 | 5.79 | 6.25 | 0.134 | -0.243 | 0.204 | 27 | 4( | 43 | 46.25 | 57 | 100 |
| (10;50] | 43.90 | 5.87 | 7 | 0.134 | -0.083 | 1.094 | 31 | 4 | 44 | 48 | 62 | 60 |
| (50;1200] | 44.52 | 6.95 | 9 | 0.156 | -0.888 | 1.666 | 22 | 4 | 44 | 50 | 57 | 54 |

Source: Authors.

Conditional distribution is shown on Figure 6.

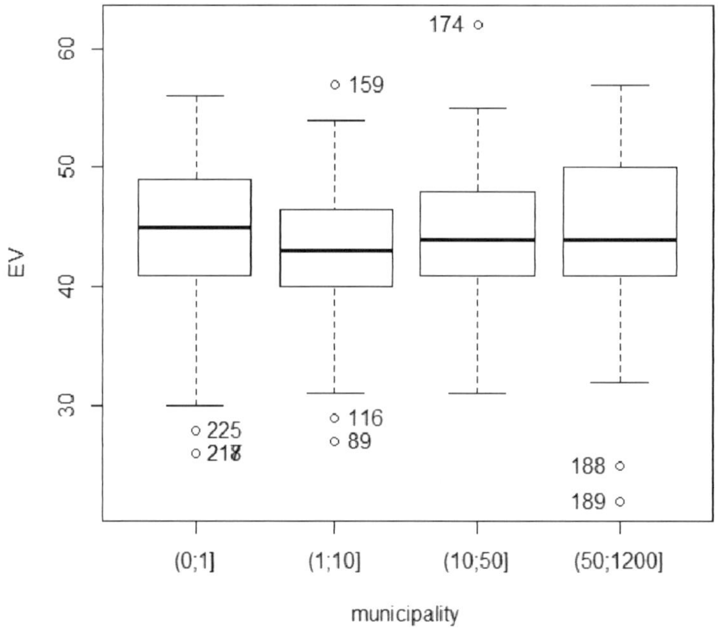


**Fig. 6** Conditional distribution of the EV evaluation variable according to the size of the respondents' municipality

Source: Authors.

The test of independence of EV evaluating EV on education is done using the Kruskal-Wallis test. The test provided these outputs: chi-squared = 2.9836, df = 3, p-value = 0.3942 A free view of Figure 5 and Table 5 gives us the assumption that EV does not depend on the size of the residence. This assumption was confirmed by the Kruskal-Wallis test, which confirmed the zero hypothesis and therefore no statistically significant dependence on the 5% significance level was confirmed.

## 4 Conclusion

The article presents partial research results that focus on customer perceptions of loyalty programs. The dependence of their attitude is tested in terms of age, gender, education, and municipality size. From the processing of the data obtained, it was found that customers mostly enjoy loyalty programs and a well-established loyalty program motivates them to buy more from specific companies. This is confirmed by the control question, which is a negation of satisfaction with loyalty programs, and customers identify with it at least. In the second tent, customers also mention areas where businesses have an area to improve their loyalty program settings. They most sympathize with the belief that loyalty programs are many and customers sometimes do not have an overview of which programs they are involved in. Here is the space to differentiate from other loyalty programs of competition and reduce the demand for its use for customers – friendly setting and ease of use for customers. For customers, the problem area is a time limit for taking advantage of loyalty programs. There is a need to find a compromise point where the program will create a permanent advantage for a trader based on time constraints, while at the same time it will be of interest to the customer so that the time limit is not the fact that the advantage is not paid to him in such a short time. Technical and functional loyalty programs do not pose a significant problem for most customers. From the testing of attitudes to loyalty programs in terms of gender, age, education and municipality, there were no statistically significant differences in the significance level of 5%. Statistically significant dependence EV values was demonstrated only in education at 10% significance level. No significant dependence on other identification variables has been demonstrated.

When planning and setting loyalty programs with respect to target customers, it is appropriate not only to make effective adjustments, but also to differentiate from loyalty programs of competitors. It follows that if a competitive program exists, then there is a greater pressure on originality and uniqueness. In this way, it is possible to motivate customers to make more frequent purchases in specific stores. An invaluable fact is to consider the time constraint of the benefits offered so that for most of the target customers they are still realistically beneficial and create a sort of compromise between the benefits of one and the other of the relationship.

## Acknowledgment

This paper is an output of the science projects IGS 8110IGS201813, TL02000559 Secure cities for pedestrians and seniors, TL02000017 Intergenerational management to support digitization in construction and TL01000349 Stabilization and development of SMEs in rural areas.

## References

1. M. Tomcik, P. Rosenlacher, Number of advertisements per day and their relevance to consumers. *Acta Oeconomica Universitatis Selye*, **7**(2), 162–174 (2018)
2. E. Kicova, M. Nadanyiova, Brand as a Tool of Company's Strategic Marketing in Practice Conference. *Proceeding of 7th ESE International Conference on Sports, Health and Management. Lecture Notes in Management Science*, **73**, pp. 29-34 (2017)
3. M. Palatkova, The Opportunities to Assess the Performance of Destination Marketing Management. *12th International Conference on Hradec Economic Days (HED 2014): Economic Development and Management of Regions Location,* pp. 362-371 (2014)
4. M. Hesková, P. Štarchoň, *Marketingová komunikace a moderní trendy v marketing [Marketing communication and modern trends in marketing]*. Prague: Oeconomica (2009)
5. L. Mura, B. Gontkovicova, E. Dulova Spisakova, Z. Hajduova, Position of Employee Benefits in Remuneration Structure. *Transformations in Business & Economics*, **2**(47), 156-173 (2019)
6. R. Benda-Prokeinová, K. Dobeš, L. Mura, J. Buleca, Engel's Approach as a tool for estimating consumer behaviour. *E & M Ekonomie a Management*, **20**(2), 15-29 (2017)
7. M. Karlíček, P. Král, *Marketingová komunikace: jak komunikovat na našem trhu [Marketing communication: how to communicate in our market]*. Prague: Grada (2011)
8. A. Krizanova, L. Gajanova, M. Nadanyiova, Design of a CRM Level and Performance Measurement Model. *Sustainability*, **10**(7) (2018)
9. L. Vilcekova, P. Starchon, Consumer Perception of Selected Brands Explored through Archetypes. *Proceeding of the 4th International Conference on Education and Social Sciences (INTCESS) Istanbul*, pp. 819-823 (2017)

10. J. Brych, Věrnostní program neprodává, ale buduje [The loyalty program does not sell but builds]. *Marketing a Media [Marketing and Media]* [online], Available at: http://mam.ihned.cz/c1-63329390-vernostni-program-neprodava-ale-buduje (2015)
11. M. Zamazalová, *Marketing obchodní firmy [Business company marketing]*. Prague: Grada (2009)
12. D. Waari, R. Bonuke, D. Kosgei, Loyalty programs benefits and customer loyalty: The mediating role of customer satisfaction. *International Journal of Commerce and Management*, **6**(4), 726-737 (2018)
13. G.J. Tellis, *Reklama a podpora prodeje [Advertising and sales promotion]*. Prague: Grada (2016)
14. F. Reichheld, T. Teal, The loyalty effect: the hidden force behind growth, profits, and lasting value. *Harvard Business School Press* (2001)
15. E. Smolkova, P. Starchon, D. Weberova, Country-of-Origin Brands from the Point of View of the Slovak and Czech Consumers. *Proceeding of conference Innovation Management and Education Excellence Vision 2020. From Regional Development Sustianablility to Global Economic Growth*, pp. 2119-2130 (2016)
16. T. Fu, C. Li, L. Guo, Y. Wang, How we view competitions between the converging Internet social platforms: Does higher user number mean the final victory. *International Journal of Modern Physics C*, **26**(8) (2015)
17. V. Nývlt, The role of managing knowledge and information in BIM implementation processes in the Czech Republic. *9th International Scientific Conference on Building Defects (Building Defects),* **146** (2018)
18. J. Lojda, K. Ventrubová, The Grant Policy of the Forestry Sector in the Czech Republic After 2013. *Reports of Forestry Research-Zpravy Lesnickeho Vyzkumu*, **60**(1), 64-72 (2015)
19. V. Mulačová, P. Mulač et al., *Obchodní podnikání ve 21. Století [Business in the 21st century]*. Prague: GRADA Publishing (2013)
20. Y. Liu, The long-term impact of loyalty programs on consumer purchase behavior and loyalty. *Journal of Marketing*, **71**(4), 19-35 (2007)
21. P. Seemann, V. Farkasova, Coaching as a tool for Improving the Quality of Managerial Work. *Proceeding of Conference: 15th International Scientific Conference on Globalization and its Socio-Economic Consequences*, pp. 661-669 (2015)
22. R. Kampf, S. Lorincova, M. Hitka, O. Stopka, Generational Differences in the Perception of Corporate Culture in European Transport Enterprises. *Sustainability*, **9**(9) (2017)
23. I. Kmecová, Analysis of the Efficiency of the Educational Processes of the Subjects Business Management, Human Resource Management, and Mathematics, and their Comparison. *12th International Technology, Education and Development Conference (INTED)*, pp. 1781-1788 (2018)
24. A. Kucharcikova, M. Miciak, Human Capital Management in Transport Enterprises with the Acceptance of Sustainable Development. *Sustainability*, **10**(7) (2018)
25. S. Lorincova, M. Hitka, Z. Balazova, Corporate Culture in Slovak Enterprises as a factor of HRM Quality – Case Study. *International Journal for Quality Research*, **10**(4), 719-732 (2016)
26. V. Cempirek, I. Drahotsky, P. Vrbova, The Concept of Prices from Antiquity to today in Comparison with Selected Current International trends. *Proceeding of the Conference: 14th International Scientific Conference of Doctoral Students and Young*

*Scientists on Economic, Political and Legal Issues of International Relations*, pp. 68-75 (2015)

27. G. Fedorko, S. Honus, S. Badiarova, Analysis of the Utilization of Machinery in the Production Process Using Computer Simulation. MATEC Web of Conferences: *18th International Scientific Conference on LOGI*, **134** (2017)
28. M. Dobrodolac, L. Svadlenka, M. Cubranic-Dobrodolac, et al., A model for the comparison of business units. *Personel Review*, **47**(1), 150-165 (2018)
29. Z. Musová, Responsible behavior of businesses and its impact on consumer behavior. *Acta Oeconomica Universitatis Selye*, **4**(2), 138-147 (2015)
30. M. Gogolová, The doorstep sales and the selling events in Slovakia. *Acta Oeconomica Universitatis Selye*, **3**(2), 33-39 (2014)

# Machine prediction of US imports from the PRC in the context of mutual sanctions

*Petr* Šuleř[1*], *Jan* Mareček[1]

[1]Institute of Technology and Business, School of Expertness and Valuation, Okružní 517/10, 37001 České Budějovice, Czech Republic

**Abstract.** The aim of this paper is to mechanically predict the import of the United States of America (USA) from the People's Republic of China (PRC). The trade restrictions of the USA and the PRC caused by the USA feeling of imbalance of trade between the two states have significantly influenced not only the trade between the two players, but also the overall climate of international trade. The result of this paper is the finding that multilayer perceptron networks (MLP) appear to be an excellent tool for predicting USA imports from the PRC. MLP networks can capture both the trend of the entire time series and its seasonal fluctuations. It also emerged that time series delays need to be applied. Acceptable results are shown to delay series of the order of 5 and 10 months. The mutual sanctions of both countries did not have a significant impact on the outcome of the machine learning prediction.
**Key words:** machine prediction, trade war, sanctions, international trade

## 1 Introduction

The business relationship between the USA and China is currently supported by roughly 2.6 million jobs in a large number of industrial sectors including jobs which Chinese corporations have created in the USA [1]. Since Chinese middle class will go on developing in the next decade, American corporations will have the opportunity to benefit from a new and lucrative consumer basecamp which may further encourage the employment and economic growth [2]. A big advantage to the US economy in going into business with China is that China has become an integral part of the global production supply chain; a considerable part of its export includes components produced abroad and provided for the final assembly in China [3]. If the value of these imported components is subtracted from Chinese export, the business deficit between the USA and China will be reduced by a half to roughly 1% of GNP. The next big advantage is the expectation that China will remain one of the fastest growing strong economies and will be creating ample opportunities for the growth of the US society; supposing that China will continue in economic improvements which will remove existing obstacles to the easy access to the market in a large number of sectors [4].

* Corresponding author: petr.suler@cez.cz

According to Chossudovsky [5] China is not dependent on the import from the USA. On the contrary, the USA strongly relies on the import from China which means that the US economy is weak in regard to industry and production. As a matter of fact, "Made in China" is a backbone of the majority of retails in the USA. These retails strongly encourage domestic consumption in all distinct categories of goods – clothes, shoes, hardware, electronics, toys, jewellery, household needs, food, TV, mobile phones etc. "Made in China" is an excellent source of profit and wealth in the USA since consumer commodities imported from low-wage Chinese economy are often sold in retails for more than a tenfold price of their production costs [6].

At the end of 2018 and in the first half of 2019 the business rivalry between PRC and the USA had a deep impact on Chinese ideologies and politics, social-economic and military development and its foreign policy [7].

The "business war" between China and the USA was one of the most important international economic events in 2018. The USA and China are the two soundest global economies which means that the current business war between the USA and China also influences other economies in the world [8, 9]. Owing to the Chinese crucial role in the global economy and geopolitical impacts, the USA referred to China as a competitor of the strategy [10] and considered China as a grave threat to American security interests. The fact that China was officially marked and treated as a major threat to the US national interests had catastrophic consequences for the USA and the rest of the world [11].

Itakura [12] assesses the impact of the business war between the USA and China using a model of global business with dynamic calculation general equality (CGE). The author carries out a simulation analysis ex ante and analyses three scenarios in order to understand the way the business war influences tariffs, investments and productivity. The escalation of the business war reduces the gross national product (GNP) in China by -1.41% and in the USA by -1.35%. The business war reduces almost the entire sectoral import and outputs in the both countries. The author provided the dynamic model CGE with requirements for the import and analyses the difference between the results of both models concerning impacts of the business war on GNP and the bilateral trade.

Chong and Li [13] summarize the key ideas of Trump's administration which stands behind Chines-American business war. For example, Trump believes that the bilateral trade deficit between the USA and China jeopardizes American interests. He argues that the main cause of the bilateral deficit lies in unfair export Chinese subventions and high tariffs as contrasted to the American import. The only way for the USA to reduce the bilateral deficit is to impose high tariffs on Chinese products. The bilateral trade inequality does not actually threaten the US economy since Chinese investors invest back in the USA and use means from the bilateral trade surplus. The convenient way how to resolve Chinese-American trade inequality would be to increase Chinese imports and to persuade the USA to lift the restriction on exporting advanced technologies to China and that the both countries should overcome mutual bilateral trade obstacles.

The recent figures have shown that the USA frequently imposes economic sanctions. The sanctions, though, usually do not meet the intended political or economic goals [14]. Li [15] gives an example of the cultural differences. Cultural standards of the target countries play an important role in determining the effectiveness of the sanctions. Using examples of American economic sanctions against China, the author demonstrates that the standards and beliefs of the target countries play an essential role in assessing the effectiveness of the imposed sanctions. The result of the study is that initiators of economic sanctions should take into account the cultural responsibility and sensitivity of goals in order to impose legal and effective economic sanctions as an appropriate means of foreign policy.

The USA has imposed a large scale of economic sanctions starting from export controls up to orders to specific imports. Yang et al. [16] found out that although American sanctions did not have a significant devastating impact on the overall Chinese economic growth and trade between the both countries with respect to macro economy, the sanctions have a damaging impact on the producer and consumer in both countries. The US economic sanctions prevented the import of technologies to China. American restrictions on the import from China caused dead weight costs for the USA as a result of higher domestic production costs of import substitutes. American export controls inhibited the export from the USA to China and contributed to huge trade deficits between the USA and China. Export controls also caused losses of highly paid jobs in the USA. Moreover, the US economic sanctions against China also had significant impacts on the third party. Chinese diversification of import to sources other than the USA may have a long-term impact on the export from the USA to China even though American economic sanctions against China will be lifted.

China declared to impose sanctions on American corporations involved in the agreement to sell tanks, rockets and other related equipment equal to 2.2 billion dollars to Thai-wan; China proclaimed that this act had violated Chinese sovereignty and national security. The spokesman of Chinese Ministry of Foreign Affairs Geng Shuang said that the USA having sold the weapons presents "a serious violation of the international law and regulations governing international relationships". In order for China to defend its national interests, the country will impose sanctions on American corporations involved in the above-mentioned sale of weapons to Tchai-wan [17].

In September 2018 American president Donald Trump announced the imposition of ten-per cent tariff on the import of products from China. However, Trump also threatened to increase the tariff to 25% as of the following year. China promptly responded by a plan aiming at punitive measures in the sector of fiscal and monetary policy [18]. The monetary policy can thereby expect growing pressure on depreciating Chinese currency with the intent to compensate for the consequences of the measures. This will be implemented by mere verbal interventions or a foreign-exchange intervention of the central bank. The fiscal policy will most likely see other punitive measures [19].

Perhaps American president Donald Trump has just done a great favour to South-Korean electronics, which became Samsung after he had blocked Chinese Huawei, which the president accuses of helping Peking with its spying activities. The main American suppliers of Huawei Corporation allegedly stated that they would immediately disrupt the supplies of the Chinese corporation. In addition, Google added fuel to the flames as it stated that it would block the use of some of Huawei's services and applications for Android on its devices. As Trump had put Huawei on the black list of American technological businesses, he effectively banned Huawei from producing smartphones and tablets to sale outside China and to sell its laptops elsewhere [20].

Besedeš and Prusa [21] estimated impacts of the reduction of anti-dumping tariffs according to the probit model of random outcomes. Their research demonstrated that effects of anti-dumping tariffs which the USA imposed on the trade are under a careful examination and provisional decision and they are even more visible than after the final decision would have been reached. They also found out that for the target countries it is hard to recover lost trade level after the anti-dumping tariff has been abolished. Lee [22] compared the effects of import reduction with the target countries and the effects after the third country has abused the import policy. It demonstrates that anti-dumping tariffs have import-restricting effects. Park [23] in his works analysed the influence of anti-dumping tariffs on import studying Chinese data from 2002 to 2004 using Generalized Method of Moments (GMM). The results show that the anti-dumping protection has far-reaching effects on the inhibition and deflection of the trade.

Choi [24] analysed effects of anti-dumping measures on imports with the intent to find out whether the effect of reducing anti-dumping tariff on the business is principal. Author's results suggest that the increase of anti-dumping tariff by 1% reduces the import of the target product roughly by 0.43 which corresponds to 0.51%. However, exact figures show that the overall import of target products increased roughly by 30% while the anti-dumping tariff was legally effective. It means that anti-dumping tariff presents only a temporary relief from import schemes.

Yu, Wang and Lai [25] aim to devise a methodology that should consider seasonal fluctuations while equalizing time series through artificial neural networks giving the import from China to the USA as an example. The difficulty to predict the volume of foreign trade is often caused by restricting a lot of conventional prediction models. In order to boost the prognostic efficiency, it is necessary to develop a method which will hybridize econometric and artificial intelligence models [26].

According to the idea of non-linear prediction of the phase space reconstruction Weijin and Yuhui [27] presented a model of delays in time series of BP neural network; the model's ability to generalize was improved by Bayesian regularization. The model is used for predicting import and export trades in one sector. In comparison with a broad prognosis assessment, the authors concluded that non-linear prognosis may not only focus on data combinations and accuracy improvement, but it can also clearly reflect non-linear characteristics of the prognosis system.

## 2 Data and methods

The mutual trade restrictions between the USA and China arising from the feeling of the USA concerning the imbalance of the trade between the two countries have significantly influenced the trade between the countries in the recent years as well as the overall atmosphere in the international trade. The United States have decided to regulate trade by imposing duty on specific kinds of goods. This naturally creates an extraordinary situation in terms of the overall trade balance, import, and export. Given that the trade war has been started by the USA, it could be expected that the first variable to be influenced will be the US import from the China. Therefore, the time series of this variable will show extraordinary fluctuations. The information on the USA import from the China thus constitutes an important data set for time series related research.

The dataset illustrates the course of the time series at monthly intervals from January 2000 to July 2019. The course of the time series is shown in Figure 1.

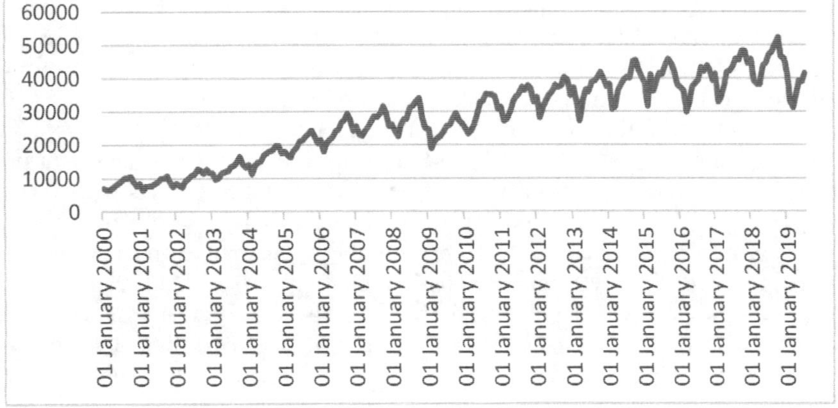


**Fig. 1.** Development of USA import from PRC between January 2000 and July 2019

Source: [28], own processing.

Note: the values in the Table and in the entire text are given in billions of American dollars (USD).

The Figure clearly shows seasonal fluctuations within the individual years of the monitored period. There was an especially significant growth in import from USD 6902 bn. in October 2018. This means more than 7 times higher USA import from the PRC in the monitored period.

Table 1 shows the basic statistical characteristics of the dataset.

**Table 1.** Basic statistical characteristics of the examined dataset

| Statistics | Date (Input variable) | Month (Input variable) | Year (Input variable) | Import (Output (target)) |
|---|---|---|---|---|
| Minimum (Training) | 36556.00 | 1.00000 | 2000.000 | 6375.60 |
| Maximum (Training) | 43646.00 | 12.00000 | 2019.000 | 48127.80 |
| Average (Training) | 40050.47 | 6.45455 | 2009.115 | 27572.76 |
| Standard deviation (Training) | 2001.16 | 3.50688 | 5.473 | 12039.78 |
| Minimum (Testing) | 36585.00 | 1.00000 | 2000.000 | 6584.40 |
| Maximum (Testing) | 43677.00 | 12.00000 | 2019.000 | 52202.30 |
| Average (Testing) | 40130.09 | 6.00000 | 2009.371 | 26146.36 |
| Standard deviation (Testing) | 2418.96 | 3.38683 | 6.691 | 12282.23 |
| Minimum (Validation) | 36646.00 | 1.00000 | 2000.000 | 7070.50 |
| Maximum (Validation) | 43373.00 | 12.00000 | 2018.000 | 50015.00 |
| Average (Validation) | 40412.86 | 6.71429 | 2010.086 | 29966.20 |
| Standard deviation (Validation) | 3403.28 | 3.44159 | 9.284 | 21320.19 |
| Minimum (Overall) | 36556.00 | 1.00000 | 2000.000 | 6375.60 |
| Maximum (Overall) | 43677.00 | 12.00000 | 2019.000 | 52202.30 |
| Average (Overall) | 40116.30 | 6.42553 | 2009.298 | 27716.79 |
| Standard deviation (Overall) | 2069.19 | 3.45140 | 5.669 | 12057.45 |

Source: Own processing.

There will be a regression carried out using neural structures. We will generate multilayer perceptron networks (MLP). Three sets of artificial neural networks will be generated. They will differ according to assumed time series lag:

1. 1-month lag,
2. 5-month lag,
3. 10-month lag.

Time series lag is the amount of data used to calculate the next value (that is in the first case, based on the value of one preceding month, and in the second case, based on five preceding values and in the third case, based on outputs of the previous 10 months). A larger time series lag can result from averaging values. A small lag can result in extreme fluctuations of equalized time series. Each time series lag is more challenging for the complexity of the artificial neural network structure, namely neurons in the input layer (the input layer of the neural network will have 3 neurons in experiment 1, and 15 neurons in experiment 2, and even 30 neurons in experiment 3). The other settings will be equal for each experiment. Time will be used as the continuous independent variable. Seasonal fluctuations will be represented by a category variable of the month and year in which the value was measured. Thus we will use possible monthly seasonal fluctuations of a time series. At the same time, we should capture the overall trend of a time series. US import from China will be used as a dependent variable. The time series will be divided into three sets – Training, Testing a Validation. The first set will comprise 70% input data. We will generate neural structures based on the Training set. We will leave 15% input information in the two remaining sets of data. Both sets will be used for reliability verification of the identified neural structure, or model. We will generate 10,000 neural networks. We will retain 5 of them that show the best characteristics. The least squares method will be used.

The process of generating networks will be finished if no improvement is achieved, i.e. the sum of squares is not decreasing. Thus we will only retain those neural structures where the sum of residual squares to the actual US import from China is the lowest possible (ideally zero). A hidden layer will have at least two neurons, but no more than 8 for experiment 1 and 9 for experiments 2 and 3. We will consider the following distribution functions in the hidden layer and in the output layer: Linear, Logistic, Atanh, Exponential, Sinus.

Other settings will remain default (according to the ANS tool – automated neural networks). In conclusion, we will compare results of the three completed experiments to determine whether the defined machine learning methodology is correct in terms of direction and to identify what time lag is closest to the correct result.

## 3 Results

### 3.1 Experiment 1 (1-month lag of time series)

Table 2 provides a summary of neural networks retained from Experiment no. 1.

**Table 2.** Neural networks retained from Experiment 1

| Network | Training perform. | Testing perform. | Validation perform. | Training error | Testing error | Validation error | Training algorithm | Error function | Activation of hidden layer | Output activation function |
|---|---|---|---|---|---|---|---|---|---|---|
| MLP 3-7-1 | 0.990716 | 0.987123 | 0.990223 | 1302564 | 1880266 | 1381484 | BFGS (Quasi-Newton) 206 | Sum of squares | Tanh | Exponential |
| MLP 3-8-1 | 0.990497 | 0.991163 | 0.989850 | 1330314 | 1293691 | 1409508 | BFGS (Quasi-Newton) 334 | Sum of squares | Logistic | Identity |
| MLP 3-4-1 | 0.986405 | 0.978384 | 0.990274 | 1916087 | 3442250 | 1383508 | BFGS (Quasi-Newton) 81 | Sum of squares | Tanh | Exponential |
| MLP 3-9-1 | 0.992081 | 0.984770 | 0.988757 | 1108202 | 2261257 | 1566143 | BFGS (Quasi-Newton) 240 | Sum of squares | Tanh | Sinus |
| MLP 3-4-1 | 0.983507 | 0.977995 | 0.988989 | 2308896 | 3318420 | 1546956 | BFGS (Quasi-Newton) 74 | Sum of squares | Tanh | Exponential |

Source: Own processing.

The table shows that neural networks with 4 to 9 neurons in the hidden layer were retained. The hyperbolic tangent function and the logistic function are used neural networks to activate the hidden layer. The output neuron layer is activated by the exponential, identity and sinus functions. Of course, the performance of neural networks is important. It is expressed by a correlation coefficient. The correlation coefficient is very high for all data sets and all networks, each time above 0.98 which is a high level of direct dependence. This parameter clearly shows that the retained neural structures constitute a highly credible model of a real time series, thus being able to truly predict future observations.

Figure 2 provides a comparison of the actual course of time series and equalized times series.

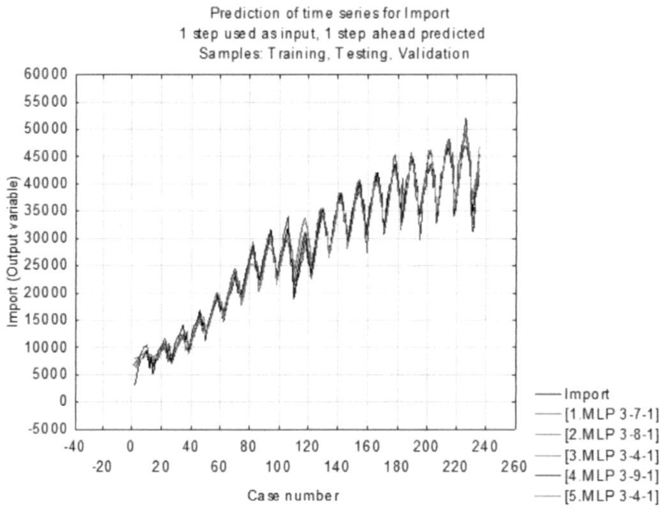


**Fig. 2.** Equalized time series of retained networks – Experiment 1

Source: Own processing.

It is also obvious from Figure 2 that all equalized times series are able to copy fairly well the course of the actual US import from China. Except for minor exceptions, each can not only capture the trend but even seasonal fluctuations. In order to establish whether networks are applicable or not, we will further examine their use in forecasting. We will predict the US import from China in the period of August 2019 to December 2020. The details are shown in Table 3.

**Table 3.** Forecast for the period of August 2019 to January 2020 by networks retained from Experiment 1

| Date | MLP 3-7-1 | MLP 3-8-1 | MLP 3-4-1 | MLP 3-9-1 | MLP 3-4-1 |
|---|---|---|---|---|---|
| August 31, 2019 | 47645.49 | 44936.4 | 47610.17 | 46645.94 | 49518.71 |
| September 30, 2019 | 49591.31 | 45182.8 | 50658.15 | 47420.89 | 52071.62 |
| October 31, 2019 | 49830.61 | 44064.5 | 52774.70 | 47654.44 | 53358.01 |
| November 30, 2019 | 48105.30 | 40009.4 | 52639.24 | 47142.82 | 53452.82 |
| December 31, 2019 | 45781.00 | 32482.2 | 51014.06 | 45928.15 | 52533.93 |
| January 31, 2020 | 45205.55 | 28064.1 | 51886.42 | 45578.78 | 50978.39 |
| February 29, 2020 | 35279.15 | -7220.4 | 40595.24 | 35745.96 | 41307.45 |
| March 31, 2020 | 36688.44 | -9904.3 | 41939.51 | 38287.74 | 42220.96 |
| April 30, 2020 | 38417.74 | -12631.8 | 43544.56 | 41572.72 | 44126.79 |
| May 31, 2020 | 40511.25 | -14734.2 | 45528.36 | 44750.52 | 47153.75 |
| June 30, 2020 | 42970.47 | -16560.3 | 48111.26 | 46893.48 | 51335.50 |
| July 31, 2020 | 45675.86 | -17818.6 | 51489.89 | 47835.32 | 56252.61 |
| August 31, 2020 | 48253.33 | -18925.5 | 55844.60 | 48096.97 | 61195.95 |
| September 30, 2020 | 49959.83 | -20026.7 | 60906.72 | 48127.50 | 65250.80 |
| October 31, 2020 | 49898.49 | -21814.3 | 65370.34 | 48126.36 | 67751.33 |
| November 30, 2020 | 47926.08 | -26244.6 | 67291.30 | 47998.17 | 68617.90 |
| December 31, 2020 | 45552.30 | -33683.7 | 67273.08 | 47393.27 | 67992.91 |

Source: Own processing.

It is apparent at first sight that network 2. MLP 3-8-1 gives a senseless prediction of the variable under review. It even shows negative values of import. However, this is not possible in real world. There is a less difference between the others. Therefore, we will use Figure 3 (Time series and forecasting trend) to choose the most successful neural network.

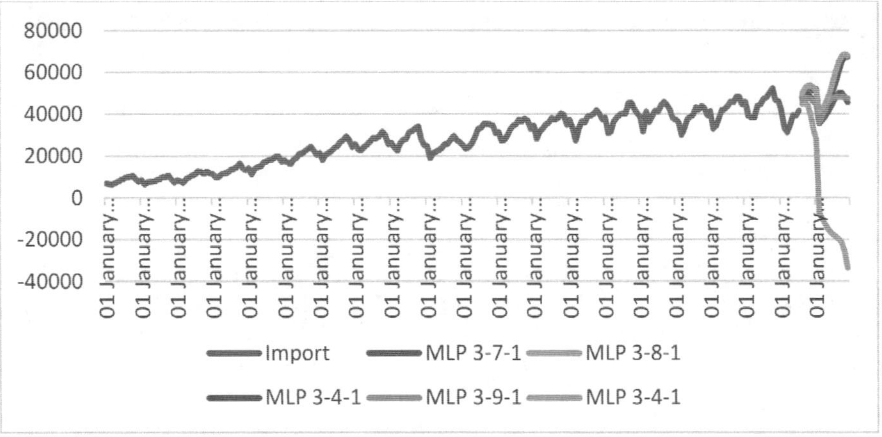


**Fig. 3.** Time series and forecasting trend by networks retained from Experiment 1

Source: Own processing.

Now the expert estimation comes into play. As a result, 1. MLP 3-7-1 and 4. MLP 3-9-1 networks are identified as close to the correct result. Considering the statistics of equalized time series, I conclude that 4. MLP 3-9- 1 is the most successful and most accurate artificial neural network.

## 3.2 Experiment 2 (5-month lag of time series)

Table 4 provides a summary of neural networks retained from Experiment no. 2 where 5-month lag of time series is expected.

**Table 4.** Neural networks retained from Experiment 2

| Network | Training perform. | Testing perform. | Valid. perform. | Training error | Testing error | Valid. error | Training algorithm | Error function | Activation of hidden layer | Output activation function |
|---|---|---|---|---|---|---|---|---|---|---|
| MLP 15-3-1 | 0.984886 | 0.969392 | 0.985259 | 2049185 | 3621980 | 1610485 | BFGS (Quasi-Newton) 64 | Sum of squares | Logistic | Identity |
| MLP 15-7-1 | 0.991430 | 0.979645 | 0.987889 | 1152035 | 2003906 | 1290950 | BFGS (Quasi-Newton) 148 | Sum of squares | Tanh | Exponential |
| MLP 15-7-1 | 0.989698 | 0.977635 | 0.986896 | 1390544 | 2312432 | 1461439 | BFGS (Quasi-Newton) 114 | Sum of squares | Tanh | Exponential |
| MLP 15-10-1 | 0.989664 | 0.977584 | 0.985530 | 1397550 | 2337306 | 1638214 | BFGS (Quasi-Newton) 150 | Sum of squares | Tanh | Exponential |
| MLP 15-4-1 | 0.982478 | 0.965299 | 0.985481 | 2377804 | 4413279 | 1725236 | BFGS (Quasi-Newton) 34 | Sum of squares | Tanh | Exponential |

Source: Own processing.

There are 15 neurons in the output layer with respect to the time series lag. The table shows that neural networks with 4 to 10 neurons in the hidden layer were retained. The hyperbolic tangent function and the logistic function are used by neural networks to activate the hidden layer. The output neuron layer is activated by the exponential function and the identity function. The correlation coefficient is very high for all data sets and all networks, each time above 0.96 which is a high level of direct dependence. This parameter clearly shows that the retained neural structures constitute a highly credible model of real time series, thus being able to truly predict future observations. The sum of least squares was used to establish the error rate. Even the error is acceptable with a view to input data (for all networks and sets).

Figure 4 provides a comparison of the actual course of time series and equalized times series.

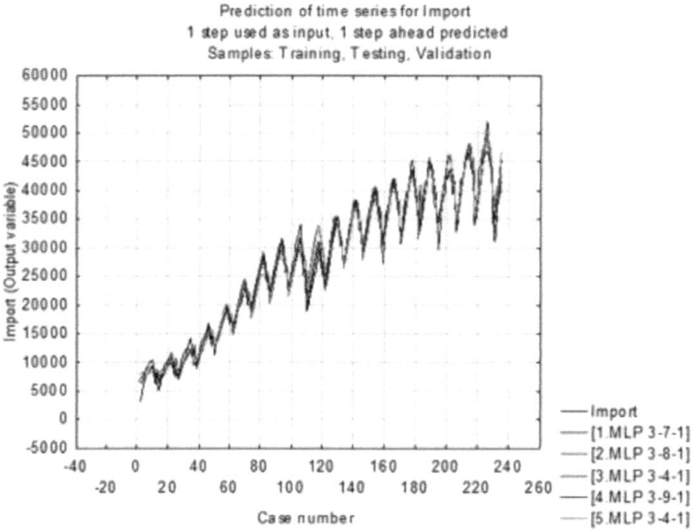


**Fig. 4**. Equalized time series of networks retained from Experiment 2

Source: Own processing.

It is also obvious from Figure 4 that all equalized times series are able to copy fairly well the course of the actual US import from China. Except for minor exceptions, each can not only capture the trend but even seasonal fluctuations.

In order to establish whether networks are applicable or not, we will further examine their use in forecasting. We will predict the US import from China in the period of August 2019 to December 2020. The details are shown in Table 5.

**Table 5.** Forecast for the period of August 2019 to January 2020 by networks retained from Experiment 2

| Date | MLP 15-3-1 | MLP 15-7-1 | MLP 15-7-1 | MLP 15-10- | MLP 15-4-1 |
|---|---|---|---|---|---|
| August 31, 2019 | 47515.57 | 47002.25 | 47040.21 | 48861.78 | 47797.37 |
| September 30, | 49604.57 | 49351.12 | 49124.91 | 51588.14 | 49334.92 |
| October 31, 2019 | 50764.12 | 51231.68 | 51143.95 | 52961.69 | 50106.41 |
| November 30, | 51112.63 | 50454.89 | 50574.42 | 50755.99 | 50462.89 |
| December 31, | 50848.47 | 47639.46 | 47206.94 | 46915.98 | 50694.26 |
| January 31, 2020 | 50248.03 | 48216.25 | 48999.36 | 46650.57 | 50964.55 |
| February 29, 2020 | 44783.80 | 35272.04 | 38268.77 | 34983.18 | 45836.01 |
| March 31, 2020 | 42825.99 | 35552.70 | 35592.07 | 37866.55 | 45078.61 |
| April 30, 2020 | 41978.92 | 41250.17 | 39894.06 | 42639.63 | 43639.14 |
| May 31, 2020 | 44659.23 | 43842.69 | 44274.29 | 43010.50 | 43976.81 |
| June 30, 2020 | 44337.75 | 47957.59 | 49797.58 | 45170.13 | 44301.28 |
| July 31, 2020 | 47481.72 | 51221.00 | 53123.88 | 52357.95 | 47687.89 |
| August 31, 2020 | 50947.31 | 54986.77 | 56840.84 | 59182.85 | 51304.98 |
| September 30, | 54119.11 | 59020.77 | 61001.23 | 65061.23 | 54268.33 |
| October 31, 2020 | 56519.55 | 62076.98 | 65109.56 | 67830.92 | 56228.17 |
| November 30, | 58001.95 | 60949.66 | 64774.65 | 64240.15 | 57365.47 |
| December 31, | 58675.62 | 57821.82 | 61761.25 | 59256.33 | 57994.40 |

Source: Own processing.

The results shown in the table represent a potential trend of the time series. It is warranted to a certain extent by the chart in Figure 5.

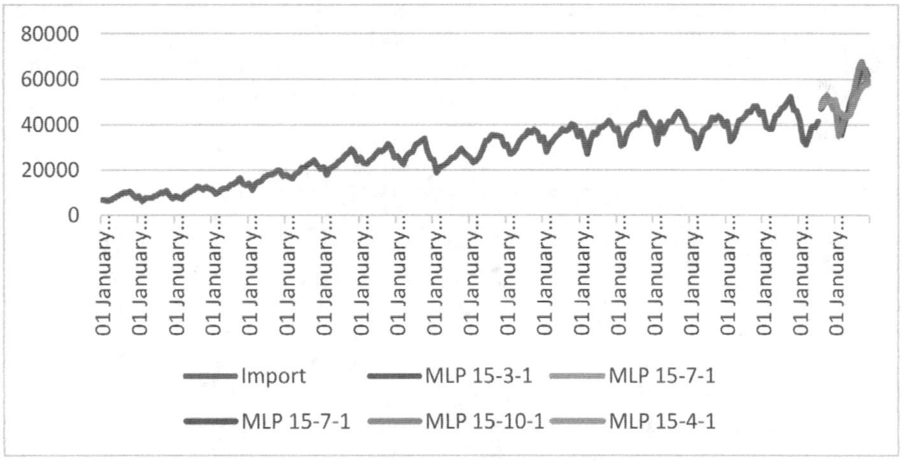


**Fig. 5.** Time series and forecasting trend by networks retained from Experiment 2

Source: Own processing.

If the actual trend of US import from China is compared to the expected time trend of the variable predicted by using the artificial neural networks retained from Experiment 2, we conclude (mainly considering the statistics table in Schedule 2) that 5. MLP 15-4-1 network is the best evaluated artificial neural structure.

### 3.3 Experiment 3 (10-month lag of time series)

There were 10,000 artificial neural networks again generated for Experiment no. 3 and five with the best characteristics were retained (more details in Table 6).

**Table 6.** Neural networks retained from Experiment 1

| Network | Training perform. | Testing perform. | Validation perform. | Training error | Testing error | Validation error | Training algorithm | Error function | Activation of hidden layer | Output activation function |
|---|---|---|---|---|---|---|---|---|---|---|
| MLP 30-3-1 | 0.982807 | 0.963986 | 0.987057 | 1875805 | 4873910 | 1391464 | BFGS (Quasi-Newton) 56 | Sum of squares | Tanh | Exponential |
| MLP 30-3-1 | 0.980938 | 0.967076 | 0.985988 | 2113288 | 4195384 | 1524165 | BFGS (Quasi-Newton) 27 | Sum of squares | Tanh | Exponential |
| MLP 30-3-1 | 0.981791 | 0.962650 | 0.985991 | 2000410 | 4995097 | 1514516 | BFGS (Quasi-Newton) 26 | Sum of squares | Tanh | Identity |
| MLP 30-4-1 | 0.980652 | 0.965798 | 0.986909 | 2160396 | 4168783 | 1443169 | BFGS (Quasi-Newton) 17 | Sum of squares | Tanh | Exponential |
| MLP 30-3-1 | 0.981757 | 0.966519 | 0.985911 | 2004951 | 4275632 | 1625783 | BFGS (Quasi-Newton) 32 | Sum of squares | Tanh | Exponential |

Source: Own processing.

For this part of experiment, there are 30 neurons in the output layer. The table shows that neural networks with 3 and 4 neurons in the hidden layer were retained. The hyperbolic tangent function is used by neural networks to activate the hidden layer. The output neuron layer is activated by the exponential function and the identity function. Of course, the performance of neural networks is important. It is expressed by a correlation coefficient. The correlation coefficient is very high for all data sets and all networks, each time above 0.96 which is a high level of direct dependence. This parameter clearly shows that the retained neural structures constitute a highly credible model of real time series, thus being able to truly predict future observations. The sum of least squares was used to establish the error rate. In this case too, the error is not significant and good-quality forecasting of time series can be expected. Figure 6 provides a comparison of the actual course of time series and equalized times series.

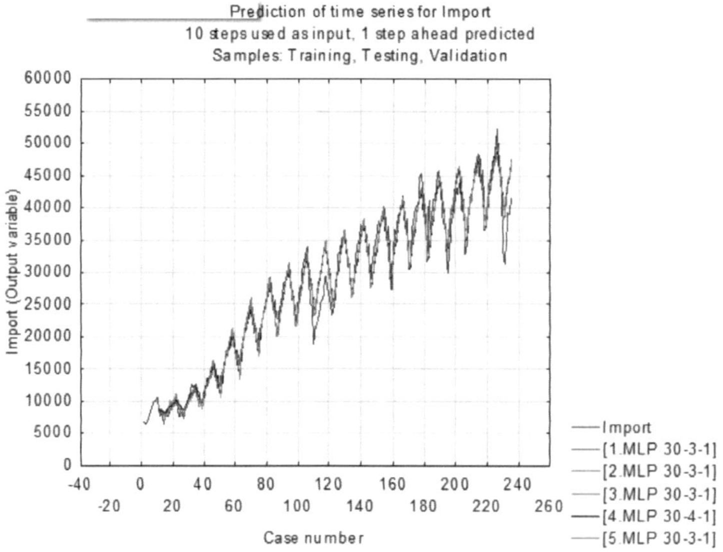


**Fig. 6.** Equalized time series of networks retained from Experiment 3

Source: Own processing.

It is obvious from Figure 6 that all equalized times series are able to copy fairly well the course of actual US import from China and moreover, the course of all equalized time series is very similar. Neural networks are able to capture the trend as well as seasonal fluctuations.

Also in the case of 10-month lag of time series, we will address future predictions of the variable under review. We will predict the US import from China in the period of August 2019 to December 2020. The details are shown in Table 7.

**Table 7.** Forecast for the period of August 2019 to January 2020 by networks retained from Experiment 3

| Date | MLP 30-3-1 | MLP 30-3-1 | MLP 30-3-1 | MLP 30-4-1 | MLP 30-3-1 |
|---|---|---|---|---|---|
| August 31, 2019 | 47628.33 | 49720.39 | 47464.98 | 48814.61 | 48596.98 |
| September 30, 2019 | 48837.86 | 51284.26 | 48327.02 | 49750.97 | 49389.40 |
| October 31, 2019 | 49959.95 | 52297.62 | 49022.30 | 50807.19 | 51060.75 |
| November 30, 2019 | 50449.75 | 50352.74 | 47706.31 | 48416.92 | 50686.28 |
| December 31, 2019 | 46867.24 | 48109.86 | 46878.11 | 47624.49 | 48046.89 |
| January 31, 2020 | 46041.69 | 45989.38 | 46020.18 | 46861.19 | 46301.59 |
| February 29, 2020 | 42208.11 | 41025.14 | 42797.71 | 41273.91 | 40974.64 |
| March 31, 2020 | 42495.83 | 41372.79 | 43634.20 | 42142.19 | 42009.30 |
| April 30, 2020 | 45026.20 | 44732.08 | 45831.48 | 45485.90 | 45045.58 |
| May 31, 2020 | 46921.96 | 47618.64 | 47618.49 | 48232.82 | 47757.49 |
| June 30, 2020 | 47976.45 | 48794.48 | 48409.38 | 49450.13 | 49140.73 |
| July 31, 2020 | 49584.24 | 52005.43 | 49645.76 | 51801.52 | 52358.90 |
| August 31, 2020 | 51077.79 | 54756.44 | 50636.59 | 53605.07 | 54543.17 |
| September 30, 2020 | 52299.32 | 56770.97 | 51352.24 | 54625.63 | 55204.93 |
| October 31, 2020 | 53389.33 | 58254.67 | 51888.43 | 55617.13 | 57312.10 |
| November 30, 2020 | 54033.93 | 56582.64 | 51327.16 | 54278.13 | 59088.72 |
| December 31, 2020 | 51211.21 | 54610.00 | 51015.99 | 53986.72 | 56393.04 |

Source: Own processing.

As previously mentioned, the forecasts are very similar. It is warranted to a certain extent by the chart in Figure 7.

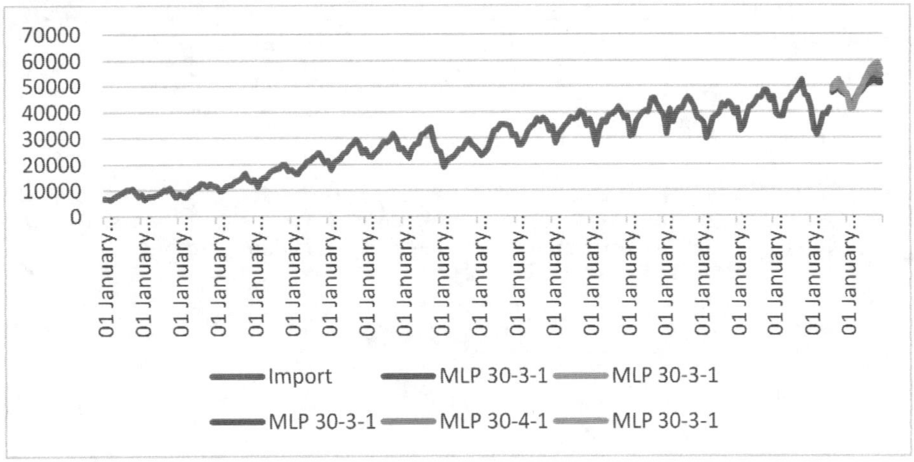


**Fig. 7.** Time series and forecasting trend by networks retained from Experiment 3

Source: Own processing.

If the actual trend of US import from China is compared to the expected time trend of the variable predicted by using the artificial neural networks retained from Experiment 3, we conclude (mainly considering the statistics table in Schedule 2) that 5. MLP 30-3-1 network provides the best prediction of future observations.

### 3.4 Discussion of results

Three experiments were conducted in research. The intention was to examine the trend of time series represented by US import from China. In particular, we examined how multilayer perceptron networks are able to equalize time series. In each experiment, 10,000 artificial neural networks were generated. Five best structures were retained from each experiment. All but one from Experiment 1 which clearly suffered from overfitting can be used for prediction of the future US import from China. In each case, the performance of networks was high and the error rate was fairly low. We defined the best network in each experiment and conducted further comparison of this network (more details in Table 8).

**Table 8.** Comparison of the most successful neural structures from all experiments

| Date | Experiment 1 | Experiment 2 | Experiment 3 |
|---|---|---|---|
| | 4.MLP 3-9-1 | 5.MLP 15-4-1 | MLP 30-3-1 |
| August 31, 2019 | 46645.94 | 47797.37 | 48596.98 |
| September 30, 2019 | 47420.89 | 49334.92 | 49389.40 |
| October 31, 2019 | 47654.44 | 50106.41 | 51060.75 |
| November 30, 2019 | 47142.82 | 50462.89 | 50686.28 |
| December 31, 2019 | 45928.15 | 50694.26 | 48046.89 |
| January 31, 2020 | 45578.78 | 50964.55 | 46301.59 |
| February 29, 2020 | 35745.96 | 45836.01 | 40974.64 |
| March 31, 2020 | 38287.74 | 45078.61 | 42009.30 |
| April 30, 2020 | 41572.72 | 43639.14 | 45045.58 |
| May 31, 2020 | 44750.52 | 43976.81 | 47757.49 |
| June 30, 2020 | 46893.48 | 44301.28 | 49140.73 |
| July 31, 2020 | 47835.32 | 47687.89 | 52358.90 |
| August 31, 2020 | 48096.97 | 51304.98 | 54543.17 |
| September 30, 2020 | 48127.50 | 54268.33 | 55204.93 |
| October 31, 2020 | 48126.36 | 56228.17 | 57312.10 |
| November 30, 2020 | 47998.17 | 57365.47 | 59088.72 |
| December 31, 2020 | 47393.27 | 57994.40 | 56393.04 |

Source: Own processing.

The differences between neural networks are already presented in Table 8. Although not visible at first sight, the differences mainly between networks selected from Experiment 2 and 3 are not significant. This situation is even better illustrated in Figure 8.

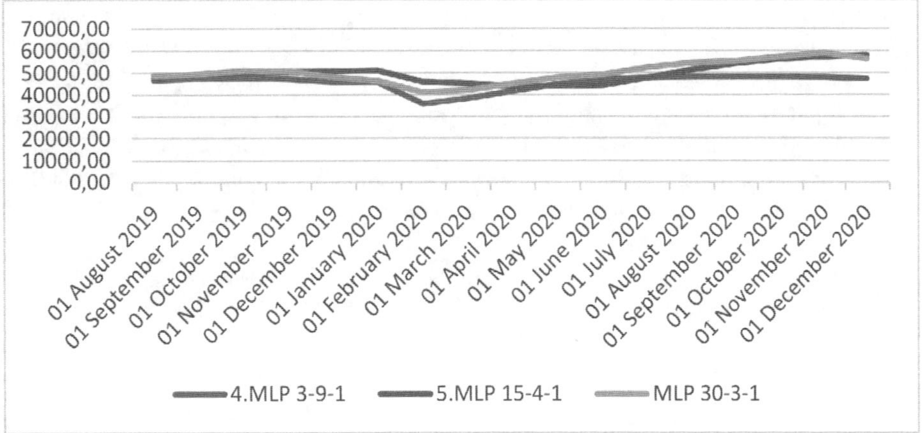


**Fig. 8.** Comparison of the most successful neural structures from all experiments

Source: Own processing.

A very similar course of all selected neural structures is apparent from the figure. Some differences, although not fundamental are identified in February 2020 and November and December of the same year. However, it should be noted that a very similar output is presented by the networks from Experiment 2 and 3.

## 4 Conclusion

The objective of this paper was to forecast the import of the United States of America from the People´s Republic of China using machine learning techniques. It can be concluded that the objective of this paper was achieved:

1. It appears as if multilayer perceptron networks will be a great forecasting tool for US import from China.
2. MLP networks are able to cover both the trend of the entire time series and its seasonal fluctuations.
3. Time-series lag must be applied. Acceptable results are established for 5- and 10-month lag of time series.
4. Mutual sanctions of both countries did not have a significant impact on forecasting using machine learning.

Further research should solely be directed to an experiment that would provide information which lag of time series would be the most appropriate for this type of task and this specific variable.

## References

1. Z. Rowland, P. Šuleř, M. Vochozka, Comparison of neural networks and regression time series in estimating the Czech Republic and China trade balance. *SHS Web of Conferences: Innovative Economic Symposium 2018 – Milestones and Trends of the World Economy*, **61** (2019)
2. H. Yaping, S. Pengfei, M. Vochozka, Pollution caused by finance and the relative policy analysis in China. *Energy & Environment*, **28**(7), 808-823 (2017)

3. M. Vochozka, J. Vrbka, Estimation of the development of the Euro to Chinese Yuan exchange rate using artificial neural networks. *SHS Web of Conferences: Innovative Economic Symposium 2018 – Milestones and Trends of World Economy,* **61** (2019)
4. Understanding the US - China Trade Relationship: Prepared fot the US - China Business Council by Oxford Economics. *The US - China Business Council: Understanding the US-China Trade Relationship* [online], Available at: https://www.uschina.org/sites/default/files/Oxford%20Economics%20US%20Jobs%20and%20China%20Trade%20Report.pdf (2017)
5. M. Chossudovsky, Trump's Trade War with China: Imagine What Would Happen if China Decided to Impose Economic Sanctions on the USA? *Global Research: Centre for Research on Globalization* [online], Available at: https://www.globalresearch.ca/imagine-what-would-happen-if-china-decided-to-impose-economic-sanctions-on-the-usa-2/5598941 (2017)
6. M. Vochozka, J. Horák, P. Šuleř, Equalizing seasonal time series using artificial neural networks in predicting the Euro-Yuan exchange rate. *Journal of Risk and Financial Management*, **12**(2) (2019)
7. V.V. Mikheev, S.A. Lukonin, China - USA: multiple vector OF "trade war". *World Economy and International Relations*, **63**(5), 57-66 (2019)
8. J. Vrbka, Z. Rowland, P. Šuleř, Comparison of neural networks and regression time series in estimating the development of the EU and the PRC trade balance. *SHS Web of Conferences: Innovative Economic Symposium 2018 – Milestones and Trends of World Economy,* **61** (2019)
9. L.D. Qiu, C. Zhan, X. Wei, An analysis of the China - US trade war through the lens of the trade literature. *Economic and Political Studies*, **7**(2), 148-168 (2019)
10. M. Yu, Introduction to the Special Issue on Understanding the Current China-U.S. 'Trade War'. *China Economic Journal*, **12**(2), 97-99 (2019)
11. E.L.C. Lai, The US - China trade war, the American public opinions and its effects on China. *Economic and Political Studies*, **7**(2), 169-184 (2019)
12. K. Itakura, Evaluating the Impact of the US - China Trade War. *Asian Economic Policy Review* (2019)
13. T.T.L. Chong, X. Li, Understanding the China - US trade war: causes, economic impact, and the worst - case scenario. *Economic and Political Studies*, **7**(2), 185-202 (2019)
14. V. Machová, J. Mareček, Estimation of the development of Czech Koruna to Chinese Yuan exchange rate using artificial neural networks. *SHS Web of Conferences: Innovative Economic Symposium 2018 – Milestones and Trends of World Economy*, **61** (2019)
15. Y. Li, US Economic Sanctions Against China: A Cultural Explanation of Sanction Effectiveness. *Asian Perspective*, **38**(2), 311-335 (2014)
16. J. Yang, H. Askari, J. Forrer, H. Teegen, US Economic Sanctions Against China: Who Gets Hurt? *World Economy*, **27**(7), 1047-1081 (2004)
17. China vows to impose sanctions on US firms supplying Taiwan military. *The Guardian* [online], Available at: https://www.theguardian.com/world/2019/jul/12/china-taiwan-sanctions-us-firms-military-sales (2019)
18. S.V. Nozdrev, China in Global Financial System. *World Economy and International Relations*, **60**(10), 29-40 (2016)

19. B. Seyoum, US foreign trade zones and import intensity examining determinants of import intensity in US foreign trade zones. *European Business Review*, **29**(1), 103-122 (2017)
20. A. Pandey, US - China trade war: Huawei's loss is Samsung's gain. *DW: Made for minds* [online], Available at: https://www.dw.com/en/us-china-trade-war-huaweis-loss-is-samsungs-gain/a-48814477 (2019)
21. T. Besedeš, T.J. Prusa, The Hazardous Effects of Antidumping. *Economic Inquiry*, **55**(1), 9-30 (2017)
22. H. Lee, Estimating the Protection Effects of Antidumping Duties. *Journal of Korea Trade*, **13**(3), 1-23 (2009)
23. S. Park, The Trade Depressing and Trade Diversion Effects of Antidumping Actions: The Case of China. *China Economic Review*, **20**(3), 542-548 (2009)
24. N. Choi, Did Anti-dumping Duties Really Restrict Import? Empirical Evidence from the US, the EU, China, and India. *East Asian Economic Review*, **21**(1), 3-27 (2017)
25. L. Yu, S. Wang, K.K. Lai, Forecasting China's Foreign Trade Volume with a Kernel-Based Hybrid Econometric-Ai Ensemble Learning Approach. *Journal of Systems Science and Complexity*, **21**(1), 1-19 (2008)
26. J. Deggans, T. Krulický, M. Kováčová, K. Valášková, M. Polliak, Cognitively enhanced products, output growth and labor market changes: Will artificial intelligence replace workers by automating their jobs? *Economics, Management, and Financial Markets*, **14**(1), 38-43 (2019)
27. J. Weijin, X. Yuhui, A novel method for nonlinear time series forecasting of time - delay neural network. *Wuhan University Journal of Natural Sciences*, **11**(5), 1357-1361 (2006)
28. World Bank [online], Available at: https://data.worldbank.org/indicator/TM.TAX.MRCH.WM.AR.ZS?end=2017&locations=CN-US&start=2014 (2019)

### *Schedule 1*
**Basic statistics of equalized time series – 1-month lag**

| Statistics | 1.MLP 3-7-1 | 2.MLP 3-8-1 | 3.MLP 3-4-1 | 4.MLP 3-9-1 | 5.MLP 3-4-1 |
|---|---|---|---|---|---|
| Minimum forecast (Training) | 7354.37 | 6559.94 | 8142.94 | 4119.17 | 7193.36 |
| Maximum forecast (Training) | 47396.03 | 48014.80 | 48288.31 | 46169.46 | 48594.69 |
| Minimum forecast (Testing) | 7254.62 | 7045.62 | 8029.47 | 3015.71 | 7079.77 |
| Maximum forecast (Testing) | 48674.80 | 49154.72 | 49613.73 | 47021.13 | 49544.61 |
| Minimum forecast (Validation) | 7478.22 | 7840.43 | 8273.81 | 5681.94 | 7347.37 |
| Maximum forecast (Validation) | 48516.18 | 48278.85 | 48569.99 | 46786.56 | 49352.04 |
| Minimum residues (Training) | -4712.82 | -5176.38 | -6518.51 | -5066.37 | -6957.79 |
| Maximum residues (Training) | 7330.04 | 7387.11 | 7219.56 | 7026.74 | 8087.28 |
| Minimum residues (Testing) | -4910.74 | -3967.62 | -7436.32 | -4426.25 | -6291.87 |
| Maximum residues (Testing) | 4514.17 | 3379.49 | 3484.47 | 5181.17 | 4175.70 |
| Minimum residues (Validation) | -2585.83 | -2612.07 | -4704.98 | -3275.40 | -4326.75 |
| Maximum residues (Validation) | 4740.17 | 5076.42 | 2623.07 | 4572.21 | 3476.45 |
| Minimum standard residues (Training) | -4.13 | -4.49 | -4.71 | -4.81 | -4.58 |
| Maximum standard residues (Training) | 6.42 | 6.40 | 5.22 | 6.67 | 5.32 |
| Minimum standard residues (Testing) | -3.58 | -3.49 | -4.01 | -2.94 | -3.45 |
| Maximum standard residues (Testing) | 3.29 | 2.97 | 1.88 | 3.45 | 2.29 |
| Minimum standard residues (Validation) | -2.20 | -2.20 | -4.00 | -2.62 | -3.48 |
| Maximum standard residues (Validation) | 4.03 | 4.28 | 2.23 | 3.65 | 2.80 |

### *Schedule 2*
**Basic statistics of equalized time series – 5-month lag**

| Statistics | 1.MLP 15-3-1 | 2.MLP 15-7-1 | 3.MLP 15-7-1 | 4.MLP 15-10-1 | 5.MLP 15-4-1 |
|---|---|---|---|---|---|
| Minimum forecast (Training) | 5752.93 | 7952.83 | 8335.37 | 8239.21 | 8144.24 |
| Maximum forecast (Training) | 48664.89 | 47750.67 | 47773.06 | 47580.68 | 48129.11 |
| Minimum forecast (Testing) | 9259.27 | 9226.17 | 9230.93 | 9160.76 | 9370.74 |
| Maximum forecast (Testing) | 48955.84 | 49144.91 | 48949.81 | 50189.47 | 48249.50 |
| Minimum forecast (Validation) | 11050.71 | 9440.66 | 9456.50 | 9598.51 | 9230.38 |
| Maximum forecast (Validation) | 48266.83 | 47781.93 | 47405.82 | 49350.75 | 47896.93 |
| Minimum residues (Training) | -7132.39 | -4323.94 | -4579.37 | -4409.89 | -7308.98 |
| Maximum residues (Training) | 7907.11 | 8082.84 | 7739.53 | 7788.84 | 8020.51 |
| Minimum residues (Testing) | -8006.76 | -3966.38 | -4111.01 | -4183.43 | -9100.39 |
| Maximum residues (Testing) | 3246.46 | 3928.01 | 3951.86 | 4231.92 | 3952.80 |
| Minimum residues (Validation) | -4409.00 | -2874.81 | -2937.65 | -4393.50 | -4758.93 |
| Maximum residues (Validation) | 2948.14 | 4569.47 | 3537.80 | 4950.74 | 3961.35 |
| Minimum standard residues (Training) | -4.98 | -4.03 | -3.88 | -3.73 | -4.74 |
| Maximum standard residues (Training) | 5.52 | 7.53 | 6.56 | 6.59 | 5.20 |
| Minimum standard residues (Testing) | -4.21 | -2.80 | -2.70 | -2.74 | -4.33 |
| Maximum standard residues (Testing) | 1.71 | 2.77 | 2.60 | 2.77 | 1.88 |
| Minimum standard residues (Validation) | -3.47 | -2.53 | -2.43 | -3.43 | -3.62 |
| Maximum standard residues (Validation) | 2.32 | 4.02 | 2.93 | 3.87 | 3.02 |

*Schedule 3*

**Basic statistics of equalized time series – 10-month lag**

| Statistics | 1.MLP 30-3-1 | 2.MLP 30-3-1 | 3.MLP 30-3-1 | 4.MLP 30-4-1 | 5.MLP 30-3-1 |
|---|---|---|---|---|---|
| Minimum forecast (Training) | 8137.69 | 8162.69 | 7379.79 | 8221.65 | 7621.39 |
| Maximum forecast (Training) | 48644.28 | 48255.74 | 46387.41 | 47328.36 | 48550.17 |
| Minimum forecast (Testing) | 9292.58 | 9020.55 | 8870.28 | 9523.50 | 8826.83 |
| Maximum forecast (Testing) | 48575.03 | 50304.35 | 48013.80 | 49276.72 | 49421.77 |
| Minimum forecast (Validation) | 9810.03 | 9189.25 | 10003.05 | 9621.15 | 9200.18 |
| Maximum forecast (Validation) | 47526.94 | 49311.92 | 47293.30 | 48260.61 | 47858.95 |
| Minimum residues (Training) | -6506.02 | -6428.79 | -6169.27 | -6466.52 | -6139.14 |
| Maximum residues (Training) | 7902.86 | 8700.19 | 7788.67 | 8921.58 | 8463.54 |
| Minimum residues (Testing) | -8618.38 | -7617.52 | -9319.86 | -7496.30 | -7656.74 |
| Maximum residues (Testing) | 3627.27 | 3661.09 | 4188.50 | 3431.46 | 3573.96 |
| Minimum residues (Validation) | -4560.79 | -5080.26 | -4431.97 | -4071.65 | -4657.13 |
| Maximum residues (Validation) | 2495.61 | 3452.72 | 2781.95 | 3265.68 | 2884.09 |
| Minimum standard residues (Training) | -4.75 | -4.42 | -4.36 | -4.40 | -4.34 |
| Maximum standard residues (Training) | 5.77 | 5.98 | 5.51 | 6.07 | 5.98 |
| Minimum standard residues (Testing) | -3.90 | -3.72 | -4.17 | -3.67 | -3.70 |
| Maximum standard residues (Testing) | 1.64 | 1.79 | 1.87 | 1.68 | 1.73 |
| Minimum standard residues (Validation) | -3.87 | -4.12 | -3.60 | -3.39 | -3.65 |
| Maximum standard residues (Validation) | 2.12 | 2.80 | 2.26 | 2.72 | 2.26 |

# The marketing value creation in the waste management sector – multi-conceptual business model

*Izabela Barbara* Sztangret[1]*

[1]University of Economics in Katowice, Department of Entrepreneurship; 1 Maja 50, 40-287 Katowice, Poland

**Abstract.** The concept of marketing value occurred interesting due to a result of the process of transformation of traditional business relationships into specific type of cooperation of enterprises to create multivalued for demanding customers. It also concerns studied companies of the sector of waste collection and recycling, as well as their partners and customers. In the studied case, marketing value is created by enterprises, their cooperants and customers for the purpose of further satisfaction of target market need, in at least three-spherical business model. This model often has multi-concept, or holistic nature while combining relationship marketing, supply marketing, integral, internal, systemic, strategic and social marketing. Companies representing the sector of waste management apply smart technological solutions which result in interactive formation of market value. Smart ways of creation of marketing value, in the sector of waste management enterprises in the environment of IT solutions is the goal of this paper. The main research methods used are case studies.
**Key words:** business model, marketing value, waste management, waste selection

## 1 Introduction

The concept of marketing value occurred interesting due to a result of the process of transformation of traditional business relationships into specific type of cooperation of enterprises to create multivalued for demanding customers. It also concerns studied companies of the sector of waste collection and recycling, as well as their partners and customers. Peculiar character of cooperation is associated with the fact of approaching information as a resource constituting relationship in the network of enterprises and their customers, which can be successfully implemented in IT environment. **Interactive contact and involvement of the stream suppliers and customers of secondary market** form the operational sub-area of the concept of creation of marketing value, in studied sector, in *subjective approach*. This is the features of creation **marketing value by integral marketing**.

* Corresponding author: izabela.sztangret@ue.katowice.pl

Technological tools enable smart management of communal waste through effective division into stream fractions, which consequently allows for separation of the material for recycling and reuse, management of the fraction obtained from mechanical-biological waste treatment, and first minimising the amount of non-productively stored waste. Knowledge about chemical composition of packaging getting into the waste stream tightens the process of precise selection. Organisational knowledge and competencies as well as skilful application of new technologies in **the processes of selection and recycling of communal waste** constitute the operational sub-area of creation of marketing value in inter-organisational level, in *process approach*. It occurs as an **extra-eco-marketing value and the symptom of strategy marketing value.**

Commercial use of information automatically registered by separator scanners of particular fractions of communal waste is another operational sub-area of creation marketing value, in this case. Information included in the barcode is important in identification and registering a particular type of product. Thereby it has **marketing value, useful in market research** in *functional approach* to management in a company.

In the studied case, marketing value (valuable relations, eco-strategy value, marketing information) is created by enterprises, their cooperants and customers for the purpose of further satisfaction of target market need, in at least **three-spherical business model**. This model often has multi-concept, or holistic nature while combining relationship marketing, supply marketing, integral, internal, systemic, strategic and social marketing. Companies representing the sector of waste management apply smart technological solutions which result in interactive formation of market value.

Therefore, smart ways of creation of marketing value, in the sector of waste management enterprises in the environment of IT solutions, raise interest. Their identification and analysis is the **goal of the paper**[2]. This will be implemented through in-depth case study analysis.

## 2 Research methods

Regional Municipal Waste Treatment Facilities are, for the needs of this paper, the subject of the research. These entities are described in the Law on waste of 14th December 2012 and the Directive 2002/83/EC that determines the minimum requirements for establishment of such an entity. The documents also describe hierarchical method of waste management through:

1. prevention of waste generation,
2. preparation of its reuse,
3. recycling,
4. other processes of recovery,

[2] This topic seems to be quite now. We haven't got comprehensive studies on this topic in the literature. This knowledge is on the first stage of its development. But it is very endurable, because we need revolution in recycling: "This means we do not recycle from material to material but from material to elements and then back to completely different materials (with new value...). This way we could take composite ingredients, break them apart, and recycle everything, not just the organic or inorganic material like we do today", by Jutta Laine-Ylijoki, senior scientist at VTT. "Extracting valuable materials from side streams, such as oil, paint, or glue, is our main concern right now. In many cases, the actual process of fractioning these streams is not the problem, increasing the production to an industrial scale is (...). Another challenge for L&T involves creating completely new products or materials from waste. Steel will be always recycled as steel, but plastic might turn into something completely different in its second life, by Mikko Paasikivi, business manager at L&T [1].

5. disposal.

The law provides the possibility to divide voivodeships, in voivodeship plans of waste management (VPWM) into regions, if they comply with the minimum requirements as determined in clause 35 section 5 of the law. It establishes that the region of municipal waste management constitutes an area of neighbouring communes that comprise in total at least 150 thousand inhabitants and is operated by facilities referred to in section 6. The area of a commune of 500 thousand inhabitants can also be the region of municipal waste management. New regulations included in the amendment to the law on waste of 22nd January 2015 define a regional facility for municipal waste processing as the waste management plant of the capacity sufficient to collect and process waste from the area inhabited by at least 120 thousand inhabitants. Also, it must comply with the requirements of the best available technology referred to in clause 207 of the law of 27th April 2001 – Environmental protection law, or technology referred to in clause 143 of this law, including those applying new available technologies of waste processing or ensuring that:

1) mechanical and biological treatment of mixed municipal waste and separation from mixed municipal waste the fractions that are suitable for recovery in whole or in part, or,

2) processing of separately collected green waste and other bio-waste, and producing from them a product of fertilising qualities, or substances supporting plant cultivation that comply with requirements as determined in separate regulations, or other material after the process of composting or fermentation admitted for recovery in the R10 recovery process that meets the requirements determined in regulations issued on the basis of clause 30 section 4, or,

3) landfilling of waste generated in the process of mechanical and biological treatment of mixed municipal waste and remains from segregation of municipal waste of the capacity that allows for reception of waste for the period of at least 15 years, in the amount not smaller than generated in the facility for mechanical and biological treatment of mixed municipal waste [2]. Furthermore, the region of municipal waste management may comprise neighbouring communes of various voivodeships if it is provided for by voivodeship waste management plans of these voivodeships [3]. Another important provision of the quoted law concerns introduction of the notion of transregional facility which can be an incineration plant for municipal waste with the capacity sufficient for reception and treatment of mixed municipal waste collected from the area inhabited by at least 500 thousand residents, and that meets the requirements of the best available technology called "transregional municipal waste incineration plants".

**Table 1.** Basic information about conducted research

| Specification | Characteristic features |
|---|---|
| Research technique | Critical analysis of the literature and trade magazines, analysis of webpages, analysis of sponsored interviews, direct interviews |
| Sample selection | Targeted selection of typical units |
| Sample size | Leader of waste management sector in region IV, by the criterion of facility machinery<br>20 representatives of RIPOK [*regional municipal waste treatment facility*] in Poland |
| | Over 10 trade webpages related to waste management sector |
| Geographical range | Regional range |
| Time range | 2014-2019 |

Source: Authors.

MASTER Odpady i Energia Sp. z o.o. enterprise [MASTER Waste and Energy limited liability company], being one of 3 similar facilities constructed in Poland that have the latest plant ensuring mechanical and biological treatment of mixed municipal waste and separation of fractions, from mixed municipal waste, that are recyclable in whole or in part, of the capacity of 120,000 Mg/year. The plant, together with the landfill constitutes an integral part of communal waste management system in Region IV.

The paper applies the methods of conceptual research, and qualitative empirical research (case study) [3]. Analysis of the literature of the subject and research based on secondary and primary sources was performed.

For the needs of the entities that have the status of RIPOK [*Regional Municipal Waste Treatment Facility*] by the criterion of indication by a leader and position according to secondary sources of the sector were selected for the analysis of expanded case study[4].

## 3 Waste selection and inter-processing as a way of integral marketing

The aforementioned directive 2008/98 / EC of the European Parliament and of the Council of 19 November 2008 on waste, and the one repealing certain directives (hereafter: the Framework Directive), focus on the prevention of negative impacts of waste generation, its management and protection of natural resources. It brings a new approach to waste management and also changes the way people think about waste, while recognizing it as a valuable resource. The directive emphasizes the definition of key concepts such as: recovery, disposal and recycling (3R)[5]. The Framework Directive stresses the need to establish a recovery and waste disposal facility using the best available technology (**BAT**-Best Alternative Technology, **BREF**- *Reference Document on Best Available Techniques for the Waste Treatment Industries*) [4], which represents appropriate use of diverse and complementary technological solutions and waste processing processes. This means that waste management system can be based on various technologies and technological processes that allow for the management of collected municipal waste, recycling, thermal

[3] Application of the method seems justified because:
1/ research concerns contemporary, dynamic phenomena and knowledge about these phenomena that is created;
2/ they concern research of actual contexts of these phenomena at rather large ambiguity of borders between their contexts and the very phenomena;
3/ the subject of research is too complicated to explain the cause and effect relationships by means of the survey method or experiment [5, 6].

[4] Sample selection was performed on the basis of leader's indication and on the basis of indications in final report of III stage expert opinion aiming at performance of waste examination in 20 waste mechanical and biological treatment plants financed from resources of the project no POPT.03.01.00-00-375/13-00, „Wsparcie na działania sieci organów środowiskowych i instytucji zarządzających funduszami unijnymi" "Partnerstwo: Środowisko dla Rozwoju"[ ["*Support for activities of the environmental networks and institutions managing EU funds." "Partnership: environment for development*"] in 2014 from Resources of 2014 Technical Assistance Operational Programme 2007-2013 within Priority III – Support for execution of structural funds operation; Action 3.1. – „Functioning of institutions involved in NSRF". Executor: Consortium: Uniwersytet Zielonogórski i Zakład Utylizacji Odpadów Spółka z o.o., [*University of Zielona Góra and Waste Disposal Facility limited liability company*] [7].

[5] The concept described more broadly in the author's publications: Zrównoważony rozwój przez zarządzanie wiedzą w eko-systemie interesariuszy, na przykładzie przedsiębiorstwa usług komunalnych [8, 9].

transformation or energy recovery. Other forms of waste recovery that require pre-treatment (separation of individual fractions) is acceptable as this allows for further neutralisation of the waste by processing. The scheme of the municipal waste treatment in studied RIPOKs [*Regional Municipal Waste Treatment Facilities*] includes most frequently four main components, described below:

1) Thermal conversion of waste involves oxidation of waste (including incineration and gasification) or decomposition of waste (including pyrolysis). These processes can be conducted in incineration plants for hazardous waste and in municipal waste incineration plants, and in the case of non-hazardous waste, in installations other than waste incineration plants or in other facilities in the way compliant with specific provisions.
2) Mechanical-biological waste processing consists of mechanical processing of waste and biological waste treatment combined into one integrated technological process of mixed municipal waste processing for the purpose of preparing them for recovery, including recycling, energy recovery, thermal conversion or storage. The mechanical processes include, among others, grinding, sifting, sorting as well as segregation of ferrous and non-ferrous metals. Once these processes have been performed, organic fraction is left, and it undergoes biological aerobic or anaerobic stabilization. Biological processing of waste in aerobic conditions is conducted in accordance with strictly defined requirements. Waste separated as a fraction 0-80 mm is processed with a transfer of waste for a total period of 8 to 12 weeks, including at least 2 weeks' process in a closed reactor or in a hall with active aeration, when impurified process air is prevented from being released into the atmosphere until AT4 value below 20 mg $O2/g$ of dry matter is obtained (perceived as breathing activity - a parameter expressing the oxygen requirement for a sample of waste for 4 days). In biological waste treatment processes in anaerobic conditions, the aforementioned waste is subject to stabilization in a two-stage process:

   a) In the first stage of mesophilic fermentation, for at least 20 days, or thermophilic fermentation, for at least 12 days, in a second oxygen stabilization stage in a closed reactor or in a hall with active aeration, with a protection preventing release of impurified process air into the atmosphere for at least the period of 2 weeks' time;

   b) in the second stage of aerobic stabilization, stabilization conducted in open-air prisms, aerated by waste moving once a week at least, for the period of at least 3 weeks is admissible.
3) Processing of selectively collected green waste and other bio-waste, i.e. composting, consists in the processing of waste by means of natural reactions of decomposition of organic substances by microorganisms. Processing takes place under controlled conditions, in the presence of oxygen and in appropriate temperature and humidity levels, and with the participation of microorganisms. The purpose of composting is to obtain organic fertilizer, at a considerable reduction in the volume of deposited waste (even by half), and a neutral chemical and biological impact of the obtained "product" on the environment.
4) Storage is a disposal method applied for waste which, for technological or economic reasons, has failed to be recovered or disposed with the use of other methods. Landfill waste is stored in a selective way according to the types specified in the landfill operating instructions.

The most modern methods for reduction of the amount of waste on stock include depolymerisation, bio-drying, hydro-mechanical segregation and autoclaving.

5) Depolymerisation is the process of plastic waste processing that takes place through the decomposition of hydrocarbon chains under the influence of temperature, lack of air, and with the use of a catalyst. As a result, a product of new properties is obtained. The

liquid fuel component, synthetic paraffin oil which is a component of diesel oil and a substitute for heating oil, is the final product of the catalytic depolymerisation process.

6) Bio-drying takes place in boxes with intensive aeration. Waste is transported in containers to the place, where piles are built and then covered with semi-permeable geomembrane. The process of intensive stabilization in aerobic conditions takes place in the piles. After about 21 days of processing, the processed waste is sorted, and it can be sent to the production line for production of alternative fuels, or transferred as the so-called. Pre-RDF28 used in energy-intensive industries such as cement plants.
7) Hydromechanical separation involves the separation of the mixed stream of waste by means of hydro-separation. Plastics (light fraction) float on water, the organic fraction moves in the water, while heavy fractions such as metals and minerals fall to the bottom. This method allows for recovery of about 70-90% of materials for further recycling. Organic fraction undergoes the process of fermentation and then is used to produce fertilizers.
8) Autoclaving involves mechanical and thermal treatment of waste. During the process, odours and pathogens are eliminated. This technology applies three physical factors, automatically controlled on the basis of waste morphology: temperature, pressure and time. Sterilized waste is subject to mechanical sorting.

Sterile organic fractions and pure nonorganic fractions that can be recycled are released from the stream of mixed municipal waste [10].

*Investigated Complex Waste Management Plant in Tychy, managed by Master Odpady i Energia Sp. z o.o. company focuses on full eco-innovation of waste management solutions. At the plant in Tychy, practically all possible solutions for efficient sorting of waste are practically applied. The main element that distinguishes the plant is the use of 8 x optical separators, 3 x separators of ferrous metals, a non-ferrous metal separators and 2D / 3D ballistic separator of hard parts for biological fraction. Thanks to smart technological solutions they can separate the individual raw materials from the waste stream with the effectiveness of over 80% on the basis of recognition of physical and chemical qualities of the waste.*

Waste selection by suppliers of the stream of waste, smart inter-processing of inter-organizational knowledge, skills and technology, relations based on information/knowledge diffusion between suppliers, companies of municipal waste sector and aims of secondary market creates the area of creation integral marketing value.

# 4 Informative value of the waste

The information contained in, or on waste, that constitutes the stream, and is obtained by an enterprise of studied sector can be used effectively to streamline the selection process, on the level of segregation of raw material to be processed or sold. Furthermore, they can provide information about buying and consumption behaviours of creators/suppliers of waste stream.

According to the Ordinance of the Minister of the Environment on packaging design patterns of 23 April 2004, **the marking of the packaging of goods** is determined by their chemical composition. The packages which have a significant environmental impact have been distinguished. This allows for identification, with the use of optoelectronic and laser separators, of waste streams for example for pyrolysis, self-depolymerisation or sales on the secondary market. The most important markings are shown in Table 2.

**Table 2.** Marking of packaging by chemical composition

| Composition | Marking |
|---|---|
| Aluminium, found in for example cans and disposable crockery | 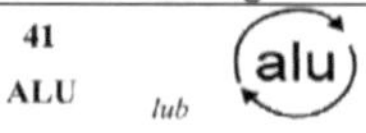  |
| Polyethylene terephthalate - pet, found for example in plastic Bottles, dishes, packages, and housings of household appliances | 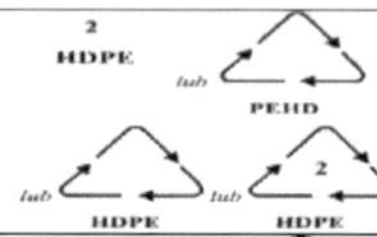  |
| High density polyethylene - pehd, hdpe, used for example for production of foils, packaging, garbage bags, sewage pipes, rainwater tanks and garbage containers | 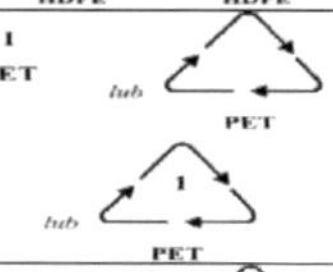  |
| Polyvinyl chloride - pvc, contained for example in syringes, liners and insulation of cables | 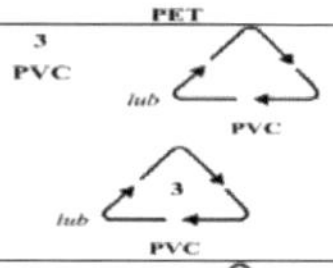  |
| Low density polyethylene - ldpe, peld, which is contained in plastic bags | 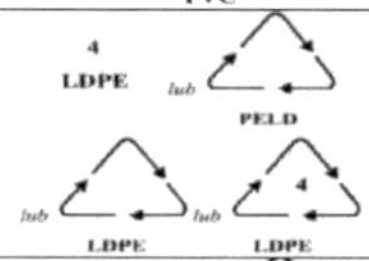  |
| Polypropylene - pp, used for example in foam insulation, floor coverings, toys, household appliances | 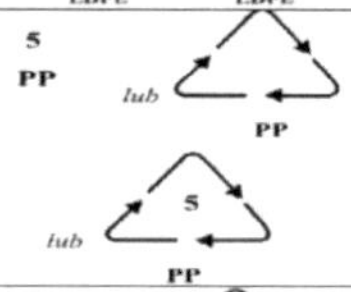  |
| Polystyrene - ps, contained for example in styrofoam, fancy goods and toys | 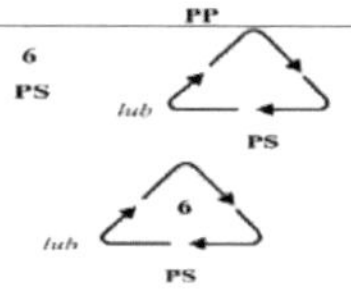  |

Source: [3].

**Waste Transfer Card** (KPO) is another source of information and intra-organisational knowledge, also offering the possibility of its commercial use. This card is the evidence for appropriate disposal of waste by the entity / supplier of the stream, while omitting a natural person, at entitled recipients, according to the waste code as specified in the Code Catalogue [Regulation of the Minister of the Environment of 27 September 2001 on the waste catalogue]. The waste catalogue divides the waste according to the source of its formation in 20 groups as shown in table 3.

**Table 3.** Catalogue of waste by groups

| Waste type | Marking |
|---|---|
| Waste arising from the search, extraction, physical and chemical processing of ores and other minerals | 01 |
| Agricultural, horticultural, hydroponic, fishery, forestry, hunting and food processing waste | 02 |
| Waste from wood processing and from the manufacture of boards and furniture, cellulose pulp, paper and cardboard | 03 |
| Waste from the leather, fur and textile industries | 04 |
| Waste from crude oil processing, natural gas purification and pyrolytic coal processing | 05 |
| Waste from the manufacture, preparation, marketing and use of nonorganic chemical products | 06 |
| Waste from the production, preparation, marketing and use of products of the organic chemistry industry | 07 |
| Waste from the manufacture, preparation, marketing and use of protective coatings (paints, varnish, ceramic enamels), putty, adhesives, sealants and printing inks | 08 |
| Waste from the photographic industry and services | 09 |
| Waste from thermal processes | 10 |
| Waste from chemical treatment and coating of metals and other materials, and from non-ferrous metal hydrometallurgical processes | 11 |
| Waste from shaping and physical and mechanical processing of metals and plastics | 12 |
| Waste oils and liquid fuel waste (excluding edible oils and groups 05, 12 and 19) | 13 |
| Wastes from organic solvents, coolants and propellants (excluding groups 07 and 08) | 14 |
| Packaging waste; sorbents, wiping cloths, filter materials and protective clothing not included in other groups | 15 |
| Waste not included in other groups | 16 |
| Waste from the construction, repair and dismantling of buildings and road infrastructure (including soil and soil from contaminated sites) | 17 |
| Medical and veterinary waste | 18 |
| Waste from installations and equipment for the management of waste, sewage treatment plants and the treatment of drinking water and water for industrial purposes | 19 |
| Municipal waste including selectively collected fractions | 20 |

Source: [2].

The waste code defining the type of waste consists of six digits. Hazardous waste is indicated in the catalogue by the top index in the form of a "*" star.

## 5 "Trash can content" as a aim of marketing research

The results of the analysis of the information contained in the barcode label of municipal waste and so-called analysis of the contents of the trash can have a marketing value and can be the basis for the marketing decisions of the suppliers of a particular type of product on the market. Therefore, they may be the subject of knowledge commercialization in relationships between enterprises operating in waste management sector and companies or research agencies. **The barcode** is a graphical representation of information through a combination of dark and light elements, determined by the code symbols of the code structure. Graphics correspond to a sequence of numbers, each of which is the carrier of specific information. Two or three initial digits are the country of origin (the country code is 590), the next five digits are the manufacturer's code assigned by the Universal Copyright Convention (UCC). The next digit is the product code assigned by the manufacturers, and the control digit confirming the correctness of the scan. The digital representation of what the code contained in the code is intended to allow to manually enter the code, if the reader is unable to read the information. The code is an informational value when the electronic reader (laser, diode or camera) is automatically read by the reader, allowing the product to be identified.

On the other hand, the information from the analysis of **"trash can content" (the so-called garbology analysis"**) [11] refers to household shopping and consumption habits, which can be the subject of comparative analysis in the category of the subject (residential and non-residential areas, including industrial areas). Geographic (international, regional, local), territorial unit (city, village), type of residential (single or multi-family buildings), or time, with the distinction of "special periods" (e.g. holidays, holidays period etc.). Analysis of the content of trash can may concern:

1) the structure of purchased and consumed goods in the household, by residents of a building, estate, district, companies of their groups, in a specific area;
2) eco-habits, manifested in purchasing behaviour and tendency to segregate waste;
3) quantity of consumed / consumed good of a category, in a given unit of time and the amount of single purchase expressed for example by the size of the package;
4) intensity of consumption, expressed as the time of filling the trash and the frequency of emptying;
5) social level of the household;
6) information about consumer habits concerning consumption of highly processed products or products / raw materials;
7) information about preferences regarding local, national or foreign product preferences;
8) the degree of waste of purchased products: food, household and TV appliances and electronics.

## 6 Conclusions. Implications for Business Marketing Practice

As shown by the results of research and analysis, the marketing value acquired in the competent, multi-faceted and multi-concept activities of waste sector entities, who establish systemic relationships with stream providers, aims of secondary markets, customers and research agencies in the process of direct and indirect relations (Figure 1).

https://doi.org/10.1051/shsconf/20207301028

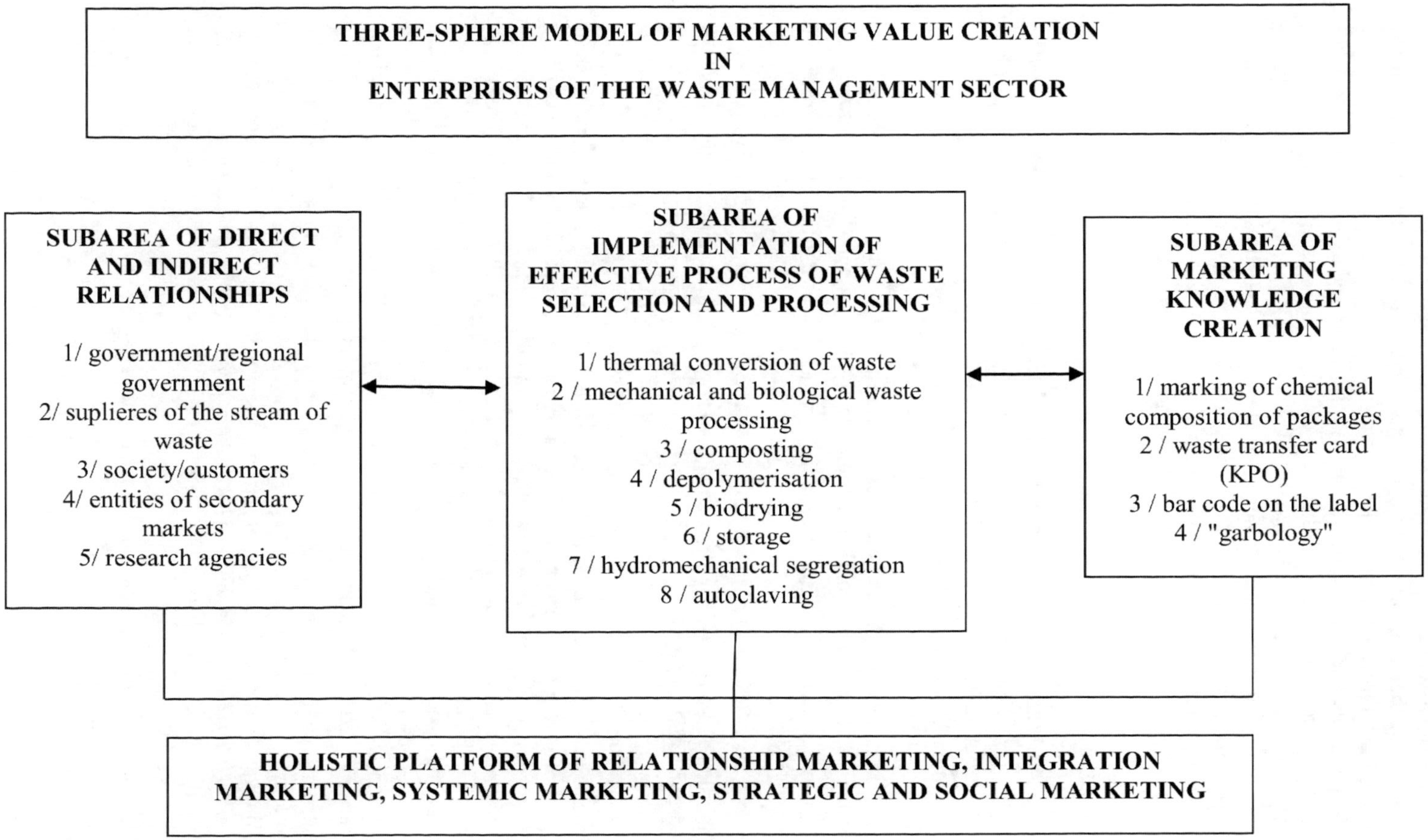


**Fig.1.** Multiconceptual business model of marketing value creation

Source: Author.

Wide range of activities of entities of waste management sector appreciates the holistic and multi-nature of model of creation of marketing value. These are the value of relations, value of wellness of society in long term perspective and social responsibility, commercial value of marketing information included in waste.

Knowledge, skills and eco-awareness are the linking elements in outside and inside relationships, but also the determinants of efficient logistics processes, including intraorganizational ones, in formulating the parameters of appropriate waste streams, desirable because of technological solutions used in the plant. In addition, information obtained based on the analysis of acquired waste starts to gain a new commercial dimension and opens a broad field of cooperation on functional basis of management, i.e. marketing research. It is not enough appreciated by entities of both sectors yet.

## References

1. P. Pekkala, Creating New Value from Waste and Industrial Side Streams. *Bioeconomy* [online], Available at: https://makingoftomorrow.com/creating-new-value-waste-industrial-side-streams/ (2017)
2. Ministry of the Environment, Environmental Protection Law, 27th April 2001 clause 143 (2001)
3. Ministry of the Environment, Reference document on Best Available Techniques for Industrial Waste Processing [online], Available at: http://ippc.mos.gov.pl (2006)
4. Ministry of the Environment Best Available Techniques for Industrial Waste Processing (2006)
5. Ch. Perry, Case Research in Marketing. *The Marketing Review,* (1), 303-323 (2001)
6. T. Żabińska, L. Żabiński, Zarządzanie marketingowe. Koncepcje marketingu a praktyki zarządzania [*Marketing management. Marketing concepts and management practices]*. Aspekty teoretyczne i badawcze, [*Theoretical and research aspects]*, Publisher: AE Katowice, University of Economics in Katowice (2007)
7. „Wsparcie na działania sieci organów środowiskowych i instytucji zarządzających funduszami unijnymi" "Partnerstwo: Środowisko dla Rozwoju" ["*Support for activities of the environmental networks and institutions managing EU funds." "Partnership: environment for development*"] in 2014 from Resources of 2014 Technical Assistance Operational Programme 2007-2013 within Priority III – Support for execution of structural funds operation; Action 3.1. – „Functioning of institutions involved in NSRF". Executor: Consortium: Uniwersytet Zielonogórski i Zakład Utylizacji Odpadów Spółka z o.o., *University of Zielona Góra and Waste Disposal Facility limited liability company*, Zielona Góra, (2015). Available at: www.ekspertyzambp.com.pl. (2016)
8. I. Sztangret, Zrównoważony rozwój przez zarządzanie wiedzą w eko-systemie interesariuszy, na przykładzie przedsiębiorstwa usług komunalnych [*Sustainable development by knowledge management in the eco-stakeholder system, on the example of a utility company*], Wydawnictwo Naukowe Uniwersytetu Szczecińskiego [Scientific Publishing House of the University of Szczecin], Studies and Research Papers of the WNEiZ US **43/2** (2016)
9. I. Sztangret, Intelligent knowledge management in the sector of waste disposal enterprises. *III. International Scientific Conference – Winter session Industry 4.0* (2018)

10. T. Styś, R. Foks, Rynek gospodarowania odpadami komunalnymi w Polsce. *Perspektywa 2030* [*Market for municipal waste management in Poland. Perspective 2030*] (2014)
11. J. Baguchinsky, Adventures in Garbology: What Trash Can Tell Us [online], Available at: http://itech.fgcu.edu/&/issues/vol2/issue2/garbology.htm (1999)

# Globalization impact on the world travel market development

*Iryna Mykhailivna* Trunina[1*], *Olena Anatoliivna* Sushchenko[2], *Viktoriia Valeriivna* Druzhynina[1] and *Olena Leonidivna* Zahorianska[1]

[1]Kremenchuk Mykhailo Ostrohradskyi National University, Pershotravneva Street 20, 39600 Kremenchuk, Ukraine
[2]Simon Kuznets Kharkiv National University of Economics, Nauky Ave 9A, 61000 Kharkiv, Ukraine

**Abstract.** The paper deals with the research of the theoretical and methodological bases of the world travel market functioning, taking into account the new trends influence on it and gives the characteristic of the forms of these trends. We provide the globalization processes dynamics assessment and their impact on the world travel market development. It enables the determination of the main features of Ukraine's participation at the world travel market and results in the indication of the Ukrainian tourist product development prospects at it. We prove that the development of tourism, as one of the basic branches of national economy, is an important factor of the social development and cultural growth of Ukraine. Tourism is the source of foreign currency income in the state treasury and also the center of new jobs creation. The interest of foreign tourists in our history and culture will grow with the Ukraine tourism development, which will enable our country to preserve its cultural and historical heritage and to be successfully placed at the world travel market.
**Key words:** globalization, tourist activity, world travel market, macroregions, integration, global trends

## 1 Problem statement

At the end of XX and the beginning of XXI centuries the processes that got the name of globalization began to manifest more distinctly and dynamically in the development of the world economy. It is these processes that significantly influence the growth of economy, the increase of the country competitiveness at the world markets, the improvement of the welfare of the population. The globalization processes at the travel market are characterized by the change of technologies, the business activity internationalization, the transport infrastructure modernization, the new regulation mechanisms creation.

The objective necessity for the further world travel market development practical bases development and improvement in the current world economy, taking into account the influence of globalization, explains the topicality of this research subject.

* Corresponding author: author@email.org

## 2 Brief literature review

The papers of such outstanding scientists as M. Bosovska, J. Nesbitt, O. Plotnikov, M. Ryza, V. Rokoch, V. Kvartalnov, G. Harris, et al. deal with the problems of the development of the world travel market. Such Ukrainian and foreign scientists as O. Bilorus [1], S. Sokolenko [2], D. Lukianenko [3], V. Sokolov [4] contributed a lot into the study of the globalization impact on the world market. However, the problems of the basic aspect's determination of current globalization processes impact on the international tourism development and the international tourism study in the global and regional processes context are insufficiently covered.

## 3 Purpose of study

Consists in the analysis of the international travel market in the context of globalization and the Ukraine efficient integration into the world travel market further prospects determination.

## 4 Results

Nowadays the most world economies develop in close cooperation. This situation is caused by the historical conditions and current relations concerning the big-scale collaboration in the fields of trade, investments, recreation trips and business. These are the direct relations between the enterprises, organizations, the bodies of local self-government, states, economies and their separate industries in their foreign activity. All the mentioned interrelations take place within the framework of such a phenomenon as globalization that essentially influences the economic and social spheres of management.

In the special literature one can come across a lot of definitions of globalization and almost all of them indicate its wide range. Table 1 contains the definitions illustrating the basis of the globalization concept.

**Table 1.** Globalization in the theoretical approach.

| Author | The definition of "globalization" | Source |
|---|---|---|
| 1 | 2 | 3 |
| P. Haggett | a series of events, decisions and reactions taking place in some part of the earth and significantly influencing the functioning of the economy and the society in the farthest parts of the world; it is a process based on the spatial issues and community independently of the location | [5] |
| M. Czerny | from the point of view of Geographical sciences, it is the change from the local, regional scale to the global one with the participation of two necessary interacting factors, i.e. man and the environment | [6] |
| The European Commission | a process including the final production and markets situated in different parts of the world, inseparably interrelated by contracts, commercial and financial-technological activity | [7] |
| Z. Bauman | a process of the acceptance of the global rules and standards and refusal from the democratic and autonomous policy | [8] |
| Zakharova M.V. Kashanina T.V. | Globalization is a multi-social phenomenon; it is a world process that covers all the spheres of the social life and joins national states into one common system | [9,10] |
| Shergin S.O. | the determination of the existent types of local communities, measures and parameters and also social relations with other communities within the global international space and the reconstruction of the model of their interrelation with the external world | [11] |

| | | |
|---|---|---|
| Kravchuk M. | the formation of the integrated general information space by means of the development of information-computer technologies, which provides a powerful impulse to the acceleration of the process of borrowing cultural and institutional acquisitions, and the creation of certain universalization of the forms of social life on this basis. This expansion results in ruining of the traditional sociocultural forms and institutions. | [12] |
| Kosolapov N.A. | the intensification of the crisis phenomena and weakening of the role of the national state the tendency to the unification of the power, the global expansion of the consumption means, the solution to the local problems in the context of the global community, the flexible and multi-center character of the international relations. | [13] |
| H. Borne, A. Doliński | New directions and tendencies in the international tourism due to geopolitical changes causing the occurrence of the phenomenon of the international mass tourism that got the name of "technified tourism" | [14] |

Source: Authors.

Analyzing Table 1 and the literature in the subject, it is possible to say that globalization is the process of the world economic, political and cultural mutual drawing together and creation of interrelations.

Unfortunately, there are definitions that object to great globalization as this phenomenon results in the loss of the state identity both economic and social dependence and even the increase of the level of unemployment and environment pollution [15]. In spite of different approaches to the globalization phenomenon, in most definitions both economy and society are the important factors underlying the basis of the globalization of the world economy. Thereby, the analysis of the main directions of globalization makes it possible to determine the human life main parts influenced by it most of all: business activity, industry, the development of new technologies, tourism, culture and recreation, social attitude, ecology, social-economic responsibility [16, 17].

Taking into account the above mentioned points, it should be noted that globalization equally concerns both economy and society [18]. The interrelation between globalization and tourism is worth thinking about as it is an important element of the social activity and, at the same time, one of the main branches of economy. From the point of view of tourism, globalization is defined as a process that unites people all over the world, minimizing the transport cost, leveling the boundaries, cultural and social barriers, facilitating the access to new fields of knowledge and work. The consequences of the globalization in tourism influence the following factors: unlimited access to information; the development and constant modernization of the means of transport, free markets and foreign trade related to them, the expansion of the hotel networks in the international markets, considerable demand for travel services, international competition.

Many specialists state that globalization positively influences the economy of tourism and tourism itself. Due to these processes the current world touristic sphere rapidly progresses. Its development promotes active economic growth of many countries of the world. However, the tourism development world trends essentially influence the tourism branch development of a particular country economy. Nowadays tourism has become an independent branch of economy. It makes almost 6% of the world gross national product, 5% of all taxes and 7% of the world investments. The tourism industry development promotes the increase of the budget income, the new jobs creation, the infrastructure development, the new tourist objects construction, the intensification of the protection and reconstruction of the existing historical-cultural monuments. In many countries the tourism has become an important source of budget and belongs to the most promising branches of the national economy. The expansion of the recreation-tourism industry and the growth of its influence on the economic development are among the world economy development priority directions.

Considerable territory irregularity is typical for the current international tourism. In the most general form it reflects different social-economic levels of the world' countries: economically developed countries obtain 57% of the profit, developing countries get 30%, the countries with transient economy – 13%. Based on such regional structure of tourism income and profit it is possible to distinguish between five tourist macroregions of the world:

1. European region (the countries of West, North, South, Central, East Europe, and also the countries of the East Mediterranean – Israel, Cyprus, Turkey).
2. American region (the countries of North, South, Central America, the island countries and the territories of the Caribbean basin).
3. Asian-Pacific region (the countries of East and South-East Asia, Australia and Oceania).
4. African region (the countries of Africa except Egypt and Libya).
5. Middle East region (the countries of West and South-West Asia, Egypt, Libya).

The progressive dynamics of development is typical of all the macroregions but the rate is different and the number of tourist arrivals is also different. The Travel &Tourism world organization makes annual rating of the world' countries as to the tourist arrivals and the tourism income and, based on these data, forms the index of competitiveness (IGC) of all world countries in the travel industry (Table 2).

**Table 2.** Analysis of the competitiveness index (IGC) of tourist macro-regions of the world in 2015-2016, (formed according to the data 19).

| No. | Touristic macroregions of the world | Number of countries | Name of the country | 2015 IGC max | 2015 IGC min | 2016 IGC max | 2016 IGC min | Share of tourist arrivals, % |
|---|---|---|---|---|---|---|---|---|
| 1 | European | 41 | Switzerland | 5.76 | - | 5.81 | - | 58 |
| | | | Bosnia and Herzegovina | - | 3.71 | - | 3.8 | |
| 2 | Asian-Pacific | 27 | Singapore | 5.68 | - | 5.72 | - | 19 |
| | | | Kirghizia | - | 3.87 | - | 3.75 | |
| 3 | American | 21 | USA | 5.61 | - | 5.7 | - | 17 |
| | | | Venezuela | - | 3.3 | - | 3.27 | |
| 4 | African | 34 | Mauritius | 4.43 | - | 3.73 | - | 4 |
| | | | Mauritania | - | 3.03 | - | 2.94 | |
| 5 | Middle East | 14 | UAE | 5.24 | - | 5.26 | - | 2 |
| | | | Yemen | - | 2.72 | - | 2.74 | |
| | Total: | 137 | - | - | - | - | - | 100 |

Source: Authors.

Analyzing the data in Table 2 one can come to the conclusion that the biggest share (58%) of tourist arrivals belongs to Europe. It is due to the availability of all the types of recreation facilities, steady economic and political situation in this macroregion, the presence of the developed infrastructure and profitable geographic location. However, the recreation in Europe is more expensive than in other macroregions [20].

Asian-Pacific macroregion is the second; the share of its tourist arrivals is 19%. It is explained by the presence of unique and exotic types of the recreation resources, the high level of service and lower prices than in Europe or America. However, unsteady political situation, frequent natural disasters such as earthquakes, tsunami, floods negatively influence this macroregion development. Singapore, being the leader in this group, took the necessary measures related to the international openness to guarantee the safe arrival of international tourists and the quality of its air transport infrastructure is high. The country invested more funds into sightseeing, the development of entertainment and games, and is at the top of the entertainment rating looked for online. Also, the country should put more efforts into environmental sustainability especially because of floods stress and the damage of the coastal ecosystems.

American macroregion is the third. A somewhat lower share of tourist arrivals in it can be explained not by the decrease of the tourist attraction of America but by the rapid tempo of the travel industry development in other macroregions. This region is characterized by the availability of most types of the recreation resources, the developed infrastructure. However, the unsteady political situation in the countries of South America, natural disasters in the USA (hurricanes, floods) cause the decrease of the tourist arrivals share. The USA is the top country in American region and the third one in the general rating. With many natural objects of the World Heritage the US occupies the third position in the rating of the natural resources, which also attracts tourists due to cultural, entertaining and sport sites. It is supplemented with a highly competitive business environment and several bilateral air traffic agreements supported by the infrastructure of the air transport of the country. However, the visa requirements are serious, which restricts the growth of tourist arrivals.

The last but one position is occupied by African macroregion, the cheapest place of recreation in the world with rich natural recreation resources, which promotes the increase of the tourist arrivals number. The insufficiently developed infrastructure and the low level of travel services negatively influence the development of the tourism industry. Besides, the availability of political problems and the irregular development of different parts of the continent are obstacles to increase of the tourist flows amount.

Middle East macroregion is characterized by the lowest share of tourist arrivals, which is explained by the unsteady political situation in the region, the absence of the developed tourist infrastructure.

However, a long tourist season, the presence of rich recreation resources, the proximity of developed European states positively influence the tourism development in the region and make a considerable share in its tourism-recreation potential. The United Arab Emirates (UAE) are the leaders in the region and occupy the 16th position in the general rating as they accepted more than 10 mln tourists in 2016. While there are no rich natural resources in the UAE, the country created the unique environment to attract tourists for business and recreation.

Analysis of the world travel market makes it possible to come to conclusion that the international tourism develops in close correlation with other spheres of international life and responds to the general economic state of the market and political situation by the increase or decrease of the growth rate.

It explains the irregularity of the international tourism development in different periods and its special features in different regions. Under these conditions it is especially important to develop the directions of the competitiveness improvement of such national travel markets, in particular, Ukrainian one, that should take into account the modern tendencies of world economy development and the special features of the country tourism-recreation potential.

Geographically, Ukraine belongs to European macroregion and has the following competitiveness indices, based on data from the world tourism organization (Travel&Tourism): in 2015 IGC was 4.03, which is the 79th position in the general rating; in 2016 IGC was 4.00, which is the 85th position in the general rating. Thus, it is obvious that the position of Ukraine at the world travel market deteriorated in comparison with 2013, one can also state that the tourism potential in Ukraine is used insufficiently; the results by the subindex "Human, cultural and natural resources" are especially low [19, 21] (Table 3).

From 2012 till 2013 the number of Ukrainian tourists who travelled abroad increased almost by 1.5 times. However, from 2013 till 2016 it decreased almost by 9 times. The number of the served tourists from 2012 till 2013 increased more than by 20 times, which is an extremely high index. From 2013 till 2016 it decreased by 16.5 times [20, 21].

**Table 3.** Rating of the world' countries according to the Travel & Tourism competitiveness index in 2015-2016 in comparison with 2013, (formed according to the data 20, 21)

| No. | Country | IGC 2013 | | IGC 2015 | | | IGC 2016 | | |
|---|---|---|---|---|---|---|---|---|---|
| | | rating | points | rating | points | Δ 2015/ 2013 | rating | points | Δ 2016/ 2013 |
| 1 | Switzerland | 1 | 5.66 | 1 | 5.76 | 0 | 1 | 5.81 | 0 |
| 2 | Singapore | 10 | 5.23 | 2 | 5.68 | ↑8 | 2 | 5.72 | ↑8 |
| 3 | USA | 6 | 5.32 | 3 | 5.61 | ↑3 | 3 | 5.7 | ↑3 |
| 4 | Netherlands | 13 | 5.14 | 5 | 5.5 | ↑8 | 4 | 5.57 | ↑9 |
| 5 | Germany | 2 | 5.39 | 4 | 5.53 | ↓2 | 5 | 5.57 | ↓3 |
| 6 | Sweden | 9 | 5.54 | 9 | 5.43 | 0 | 6 | 5.53 | ↓3 |
| 7 | Great Britain | 5 | 5.38 | 10 | 5.43 | ↓5 | 7 | 5.49 | ↓2 |
| 8 | Japan | 14 | 5.13 | 6 | 5.47 | ↑8 | 8 | 5.48 | ↑6 |
| 9 | Finland | 17 | 5.1 | 8 | 5.45 | ↑9 | 10 | 5.44 | ↑7 |
| 10 | Norway | 22 | 4.95 | 11 | 5.41 | ↑11 | 11 | 5.44 | ↑11 |
| 11 | UAE | 28 | 4.86 | 17 | 5.24 | ↑11 | 16 | 5.26 | ↑12 |
| 12 | Spain | 4 | 5.38 | 33 | 4.59 | ↓29 | 32 | 4.68 | ↓28 |
| 13 | Ukraine | 76 | 3.98 | 79 | 4.03 | ↓3 | 85 | 4 | ↓9 |
| 14 | Yemen | 133 | 2.96 | 138 | 2.73 | ↓5 | 138 | 2.74 | ↓5 |

Source: Authors.

Fig.1 allows the analysis of the state of Ukraine at the world travel market in more detail. Switzerland was chosen the leader as this country has been occupying the first position in the world competitiveness rating for many years.

Analyzing the information above, one can state that Ukraine's indices of the travel services development are below the average among the countries of European macroregion. It especially concerns the macroeconomic indices of the state development; the weak development of the financial market that hampers the investment into the travel industry and the innovation development

Ukraine has all the conditions for the intensive development of the incoming and outcoming tourism: the special features of the geographic location and relief, a favorable climate, the wealth of the natural, historical-cultural and tourism-recreation potentials. Unfortunately, Ukraine looks rather modest against the background of the world tourism that advances with extreme rapidity.

Now Ukrainian citizens are kind of investors of the foreign countries, which is one of the most painful problems of the whole tourism industry. There is no doubt that the causes of this state are to be looked for in the complicated social-economic situation in the country, in the unadjusted mechanisms of stimulation in the travel industry, in the absence of the efficient strategy of the development of this industry both at the national and regional level, in the inaccessibility of financing sources, in the taxation system imperfection, in the innovation sphere insufficient development.

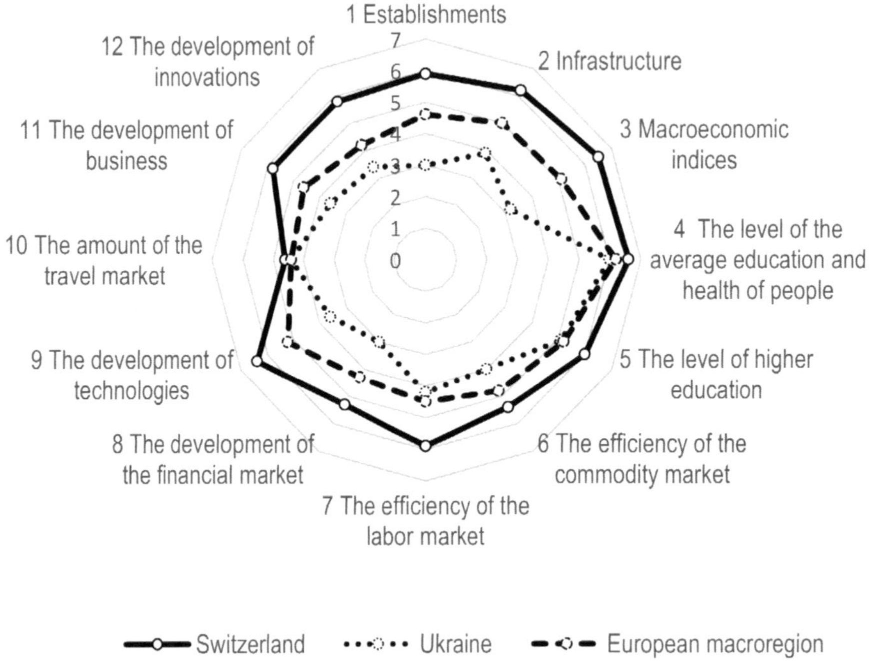


**Fig. 1.** The state of Ukraine at the European travel market, (formed according to the data 20, 21).

Source: Authors.

Obviously, Ukraine is extremely rich in almost all the types of tourism resources and can be rather attractive for foreign investors. However, unfortunately, the country tourist infrastructure is insufficiently developed for it, so, first of all, it is necessary to completely modernize all the hotels and restaurant enterprises, and roads. Of course, such measures require considerable funds that our country is not able to provide, especially under the conditions of economic crisis.

There is no complete examination of the whole tourism industry of Ukraine with the profound information, statistic data, analytical conclusions and remarks.

Despite the fact that nowadays there are general principles of the international tourism organization, it's functioning and development, every country that pays special attention to this branch of economy applies its own models of tourism. It results from different reasons based on historical, geographical, climatic, economic and social specific features of each country or region. Every country, that develops the tourism industry, tries to completely use its national resources aiming at obtaining considerable currency income to the budget.

## 5 Conclusions

Among the current tendencies of the world market development on the whole and the travel market in particular, the globalization processes deserve special attention. The present paper contains the theoretical substantiation of these processes. They are characterized by the following features: the change of technologies, the modernization of the transport infrastructure, the business activity internationalization, the international tourism regulation mechanism creation. With the development of these processes one can observe active economic growth of many countries of the world, in particular, due to the tourist flows amount increase.

Taking into account the performed analysis of the main world macroregions development, we revealed their development irregularity by means of the travel industry competitiveness index calculation. We have found out that the international tourism rapidly responses to the changes of the general economic state of the market and political situation and under such conditions the assessment of Ukrainian travel market competitiveness, taking into account the world economy development trends and the country tourism-recreation potential features, is of special significance. The results of this assessment certify that Ukraine demonstrates considerable lag in the average indices of European macroregion development as to the country development macroeconomic indices, the financial market development indices, which hampers the investment into the tourism industry and the innovations development.

Taking into account the above results of the research, it is possible to formulate the main strategic imperatives for the Ukrainian tourism industry development both at the national and world level. Firstly, the intensive information and Ukraine tourism potentials advertising by means of the circulation of the whole tourism industry complete examinations with the profound information, statistic data, analytical conclusions and remarks. Secondly, the travel activity stimulation mechanisms creation of both incoming and outcoming tourists via the tourist infrastructure modernization, in particular, hotels and catering enterprises. Thirdly, the innovations in tourism local programs development working out, their efficiency assessment, the innovation financial-credit support substantiation, the travel enterprises innovation policy strategies and tactics determination with the further involvement of highly qualified guides, hotel business employees, marketing specialists, managers and other specialists into the travel market.

The formed database can be used in the further research to create the economic-mathematical models of the Ukrainian tourism industry development taking into account its regions development potential.

## References

1. O.H. Bilorus, *Ekonomichna systema hlobalizmu [The economic system of globalism]*. KNEU, Kyiv, 360 (2003)
2. S.I. Sokolenko, *Hlobalizatsiia i ekonomika Ukrainy [Globalization and the economy of Ukraine]*. Kyiv, 568 (2009)
3. D.H. Luk'ianenko, *Hlobalna ekonomichna intehratsiia [Global Economic Integration]*. Kyiv, 220 (2008)
4. V. Sokolov, Konturyi buduschego mira: natsii, regionyi, transnatsionalnyie obschnosti [The contours of the future world: nations, regions, transnational communities]. *Moscow University Bulletin*, **3**, 23-67 (2004)
5. P. Haggett, *Geography*. A global synthesis. Pearson Education Limited, Prentice Hall, Harlow, UK (2001)
6. M. Czerny, Globalizacja a rozwój. *Wybrane zagadnienia geografii społecznogospodarczej świata* [Globalization and development. *Selected issues of socio-economic geography of the worl*d] (2005)
7. P. Dominiak, J. Wasilczuk, N. Daszkiewicz, Small and medium-sized enterprises in the face of internationalization and integration of European economies. *Examples of Italy, France, Poland and the Czech Republic*, *Scientific Publishing Group*, **7**, 42-53 (2005)

8. Z. Bauman, *Globalizacja*. Państwowy Instytut Wydawniczy, Warszawa, 67-83 (2000)
9. M.V. Zaharova, *Sravnitel'noe pravovedenie: voprosy teorii i praktiki* [*Comparative Law: Theory and Practice*]. Moskow: Prospekt, 160 (2013)
10. T.V. Kashanyna, *Struktura prava* [*Right structure*]. Moskow: Prospekt, 584 (2013)
11. S.O. Sherhin, Kontseptualno-prakselogichni aspekty globalizatsiinogo rozvytku [Conceptual-praxeological aspects of globalization development]. *Globalizatsiia i suchasnyi mizhnarodnyi protsess – Globalization and modern international process*, 104-105 (2009)
12. M. Kravchuk, Kontseptualna evoliutsiia teorii globalizatsii [The conceptual evolution of the globalization theories]. *Politychnyi menedzhment – Political management*, **2**, 122-137 (2009)
13. N.A. Kosolapov, Mezhdunarodno-politicheskaia organizatsiia globaliziruiuzhegosia mira [The international political organization of the globalized world: models for a medium-term prospect]. *Obshchestvennye nauki i sovremennost – Social sciences and modernity*, **6**, 141-143 (2001)
14. H. Borne, A. Doliński, *Organizacja turystyki*. WSiP, Warszawa, 23-26 (1998)
15. V. Druzhynina, G. Likhonosova, Strategic imperatives ensuring population welfare under transformation exclusion conditions. *Baltic Journal of Economic Studies*, **3**(5), 135-146 (2017)
16. I.M. Trunina, O.A. Sushchenko, G.S. Likhonosova, The territory recreation ensuring management system design. Economic *Theory and Practice*, **1**, 635-647 (2017)
17. I.M. Trunina, O.A. Sushchenko, Creation of innovation clusters as a line of enterprise competitiveness improvement in the field of foreign economic activity. *Actual Problems of the Economy,* **3** (177), 191-198 (2016)
18. V. Druzhynina, G. Likhonosova, G. Lutsenko, Assessment welfare of the population in the synergetic system of socio-economic exclusion. *Marketing and Management of Innovations*, **2**, 54-68 (2018)
19. A.Y. Dubeniuk, Otsinra konkurentospromozhnosti turystychnoi galuzi Ukrainy [The assessment of the competitiveness of the tourist industry of Ukraine]. Problemy razvitiia vneshneekonomicheskih sviazei i privlechenie inostrannyh investitsii: regionalnyi aspect – The problems of the development of the foreign economic relations and attraction of foreign investments: regional aspekt, Donetsk: DonNU, **7**, 754-761 (2007)
20. The Travel & Tourism Competitiveness Report [online], Available at: http://www3.weforum.org/docs/WEF_TT_Competitiveness_Report_2013.pdf. (2013)
21. The Travel & Tourism Competitiveness Report [online], Available at: http://www3.weforum.org/docs/GCR20162017/05FullReport/TheGlobalCompetitivenessReport2016-2017_FINAL.pdf (2017)

# Human factors errors and identification of causes of successful/unsuccessful completion of business projects

*Radka* Vaníčková[1*]

[1]Institute of Technology and Business in České Budějovice, Faculty of Corporate Strategy, 37010 České Budějovice, Czech Republic

**Abstract.** Managers often deal with uncertainty of a different nature in their decision processes. The objective of the contribution is to confirm or refute the facts published in professional literature including the opinions resulting from the erudition of the author of the paper claiming that human factor errors and identifying causes of successful / unsuccessful completion of business projects is a significant explicit factor. With regard to this finding, the paper will deal with the quintessence of the human factor positive/negative impact on the business projects success, as well as recognizing and prevention of negative impacts on the implementation of success/failure a business project. The theoretical part defines the influence of the human factor and correlation of a business project success/failure with the SMART method. The analytical part of the paper interprets the theoretical findings into practical models of project teams by means of comparison and erudition of the author of the paper. The final part proposes measures, suggestions and recommendations concerning the optimal setting of the human factor key values so that the personality positive/negative impact on the business project success is evaluated and from the socio-economic perspective in terms of invested financial resources and the return of investment, from a substantive and time point of view. For the purposes of the contribution, the term "influence" can be divided into exogenous and endogenous, while "success" can be defined as a determinant of positive/negative fulfilment of expected vision and business strategy entities. In terms of project management, "project" is a set of individual and unique activities to implement a planned change or innovation in a given period of time using specific sources.
**Key words:** human factors errors, interest groups, project team, team roles, behavioural competence

## 1 Introduction

Project is a set of concrete activities aiming to achieve the objective pursued. It is determined by time, financial resources, and human and material resources. Most projects have a unique objective. Business projects are generated by project teams. Project objective is defined with regard to the originality and uniqueness of partial tasks, necessary flexibility, creativity,

* Corresponding author: vanickova@mail.vstecb.cz

expertise, practical experience and professional skills, ambitions and self-realization, knowledgeable project team is build up with the aim to achieve the objective with higher value added and synergic effect of economic and social benefit. The basic attributes of the influence of individual personalities and interest groups affecting a business project success are defined [1].

# 2 Data and methods

## 2.1 Human factor as characteristic

The bearer of project activities, productivity and prosperity are people, that is, core employees and co-workers. The activity in a business and in a micro-enterprise business shows analogy of the human factor – project management [2], [3]. Project manager has necessary professional qualifications and specific knowledge in the field / discipline in the form of so-called hard skills, i.e. technical, methodical, language, knowledge, etc. [4]. In the case of manual workers´ positions, this refers to e.g. service and maintenance of numerically controlled machines, setting standards, regulations and internal rules of protection of safety and health at work, etc. The examples of soft skills are e.g. the development of verbal and on-verbal communication skills, individual and team work, project team leadership [5], [6], conceptual and entrepreneurial thinking supporting recognition, understanding, critical thinking, constructive criticism and the willingness to accept criticism, willingness to risk taking, resistance to stress, conflict resolution and business negotiation in a way of team partner cooperation, the ability to adapt to a situation and to accept innovations [7], learning new things, assertiveness, self-reflection, recognizing moral and ethical values, tolerance, networking, etc. [8]. Techniques of mental work and hygiene are hard to measure, they are less tangible, but equally important. The ability to acquire and develop professional knowledge and practical skills is influenced mainly by the temperament including the environment in family. An example for a leadership position is assessment of the soft skills influence in relation to employees / co-workers engagement, motivation and satisfaction, or leadership and facilitation of work meetings [9].

**Table 1.** Basic specification of project manager competencies

| | |
|---|---|
| **Communication skills** | Listening, negotiation, presentation |
| **Organization skills** | Planning, forecasting, controlling, setting KPI |
| **Team work** | Motivation, empathy, creativity, emotional stability |
| **Leadership** | Vision, delegating tasks, setting a good example, positive attitude |
| **Art of overcoming obstacles** | Flexibility, ambition, perseverance |
| **Expertise** | Technical, technological, methodical, hard and soft skills, practical experience |

Source: Own processing.

## 2.2 Influence of interest groups

Interest groups are individuals or organizations actively involved in a project or whose interests are positively or negatively influenced by the project implementation or the result of a sub-project. Successful preparation, implementation, and completion of a project depends on the approach / behaviour / expected result of an individual as a bearer of a personality, but also as personalities of so-called interest groups. Significant determinants of success / failure are necessary for the interpretation at an early stage of a business project management, in a feasibility study with the support of the SWOT analysis with the aim to eliminate threats / limitations / milestones / risks [10] and to enhance the functionality of selecting suitable alternatives to strategic and business opportunities and development potential of the successful implementation of a business project – see Table 2 Important influential subjects.

**Table 2.** Important influential subjects

| **Project organization** | **Defines assignment, influences employees in project** |
|---|---|
| Strategic management | Executive and strategic management approve project parameters and sources. |
| Business owners | Propose and formulate a strategy. In the case of inconsistent owner structure, there is increased risk of negative influence |
| Project team | Choosing project team members, assignment of roles to individual team members together with an authorised person in the position of a project manager is a strong motivator and stabilizer of fulfilment / successful completion of a business project. |
| Customer – buyer | Customer/buyer has a unique influence on the usefulness and economic return of investment in a business project. |
| Investor | Banking institutions, shareholders, third subjects requiring prediction of appreciated investment with setting business conditions and acceptable risks of a business project. |
| State institutions | Government, legislative and control institutions influence compliance with legal, tax and legislative conditions that each business entity takes into account and calculates their amount. |
| Other public and media | Media, social, cultural, interest, and environmental centres can have different preferences and personal interests. It is therefore essential to find common solutions or acceptable compromise in order to enhance open communication, conceptual and creative thinking, logical thinking, and meaningful decision-making in favour of a clearly defined business project objective. |

Source: Own processing.

## 2.3 Project team influence

Each person is different, has different dispositions, approaches, abilities, and skills. However, even people with the same skills and competencies can act differently in a community [11]. Human factor has a significant impact on success/failure business project [12], since the basic management requirement is to choose a team, assign key competences, sub-competences and roles in the project with regard to professional and specific experience [13], period of implementation of successfully/unsuccessfully completed projects, personal traits and characteristics of individual project team members including the conformity.

**Table 3.** Team roles according to M. R. Belbin

| | |
|---|---|
| **Chairman, project manager** | dominant, stable, with a great share of soft skills; project leader, leader at a social level, coordinator; his contribution to discussions is properly formulated questions and answers |
| **Proposer, director** | active, impulsive, often dominant member of a project team; manages team work, requires discipline and productivity |
| **Innovator** | intelligent individualist, opinion maker; foresighted, inspiring for team members, brings new ideas |
| **Whistle-blower, controller** | rational and emotionally stable person; analyst with conservative opinions |
| **Organizer, implementer** | conscientious, disciplined person; translates team decisions into concrete tasks assigned to team members; works systematically |
| **Discoverer** | gathers ideas, contributes with his own invention; establishes business contacts, gains external support, improvises |
| **Supporter** | team worker, team builder; enhances team work |
| **Finisher** | team member, whose main characteristics are diligence, precision, sense of order, time management with a high share of control and legal awareness |
| **Specialist** | professional in a given field, with a specific specialization; depending on the nature of a project, he has technical, economic, and legal knowledge; often works as an expert from practice |

Source: Own processing.

## 2.4 Influence of project manager behavioural competence

Project manager shall be chosen based on the interest in people and the awareness of interpersonal relationships. The term "behavioural competence" (hereinafter referred to as BC) in relation to the human factor influence on the business project success/failure is a paradigm of the project manager´s competencies and skills reflected in leading working groups and motivating them to work performance [14]. IPMA standards define sets of competencies including project management topics in the theoretical and conceptual, technical, empirical, and behavioural areas. The basic fragments are further divided into competency elements, while the behavioural competencies in the area of soft skills listed below consider fifteen basic elements.

**Table 4.** Behavioural competencies elements

| | | |
|---|---|---|
| 1. | **Leadership** | Leadership consists in motivating and leading the team to fulfilling the potential and achieving project objectives. |
| 2. | **Involvement and motivation** | personal contribution of a project manager; due to his involvement, people align themselves with the implementation of the project and assigned roles |
| 3. | **Self-control** | systematic and positive approach to dealing with everyday routine and stress situations in a project team |
| 4. | **Assertiveness** | ability to present and defend own opinion, adopt an objective view; the art of persuasion and factual argumentation; protection against conditional manipulation; ability to say NO |
| 5. | **Relaxation** | ability to alleviate stress and tension; personal and team prevention against the burn-out syndrome |
| 6. | **Openness** | Openness is defined as the ability to motivate other team members so that they can see the value of their own opinion, loyalty in relation to self-expression and self-realisation. |
| 7. | **Creativity** | Creativity is the ability of logical and conceptual thinking, the ability to think originally, react flexibly with the use of imagination, tasks and plans visualisation, association, constructional and creative fantasy in terms of increasing performance, sensitivity, and intuition. |
| 8. | **Result orientation** | managerial skill – ability to focus on key objectives, set personal and team priorities using the ABC methods and time management |

| | | |
|---|---|---|
| 9. | **Performance** | managerial skill; the ability to effectively use time, resources, and human capital; performance is a set of characteristics and disposition, preparedness to deliver expected performance / outputs |
| 10. | **Discussion** | a form of work meeting, an efficient form of collective problem solution during organizational, conceptual, and informative meetings |
| 11. | **Negotiation** | Negation between team members within a project and supplier-customer relations is desirable for the purpose of successful business relations of the win-win type. Managerial skills shall include mainly the negotiation strategy |
| 12. | **Conflicts and crisis** | Conflict can be defined as a centre of conflicting interests or incompatible personalities. The task of a project manager is to operatively tackle a crisis or misunderstanding by means of motivating team using leadership to achieve excellence. |
| 13. | **Reliability** | To be a reliable person means to build close and friendly working relationships with the environment based on reliability, fair behaviour, straightforwardness, decisiveness, and self-confidence. |
| 14. | **Understanding values** | A manager sees the value of people, organizations, and societal values, see the quality of the people and the participants´ points of view. |
| 15. | **Ethics** | Ethical behaviour is a basis of each business; it is a part of advanced companies and corporation's business culture. Ethics is a personal and professional autonomy of conscience and human freedom. |

Source: Own processing.

In the theoretical part of the contribution on the topic of "Human factors errors and identification of causes of successful/unsuccessful completion of business projects", the author defined significant correlations of abilities and skills with which the human factor influences the implementation of the project. The more these paradigms are applied in user practice, the more the personality of a project manager/ leader is important in the interaction with the defined objectives and expected outputs [15]. Conversely, the more the competency model of a project manager / leader differs from the characteristic nature of behavioural competencies and leadership, the greater the risk of adverse influence on the success/failure and conduct of a business project is [16].

# 3 Results

Business projects management requirements for speed and efficiency of work, and conducting the project activities, minimization of overhead and wage costs often increase with the growing project life cycle and often move beyond higher demands of human admissibility and resistance mainly in the area of hard skills, specific and professional specializations, practical skills, and managerial competencies, mainly the technical and methodological, conceptual, empirical and behavioural ones, with a lack of limits to the development of human factor competencies. Only the ability of a concrete personality is limited. The erudition of the author† is the fact that there are no limits of the personality development possibilities. Each person sets their own alternatives acceptable without limitations at different stages of life and development influenced by their own ambitions reflected in the self-evaluation of the personality potential.

Another explicit finding of the author is the analysis of the degree of the human factor impact on business project success and comparison of the needs of hard versus soft skills. Both practical experience of the author and publicly accessible studies and documentation in various business sectors indicate higher share of soft skills in behavioural competencies compared to hard skills, e. g. technical, professional, specific, etc.

Briefly, it can be said that the higher influence of behavioural competencies is able to compensate a potentially lower technical and professional competence; or that the personality behavioural competencies enable rapid acquisition of technical knowledge applicable in the

† The author has been active in business development and project management for a long time.

business project. Based on the findings specified above it can be stated that a competent project manager / leader is able to conduct a business project successfully in various sectors, even in completely different business sectors.

### 3.1 Positive influence of personality on business project success

In order to demonstrate the positive influence of human factor on business project success/failure, we will define a completed Czech project. In order to protect personal data of the corporation selected and in compliance with the European General Data Protection Regulation (GDPR) for protecting personal data of a concrete manager, only general, not publicly available parameters of the selected project sector and management of the selected company.

**Table 5.** Basic definition of business project

| | |
|---|---|
| **Sector** | Civil engineering and constructing |
| **Name of project** | Integrated parking system |
| **Final consumer** | Municipality |
| **Position of project manager** | Team leader |
| **Location** | East Bohemia |
| **Project submitter** | Large constructing and design company operating in the given field in the entire Czech Republic |
| **Number of project team members** | 4 (Project leader, specialists and implementer) |
| **Method of implementation** | Acceptance of the contract by the consumer, implementation of business project |
| **Sources** | CZK 5,000,000 |
| **Time** | 36 months |

Source: Own processing.

The starting position of the selected business project – individual phases: Start-up. (Know-how, business meetings with municipalities and political clubs and associations, business lobbying, contracting documents, expert case and economic studies of return of resources, finding a business project investor). This part of a project, from vision to signing the implementation phase of the project embedded in the contract, handing over the successfully completed project to the investor, took 36 months. The management of the project was at its early stage intuitively led by common sense, since no similar business project had been implemented in the Czech Republic by then. It was necessary to create a compact development and professional project team that was able to cooperate flexibly with the maximum workload, integrity, ambitions, patience, but also possesses demonstrable negotiation skills.

The start-up phase was carried out by a four-member team:

Project manager: leader, visionary, innovator, author of the business project concept. This assertive person had predominant behavioural competitions, leadership, negotiation and presentation skills, creativity, and high share of the ability to involve and motivate the project team members. The expert part of the project and legal know-how was delegated to the other team member with a partial support of outsourcing service of a design office, which had been evaluated as the optimal alternative of cooperation in terms of resources and quality of the project.

Project specialist 1: Expert in the field of civil and labour law with overlapping corporate and contract law dealing with the issue of public procurement in an explicit team role dealt with the creating the legal framework of know-how, streamlining the corporate visions with the emphasis on functionality, efficiency, and viability of the business project in the medium and long terms, and with the elimination of risks to prevent later legal failure. This person had a majority share of legal know-how and decision-making competence in technical areas of assessment in terms of risk management, implementation of outputs / impacts on the individual solutions to partial problems, the issue of contractual relations and the quality of business project documentation. In terms of conceptual competence, the specialist 1 excelled in the area of result orientation related to the issue of corporate and contract law. In terms of behavioural competencies, the emphasis was put on reliability, responsibility, ethics of moral behaviour and mutual communication, involvement, and devotion.

Project specialist 2: Expert in the field of economics with a focus on planning, controlling, budgeting, tax records and reporting to project submitter. Specialist 2 had highly specialized technical skills in the field of business economics and economy. Specialist 2 was an owner of a consultancy and advisory centre, an external lecturer at the University of Economics in Prague, the Czech Republic, and lecturer at several educational centres of renowned companies in the Czech Republic. This specialist had a minority share of soft skills in the area of behavioural competencies, e. g. the art of open and shared communication, listening, negotiating, communicating with people, resolving conflicts, and dealing with stress and crisis situations. The members of the project team management actively worked with specialist 2, but the lack of soft skills had to be compensated by other team members, not by the expert in economics.

Project implementer: Executive member of the team with a sense of meeting deadlines, detail, stickler, team builder and coordinator, with the ability to complete the tasks assigned and project activities, diplomat of social intercourse and protocol, harmonized time periods set, convened meetings, coordinated written project documents, reported obtained and achieved internal results / outputs of member teams with outsources outputs into a complex project manual. The project implementer accepted the visions and integrated them into the business plan. It was a person with a high share of conceptual and technical competencies, behavioural competencies with a predominance of creativity, work performance, reliability, negotiation skills, and the ability to deal with conflicts and crisis and stress situations.

As a team, the selected team members created, implemented and finally handed over one of the most successful business projects in the Czech Republic. In terms of evaluation and erudition of the author of the contribution in the implementation of the start-up phase of the above-mentioned business project, it can be stated that the human factor impact on the business project success/failure is an equivalent of success/failure, where the human factor is not considered in terms of an individual, but of the project team with optimally integrated personalities.

## 3.2 Negative influence of personality on business project success

**Table 6.** Basic project definition II

| | |
|---|---|
| **Sector** | Civil engineering. |
| **Name of project** | Improvement a restructuring of company |
| **Final consumer** | Project submitter |
| **Position of project manager** | Team leader and Interim manager |
| **Location** | East Bohemia |
| **Project submitter** | Regional producer of steel constructions |
| **Number of project team members** | 6 (Project leader, specialist, implementer, coordinator) |
| **Method of implementation** | Achievement of required economic indicators over time |
| **Sources** | CZK 3,500,000 |
| **Time** | 12 months |

Source: Own processing.

The company breaking into market in 1992 was directed by the management of owners, i.e. members of a private limited company. Its added value has year-to-year lowered since 2008 while the negative aberration of the added value was higher than the decrease of its total revenues. The company owners decided to restructure the company in a project in 2011. The project management was entrusted to an external manager – a restructuring expert who won the tender. The project team was picked in accordance with owners from internal resources of the company management. The primary task confronted the sequences in individual segments of the business economy and processes, e.g. cash-flow, expenses, process transparency, personnel, business revival and update of system procedures.

The project manager: is an external specialist with great professional and technical competences in the area of mechanical engineering, business economy and the economics of legal awareness who applied a directive style of the management of the project team with one-way communication. Its members called him an authoritarian since he had not shown any signs of behavioural competence in negotiations, discussions, mutual understanding and communication, openness and self-control. What was appreciated were personal traits such as ambition, endurance, independence, goal-mindedness, the trait of which initiated the restructuring process and crisis project management (the author marked him as 'Archer').

The project specialist: is an internal worker/expert – the current chief financial officer. He is an important person with a university education in the relevant field of study. He is on the long term basis associated with the company disposing of technical and professional competences in the area of financial management and planning, controlling and tax records. He is the general manager appointed to the project team by owners of the company. Unfortunately, the role of the chief financial officer and the project specialist were not identical throughout the implementation of the business project. The chief financial officer considered the position held as a threat and degradation to his present job. In regard to erudition and author's own terminology, he is marked as 'Teddy'.

The project implementer: is an internal worker/tuner, present HR manager familiar with the social environment of the company who has developed behavioural competences and disposes of amication elements and who also knows individual internal workers. The project implementer has professional and technical competences in Labour Law and he identifies himself with the business project. With respect to a long-term influence of the labour market and employment rate, he considered the project implementation as necessary and he did not impose any measures for reducing jobs.

The project coordinator: is a member of the project team of the internal environment of the secretariat of the company owner; the project coordinator was also a coordinator, administrator, tuner and cementer (a person that glues the team together). The personality of the project coordinator can be described as well-balanced with the equal proportion of participation in various areas of competence. Her pros were prioritizing the team cooperation, quality and carefulness in drafting written documents, knowledge of business processes, basic legal awareness, and knowledge of IT, issuing internal regulations and workflows. Her behavioural competences included the art of communication, presentation and negotiation.

The project implementer 1 and 2: are internal technical workers who worked on the business project, as had been required by the project call and assignment, simultaneously, having their own job outside the project by contract based on an agreement to complete a job. In regard to competences and skills, they mostly focused on the implementation of business systems and products in the area of health and safety at work. They display behavioural competences such as reliability, understanding of true values, effectiveness and usefulness of the business project. These implementers did not examine larger context of the project; however, the supported and stimulated operational activities of the business project.

The personnel of project team II and its own implementation failed to achieve required parameters and thus to implement the project itself. During the project's life cycle, the cooperation with the project manager/team leader was discontinued and an outsourcing external restructuring company was admitted to complete the project.

The causality of the business project might be defined as follows:

1. A complete lack of behavioural communication and presentation competences and social skills of the project manager/team leader; a considerable fluctuation of continuous and direct communication with members of the project team; the explanation of necessary restructuring measures-to-be caused a negative attitude of the team which negatively influenced the business project which resulted in reducing the personnel in the company.
2. The choice of project team members from managers of internal resources who had been internal workers even in pre-project phases was not optimal. In order to make necessary mends, i.e. systematic changes according to global trends and future expectations, it was necessary to cooperate with managers/team leaders who are interested and ambitious to promptly change the existing state and swiftly respond to it; at the same time, they should not be associated with the existing organizational structure of the company.
3. Company owners' interventions in the project management in the phase of team formation, which preferred members from internal personal resources in order to reduce overheads and labour costs and which admitted an external restructuring company, completed the project without a project manager/team leader.
4. The lack of behavioural competences of the project manager/team leader resulted in so called 'stagnation' of activities within the project team: i.e. individual team members focused only on themselves and not on the team and its close cooperation. In addition, open communication broke down which resulted in a breakdown in communication and barriers.

With the view to the depicted project management in relation to examining the influence of the human factor on the success/failure of a business project we might argue that neither great managerial competence counterbalances the lack of behavioural competences in communication skills, negotiation, leadership, assertion, conflict settlement and resolving crisis and stress situations. If the project team had been well-balanced with the focus on integrity and team members had developed behavioural competences and motivation stimuli, the likelihood of the successful completion of the business project would have been strong, although the project manager/team leader and the contracting authority is responsible for the failure of the project.

## 4 Conclusions

The analytical part compared two different project teams in different branches of entrepreneurship. Although the requirement for the project management is set out in instructions, resources and time according to the project triple constraint, each project is original and unique in its implementation and satisfying its expectations. The project resources include the most precious business resource – human capital. Unfortunately, this unique capital quantity is the hardest for the sociological measuring and persistence; therefore, it exerts the biggest influence not only on the success of the business project in regard to the successful completion of the project, but also a negative impact on the failure of the project.

The summary evaluation of findings provides a direct answer to causal stimuli of the influence of the human factor on the success/failure of the business project in these segments:

1. The influence of external interested subjects including individuals from enterprises, banks, interested groups of investors, competing companies and other strategic subjects and institutions; the whole number of these individuals

have their particular interests which may influence the success/failure of the business project in both ways. A systematically managed project must not only know how to exercise positive influences and convert them into success, but also predict negative influences and reduce them to minimum.

2. Internal subjects of the interested organization as a contracting authority and important persons of the business project are of a strategic significance, as the company not only funds the project, but also changes the approach and comfort of involved persons who may participate in influencing the project in the both ways. The project team and contracting authority are responsible for the risk management of the project.
3. The influence of the project manager/team leader as a human factor on the success of the project is the key determinant to the expected success of the business project.

The summary evaluation in the theoretical and practical part of the article defines and recommends necessary competences and personal traits as follows:

- Qualification standards in technical, behavioural and conceptual competences.
- Personal charm, due respect and natural traits of the person.
- Ambitions, discipline and determination to achieve sth.
- To be a leader to yourself and also to the project team.
- Promptly respond to partial changes.
- Use the 'common sense'.

These coherent parameters of human factor significantly influence the success/failure of the business project.

## References

1. A. Gorod, L. Hallo, T. Nguyen, A Systemic Approach to Complex Project Management: Integration of Command-and-Control and Network Governance. *Systems Research and Behavioral Science*, **35**(6), 811-837 (2018)
2. H.T. Nguyen, B. Hadikusumo, Impacts of human resource development on engineering, procurement, and construction project success. *Built Environment Project and Asset Management*, **7**(1), 73-85 (2017)
3. J. Doležal, P. Máchal, B. Lacko, *Projektový management podle IPMA [Project management according to IPMA]*. Prague: Grada (2012)
4. E. Mtsweni, Effect of technical issues and soft issues towards team performance and project success. *2017 IEEE AFRICON: Science, Technology and Innovation for Africa, AFRICON 2017*, pp. 957-962 (2017)
5. A.R. Gilal, J. Jaafar, M. Omar, S. Basri, A. Waqas, A rule-based model for software development team composition: Team leader role with personality types and gender classification. *Information and Software Technology*, **74**, 105-113 (2016)
6. M. Taghavi, H. Taghavi, M. Taghavi, Research project leadership stipulation system. *Campus-Wide Information Systems*, **30**(4), 266-287 (2013)
7. E. Ambrozová, J. Koleňák, D. Ullrich, V. Pokorný, Corporate environment and the abilities required of professional managers. *Proceedings of the 25th International Business Information Management Association Conference - Innovation Vision 2020: From Regional Development Sustainability to Global Economic Growth, IBIMA 2015*, pp. 753-769 (2015)

8. K.L. Cullen, W.A. Gentry, F.J. Yammarino, Biased self-perception tendencies: Self-enhancement/self-diminishment and leader derailment in individualistic and collectivistic cultures. *Applied Psychology*, **64**(1), 161-207 (2014)
9. R. Bautista-Mesa, H. Molina Sánchez, J.N. Ramírez Sobrino, Audit workplace simulations as a methodology to increase undergraduates' awareness of competences. *Accounting Education*, **27**(3), 234-258 (2018)
10. R. Štefánek, K.H. Bočková, How to manage risks in Czech and Slovac engineering projects. *E a M: Ekonomie a Management*, **14**(4), 67-77 (2011)
11. M. Taylor, *Začínáme řídit projekty [We are starting to manage projects]*. Brno: Computer Press (2007)
12. H.S. Ng, D.M.H. Kee, T. Ramayah, The role of transformational leadership, entrepreneurial competence and technical competence on enterprise success of owner-managed SMEs. *Journal of General Management*, **42**(1), 23-43 (2016)
13. M.Y.A. Alsabbah, H.I. Ibrahim, Employees' competence and job performance. *Advanced Science Letters*, **23**(1), 313-316 (2017)
14. N. Gruden, A. Stare, The Influence of Behavioral Competencies on Project Performance. *Project Management Journal*, **49**(3), 98-109 (2018)
15. J.R.S.C. Mateo, E.D.R. De Navamuel, M.A.G. Villa, Are project managers ready for the 21thchallenges? A review of problem structuring methods for decision support. *International Journal of Information Systems and Project Management,* **5**(2), 43-56 (2017)
16. Q. Wen, M. Qiang, Project Managers' Competences in Managing Project Closing. *Project Management Journal,* **50**(3), 361-375 (2019)

# Increasing the availability of financial services in remote, sparsely populated and hard-to-reach regions through the development of an entrepreneurial initiative in rural tourism

*Lyudmila* Vasyukova[1], *Natalya* Masyuk[1*], *Marina* Bushueva[1], *Anastasiya* Bessonova[1], *Nina A.* Mosolova[1]

[1]Vladivostok State University of Economics and Service, Ulitsa Gogolya 41, 690014 Vladivostok, Russia

**Abstract.** Activities to increase the availability of financial services are the primary task of the state economic policy. The formation of an active consumer of financial services is possible through the development of entrepreneurial activity among residents of remote areas, whose sparse population can become a competitive advantage. World experience in rural tourism development has shown that it is the sparsely populated, remote rural settlements from industrial centers that make them attractive for potential users of tourist services that are supplied by local people: tourist accommodation, catering, souvenir production, etc. This article presents the results of a study of the level of physical accessibility of financial services for residents of the Pozharsky district of Primorsky Krai, as well as their level of satisfaction with financial products and services. The study was conducted in 2018-2019. The main goal of the article is to examine the problems of ensuring financial affordability in remote, hard-to-reach and sparsely populated regions by the example of Pozharsky District, Primorsky Region of the Russian Federation. Measures are proposed to increase the availability of financial services through the development of entrepreneurial initiative of the inhabitants of the region.
**Keywords:** financial affordability, financial literacy, entrepreneurial initiative, financial services

## 1 Introduction

The most important factor in increasing the efficiency of economic development and the quality of life of the population in any region is the availability of modern financial services. The World Bank considers financial affordability to be the main, key factor in increasing overall well-being and reducing extreme poverty in the world. For the implementation of 7 of 17 goals in the field of sustainable development of society, the availability of financial

* Corresponding author: masyukn@gmail.com

services is defined as a factor contributing to the achievement of goals in the field of education, gender equality, creating conditions for economic growth, reducing inequality [16]. Issues of cash flows in the regions and their spatial migration are considered in the early works of [15].

The global goal of the World Bank Group (World Bank Group) is to implement the Universal Financial Access (UFA) project by 2020 [8]. Russia, as a member of the G20 countries, has committed itself to helping ensure the availability of financial services for citizens and entrepreneurs of our country. The Bank of Russia has developed a "Strategy for improving financial affordability in the Russian Federation" [22] currently, the provisions of this document regulate measures to ensure financial affordability, the implementation of which is carried out in the period 2018-2020.

However, none of the strategies will be implemented without the development of digital technologies, which in the conditions of low digital literacy of the population often has not only a positive, but also a negative impact. The lack of modern scientific knowledge hinders the development of optimal solutions (strategies and programs) that would offset the negative effects and contribute to the manifestation of the fully positive consequences of the digitalization process. The reason why digitalization is such a hot topic today is that it is one of the first true revolutions in business history. If the industrial revolution of 1760-1840. It took 80 years; the digital revolution is currently taking place rapidly [17]. Digitization is constantly happening around us much faster than any other changes we face on this scale. That's why we call it a revolution. Digitalization is defined as the integration of digital technology in everyday life. Since business and work make up a large part of everyday life for most people, perhaps this is the main thing that needs to be changed. It became possible to automate processes, measure aspects of a business that were previously immeasurable, and even cross-referenced and used information from a variety of sources to provide an understanding of the decision-making process [16].

Despite significant positive changes in the field of ensuring financial accessibility for citizens and business entities, there are still a number of unresolved problems of financial accessibility in Russia. One of these problems is to ensure financial accessibility for the population living and conducting economic activities in remote, sparsely populated, hard-to-reach areas, in small rural settlements. Meanwhile, about 26% of Russian citizens live in a countryside and villages. In total, there are more than 134 thousand rural settlements in the country [6]. The work on ensuring the priority of scientific research on the issues of ensuring the availability of financial services, summarizing the best practices, searching for new methods, techniques, forms, tools to improve the financial literacy of Russian citizens seems to be relevant and practically significant.

Many modern scientists pay great attention to the problems to which this study is dedicated. Examples and analyses of successful digital transformations are described in the works of M.A. Bushueva, V.K. Vasyukova, A.P. Dobrynin, A.E. Kiryanov, V.P. Kupriyanovsky, N.N. Masyuk, D.E. Namiot, S.A. Sinyagov, B. Benlian, B. Bloching, T. Hess, P. Leutiger, C. Matt, T. Oltmanns, G. Remain, C. Rossbach, J. Sal., T. Schlick, D. Schwir, O. Shefranyuk, F. Visbek, and others. Studies on the availability of financial services are presented in the works of Russian economists Krivoruchko S.V., Ponomarenko V.E. [12], Mamuta M.V. [14]; [9] etc. A summary of the best practices in the use of tools for financial accessibility is presented in the works of N.V. Baidukova, V.L. Dostov [2]. E.A. Fedulova, N.A. Vakutin [28] etc. The role of universities in the implementation of the strategy for improving financial literacy is presented in the works of I.G. Teleshova, E.N. Kudryashova, R.A. Kokoreva, S.A. Trukhacheva [13]. The interrelation of the system of continuous education and the concept of improving the financial literacy of the population is covered in the works of S.M. Vdovin [26] and others. Meanwhile, an active discussion of the availability of financial services for people in remote and sparsely populated areas is of scientific and practical interest. We believe that the problems of ensuring financial affordability in remote, hard-to-reach and sparsely populated regions should be solved through the development of entrepreneurial initiative among citizens who not only live but also conduct economic activity in these areas, very often being the main participants in entrepreneurial activity in their territories.

## 2 Method and methodology

In developing the research methodology, general scientific research methods, applied economic and statistical methods, analysis and synthesis methods have been applied. When constructing a heat map, sociological and expert polls, interviews were used, and a complex of empirical methods was used, including content analysis, methods of comparative analysis, graphic modeling, algorithmizing, and rating.

In order to investigate the essence of an economic phenomenon, it is necessary to determine its essential or attributive characteristics.

The modern definition of the availability of financial services has evolved over the past decade. In 2009, analysts of the Center for Financial Inclusion (CFI) formulated the main parameters of financial affordability:

1) the ability of the entire working-age population to receive a full range of high-quality financial services;
2) the affordability of financial services;
3) the availability of a convenient way to obtain financial services;
4) respect by a financial service provider for the human dignity of clients [14].

In 2010, World Bank experts identified three main components of financial affordability as an economic category: products (payments, savings, insurance, loans), characteristics (price affordability, physical accessibility, convenience, quality), channels (access points, institutions, customers, infrastructure) [9]. Alliance for Financial Inclusion (AFI) is the world's leading organization for regulating and shaping policies in the field of financial accessibility, in 2011 identified three components of financial accessibility: physical accessibility, relevance, and quality of financial services [18].

In 2012, the Organization for Economic Cooperation and Development (OECD) proposes to consider financial affordability combined with financial literacy, identifying opportunities to use innovative ways of providing financial services as a key characteristic of financial accessibility [14].

The Bank of Russia in the definition of financial affordability adopted in 2015, taking into account international experience and the specifics of the development of Russian financial markets, identifies the following components of financial accessibility:

1) the physical ability to receive financial services, i.e. availability of financial services infrastructure;
2) the demand for financial services, i.e. availability of sustainable solvent demand for financial services;
3) the quality of financial services, i.e. uninterrupted and safe provision of financial services at an affordable price, as well as an effective system for protecting the rights of consumers of financial services;
4) the usefulness of financial services, including the consumer's understanding of the properties of acquired financial services, and the assessment of the extent to which the service is provided on the quality of life.

According to these four components of financial accessibility, the Bank of Russia has developed indicators that allow a quantitative assessment of the current state in the area of availability of various types of financial services [11]. In determining the financial accessibility of the Bank of Russia, target groups of individuals are identified whose security with financial services is the main goal of the Strategy developed by the Bank: the country's capable population, as well as small and medium-sized businesses.

The complexity of the implementation of the Strategy for improving financial affordability is determined by the unique geographic characteristics of our country: a huge Region, the presence of large areas with a low population density, a large number of hard-to-reach places, etc.

Financial accessibility measures are divided into the following groups: physical accessibility; price accessibility; assortment availability; mental accessibility. As follows from the analysis of indicators of financial affordability [11] in general "the general state of financial affordability in Russia can be characterized as satisfactory" [22].

A team of authors initiated a project to study the issues of increasing the availability of financial services in remote, sparsely populated or difficult-to-reach regions through the development of an entrepreneurial initiative in rural tourism. The study of the development of accessibility of financial services was carried out on the example of Pozharsky District of Primorsky Region (Figure 1) with the involvement of students from the School of Economics and Management of the Far Eastern Federal University (Russia).

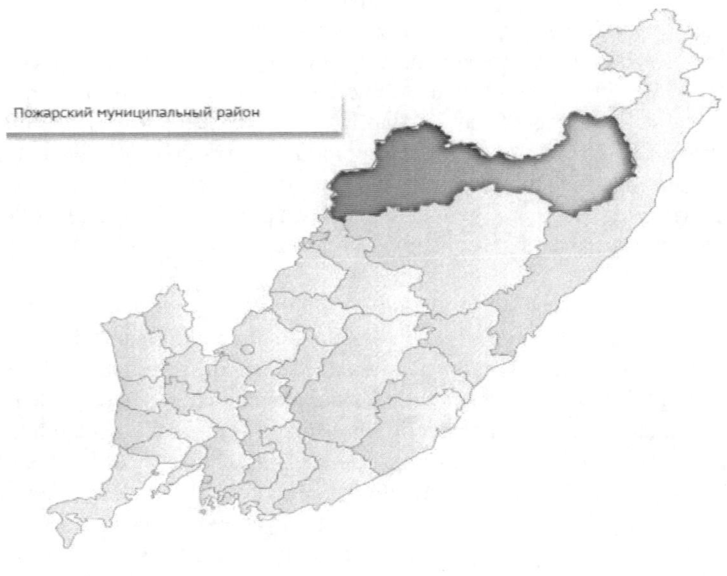


**Fig. 1.** Pozharsky municipal district on the map of Primorsky Region

Source: [30].

## 3 Results

Studies have shown that indicators characterizing the physical ability to receive financial services, including in the Far Eastern Federal District (FEFD) far from the center of the country, are constantly growing. The number of units operating per 1 million people of the adult population in Russia as a whole is 299 units, in the FEFD this figure is higher than the national average - 355 units. The number of ATMs of credit institutions with the function of issuing and receiving cash using payment cards in Russia as a whole is 1,669 units per 1 million people in the adult population, and in the Far-Eastern Federal District a similar indicator is 1,806 units per 1 million people in the adult population. At the same time, the number of ATMs in the Far Eastern Federal District is 141 units per 100 thousand square kilometers, which is 8 times less than the average Russian indicator - 1,140 units per 100 thousand square kilometers.

The most universal organizations providing payment services in sparsely populated and inaccessible areas are branches of the federal postal service. On average, in Russia, there are 326 postal offices per 1 million people in the adult population, in the Far Eastern region, this figure is higher than the average Russian – 396 offices. But in terms of the density of placement of offices of the federal postal service, the indicator for the DFO is 31 units per 100 thousand square kilometers, 7 times lower than the average - 223 units per 100 thousand square kilometers, and almost 49 times lower than in the Central Federal District - 1 508 units per 100 thousand square kilometers.

For all regions, financial market entities offer a fairly diverse range of financial products. It is possible to open a bank account or purchase an insurance policy from the general line of financial products in any region of the country, and it is unlikely that anyone will be able to present reliable facts of usurping their interests in connection with a gender mark, territorial location of the office or belonging to a particular nationality.

The affordability of financial services remains a question, which cannot be resolved using only administrative resources. Strengthening the monopolization of individual segments of the financial market in the regions, undoubtedly, adversely affects the development of competition. A quick analysis of the Russian insurance market showed that the 20 largest insurance companies (7.2% of all insurance market participants) collected 80.55% of insurance premiums in 2018. In the regions, the concentration of the insurance market is even higher. So, in Primorsky Region, in 2018, the 20 largest insurance companies collected 92.6% of insurance premiums, and this list contains only one regional insurance company – "Sterkh" from Yakutia.

The Russian insurance market, a market with a very high level of monopolization, an oligopolistic market [29], requires an active position, both on the financial market regulator, the Bank of Russia, and on the antimonopoly authorities, on the formation of fair competition principles in the market. Convergence in the financial market creates a struggle for resources - a solvent client, among representatives of other financial institutions - commercial banks. Opportunistic behavior of the overwhelming majority of financial organizations - overpricing of financial services tariffs, imposing additional services (refusal of credit when a client disagrees with buying insurance from a bank, imposing additional insurance contracts when purchasing an OSAGO policy, etc.) cause a negative reaction from consumers of financial services, and become a strategic obstacle to the formation of the most difficult component of financial affordability - mental accessibility.

It should be noted that the state, realizing the need to expand the availability of financial services in the peripheral regions of the country, initiates projects to create special administrative regions with preferential tax regimes for investors with a low administrative burden. One of such areas should be Russky Island, which has all the necessary infrastructure and geographical position in order to become a significant business and financial center that can attract Russian and foreign capital. The creation of an offshore can give a cumulative effect, attract the necessary financial resources to the region for the implementation of business projects, make financial resources more accessible for entrepreneurs in the region, and, as a result, turn the island into a serious center of business life in the Asia-Pacific region.

An obvious compromise to resolve the contradictions between suppliers and consumers of financial services in terms of affordability is to launch a digital financial marketplace under the auspices of the Bank of Russia. The peculiarity of this digital financial instrument is that it is a superstructure above the already existing mono-brand financial marketplaces, on which financial products from different suppliers will be placed. The main advantage of the digital marketplace under the auspices of the Bank of Russia is the independence of this electronic platform, and, most importantly, guaranteed protection of the rights of consumers of financial services.

The mental accessibility of financial services, in our opinion, is determined not only by the ability to consume various financial products, simple or complex, but, most importantly, by the consumer's understanding of the usefulness of this product, the realization that this financial product changes the quality of life of the consumer - a citizen or an entrepreneur.

According to a survey conducted by the NAFI Analytical Center on November 2018, more than 87% of the country's adult population uses various types of financial services [21]. The vast majority - 62% of these consumers of financial services use plastic cards to receive pensions, salaries, scholarships, etc., i.e. are "passive" users of financial services. As a rule, the decision on the method of receiving wages, scholarships, pensions and other payments is made by third parties - employers, school management, etc. Consumer plays an extremely passive role in deciding the form and quality of consumption of financial services. Therefore, in those matters when the decision is made by the consumer of financial services, he must have certain competencies: know the content of financial services; be able to assess the usefulness of the selected financial services; possess the skills to use financial services in order to get the maximum value from this service. The main task at the same time is to make the consumer of financial services an active, initiative consumer who is able to critically evaluate the content and quality of the service, to choose the necessary services for themselves.

For residents of the sparsely populated, located far from the administrative centers, the benefits from the consumption of financial services become even more unclear: to withdraw money from a bank card, you need to go to the district center; there is nothing to put in the cell, the postman will bring a pension, etc. Studies carried out by undergraduates of the Far Eastern Federal University, while carrying out a project to research the microfinance market in the Far Eastern Federal District, showed that residents of remote areas do not have a position of active consumption of financial services, in particular, microfinance services - microloans, micro insurance products. In this situation, the unique experience of Tomsk region in issuing regional bonds [5] seems to be inapplicable not only because the consumer does not have knowledge of this experience, skills, and abilities to acquire and use such financial instruments, but also the lack of an active position of an investor who is interested in obtaining experience in shaping the right, profitable and safe financial behavior.

The first stage of the study of the availability of financial services was to assess the satisfaction of the residents of the area:

- the work of financial organizations;
- financial services (products) offered by financial organizations;
- channels of providing financial services data.

It should be noted that at present there is no uniform methodology for assessing the level of satisfaction with financial products and services, both in international and in Russian practice.

In international practice, assessing the level of satisfaction with financial products and services presented by the World Bank in the Financial Capability and Consumer Protection Surveys reviews [7] highlight the following criteria for calculating financial services satisfaction indicators:

- the proportion of adults who are not satisfied with the level of any service;

- the proportion of the adult population that has ceased to use any financial service due to dissatisfaction with its level.

The project conducted a study of the level of physical accessibility of financial services for residents of the Pozharsky district of the Primorsky Region, as well as their level of satisfaction with financial products and services.

The population of Pozharsky District is 28,109 people living in the area of 2,257 thousand square kilometers. The population of Pozharsky District is extremely uneven. The most populated places are located along the federal highway Khabarovsk-Vladivostok. Most of the area is characterized by focal settlement or no resident population at all. The average population density is 0.12 people per 1 square kilometer, which is one of the lowest indicators of Primorsky Region (Figure 2).

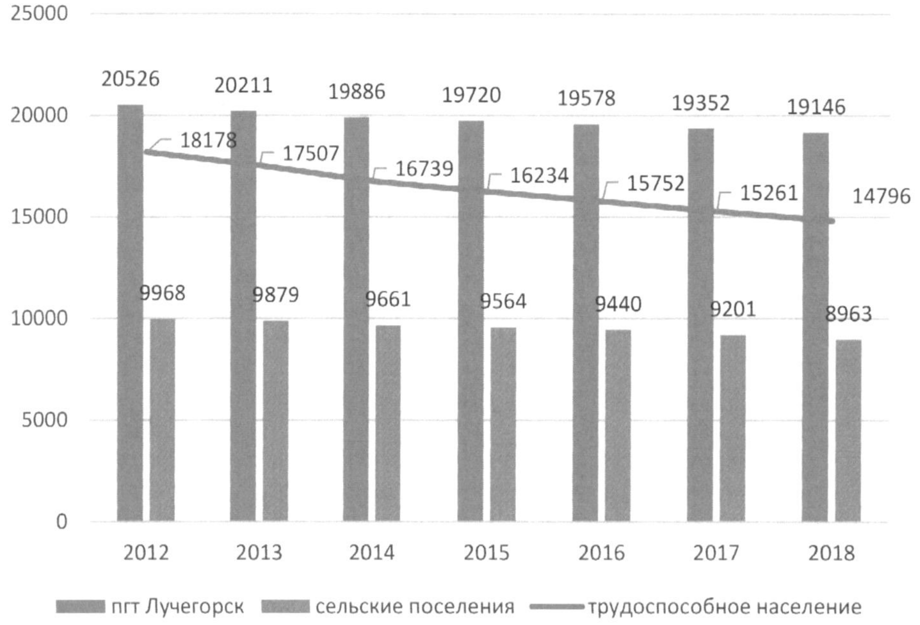


**Fig. 2.** Population dynamics of the Pozharsky district of the Primorsky Region in 2012-2018

Source: [6].

The length of local public roads owned by municipalities has not changed since 2015 and is only 169.3 km. This indicates a very low level of development of the transport infrastructure of the region. The problems of transport accessibility of rural settlements, farms, and places of residence of indigenous peoples are the most acute and significantly impede the socio-economic development of remote areas of the region, including through the promotion of business initiatives of citizens in the field of tourism.

The level of physical accessibility of financial services was assessed on the basis of data on the actual availability of various types of financial services in different localities of the district.

On the 10th of March 2019, there are 425 divisions of forty-four commercial banks, including 8 head offices, 11 branches, and 249 additional offices in the Primorsky Region. According to study Pozharsky district, there are 10 branches and additional and offices of commercial banks ("Sberbank", "Sovcombank", ROSBANK, "MTS-Bank", and "Vostochny"), five of which are located in the urban-type settlement of Luchegorsk. The location of payment terminals and ATMs is also concentrated in the district center, in some rural settlements there are none at all. Post offices in the area of ten, which is a negligible relative value of 0.44 units per 100 thousand square kilometers.

Non-bank credit and financial organizations in the Pozharsky district are represented only by one microfinance organization, which has been in the process of reorganization since January 2019. Insurance companies in Pozharsky district are present only in the district center ("Eastern Insurance Alliance" and "Rosgosstrakh"). The result of the analysis of the physical accessibility of financial services was the "heat map" of the physical accessibility of financial services in the settlements of Pozharsky district of the Primorsky Region.

To assess the level of satisfaction of the population with financial products and services, the principal component method was used, the essence of which is as follows: if the characteristics described using the initial variables change consistently, then we can assume the existence of certain general causes of this variability.

The composite index of population satisfaction with financial products and services in the Pozharsky district of the Primorsky Region, calculated on the basis of a survey of residents living in 7 of 9 rural settlements of the district, as well as in the urban-type settlement Luchegorsk, formed at the level of 39.5, and entrepreneurs - representatives of small business, developed at the level of 42.7. It should be noted that the value of the composite index of satisfaction with the performance of financial organizations, financial products (services) and channels of financial services, presented in the analytical report of the Bank of Russia [3] reached 51.1 by May 2018.

The main factor determining the difference in the assessment of satisfaction with financial services by residents of the Pozharsky District and the average Russian indicator is the limited distance channels for providing financial services in the Primorsky Region.

When calculating the composite index of population satisfaction with the work of financial organizations, the quality of financial products (services) and channels of financial services, the following variables were considered:

- the level of job satisfaction of banks, microfinance organizations (MFO), consumer credit cooperatives (CCC), agricultural consumer credit cooperatives (SKPK), insurance subjects (SSD), non-state pension funds (NPF), brokers;
- the level of satisfaction with the products and services of banks of microfinance organizations (MFO), consumer credit cooperatives (CPC), agricultural consumer credit cooperatives (ACCC), insurance subjects (IS), non-state pension funds (NPF), brokers;
- level of trust in financial organizations;
- ease of access to financial products: physical accessibility ("accessible everywhere"), waiting time for access, the security of access.

Practice shows that in the country as a whole, online financial services are becoming more and more common among consumers. For example, Yandex Money, Web Money, PayPal, MasterCard, Visa payment services, mobile applications of banks, microfinance organizations and even investment online securities and currency trading services. To use them, two factors are necessary: high-speed Internet access and basic financial literacy for making independent decisions.

What ways of the first technical factor: there are three Internet service providers in Pozharsky district? The coverage area of the Internet is limited to the urban-type settlement of Luchegorsk and the territories located close to cell phone towers. In 5 of 9 rural settlements: Verkhneperevalsky, Ignatievsky, Svetlogorsk, Sobolinsky, Krasnoyarovsky there is no cellular communication, and, accordingly, there are no Internet services. The need for the development of the Internet in remote settlements of Primorye has been discussed at different levels for more than 10 years now the development plans for this area are outlined in the national program Digital Economy of the Russian Federation, on the basis of which Rostelecom and the Administration of Primorsky Region will jointly implement the Smart Region project ". It is planned that by 2024 the construction of fiber-optic communication lines to most of the settlements will be completed, after which municipalities with more than 250 people will be provided with ubiquitous Internet access.

According to the second factor, financial literacy results are as follows: the first studies conducted within the framework of the Far Eastern Federal University (School of Economics and Management) project showed that the local population as a whole has little information about the tools of financial support for small businesses, almost none of the respondents have information about the national project “Small and Medium Enterprises” and support for individual entrepreneurial initiative , but the main thing is that people do not feel that they are subjects of relations whose financing is determined within the framework of this national project.

## 4 Conclusion

At the present stage of development of the economy, the problem of financial affordability is acute for both the population and entrepreneurs. The key role in meeting the unrealized demand for financial services is given to innovative ways of providing them on the basis of modern payment and communication technologies, the development of remote service channels, i.e. electronic financial services [3, 4]. However, these measures alone will not lead to an increase in demand for financial products in remote regions and single-industry towns. A necessary condition is the formation of the entrepreneurial activity of residents of the Region. The main line of work on the development of the availability of financial services should be focused on the formation of the mental accessibility of financial services, increasing the population’s motivation to acquire knowledge and skills in the consumption of financial services through the development of the entrepreneurial initiatives. The active inclusion of university students in research programs on the development of accessibility of financial services, the development of entrepreneurial initiatives of the population of sparsely populated, remote areas, participation in financial literacy training programs for local residents, entrepreneurs, and schoolchildren will help develop new approaches to solving accessibility issues financial services.

At the next stage of the study, it is supposed to assess the current state of affordability of retail financial services on the regional market in the context of territorial and technological accessibility and to determine the main directions of its increase; clarification of target client groups that need improved access to funding; development of proposals for the development of financial intermediaries, designed to increase the availability of retail financial services; study of a set of measures of state and municipal support of entrepreneurial initiatives and the development of proposals for their further improvement; development of the main directions of further integration of the non-banking and banking sector into the economy of the Region in order to increase the availability of retail financial services aimed at the development of entrepreneurship; development of methodological materials for conducting training and consulting classes to improve the financial literacy of different age groups of the rural population of the Pozharsky district; development of a plan for consulting and training sessions on familiarizing and educating the public in the use of digital tools to increase the availability of financial services.

## References

1. S. Alessandrini, F. Vandenberghe, J.P. Hacker, Definition of typical-day dispersion patterns as a consequence of a hazardous release. *International Journal of Environment and Pollution*, 62(2,3,4), 305-318 (2017)
2. N.V. Baidukova, V.L. Dostov, Analysis and evaluation of best practices in the use of tools to improve financial affordability (Part 1). *News of St. Petersburg State University of Economics,* 6-114, 48-58 (2018)
3. Composite index of population satisfaction with the work of financial organizations, financial products (services) and channels for the provision of financial services in 2018. Bank of Russia [online], Available at: http://cbr.ru/finmarket/development/development_affor / indicators/ (2018)
4. M. Eling, M. Lehmann, The Impact of Risks. *Geneva Pap Risk Insurance Issues Practice,* **43**(3), 359-396 (2018)
5. A. Fedenyov, E. Mekhtiev, To government bonds - through trust in the authorities. *Bank review* [online], Available at: https://bosfera.ru/bo/k-gosobligaciyam-cherez-doverie-k-vlasti (2019)
6. Federal State Statistics Service. Official site [online], Available at: http://www.gks.ru/ (2019)
7. Financial Capability and Consumer Protection Surveys. *World Bank Group* [online], Available at: http://responsiblefinance.worldbank.org/surveys/demand/exploreeconomies/russia/2008#consumer-protection (2019)
8. Financial Inclusion. Universal Financial Access 2020 initiative. *World Bank Group* [online], Available at: http://www.worldbank.org/en/topic/financialinclusion/overview (2019)
9. Financial Literacy Guide Moscow, Russia: KnoRus (2010)
10. Financial organizations positively assess the impact of digitalization on the level of financial literacy of clients. Representative All-Russian Survey. *NAFI Analytical Center* [online], Available at: https://nafi.ru/analytics/finansovye-organizatsii-polozhitelno-otsenivayut-vliyanie-tsifrovizatsii-na-uroven-finansovoy-gramot/ (2019)
11. Indicators of financial availability of the Bank of Russia [online], Available at: http://cbr.ru/finmarket/development/development_affor/indicators/ (2017)
12. V.Y. Krivoruchko, V.A. Ponomarenko, M.V. Lopatin, R. Maklakova, T.S. Bragina, P.M. Shust, Increasing the availability of payment services due to the development of user identification systems. Moscow, Russia: Infra-M (2019)
13. R.A. Kokorev, I.G. Teleshova, E.N. Kudryashova, S.A. Trukhachev, New tasks of universities in improving financial literacy. *Materials of the International Scientific Conference "Lomonosov Readings*, 1174-1182 (2016)
14. M.V. Mamuta, Financial Accessibility and Financial Literacy. *Proceedings of the International Conference on Financial Literacy and Financial Availability* [online], Available at: https://www.youtube.com/watch?v=uEHmDsonXtg (2017)
15. R. Martin, R. Minns, Undermining the Financial Basis of Regions: The Spatial Structure and Implications of the UK Pension Fund System. *Regional Studies: The Journal of the Regional Studies Association,* **29**(2), 125-144 (1995)

16. N.N. Masyuk, M.A. Bushueva, L.K. Vasyukova, A.E, Kiryanov, Platforms of digital experience and digital transformation in an innovative economy. *Proceedings of the III International Scientific and Practical Conference "Resonances Science"* (2018)
17. N.N. Masyuk, C. Zhao, L.K. Vasyukova, Change management and digital transformations in the era of business digitalization. *Proceeding of the All-Russian Scientific and Practical Conference of teachers, graduate students, undergraduates of the Ivanovo branch of the Russian Economic University named after G.V. Plekhanov*, pp. 253-258 (2019)
18. Maya Declaration. Alliance for Financial Inclusion (AFI) [online], Available at: https://www.afi-global.org/ (2011)
19. Methodology of development of the economy, industry and services in the conditions of digitalization. Russia: Polytech-Press (2018)
20. E. Rhyne, Financial Inclusion from a Consumer Perspective. *Center for Financial Inclusion (CFI)* [online], Available at: https://www.centerforfinancialinclusion.org/financial-inclusion-from-a-consumer-perspective (2017)
21. Russians began to use credit cards more often. Initiative All-Russian Poll. *NAFI Analytical Center* [online], Available at: https://nafi.ru/analytics/pervyy-reyting-finansovoy-gramotnosti-regionov-predstavili-na-sessii-v-ramkakh-rossiyskogo-ekonoiche/ (2018)
22. Strategy of increasing financial affordability in the Russian Federation for the period 2018-2020. *Bank of Russia* [online], Available at: http://cbr.ru/finmarket/development/development_affor/strategy/ (2018)
23. Strategy of increasing financial literacy of the population in the Russian Federation for 2017-2023. Order of the Government of the Russian Federation of 25.09.2017 [online], Available at: http://www.consultant.ru/document/cons_doc_LAW_278903/ (2017)
24. The main directions of development of financial technologies for the period 2018-2020. *Bank of Russia* [online], Available at: http://cbr.ru/StaticHtml/File/36231/ON_FinTex_2017.pdf (2017)
25. The strategy of the state policy of the Russian Federation in the field of consumer protection for the period up to 2030: order of the Government of the Russian Federation of 28.08.2017 [online], Available at: http://docs.cntd.ru/document/436763261 (2017)
26. S. M. Vdovin, et al., Financial literacy of the population in the aspect of the development of lifelong education. *Education Integration,* 22-3, 441-459 (2018)
27. United Nations. Sustainable Development Goals [online], Available at: https://www.un.org/sustainabledevelopment/sustainable-development-goals (2019)
28. N.A. Vakutin, E.A. Fedulova, Formation of a methodology for assessing the financial availability of leasing services. *Finance and Credit,* 23-39, 2329-2345 (2017)
29. L.K. Vasyukova, N.A. Mosolova, E.V. Shvetsova, Improving the antimonopoly regulation of the insurance market in Russia. *Science of Krasnoyarsk,* **4**(37), 181-190 (2016)
30. Official website of the Primorsky Territory and the executive authorities of the Primorsky Territory [online], Available at: https://www.primorsky.ru/primorye/munitsipalnye-obrazovaniya/

# Forecasting trade balance of Czech Republic and People´s Republic of China in equalizing time series and considering seasonal fluctuations

*Marek* Vochozka[1*], *Zuzana* Rowland[1]

[1]Institute of Technology and Business, School of Expertness and Valuation, Okružní 517/10, 37001 České Budějovice, Czech Republic

**Abstract.** The objective of the contribution is to introduce a methodology for considering seasonal fluctuations in equalizing time series using artificial neural networks on the example of the Czech Republic and the People´s Republic of China trade balance. The data available is the data on monthly balance for the period between January 2000 and July 2018, that is, 223 input data. The unit is Euro. The data for the analysis are available on the World Bank web pages etc. Regression analysis is carried out using artificial neural networks. There are two types on neural networks generated, multilayer perceptron networks (MLP) and radial basis function networks (RBF). In order to achieve the optimal result, two sets of neural structures are generated. There are generated a total of 10,000 neural structures, out of which only 5 with the best characteristics are retained. Finally, the results of both groups of retained neural networks are compared. The contribution this paper brings is the involvement of variables that are able to forecast a possible seasonal fluctuation in the time series development when using artificial neural networks. Moreover, neural networks have been identified that achieve slightly better results than other networks, specifically these are the neural networks 1. MLP 13-6-1 and 3. MLP 13-8-1.
**Key words:** trade balance, Czech Republic, People´s Republic of China, time series, seasonal fluctuations

## 1 Introduction

Exchange equalization of a nation is the net whole of a nation's exports and imports of products without considering every single budgetary exchange, ventures and other money related parts. On one hand, when a nation's exchange equalization is in positive implies that it enlists an excess if the estimation of export surpasses the estimation of imports. On the other hand, a nation's exchange parity is negative, or registers a shortfall, if the estimation of imports surpasses that of exports. Hence the term trade balance is the official term that is realistic for net exports in the account record.

Czech Republic (CR) and the People's Republic of China (PRC) are seriously involved in international trade. According to Koszek [1], Chinese trade relation is at highest level in

* Corresponding author: vochozka@mail.vstecb.cz

Czech Republic. CR exports mainly manufactured goods that are used in production of furniture, electrical appliances, and automobiles and imports raw materials, fuel and chemicals, and machinery and transportation equipment.

The big question to ask is the trade between CR and PRC healthy? This review hopes to answer the research question and to pinpoint trade balance prediction methods while equalizing time series and taking into account seasonal fluctuation.

## 2 Review of related literature

Rowland, Šuleř and Vochozka [2], compared accuracy of time series by means of regression analysis and neural networks using trade balance data between CR and PRC in a certain period. They used statistical methods which gave them a basis to determine trade balance of the two countries. Although their findings cannot be generalized due to the fact that the ignore variables which often have important effect on trade balance forecasting, their findings are applicable in forecasting short term trade balance. Vrbka, Rowland and Šuleř [3] again explored the topic from a broader perspective and put forward that China´s focal business partner is the EU and CR is a member country of the union, which means that China import in the union is large. Export and import data between the two business partners have been utilized by researchers using regression and neural network models to predict future development of trade balance while taking into account seasonal fluctuation. The outcome evidently shows that trade balance between the two partners have amplified over the years.

Evans [4] projected that variations in external positions of most countries reflect changing expectations about trade conditions far into the future. Established is that the changing forecast for the future path of the world stochastic markdown factor are mirrored in the dynamics of the U.S. external position.

Garlick [5], recommended that regardless of diversities in Central and East European (CEE) countries, there is a need to coordinate instead of contend concerning trade with China because they are all struggling with trade balance shortfalls with the Chinese. Matura [6], lamented on China-CEE investment goals and stated that China is using the CEE countries to gain political influence in the EU for economic reliefs. The rising collaboration of the Chinese and CEE nations including CR has spawned far-reaching criticism by other western nations who thinks that it not a healthy relation. However, Šteinbuka, Muravska and Kuznieks [7] concluded that China-CEE cooperation does not contradict the EU investment plan, and that none of the two parties is aiming to abate the union. They believed that the EU to which CR belong and China have a great interest in each other's top development plans. The Silk Road Economic Belt and 21st-Century Maritime Silk Road Initiatives and the Investment Plan for EU are complementing each other.

CR has prevailed regards to building its economy by maintaining the Czech Koruna (CZK). Nonetheless, PRC runs a big trade surplus with CR. Chinese imports into the CR are large and over the years, there is a drop in Czech companies with comparative benefit [8].

From the literature reviewed it can be perceived that CR will undoubtable experience proceeding trade imbalance with China. However, geographical trade concentration in China is expected to relax in the future. Rapid economic growth and trade surpluses enables China to amass capital and prompted expanding capital streams abroad from Chinese firms [9]. Chinese investment increase greatly in the EU during the economic crisis in 2008. During this period, Chinese investors invested heavily in the member countries thereby helping to savage the economic gap, which was as a result of a decrease in domestic investment [10]. The 2008 crunch help widened the trade balance between CR and PRC. Pikhart [11], Patterned change of external balance of trade give a valuable device for assessment of the sustainability of trade development and subsequently help to analyze time series and predict future advancement.

Bayesian method has been used by other researchers to estimate a sign restricted structural vector autoregressive model with quarterly time series data. Updating is particularly important for dynamic analysis. Result from such research demonstrate that asset prices have direct impact on trade balance. It is confirmed so far when exchange rate appreciate trade balance depreciates [12].

While there are a wide range of trade balance prediction methods; autoregressive moving average (ARMA), autoregressive integrated moving average (ARIMA), multiple linear regression (MLR), and artificial neural network (ANN) are frequently used tools in predicting trade balance. In recent years, ANN have proven to be more accurate because it can deal with non-linear problems and has higher prediction accuracy rate in empirical research [13]. Similarly, time series data can effectively increase the accuracy of trade balance prediction. However, the accuracy of prediction methods depends on design of the method itself [14].

CR-PRC trade has tremendously grown to the detriment of CR. Czech local firms export to China is very low while imports from PRC is at a record high.

China had been adopting a conservative policy toward trade balance. China have accumulated surplus in many periods while CR ran a deficit. This is as a result of China exporting more to CR instead of the other way round.

Nowadays, the modernization of china means more imports from developed countries, which tends to widen the trade gap. Due to its location and moderately cheap and high quality work force, economic growth and political stability CR has a benefiting advantage over the other EU countries, especially in the most recent two decades. Chinese value added goods are process in CR in other to easily reach EU western market. China´s economy is the second largest in the world which makes them significant actors in economic global relations [15].

CR and PRC have benefited from a well-established formal relation. The long time relations have led to the signing of trade agreement in 2010 for thorough economic improvement between the two allies. CR has an open economy framework and it fused into the EU financial outline. CR economy is strongly influenced by innovation and for the most part relies on remote capital joined with a defenseless capacity to hold it and the low conveyed cost [16].

Seasonal fluctuation of time series data enhance researchers to understand trade balance behavior in the past so that the patterns could be helpful for future predictions. However, conclusions made from analyzing time series is not always perfect, and factors that affect seasonal fluctuations of a time series cannot be fully adjusted by the time series analysis [17].

ARMA, ARIMA, MLR, ANN are trade balance prediction methods. Although, ANN is proven to be more suitable and accurate for predicting trade balance using time series data.

The need for a healthier reciprocated balance economic cooperation can be beneficial for the two sides as long as the two allies stop misappropriation of public funds, offset unfortunate behavior in property rights, link social vacancies and manage turn over [18].

The need for further research on the topic is essential. Forecast of trade balance of the Czech Republic and the People's Republic of China while equalizing time series and taking into account seasonal fluctuation can be used to understand the past as well as predict the future trade balance of the two parties.

## 3 Data and methods

For the purposes of the contribution, the data on the trade balance between the Czech Republic and the People´s Republic of China. The focus will be on the difference between the overall export and import between both countries from the point of view of the Czech Republic. The data available will be the data on monthly balance for the period between January 2000 and July 2018, that is, 223 input data. The unit will be Euro. The data for the analysis are available on the World Bank web pages [19], etc.

Table 1 shows descriptive characteristics of the data.

**Table 1.** Descriptive characteristics of data sets

| Samples | Month (Input variable) | Balance (Output (target)) |
|---|---|---|
| Minimum (Training) | 36526.00 | -1.050893E+09 |
| Maximum (Training) | 43252.00 | -4.054250E+07 |
| Median (Training) | 39916.52 | -3.661027E+08 |
| Standard deviation (Training) | 1949.82 | 2.289766E+08 |
| Minimum (Testing) | 36586.00 | -9.687806E+08 |
| Maximum (Testing) | 43282.00 | -5.088367E+07 |
| Median (Testing) | 39702.30 | -3.679126E+08 |
| Standard deviation (Testing) | 2174.13 | 2.719910E+08 |
| Minimum (Validation) | 36951.00 | -1.168102E+09 |
| Maximum (Validation) | 43040.00 | -6.679944E+07 |
| Median (Validation) | 40047.88 | -3.945945E+08 |
| Standard deviation (Validation) | 3096.16 | 3.098454E+08 |
| Minimum (Overall) | 36526.00 | -1.168102E+09 |
| Maximum (Overall) | 43282.00 | -4.054250E+07 |
| Median (Overall) | 39904.26 | -3.705868E+08 |
| Standard deviation (Overall) | 1963.77 | 2.400326E+08 |

Source: Own processing.

The development of the trade balance over time is very interesting. Therefore, Figure 1 is used to show selected statistical characteristics in a graphical form including the histogram of the input data.

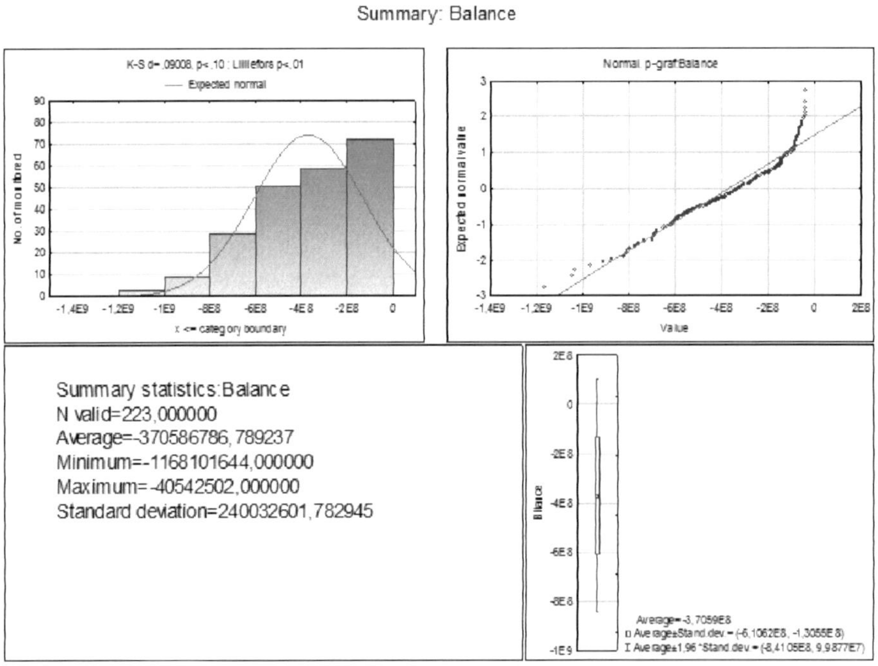


**Fig. 1.** Basic statistical characteristics

Source: Own processing.

The histogram corresponds to normal division. The data will be processed using DELL´s Statistica software, version 12. Regression analysis will be carried out using artificial neural networks. There will be two types on neural networks generated, multilayer perceptron networks (MLP) and radial basis function networks (RBF). In order to achieve the optimal result, two sets of neural structures will be generated:

1. The dependent variable will be trade balance of the Czech Republic and the People´s Republic of China, while the independent variable will be time.
2. The dependent variable will be trade balance of the Czech Republic and the People´s Republic of China. The categorical variable will be represented by a seasonal fluctuation in the form of the month in which the value was measured. The continuous variable will be time. Within this approach, it will be possible to work with a monthly seasonality of the time series.

Working with data sets will be analogous. The time series will be divided into three sets: training, which will be used for generating neural networks and will contain 70 % of the input data; testing, which will be used to verify the reliability of the neural network (model) found and it will contain 15 % of the input data; validation, which will be used for the same purpose as the testing data set and will also contain 15 of the input data. The delay of the time series will be 1. There will be generated a total of 10,000 neural structures, out of which only 5 with the best characteristics will be retained. For this purpose, the least squares method will be used. Networks generating will be finished when there is no improvement, that is, if the value of the sum of the squares does not decrease. Only the neural networks whose sum of the squares to the actual development of the Czech Republic and the People´s Republic of China trade balance is as low as possible (zero in ideal case) will be retained. The hidden layer of the MLP networks will contain not fewer than 2 and not more than 50 neurons. The RBF networks will contain at least 21 and a maximum of 30 neurons in the hidden layer. In the hidden and output layers, the following distribution functions will be considered: linear, logistic, atanh, exponential, and sinus. Other settings will remain according to the setting of the ANN (automated neural networks) tool. Finally, the results of both groups of retained neural networks will be compared.

## 4 Results

### 4.1 Neural networks A

In accordance with the methodology, a total of 10,000 neural networks will be generated, out of which 5 with the best parameters will be retained. Table 2 shows the overview of the retained neural networks including their performance and further information.

**Table 2.** Retained neural networks – neural networks A

| Network | Train. perform. | Test. perform. | Valid. perform. | Train. error | Test. error | Valid. error | Train. algorit. | Error function | Activation of hidden layer | Output activation function |
|---|---|---|---|---|---|---|---|---|---|---|
| RBF 1-22-1 | 0.943854 | 0.942643 | 0.953409 | 2.803986E+15 | 4.437812E+15 | 4.527628E+15 | RBFT | Sum of squares | Gaussian | Identity |
| RBF 1-24-1 | 0.893296 | 0.855999 | 0.951235 | 5.208266E+15 | 1.004419E+16 | 3.610412E+15 | RBFT | Sum of squares | Gaussian | Identity |
| RBF 1-30-1 | 0.917891 | 0.870478 | 0.950481 | 4.046748E+15 | 8.965170E+15 | 5.688761E+15 | RBFT | Sum of squares | Gaussian | Identity |
| RBF 1-27-1 | 0.925175 | 0.904176 | 0.959754 | 3.701526E+15 | 6.905686E+15 | 4.272654E+15 | RBFT | Sum of squares | Gaussian | Identity |
| RBF 1-25-1 | 0.942461 | 0.928905 | 0.949845 | 2.871419E+15 | 5.767100E+15 | 4.758436E+15 | RBFT | Sum of squares | Gaussian | Identity |

Source: Own processing.

As the table clearly shows, only the RBF contain one variable (time) in the input layer. The hidden layers contain between 22 and 30 neurons, while the output layers contain only one variable, the Czech Republic and the People´s Republic of China trade balance. For all retained networks, the same RBFT training algorithm was used, as well as the same error function (the sum of the least squares). Similarly, as the output layer activation function, the Gaussian curve was used. The output activation function was also the same for all retained network, it was the Identity function.

Generally, the focus shall be on the training, testing and validation performance of the individual retained networks. We are looking for a network whose performance in all data sets is ideally the same, and the error is as small as possible. It shall be noted that the division of the data into data sets was random. The performance of these data sets is given in the form of the correlation coefficient. Therefore, Table 3 shows the values of the individual data sets by the specific retained neural networks.

**Table 3.** Correlation coefficients of individual data sets – neural networks A

| | Balance (Training) | Balance (Testing) | Balance (Validation) |
|---|---|---|---|
| RBF 1-22-1 | 0.943854 | 0.942643 | 0.953409 |
| RBF 1-24-1 | 0.893296 | 0.855999 | 0.951235 |
| RBF 1-30-1 | 0.917891 | 0.870478 | 0.950481 |
| RBF 1-27-1 | 0.925175 | 0.904176 | 0.959754 |
| RBF 1-25-1 | 0.942461 | 0.928905 | 0.949845 |

Source: Own processing.

The table indicates that the performance of all retained neural networks is very similar. The slight differences do not have any significant influence on the networks performance. In the case of the training data set, the correlation coefficients values are between 0.89 and 0.94, in the case of testing data set, the range is slightly wider, with networks achieving the values between 0.85 and 0.94. The coefficients of the validation data set have the smallest range, achieving the values of about 0.95. However, it is clear that the second retained neural network, RBF 1-24-1, achieve the worst correlation coefficients. The best and constant results in all data sets are achieved in the case of the first retained network, RBF 1-22-1. However, in order to choose the most suitable retained network, it is necessary to carry out a more detailed analysis of the results obtained. The basic statistical characteristics of the individual data sets for all retained neural networks are listed in Table 4.

**Table 4** Statistics of individual data sets by retained neural structures – neural structures A

| Statistics | 1.RBF 1-22-1 | 2.RBF 1-24-1 | 3.RBF 1-30-1 | 4.RBF 1-27-1 | 5.RBF 1-25-1 |
|---|---|---|---|---|---|
| Minimal forecasts (Training) | -898391801 | -821663096 | -857909633 | -914757095 | -905237832 |
| Maximal forecasts (Training) | -49086981 | -40172975 | -13537671 | -4452348 | -41323566 |
| Minimal forecasts (Testing) | -789576394 | -804782911 | -828736787 | -706475950 | -744790331 |
| Maximal forecasts (Testing) | -49635722 | -66639725 | -23270632 | -73460448 | -55363423 |
| Minimal forecasts (Validation) | -803294038 | -842962423 | -777155961 | -816475698 | -807210541 |
| Maximal forecasts (Validation) | -79169241 | -72858834 | -16165433 | -68026809 | -85082148 |
| Minimal residuals (Training) | -336438354 | -427989158 | -399848725 | -377532214 | -373817164 |
| Maximal residuals (Training) | 258520249 | 277584103 | 188100987 | 225393996 | 210531734 |
| Minimal residuals (Testing) | -252876501 | -563851527 | -352153219 | -283794827 | -333348153 |
| Maximal residuals (Testing) | 208414123 | 238241010 | 163190692 | 208652287 | 135282887 |
| Minimal residuals (Validation) | -364807606 | -325139221 | -390945683 | -351625946 | -360891103 |
| Maximal residuals (Validation) | 146675635 | 131088493 | 100813927 | 75309342 | 141610295 |
| Minimal standard residua (Training) | -6 | -6 | -6 | -6 | -7 |
| Maximal standard residuals (Training) | 5 | 4 | 3 | 4 | 4 |
| Minimal standard residuals (Testing) | -4 | -6 | -4 | -3 | -4 |
| Maximal standard residuals (Testing) | 3 | 2 | 2 | 3 | 2 |
| Minimal standard residuals (Validation) | -5 | -5 | -5 | -5 | -5 |
| Maximal standard residuals (Validation) | 2 | 2 | 1 | 1 | 2 |

Source: Own processing.

In ideal case, the individual characteristics of the neural network should be horizontally the same in all data sets. This refers to the minimal and maximal forecasts, minimal and maximal residuals, and minimal and maximal standard residuals. As for the equalized time series, there are minimal differences only. The differences are bigger in the case of the residuals characteristics. However, even based on this information, it is not possible to determine unambiguously which of the retained structures achieves the best results.

Therefore, Figure 2 was used to show a line graph representing the actual development of the Czech Republic and the People´s Republic of China trade balance and also the forecasts made using individual generated and retained networks.

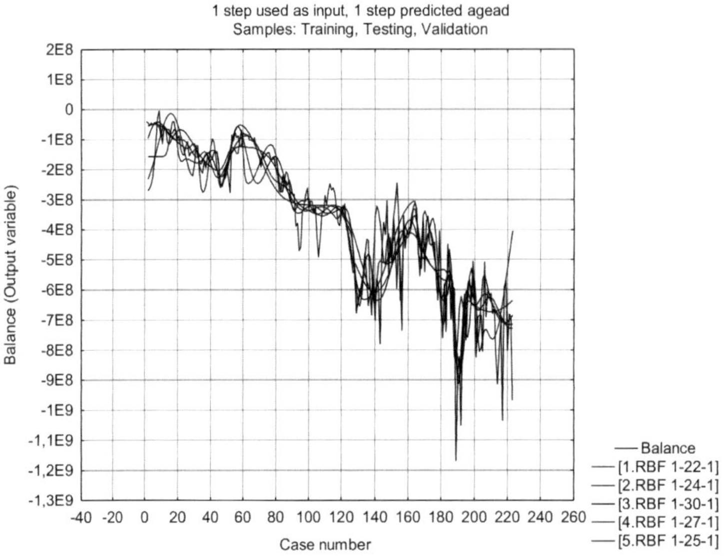


**Fig. 2.** Line graph – development of Czech Republic and People´s Republic of China trade balance forecast by neural networks compared with actual trade balance in monitored period – neural structures A

Source: Own processing.

As can be seen from the graph, the retained neural networks forecasts of the trade balance development are partially different at the individual intervals. However, what is important at the moment is not the similarity between the individual networks forecasts but their similarity (consistency) with the actual development of the Czech Republic and People´s Republic of China trade balance. In this respect, all networks appear to be interesting, following the curve assessing the trade balance development, but at the same time they are often able to capture the local minimum and maximum, that is, the extremes of the curve. The best neural network appears to be the first retained network, RBF 1-22-1.

## 4.2 Neural networks B

Based on the methodology mentioned above, the second set of 10,000 neural networks was generated, out of which 5 with the best characteristics were retained. Table 5 shows the overview of the retained networks.

**Table 5.** Retained neural networks – neural networks B

| Network | Train. perform. | Test. Perform. | Valid. perform. | Train. error | Test. error | Valid. error | Train. algorit. | Error function | Activat. of hidden layer | Output activat. function |
|---|---|---|---|---|---|---|---|---|---|---|
| MLP 13-6-1 | 0.876325 | 0.943756 | 0.876818 | 6.514775E+15 | 4.128195E+15 | 7.879408E+15 | BFGS (Quasi-Newton) 5 | Sum of squares | Identity | Tanh |
| MLP 13-3-1 | 0.879131 | 0.934377 | 0.878464 | 7.371077E+15 | 4.822871E+15 | 8.273858E+15 | BFGS (Quasi-Newton) 5 | Sum of squares | Sinus | Tanh |
| MLP 13-8-1 | 0.871402 | 0.932727 | 0.876998 | 6.755701E+15 | 4.733471E+15 | 7.776348E+15 | BFGS (Quasi-Newton) 5 | Sum of squares | Identity | Tanh |
| MLP 13-4-1 | 0.876096 | 0.943739 | 0.877163 | 7.745479E+15 | 4.489302E+15 | 8.095790E+15 | BFGS (Quasi-Newton) 5 | Sum of squares | Identity | Sinus |
| MLP 13-9-1 | 0.870916 | 0.947858 | 0.880312 | 6.884011E+15 | 3.695509E+15 | 8.112908E+15 | BFGS (Quasi-Newton) 5 | Sum of squares | Identity | Logistic |

Source: Own processing.

As the table clearly shows, the networks are only multilayer perceptron networks. The input layer contains only two variables – continuous variable (time) and categorical variable (the month of measurement). The continuous variable is represented by one neuron, while the categorical variable by twelve neurons. Logically, the input layer of all networks contains 13 neurons. The hidden layer of each retained network contains a different number of neurons – between 3 and 9 neurons. The output layer of all retained networks contains one neuron, that is, one output variable – the Czech Republic and the People´s Republic of China trade balance. All retained networks use the same alternative of the Quasi-Newton training algorithm. All networks also use the same error function – the sum of the least squares. For the activation of the hidden layers, the retained networks use the Identity function in 4 cases, and the sinus function in one case. As the output activation function, three networks used the function of hyperbolic tangent, one the sinus function and one the logistic function. The values of the individual data sets, that is, the correlation coefficient by the specific neural networks are given in Table 6.

**Table 6.** Correlation coefficients of individual data sets – neural networks B

| | Balance (Training) | Balance (Testing) | Balance (Validation) |
|---|---|---|---|
| 1.MLP 13-23-1 | 0.876325 | 0.943756 | 0.876818 |
| 2.MLP 13-4-1 | 0.879131 | 0.934377 | 0.878464 |
| 3.MLP 13-6-1 | 0.871402 | 0.932727 | 0.876998 |
| 4.MLP 13-13-1 | 0.876096 | 0.943739 | 0.877163 |
| 5.MLP 13-8-1 | 0.870916 | 0.947858 | 0.880312 |

Source: Own processing.

The table indicates that the performance of all retained networks is very similar. The differences are not big, and therefore do not have any influence on the retained networks performance. In the case of the training data set, the value of the correlation coefficient ranges between 0.87-0.88, as in the case of the validation data set. The correlation coefficients of the testing data set achieve the value of up to 0.948. All correlation coefficients achieve relatively high values and are similar to each other in the individual data sets. Next step can thus follow, that is, selection of the most suitable neural structure. In order to find the most suitable one, it is necessary to carry out a detailed analysis of the results obtained. Table 7 shows the basic statistical characteristics of the data sets for all retained neural structureS.

**Table 7.** Statistics of individual data sets by retained neural structures – neural structures B

| Statistics | 1.MLP 13-6-1 | 2.MLP 13-3-1 | 3.MLP 13-8-1 | 4.MLP 13-4-1 | 5.MLP 13-9-1 |
|---|---|---|---|---|---|
| Minimal forecast (Training) | -9,426152E+08 | -886640266 | -914766256 | -956908379 | -856459363 |
| Maximal forecast (Training) | -1,066437E+08 | -104264109 | -102226019 | -53088892 | -82724549 |
| Minimal forecast (Testing) | -1,021248E+09 | -878208259 | -923598942 | -896907808 | -865647214 |
| Maximal forecast (Testing) | -1,083489E+08 | -103406677 | -123442616 | -63808204 | -91605519 |
| Minimal forecast (Validation) | -9,206647E+08 | -814400994 | -890733429 | -939409841 | -795408474 |
| Maximal forecast (Validation) | -1,256759E+08 | -134458650 | -136309325 | -78432107 | -119006283 |
| Minimal residuals (Training) | -3,400883E+08 | -313803029 | -316749726 | -258840713 | -354397024 |
| Maximal residuals (Training) | 2,939457E+08 | 292481769 | 344142148 | 361537500 | 280016174 |
| Minimal residuals (Testing) | -2,571735E+08 | -270129973 | -271096095 | -163280841 | -241846698 |
| Maximal residua (Testing) | 1,697501E+08 | 218590947 | 143661438 | 252434082 | 201593398 |
| Minimal residuals (Validation) | -4,709871E+08 | -402178184 | -421508582 | -413710569 | -410637784 |
| Maximal residuals (Validation) | 2,357501E+08 | 325578252 | 282765747 | 312819686 | 337110906 |
| Minimal standard residua (Training) | -4,213490E+00 | -4 | -4 | -3 | -4 |
| Maximal standard residuals (Training) | 3,641811E+00 | 3 | 4 | 4 | 3 |
| Minimal standard residuals (Testing) | -4,002636E+00 | -4 | -4 | -2 | -4 |
| Maximal standard residuals (Testing) | 2,641982E+00 | 3 | 2 | 4 | 3 |
| Minimal standard residuals (Validation) | -5,305939E+00 | -4 | -5 | -5 | -5 |
| Maximal standard residuals (Validation) | 2,655860E+00 | 4 | 3 | 3 | 4 |

Source: Own processing.

As for the individual data sets statistics, it is desirable that the individual statistics conformed to each other in all data sets (minimum, maximum, residuals, etc.). In the case of equalized time series, there have been minimal differences detected in minimal and maximal forecasts, minimal and maximal residuals, and standard residuals. Despite this detailed specification, it is not possible to determine unambiguously, which of the retained networks achieves the best results.

For illustrative purposes and determining which of the retained networks shows the best results, Figure 3 shows a line graph representing the actual development of the Czech Republic and the People´s Republic of China trade balance, together with the development of forecasts using individual retained neural networks.

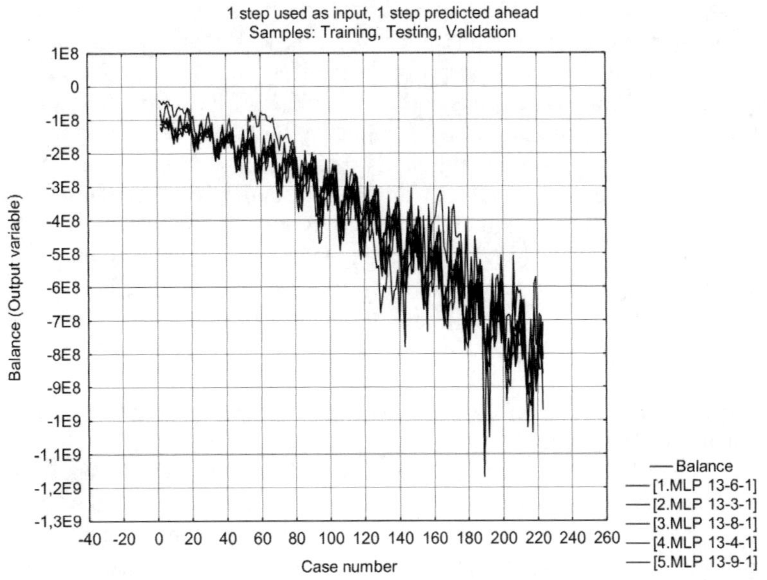


**Fig. 3.** Line graph – development of Czech Republic and People´s Republic of China trade balance forecast by neural networks in comparison with actual trade balance in monitored period – neural structures B

Source: Own processing.

The Figure clearly indicates that all retained networks forecasts of the trade balance development at the individual intervals are slightly different. However, what is important is not the similarity of the individual forecasts but the extent to which they correspond with the actual development of the trade balance (represented by blue curve). Even in this case, it can be stated that the retained networks appear to be interesting. They are able to forecast the basic trend of the trade balance development very well, but very often, they are also able to capture the local extremes – minimum and maximum. A closer look indicates that mainly two networks are applicable: the 1st retained network 1. MLP 13-6-1 and the 3rd retained network 3. MLP 13-8-1. Nevertheless, as it has already been mentioned, all networks performance and forecasting ability are very similar and can be used in practice.

## 4.3 Comparison of results of neural networks A and B

The examined time series, the Czech Republic and the People´s Republic of China trade balance, could be equalized by all generated and retained neural networks.

The results are very interesting. Based on the individual networks performance, it shall be stated that the RBF networks show significantly better results than the MLP networks. Specifically, it is the training and validation data sets performance. The performance of both neural networks data sets are the same in terms of the testing data set. However, the values do not clearly correspond to the graphical representation of the equalized time series course. From the course of the actual time series (marked blue in Figure 2 and Figure 3) it results clear that there are seasonal fluctuations. Nevertheless, those are not captured by the RBF networks. The individual RBF networks are capable of following the course of the time series only at certain intervals, but they still show relatively significant deviations. On the other hand, the time series equalized by means of the MLP networks are able to follow the course of the actual development of the Czech Republic and the People´s Republic of China trade balance quite well. However, they achieve worse performance in the training and validation data sets, at least in terms of the correlation coefficient. Basically, there are two reasons identified. Firstly, it can be due to inappropriate division of the original data set into the training, testing and validation data sets in the case of the MLP networks. Secondly, it can be due to distortion caused by division of the residuals of both data sets.

As for dividing the data into the individual data sets, it shall be noted that the division was random – as a random sample. Although the author´s conscious or unconscious preferences were not reflected in the division, there is still a possibility that the data set used for the training of the MLP networks had been created incorrectly, unbalanced. All RBF networks were generated on the same data sets – training, testing, and validation. Thus, they show approximately the same results (at a large number of iterations). The MLP networks were also generated on the same data sets, but different from the data sets used for the RBF networks.

Regarding the division of the residuals, this could also be caused accidentally. However, even such a situation may occur.

On the other hand, it is obvious that the MLP networks capture the course of the time series better. Therefore, it must be concluded that the MLP networks captured the seasonal fluctuations of the time series (which are obvious), thus actually achieving much better results that the RBF networks. The most suitable structure is therefore supposed to be found among the MLP networks according to the performance and errors of the individual neural networks. As stated above, two retained MLP networks show slightly better performance than the others – 1. MLP 13-6-1 and 3. MLP 13-8-1.

## 5 Conclusion

The objective of the contribution was to introduce a methodology for considering seasonal fluctuations in equalizing time series using artificial neural networks on the example of the Czech Republic and the People´s Republic of China trade balance.

Forecasting based on the past data always means understanding past events, their simplification and subsequent application of the simplified model for the purpose of future development. This means the view to the future is also simplified. However, it is essential in many cases to be able to forecast at least the approximate development trend [20]. This way it is possible to make plans, to implement business plans at reduced risk.

Forecasting definitely does not offer an opportunity to forecast the occurrence of extraordinary situations. The method is therefore suitable for planning for days. The contribution this paper brings is the involvement of variables that are able to forecast a possible seasonal fluctuation in the time series development when using artificial neural networks. It was involvement of the variables at the level of time, which was not seen as a continuous but categorical variable, specifically year, month, and day when we tried to identify seasonal fluctuations in the individual years and months. In this case, the classification was sufficient to capture the development of the time series examined – the CR and the PRC trade balance. The analysis can be simple explained by the fact that the continuous variable "date" was able to equalize time series, while the categorical variables corrected this time series according to the seasonal fluctuations identified. For future forecasts it is therefore very important to work with the time series delay. This way we would be able to find out the network optimal performance for future period – the network predictive power would increase. However, it should be noted that in such a case, the demand for computing performance in neural network training would be higher. Each delay means several extra neurons in the input layer for each variable (one in the case of continuous variable, in the case of categorical variable up to the maximum values possible). However, this is a subject of possible further investigation.

In any case, there have been obtained neural networks which are able to "understand" the seasonal fluctuations in equalizing time series. Moreover, neural networks have been identified that achieve slightly better results than other networks, specifically these are the neural networks 1. MLP 13-6-1 and 3. MLP 13-8-1.

The objective of the contribution has been achieved.

## References

1. R. Koszek, Central and Eastern Europe Countries in View of Chinese Economic Expansion. *Studies of the Industrial Geography Commission of the Polish Geographical Society*, **30**(1), 176-191 (2016)
2. Z. Rowland, P. Šuleř, M. Vochozka, Comparison of neural networks and regression time series in estimating the Czech Republic and China trade balance. *SHS Web of Conferences: Innovative Economic Symposium 2018 – Milestones and Trends of the World Economy*, **61** (2019)
3. J. Vrbka, Z. Rowland, P. Šuleř, Comparison of neural networks and regression time series in estimating the development of the EU and the PRC trade balance. *SHS Web of Conferences: Innovative Economic Symposium 2018 – Milestones and Trends of the World Economy*, **61** (2019)
4. M.D. Evans, External balances, trade flows and financial conditions. *Journal of International Money and Finance*, **48**, 271-290 (2014)
5. J. Garlick, China's Trade with Central and Eastern European EU Members: An Analysis of Eurostat Data, 2004-2014. *Acta Oeconomica Pragensia*, **2015**(4), 3-22 (2015)
6. T. Matura, China–CEE Trade, Investment and Politics. *Europe-Asia Studies*, **71**(3), 388-407 (2019)
7. I. Šteinbuka, T. Muravska, A. Kuznieks, Cooperation formats of China and Europe: Synergies and divergences. *Baltic Journal of European Studies*, **7**(1), 98-118 (2017)
8. T. De Castro, Z. Stuchlíková, China-V4 trade relations–a Czech perspective. *Current Trends and Perspectives in Development of China – V4 Trade and Investment*, pp. 46-60 (2014)
9. T. Dudas, M. Dudasova, Growth of Chinese investments in Europe after the global economic crisis of 2008-2009. *Economic Annals - XXI*, (160), 9-14 (2016)
10. T. Dudas, Dynamic growth of Chinese investments in Europe after the global economic crisis of 2008/2009. *16th International Scientific Conference on International Relations - Current Issues of World Economy and Politics, pp.* 236-244 (2015)
11. Z. Pikhart, Cyclical-Adjusted External Balance of Goods and Services in the Czech Republic. *Statistika-Statistics and Economy Journal*, **98**(2), 103-112 (2018)
12. W. Kim, Asset Price, the Exchange Rate, and Trade Balances in China: A Sign Restriction VAR Approach. *East Asian Economic Review*, **22**(3), 371-400 (2018)
13. M.Y. Chen, M.H. Fan, Y.L. Chen, H.M. Wei, Design of experiments on neural network's parameters optimization for time series forecasting in stock markets. *Neural Network World*, **23**(4), 369-393 (2013)
14. F. Ludbrook, K. Frajtová Michalíková, Z. Musová, P. Šuleř, Business Models for Sustainable Innovation in Industry 4.0: Smart Manufacturing Processes, Digitalization of Production Systems, and Data-driven Decision Making. *Journal of Self-Governance and Management Economics*, 7(3), 21-26 (2019)

15. V. Machová, J. Mareček, Estimation of the development of Czech Koruna to Chinese Yuan exchange rate using artificial neural networks. *SHS Web of Conferences: Innovative Economic Symposium 2018 - Milestones and Trends of World Economy*, **61** (2019)
16. O. Krpec, V. Hodulák, The Czech economy as an integrated periphery: The case of dependency on Germany. *Journal of Post Keynesian Economics*, **42**(1), 59-89 (2019)
17. T. Kliešťik, J. Vrbka, Z. Rowland, Bankruptcy prediction in Visegrad group countries using multiple discriminant analysis. *Equilibrium-Quarterly Journal of Economics and Economic Policy*, 13(3), 569-593 (2018)
18. M. Vochozka, J. Vrbka, Estimation of the development of the Euro to Chinese Yuan exchange rate using artificial neural networks. *SHS Web of Conferences: Innovative Economic Symposium 2018 - Milestones and Trends of World Economy*, **61** (2019)
19. World Bank [online], Available at: http://www.worldbank.org/ (2019)
20. M. Kováčová, T. Kliešťik, Logit and Probit application for the prediction of bankruptcy in Slovak companies. *Equilibrium-Quarterly Journal of Economics and Economic Policy*, **12**(4), 775-791 (2017)

# Considering seasonal fluctuations on balancing time series with the use of artificial neural networks when forecasting US imports from the PRC

*Jaromír* Vrbka[1]*, *Marek* Vochozka[1]

[1]Institute of Technology and Business, School of Expertness and Valuation, Okružní 517/10, 37001 České Budějovice, Czech Republic

**Abstract.** The paper's objective is to propose a particular methodology to be used to regard seasonal fluctuations on balancing time series while using artificial neural networks based on the example of imports from the People's Republic of China (PRC) to the USA (US). The difficulty of forecasting the volume of foreign trade is usually given by the limitations of many conventional forecasting models. For the improvement of forecasting it is necessary to propose an approach that would hybridize econometric models and artificial intelligence models. Data for an analysis to be conducted are available on the World Bank website, etc. Information on US imports from the PRC will be used. Each forecast is given by a certain degree of probability which it will be fulfilled with. Although it appeared before the experiment that there was no reason to include the categorical variable to reflect seasonal fluctuations of the USA imports from the PRC, the assumption was not correct. An additional variable, in the form of monthly value measurements, brought greater order and accuracy to the balanced time series.
**Key words:** forecasting models, artificial neural networks, time series, development

## 1 Introduction

The difficulty of forecasting the volume of foreign trade is usually given by the limitations of many conventional forecasting models. For the improvement of forecasting it is necessary to propose an approach that would hybridize econometric models and artificial intelligence models [1]. The given topic has already been addressed, although not widely, but the example of the USA and China as two dominant economic powers is very interesting.

Export is often seen as a way to increase economic growth. Bernard and Jensen [2] examines the interaction between export the productivity growth in the USA manufacturing. Export involves the redistribution of resources from less efficient facilities into more efficient ones. The overall redistributions are rather large, accounting for more than 40% of the overall productivity growth of the industries in the manufacturing sector. Yu, Dong and Feng [3]

* Corresponding author: vrbka@mail.vstecb.cz

state that the key policy of the industry support has resulted in a 0.35% increase compared to import since China´s accession to the WTO, while in the USA this has decreased to 0.35%. With regard to foreign export control, the impact of the key industry´s support policy on China´s trade surplus, indicating that the foreign export control to China has increased the impact of the key industries support on the trade balance.

At the end of 2018, in the first half of 2019, the trade rivalry between the People´s Republic of China and the USA had a significant impact on Chinese ideological and political component, socio-economic, and military development, as well as on the development of its foreign policy. The impact of the "trade war" on the economic development of China still remains slightly negative but the "survival potential" of the country´s economy is quite strong. Coincidentally, China´s "development potential" can suffer considerable losses if the USA continues to maintain a strict restrictive line – not in terms of the tariff-related topics, such as the expansion of Chinese capital into the US market and the access of China to the new American technology market [4]. Given the complexity of the economic system, the interactive effects of the economic variables of factors on Chinese foreign trade make forecasting the state of Chinese foreign trade extremely difficult. In terms of analysing the relationship between the economic variables and foreign trade, this study proposes a new non-linear set of methodology for hybridizing non-linear econometric model and artificial neural networks (ANN) learning in order to forecast Chinese foreign trade. Yu, Wand and Lai [5] create an auto-regression model (EC-VAR) based on the correction of the errors on the basis of co-integration, first used to capture the impacts of economic variables on Chinese foreign market in terms of multidimensional analysis. To capture non-linear patterns hidden between foreign trade and economic factors, the ANN-based EC-VAR model is used. All economic variables, outputs of linear and non-linear EC-VAR models and assessment model are used as other neural networks input for the forecasting purposes. For illustration, the proposed methodology of the set integrating teaching econometric technics and artificial intelligence (AI) methods is applied to the problem of forecasting Chinese export trade.

In order to investigate inefficiency of the market resulting from fast developing software and technological progress all over the world, technology prejudice has been introduced in the foreign Exchange market. The idea of technology distortion arises from the fact that innovative approaches have been used to address business challenges and to find the best business strategies [6]. The situation described above indicates market inefficiency, which casts doubt on the idea of an effective market theory, but is in line with the adaptive market hypothesis [7]. Galeshchuk [8] uses shallow neural network that surpasses a random walk model which confirms the idea of market inefficiency. Zhang [9] finds out that the indicators of GDP, direct foreign investment, and research and development affects Chinese export of goods by means of sensitivity analyses. This method along with the analysis of policy shall be systematically analysed and forecast by Chinese export, thus providing theoretical background for the relevant decision-making departments.

Horák, Šuleř and Vrbka [10] aimed to compare the accuracy of equalizing time series using regression analysis and neural networks on the example of the USA export to Chine. The objective is to show possible application and advantages of using neural networks in practice. The study compares two statistical methods; artificial intelligence is not used in these applications. In many economic sectors, however, better results have been shown. It has been found out that artificial neural networks are able to learn effectively the dependence within time series and between them in the form of export development data [11]. Vrbka, Rowland and Šuleř [12] compare the accuracy of equalizing time series by means of regression analysis and neural networks on the example of the trade balance between the EU and the People´s Republic of China. In general, each forecast is given by certain degree of probability of its fulfilment.

Rowland, Šuleř and Vochozka [13] aimed to compare the accuracy of equalizing time series by means of regression analysis and neural networks on the example of the trade balance between the Czech Republic and the People´s Republic of China. They use the data on the trade balance between the Czech Republic and the People´s Republic of China. First, linear regression was carried out, followed by regression using artificial neural networks. The compared both methods at the expert level and the level of the experience of an evaluator, economist. Optically, the LOWESS curve appears to be the most successful from the linear regression, and the fifth retained network (RBF 1-24-1) appears to be a suitable one from the retained neural networks. Neural networks also appear to be a method of optimizing company inventory. They are useful in forecasting time series [14]. Šuleř [15] strives to verify the possibility of using neural networks for inventory management on the example of a concrete company. They are useful for forecasting time series. Chang [16] deals with artificial networks, decision trees, and a hybrid model of ANN and decision trees. The author compares the methods of forecasting stock price derived from the three methods mentioned above. The study has also found that compared to the two methods, ANN is a more stable forecasting method.

Based on non-linear prediction of phase space reconstruction, Weijin and Yuhui [17] presented a model of time lag of a BP neural network, whose generalisation ability had been improved by Baesian regularisation. The model is used to forecast import and export trade within one sector. Compared to conventional evaluation of forecasts, it has been concluded that non-linear forecasting can be focused not only on combining data and improving accuracy but can also reflect non-linear characteristics of the forecasting system.

Zhao et al. [18] examine the impacts of China´s industrial plans for export and trade balance, as well as on how these countries are enlarged by controlling export to China from foreign countries. Empirical results show that China´s political support of the key industries through five-year plans has pad a positive impact on its export and trade balance. Specifically, this key policy of supporting industry has resulted in a 0.35% increase in Chinese export compared to the import since China´s admission to the WTO, while in the USA the increase was only 0.3%. With regard to foreign trade control, the impact of the key industries support policy on China´s surplus decreases, which indicates that foreign controls of export to China have increased the impact of the key industries support policy on the trade balance.

## 2 Data and methods

The Data for an analysis to be conducted are available on the World Bank website, etc. Information on US imports from the People's Republic of China (PRC) will be used. A time interval for which the data will be available is the monthly balance beginning January 1985 and ending August 2018, i.e. 404 pieces of input data, with the main unit being billions of US dollars.

Descriptive characteristics of the data are given in Table 1.

**Table 1.** Characteristics of data sets

| Data Sets | Month (Input Variables) | Imports (Output) (Target) |
|---|---|---|
| Minimum (Training) | 31048,00 | 264,90 |
| Maximum (Training) | 43313,00 | 48167,70 |
| Average (Training) | 37316,95 | 16718,52 |
| Standard deviation (Training) | 3549,13 | 14981,52 |
| Minimum (Test) | 31138,00 | 283,30 |
| Maximum (Test) | 42948,00 | 45817,80 |
| Average (Test) | 36651,48 | 14618,18 |
| Standard deviation (Test) | 3758,45 | 16187,18 |
| Minimum (Validation) | 31199,00 | 348,70 |
| Maximum (Validation) | 42979,00 | 45429,70 |
| Average (Validation) | 37060,87 | 15188,48 |
| Standard deviation (Validation) | 5186,49 | 14723,70 |
| Minimum (Overall) | 31048,00 | 264,90 |
| Maximum (Overall) | 43313,00 | 48167,70 |
| Average (Overall) | 37180,08 | 16179,36 |
| Standard deviation (Overall) | 3554,16 | 15075,35 |

Source: Own processing.

Certain development of the US imports from the PRC over the course of time is indeed an interesting topic. Figure 1 therefore presents graphically selected statistical characteristics, including a histogram of the input data.

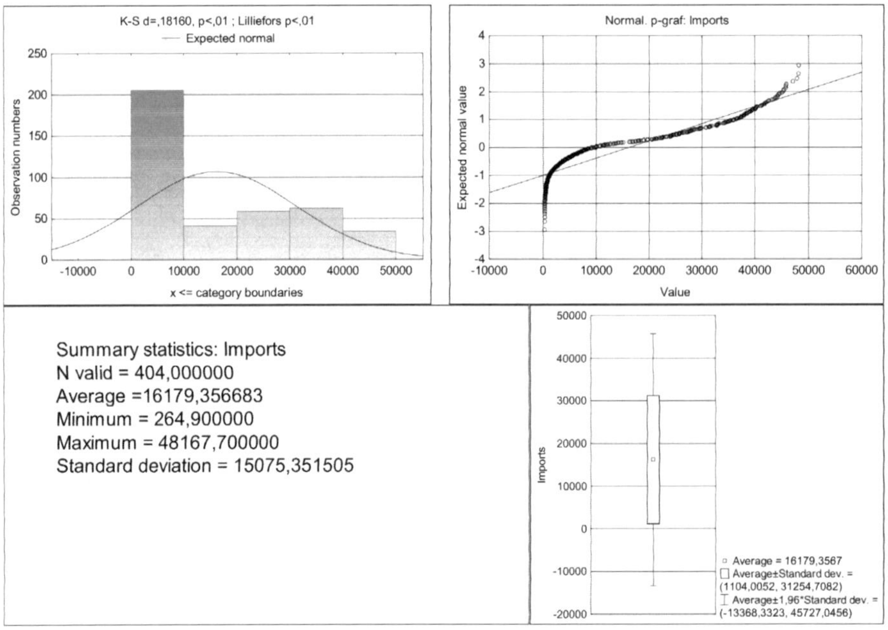


**Fig. 1.** Graph of basic statistical characteristics

Source: Own processing.

Furthermore, the histogram does not correspond to normal distribution. Data will be processed using DELL Statistica software version 12.

A specific regression using neural structures will be performed. Multilayer perceptron networks and neural networks of a basic radial function will be generated along with two sets of artificial neural networks:

1. Time shall be an independent variable, with the US imports from the PRC being a dependent variable.
2. Time shall be a continuous independent variable. The seasonal fluctuations shall be represented by a categorical variable in the form of a month in which a certain value was measured. Thus, possible monthly seasonality of the time series will be regarded, with the US imports from the PRC being a dependent variable.

Next, the time series shall be divided into three sets – training, test and validation. The first group will include 70% of the input data. Based on the training data set, neural structures will be generated. The remaining two data sets will always include 15% of the input information. Both of the groups will serve to verify reliability of a particular neural structure, or model, found. The time series delay shall be 1, and 10,000 neural networks are to be generated. Of these, 5 networks, showing the best characteristics[1], will be retained. In the hidden layer, there will be at least two neurons, with the maximum being 50. In relation to the radial basic function, there will be at least 21 neurons in the hidden layer, with the maximum being 30. As for the multiple perceptron network, the following distribution functions in the hidden layer and in the output layer shall be considered:

The method of least squares is to be applied. Network generating shall be terminated unless there is an improvement, i.e. a decrease, in the sum of the squares. There will therefore be retained those neural structures whose sum of squares of residuals related to the actual development of the US imports from the PRC will be as low as possible – ideally zero).

- Linear,
- Logistic,
- Atanh,
- Exponential,
- Sinus.

Other settings shall remain as default (according to ANS – Automated Neural Network). As part of conclusion, results of both groups of retained neural networks will be compared.

# 3 Results

## 3.1 Neural structures A

Based on the procedure established, 10,000 neural networks were generated. Of these, 5 networks, showing the best parameters, were retained. They are given in Table 2.

**Table 2.** Overview of retained neural networks

| Network | Train. Perform. | Test. Perform. | Valid. Perform. | Train. Error | Test. Error | Valid. Error | Train. Algorit. | Error Function | Activation of Hidden Layers | Output Activation Function |
|---|---|---|---|---|---|---|---|---|---|---|
| RBF 1-28-1 | 0.985565 | 0.993687 | 0.988204 | 3199247 | 2234000 | 2510302 | RBFT | Sum of squares | Gaussian | Identity |
| RBF 1-24-1 | 0.981934 | 0.994521 | 0.988715 | 3985467 | 2272140 | 2446088 | RBFT | Sum of squares | Gaussian | Identity |
| RBF 1-25-1 | 0.974559 | 0.988826 | 0.988329 | 5593698 | 2897250 | 2473482 | RBFT | Sum of squares | Gaussian | Identity |
| RBF 1-30-1 | 0.988410 | 0.992897 | 0.988280 | 2564933 | 2679733 | 2424616 | RBFT | Sum of squares | Gaussian | Identity |
| RBF 1-28-1 | 0.979283 | 0.989343 | 0.987745 | 4565578 | 3029347 | 2526599 | RBFT | Sum of squares | Gaussian | Identity |

Source: Own processing.

There are only neural networks of the basic radial function with only one variable in the input layer – time. In the hidden layer, the neural networks contain 24-28 neurons. Consequently, there is a single neuron in the output layer and a single output variable, i.e. the US imports from the PRC. The RBFT training algorithm was applied to all networks. In addition, all neural structures used the same function, namely the Gaussian curve, to activate the hidden layer of neurons. The same function was used to activate the outer layer of neurons (see Table 2).

Training, test and validation performance is worth mentioning as well. Generally, the goal is to search for a network that performs equally well across all data sets (please note that the data were randomly distributed). Errors should be as small as possible.
The performance of individual data sets is given in the form of a correlation coefficient. Values of individual data sets according to specific neural networks are shown in Table 3.

**Table 3.** Correlation coefficients of individual data sets

| | Imports (Training) | Imports (Test) | Imports (Validation) |
|---|---|---|---|
| 1.RBF 1-28-1 | 0.985565 | 0.993687 | 0.988204 |
| 2.RBF 1-24-1 | 0.981934 | 0.994521 | 0.988715 |
| 3.RBF 1-25-1 | 0.974559 | 0.988826 | 0.988329 |
| 4.RBF 1-30-1 | 0.988410 | 0.992897 | 0.988280 |
| 5.RBF 1-28-1 | 0.979283 | 0.989343 | 0.987745 |

Source: Own processing.

The table shows that the performance of all retained neural structures is approximately the same, with slight differences not affecting each network. The correlation coefficient value of all training data sets ranges from more than 0.974 to more than 0.988. The correlation coefficient value of the test data sets reaches more than 0.988 for all neural networks. The correlation coefficient value of the validation data set of all neural networks is above 0.987. In order to select the most suitable neural structure, it is necessary to carry out a closer analysis of the results obtained. Table 4 gives the basic statistical characteristics of each data set for all neural structures.

**Table 4.** Statistics of individual data sets according to neural structures retained

| Statistics | 1.RBF 1-28-1 | 2.RBF 1-24-1 | 3.RBF 1-25-1 | 4.RBF 1-30-1 | 5.RBF 1-28-1 |
|---|---|---|---|---|---|
| Min forecasts (Training) | -105.81 | 580.89 | 121.1 | -246.8 | -297.8 |
| Max forecasts (Training) | 45281.06 | 46726.83 | 46002.4 | 44572.1 | 44832.3 |
| Min forecasts (Test) | -17.55 | 875.04 | 204.9 | -236.1 | -66.6 |
| Max forecasts (Test) | 43523.94 | 43272.06 | 45755.1 | 41867.8 | 44650.5 |
| Min forecasts (Validation) | -115.73 | 782.33 | 664.4 | -228.6 | -361.7 |
| Max forecasts (Validation) | 43676.46 | 43430.95 | 45447.6 | 41848.5 | 44800.2 |
| Min residues (Training) | -9388.83 | -9090.02 | -12440.4 | -11451.6 | -10678.7 |
| Max residues (Training) | 9830.06 | 11453.48 | 13102.8 | 8460.7 | 10527.1 |
| Min residues (Test) | -2929.57 | -4211.16 | -4982.3 | -3175.6 | -6051.8 |
| Max residues (Test) | 8548.44 | 8099.91 | 9591.5 | 7255.6 | 9059.8 |
| Min residues (Validation) | -8438.26 | -7010.75 | -6153.3 | -8573.7 | -6018.4 |
| Max residues (Validation) | 4860.34 | 3833.17 | 7097.7 | 6317.5 | 6693.0 |
| Min standard residues (Training) | -5.25 | -4.55 | -5.3 | -7.2 | -5.0 |
| Max standard residues (Training) | 5.50 | 5.74 | 5.5 | 5.3 | 4.9 |
| Min standard residues (Test) | -1.96 | -2.79 | -2.9 | -1.9 | -3.5 |
| Max standard residues (Test) | 5.72 | 5.37 | 5.6 | 4.4 | 5.2 |
| Min standard residues (Validation) | -5.33 | -4.48 | -3.9 | -5.5 | -3.8 |
| Max standard residues (Validation) | 3.07 | 2.45 | 4.5 | 4.1 | 4.2 |

Source: Own processing.

Ideally, individual statistics of the given neural network are more or less identical across all data sets (minimums, maximums, residues, etc.). As for the balanced time series, however, the differences are minimal, and slightly larger differences are reflected in the residue characteristics. Yet, it is not possible to clearly determine which of the retained neural networks shows the best results.

Figure 2 illustrates a line chart representing the actual development of the US imports from the PRC, as well as particular development of forecasts by each generated and retained network.

The graph shows that forecasting of all neural networks related to the imports' development is slightly different at individual intervals. What is important, however, is not the similarity of forecasts of individual networks, but the similarity (or a degree of consistency) to the actual development of the US imports from the PRC. In that respect, it may also be stated that all retained neural networks are worth noting. They respect guidelines of the curve assessing the development of the imports, and, at the same time, tend to (albeit not exactly) perceive extremes of the curve.

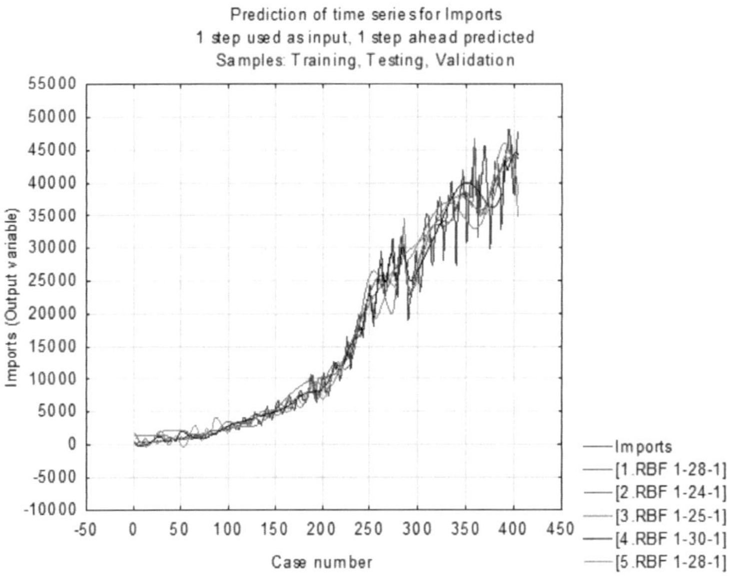


**Fig. 2.** Line graph – development of US imports from the PLC as forecast by neural networks compared to the actual imports in the period under review

Source: Own processing.

## 3.2 Neural structures B

Based on the procedure established, additional 10,000 neural networks were generated. Of these, 5 networks, again showing the best parameters, were retained. They are given in Table 5.

**Table 5**. Overview of retained neural networks

| Network | Train. Perform. | Test. Perform. | Valid. Perform. | Train. Error | Test. Error | Valid. Perform. | Train. Algorit. | Error Function | Activat. of Hidden Layers | Output Activation Function |
|---|---|---|---|---|---|---|---|---|---|---|
| MLP 13-16-1 | 0.996152 | 0.995924 | 0.994736 | 853763.9 | 1195338 | 1163517 | BFGS (Quasi-Newton) 208 | Sum of squares | Logistic | Exponent. |
| MLP 13-6-1 | 0.996152 | 0.996323 | 0.994639 | 853979.1 | 1056404 | 1170950 | BFGS (Quasi-Newton) 227 | Sum of squares | Logistic | Logistic |
| MLP 13-8-1 | 0.996010 | 0.996159 | 0.994616 | 886919.4 | 1151179 | 1198310 | BFGS (Quasi-Newton) 94 | Sum of squares | Logistic | Exponent. |
| MLP 13-9-1 | 0.995902 | 0.995885 | 0.994573 | 912038.0 | 1152619 | 1205319 | BFGS (Quasi-Newton) 93 | Sum of squares | Logistic | Exponent. |
| MLP 13-13-1 | 0.996205 | 0.995917 | 0.994870 | 841885.8 | 1200477 | 1133837 | BFGS (Quasi-Newton) 186 | Sum of squares | Logistic | Exponent. |

Source: Own processing.

There are only multilayer perceptron networks. In the input layer, there are two variables – time (a continuous variable) and a month of measurement (as a categorical variable). Time is represented by one neuron in the input layer, whereas months are represented by twelve neurons. In the hidden layer, the neural networks contain 8-16 neurons. In the output layer, there is a single neuron and a single output variable, i.e. the US imports from the PRC. For all networks, the Quasi-Newton training algorithm was applied (yet always in a different variant). Neural structures used a single function, i.e. the logistic function, to activate the hidden layer of neurons. To activate the outer layer of neurons, the exponential function and the logistic function are used (see Table 5). All retained neural structures used the sum of the least squares as an error function.

Performance values, or correlation coefficients of individual data sets by particular neural networks, are given in Table 6.

**Table 6.** Correlation coefficients of individual data sets

| | Imports (Training) | Imports (Test) | Imports (Validation) |
|---|---|---|---|
| 1.MLP 13-16-1 | 0.993132 | 0.994897 | 0.992814 |
| 2.MLP 13-6-1 | 0.992884 | 0.994211 | 0.992703 |
| 3.MLP 13-8-1 | 0.993308 | 0.995170 | 0.992898 |
| 4.MLP 13-9-1 | 0.993615 | 0.994443 | 0.992741 |
| 5.MLP 13-13-1 | 0.994173 | 0.994883 | 0.993739 |

Source: Own processing.

As may be gathered from the above table, the performance of all retained neural structures is approximately the same, and slight differences do not affect each network. The correlation coefficient value of all training data sets ranges from more than 0.992 to more than 0.994. The correlation coefficient value of the test data sets reaches more than 0.994 for all neural networks. The correlation coefficient of the validation data set of all neural networks is above 0.992. In order to select the most suitable neural structure, the obtained results need to be analyzed more closely. Table 7 provides basic statistical characteristics of each data set for all neural structures.

**Table 7.** Statistics of individual data sets according to retained neural structures

| Statistics | 1.MLP 13-16-1 | 2.MLP 13-6-1 | 3.MLP 13-8-1 | 4.MLP 13-9-1 | 5.MLP 13-13-1 |
|---|---|---|---|---|---|
| Min forecasts (Training) | 276.69 | 268.76 | 265.20 | 264.93 | 265.88 |
| Max forecasts (Training) | 48303.09 | 46260.01 | 47565.49 | 47652.57 | 48439.01 |
| Min forecasts (Test) | 366.03 | 274.49 | 265.18 | 265.00 | 273.38 |
| Max forecasts (Test) | 46552.65 | 45450.92 | 46248.88 | 46161.99 | 46629.56 |
| Min forecasts (Validation) | 291.68 | 273.66 | 267.87 | 267.24 | 266.58 |
| Max forecasts (Validation) | 47439.21 | 45750.86 | 46902.02 | 46386.91 | 47590.03 |
| Min residues (Training) | -5370.26 | -5077.31 | -5378.34 | -5404.66 | -5327.13 |
| Max residues (Training) | 8343.72 | 8620.20 | 8461.27 | 8086.81 | 8384.53 |
| Min residues (Test) | -6410.10 | -6259.24 | -6288.37 | -6174.76 | -6395.05 |
| Max residues (Test) | 2755.11 | 2264.54 | 2071.95 | 3199.24 | 2627.26 |
| Min residues (Validation) | -5098.13 | -4958.96 | -5168.12 | -5277.37 | -4980.81 |
| Max residues (Validation) | 3665.87 | 4106.84 | 3653.21 | 3628.02 | 3667.19 |
| Mini standard residues (Training) | -5.81 | -5.49 | -5.71 | -5.66 | -5.81 |
| Max standard residues (Training) | 9.03 | 9.33 | 8.98 | 8.47 | 9.14 |
| Min standard residues (Test) | -5.86 | -6.09 | -5.86 | -5.75 | -5.84 |
| Max standard residues (Test) | 2.52 | 2.20 | 1.93 | 2.98 | 2.40 |
| Min standard residues (Validation) | -4.73 | -4.58 | -4.72 | -4.81 | -4.68 |
| Max standard residues (Validation) | 3.40 | 3.80 | 3.34 | 3.30 | 3.44 |

Source: Own processing.

Ideally, individual statistics of the given neural network coincide across all sets (minimums, maximums, residues, etc.). As for the balanced time series, however, the differences are minimal. Essentially, there are minimal differences in residue characteristics as well. Nonetheless, it is not possible to clearly determine which of the retained neural networks shows the best results. According to Table 4, this may involve 3.MLP 13-8-1 and 4.MLP 13-9-13 networks.

Figure 3 is a line chart representing the actual development of the US imports from the PRC, as well as particular forecasts by each generated and retained network.
The graph shows that all neural networks forecast the development of the imports in a very similar way. At the same time, the balanced time series´ shape is very similar to the imports' actual development. All retained neural networks are again worth noting. They all forecast not only the basic direction in the development of the imports, but can also forecast local minimums and maximums.

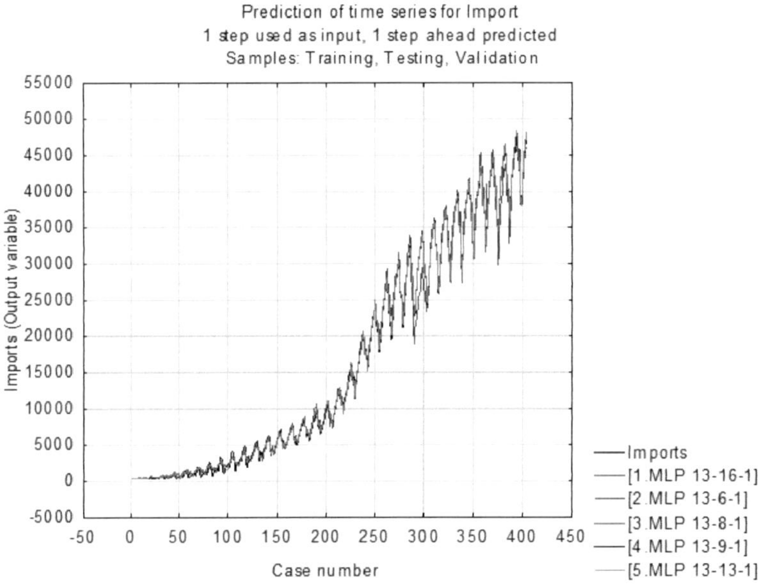


**Fig. 3.** Line graph – development of US imports from the PRC as forecast by neural networks compared to the actual imports in the period under review

Source: Own processing.

It may be seen from the above figure that all retained multilayer perceptron networks are usable.

### 3.3 Comparison of A and B results

All generated and retained artificial neural networks were able to match the discussed time series – the US imports from the PRC. Having compared the correlation coefficients (Tables 3 and 6), one may observe higher performance in relation to the B variant, i.e. retained MLP networks (when an additional categorical variable is included). This is also reflected in the evaluation of basic statistics of the forecasts, balanced time series and in Tables 4 and 7. The MLP networks, or their balanced time series, show less differences in the training, test and validation data sets than in the RBF networks (without an additional variable). The aforementioned is eventually confirmed by Figures 2 and 3. It is clear that only the MLP networks in the B variant are able to capture the time series according to its actual direction. All retained perceptron networks can cover the entire course of the US imports from the PRC and are applicable to forecasting the variable´s future development.

## 4 Conclusion

The objective of this paper was to propose a methodology for considering seasonal fluctuations on balancing time series by the means of artificial neural networks on the example of US imports from the People's Republic of China.

Generally, it follows that each forecast is given by a certain degree of probability which it will be fulfilled with. When forecasting future development of a certain variable, there is an attempt to estimate the variable´s development on the basis of data from previous periods under review. Although one can include most factors affecting the target variable into a given model, there is always a simplification of reality and a certain degree of probability that any scenario forecast will be fulfilled [19].

In the paper, a particular application of the same instrument with different initial assignments was compared. Although it appeared before the experiment that there was no reason to include the categorical variable to reflect seasonal fluctuations of the US imports from the PRC, the assumption was not correct. An additional variable, in the form of monthly value measurements, brought greater order and accuracy to the balanced time series.

It may be argued that the above imports may be determined on the basis of statistical, causal and intuitive methods, with the statistical methods being compared here. However, they provided only a possible framework for the development of the observed variable. Subsequently, it is important to work with information related to possible future development of the economic, political or legal environment. If one is able to forecast such development, it may then be projected into the monitored variable. At the same time, however, comes a standpoint of an evaluator – an economist, who, on the basis of his/her knowledge and experience, corrects prices determined by the framework statistical methods and specified on the basis of causal links. In this case, however, it seems possible to test forecasting only with the use of the B variant that brings a relatively high degree of accuracy. It may be stated that all retained perceptron networks are usable.

The paper´s objective was fulfilled.

## References

1. L. Yu, S. Wang, K.K. Lai, Forecasting China's Foreign Trade Volume with a Kernel-Based Hybrid Econometric-Ai Ensemble Learning Approach. *Journal of Systems Science and Complexity*, **21**(1), 1-19 (2008)
2. A.B. Bernard, J.B. Jensen, Entry, Expansion, and Intensity in the US Export Boom, 1987-1992. *Review of International Economics*, **12**(4), 662-675 (2004)
3. Z. Yu, J. Dong, Y. Feng, The Impacts of the Government Industrial Plans on China's Exports and Trade Balance. *Journal of Contemporary China*, **20**(3), 1-18 (2019)
4. T.S. Chang, A comparative study of artificial neural networks, and decision trees for digital game content stocks price prediction. *Expert Systems with Applications*, **38**(12), 14846-14851 (2011)
5. L. Yu, S. Wang, K.K. Lai, A hybrid econometric-AI ensemble learning model for Chinese foreign trade prediction. *Computational Science - ICCS 2007 Proceedings*, **4490**, pp. 106+ (2007)
6. K. Valášková, T. Klieštik, L. Švábová, P. Adamko, Financial risk measurement and prediction modelling for sustainable development of business entities using regression analysis. *Sustainability*, **10**(7), 2144 (2018)
7. V. Machová, J. Mareček, Estimation of the development of Czech Koruna to Chinese Yuan exchange rate using artificial neural networks. *SHS Web of Conferences*:

*Innovative Economic Symposium 2018: Milestones and Trends of World Economy*, **61** (2019)

8. S. Galeshchuk, The role of technological changes in foreign-exchange market inefficiency. *Commemoration of the Birth Centennial of Herbert A. Simon 1916-2016 (Nobel Prize in Economics 1978),* 27-34 (2016)
9. Q. Zhang, Prediction on China's merchandise exports based on BP neural network associated with sensitivity analysis. *Modern Computer Science and Applications*, pp. 351-356 (2016)
10. J. Horák, P. Šuleř, J. Vrbka, Comparison of neural networks and regression time series when predicting the export development from the USA to PRC. *Proceedings of 6th International Scientific Conference Contemporary Issues in Business, Management and Economics Engineering '2019*, pp. 170-180 (2019)
11. Ch. Bolton, V. Machová, M. Kováčová, K. Valášková, The power of human-machine collaboration: Artificial intelligence, business automation, and the smart economy. *Economics. Management, and Financial Markets*, **13**(4), 51-56 (2018)
12. J. Vrbka, Z. Rowland, P. Šuleř, Comparison of neural networks and regression time series in estimating the development of the EU and the PRC trade balance. *SHS Web of Conferences*: *Innovative Economic Symposium 2018: Milestones and Trends of World Economy*, **61** (2019)
13. Z. Rowland, P. Šuleř, M. Vochozka, Comparison of neural networks and regression time series in estimating the Czech Republic and China trade balance. *SHS Web of Conferences: Innovative Economic Symposium 2018: Milestones and Trends of World Economy*, **61** (2019)
14. J. Horák, T. Krulický, Comparison of exponential time series alignment and time series alignment using artificial neural networks by example of prediction of future development of stock prices of a specific company. *SHS Web of Conferences*: *Innovative Economic Symposium 2018: Milestones and Trends of World Economy*, **61** (2019)
15. P. Šuleř, Using Kohonen´s neural networks to identify the bankruptcy of enterprises: Case study based on construction companies in South Bohemian region. *Proceedings of the 5th International Conference Innovation Management, Entrepreneurship and Sustainability*, pp. 985-995 (2017)
16. T.S. Chang, A comparative study of artificial neural networks, and decision trees for digital game content stocks price prediction. *Expert Systems with Applications*, **38**(12), 14846-14851 (2011)
17. J. Weijin, X. Yuhui, A novel method for nonlinear time series forecasting of time-delay neural network. *Wuhan University Journal of Natural Sciences*, **11**(5), 1357-1361 (2006)
18. Y. Zhao, Y. Liu, X. Qiao, S. Wang, Z. Zhang, Y. Zhang, H. Li, Tracing value added in gross exports of China: Comparison with the USA, Japan, Korea, and India based on generalized LMDI. *China Economic Review*, **49**(5), 24-44 (2018)
19. M. Kováčová, T. Klieštik, Logit and Probit application for the prediction of bankruptcy in Slovak companies. *Equilibrium-Quarterly Journal of Economics and Economic Policy*, **12**(4), 775-791 (2017)

# The factors effecting entrepreneurial intention of university students: case of Mongolia

*Altanchimeg* Zanabazar[1*], *Sarantuya* Jigjiddorj[1]

[1]Management Department, Business School, National University of Mongolia Ulaanbaatar, University Street 1, Ulaanbaatar, Mongolia

**Abstract.** Small and medium sized businesses with entrepreneur orientation have been developing rapidly creating wealth for the nations and often become large global enterprises. The researches on the entrepreneur intention, entrepreneur education, attitude as well as the factors affecting to them are getting popular. Simultaneously, the study on whether the university is nurturing the entrepreneur intention, attitude or boosts students' willingness starting their own businesses has been remaining an interesting topic. The current study aimed at researching various factors affecting the entrepreneur intention, attitude of the students, subjective norms, entrepreneur education, and perceived behaviour control using sample survey involving 568 university students. The survey results demonstrated that the personal attitude had an influence on entrepreneur intention and the participants expressed their willingness starting their own businesses having an awareness on prospective challenges and opportunities. The determinants such as subjective norms and perceived behaviour control had both positive impacts on becoming an entrepreneur; however, an entrepreneur education demonstrated a slight impact in entrepreneur intention.
**Keywords:** small and medium sized businesses, entrepreneurial intention, university students, entrepreneur education

## 1 Introduction

An entrepreneurship is valued as a respectable strategy for the national development of a country and the start-up businesses have been supported and reflected in the government policies of the countries around the world. Researchers considered an entrepreneurship as a necessary component for development of country that enables achieving an adequate economic growth and serves as a driving force of the development [1].

An encouragement of entrepreneurship has numerous benefits like an increase of job opportunities, the reduction of the dependence on a single industry, an enhancement of capital flow that improves the business environment, an increase of prospective businesses, use of outsourcing to provide effective operations and development of new ideas and innovation. Moreover, it is regarded as a booster of social changes through creating new products and

* Corresponding author: sarantuyaj@gmail.com, altanchimegz@num.edu.mn

services that contribute to changing old technologies and promoting green technologies that is compatible for the millennium we live at the present time.

Venture creation is a significant source of employment for all countries regardless the development level [2]. While technology and innovation-based start-up companies have been established in recent years in Mongolia, unemployment among higher education graduates still exist and become one of the unsolved social problem. Hence, the universities and colleges' role in addressing emerging needs for improvement of entrepreneur education and teaching curriculum at colleges, preparation of human resources with entrepreneurship mind is crucial.

Objective of the current study aimed at studying correlation between entrepreneur intention and attitude, subjective norm, perceived behaviour control, and entrepreneur education.

## 2 Literature Review

The studies provided in different countries to identify the factors affecting an entrepreneurial intention.

According to the earlier studies it is frequently studied from the perspectives of personal psychological factors in relation to personal traits and behaviour, later scholars started giving a significance inclusion of other demographic factors like age, origin, religion, level of education and employment.

According to Ajzen [3] intentions to perform behaviours can be predicted and attitudes toward behaviour. Therefore, it can be concluded that personal outlook, attitude and close personal associates have a significant influence on an entrepreneur intention. An entrepreneur intention is a personal intent and an intellectual process towards to implementation of the goals [4].

Becoming an entrepreneur is a voluntary, sensible decision and the entrepreneur decision making is explained by entrepreneur intention model. Thus, an entrepreneur intention can be considered as an element of recognizing an entrepreneur behaviour performing and an intention to perform the behaviour depends on the personal attitude.

Historically, various theories and models have proposed in by different scholars like 'entrepreneur event' [5], 'implementing entrepreneur ideas' [6], "theory of planned behaviour' [3], 'an attitude approach to the prediction of entrepreneurship' [7], 'Entrepreneur intentions' [8], and " Competing model of entrepreneurial intentions" [9].

"Theory of planned behaviour" proposed by Azjen in 1991, applied in similar studies more frequently, explained thoroughly factors affecting behavioural intentions intensified researches in entrepreneur intentions and has been applied as the key theory.

Theory of planned behaviour. As stated previously the current concept is widely applied in entrepreneur intention research that is based on the concept 'theory of reasoned action' proposed by Ajzen and Madden. An intention to perform and actual performance have been considered as factors that are strongly correlated and indicates the behaviour. An intention expresses the motivation to perform the certain behaviour and a strong intention increases the probability of performing the behaviour. Three interrelated determinants are used for explaining the intention.

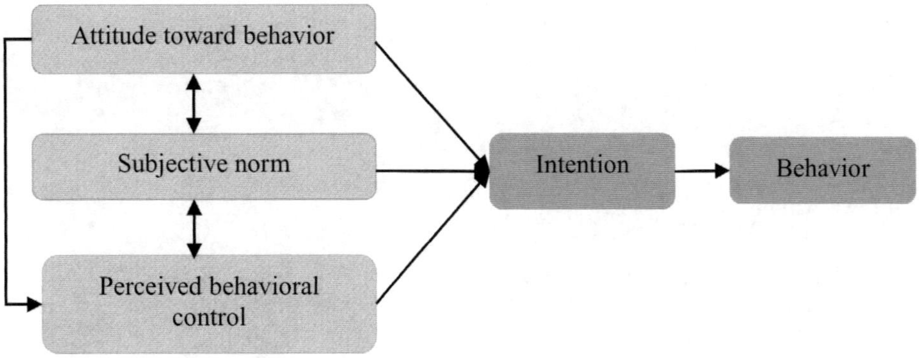


**Fig. 1.** Model by theory of planned behaviour

Source: [3].

**Attitude toward behaviour**- the degree to which performance is positively or negatively valued. The outcome of the performance can be determined by the positive or negative experience [10]. An attitude is determined by a certain outcomes, behaviors towards to a certain performance and personal skill, personality can influence it. Personality includes numerous factors like risk taking skill, and goal orientation etc,

**Subjective norm** – the state can be explained by attitude, beliefs and perceptions of people surrounding an individual. People develop beliefs on what is acceptable or not acceptable and these beliefs shape one's perception of behaviour and intention to perform or not. If the social view is positive towards to the entrepreneur intention increases or if the perception is negative the intention declines [10].

**Perceived behaviour control** – refers to the person's belief that the performance of behavior is under his control. In other words, it is a perception on whether it is easy or complicated performs certain behaviour. 'Perception' can be determined by the past experiences, ability to overcome forthcoming challenges [11]. More people get experienced and obtain essential knowledge, skills become motivated in performing the behavior.

**Entrepreneurial education** – Bakri and Mehrez [12] and del Rio-Rama et al. [13] discussed that college education played a significant influence in attitudes toward entrepreneur intention over roles models. Entrepreneurship education is also assumed enhancing among students increasing awareness of entrepreneurship as another career path to employment [14], Laspita, Patzelt, as they have a solid knowledge on business administration.

In the current research we aimed at revealing the interrelationship between entrepreneur education background and entrepreneur intention in relation to the theory of planned behaviour. Entrepreneur education supports the entrepreneurs to perform their intentions to do business and develop necessary skills [15]. According to Linan [16] an entrepreneur education increases the intent of becoming entrepreneur. Entrepreneur education enables entrepreneurs developing key skills for managing the business. In selecting the model of the current study both theory of planned behaviour by Ajzen and model of entrepreneurial intention by Shapero and Sokol are considered with inclusion of a new determinant entrepreneur education as it is demonstrated below.

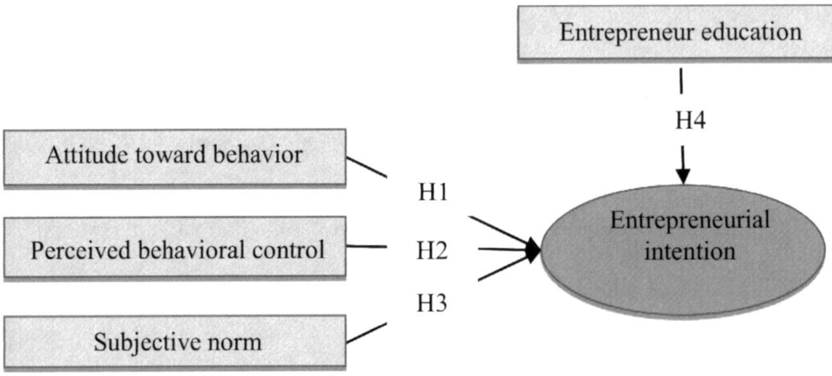


**Fig. 2.** Model for the study

Source: Authors.

**Research hypothesis:**

H1: An attitude toward behaviour positively influences to the entrepreneur intention.

H2: Subjective norm positively impacts in the entrepreneur intention.

H3: Perceived behavioural control has a positive impact to the entrepreneur intention.

H4: Entrepreneur education has a positive effect on the entrepreneur intention of the students.

## 3 Research Methodology

The research embraces a quantitative method to empirical support for the hypothesis. To study the how attitude toward behaviour, perceived behavioural control and subjective norms and entrepreneur education have an impact in entrepreneurial education we carried out survey among the Business School students who attended the subject in entrepreneurship in the academic year 2018-2019 between May-June, 2019. A total of 577 questionnaires were collected and 568 questionnaires considered valid for the analysis. The survey instrument included 22 questions that are grouped in to 22 questions along with the four questions on personal background of the respondents.

### 3.1 Background information on the survey participants

Out of 568 participants 354 or 62.3% are female and 214 or 37.7% male students studying in different years of their study. By the year of study, 40 students are studied in their first year, 186 students in their second year, 196 of them continued studying in the third year and 146 of them were the fourth-year students.

For testing relevance for factor analysis for data set, Kaiser-Meyer-Olkin (KMO) measure of sampling adequacy is used. According to KMO indicated .845 which means the size of the sample is adequate for the study. In addition, it considered essential to apply the Barlett test of Sphericity to assure that factor analysis can be applied to the data set and the results are shown below in table 1.

**Table 1.** KMO and Bartlett's Test

| | | |
|---|---|---|
| Kaiser-Meyer-Olkin Measure of Sampling Adequacy. | | 0.845 |
| Bartlett's Test of Sphericity | Approx. Chi-Square | 508.535 |
| | df | 10 |
| | Sig. | 0.000 |

Source: Authors.

Sig. value .000 demonstrated that we can apply the proposed analysis.

Reliability test. For the internal reliability of each factor Cronbach Alpha coefficient is used, which according to [17] a value greater than 0.7 to be considered acceptable. In our case, cronbach α demonstrated value of .880 and considered acceptable (table 2).

T test is applied to test whether the students' year of study or academic experience has an influence in an entrepreneur intention of the students.

**Table 2.** Questionnaire Reliability Analysis result

| Factors | Number of questions | Cronbach's α |
|---|---|---|
| Entrepreneurial intention- EI | 4 | .778 |
| Attitude toward behavior - EA | 4 | .839 |
| Subjective norm - SN | 3 | .852 |
| Perceived behavioral control - TPB | 6 | .809 |
| Entrepreneur education- EE | 5 | .810 |
| | 22 | .880 |

Source: Authors.

The test results are shown in the table 3.

**Table 3.** Influence of students' year of study in an entrepreneur intention

| | N | Mean | Std. Error | t | Sig. (2-tailed) |
|---|---|---|---|---|---|
| I year students | 40 | 3.05 | .1285 | 23.671 | .000 |
| II year students | 186 | 3.23 | .0671 | 49.426 | .000 |
| III year student | 196 | 3.72 | .096 | 63.588 | .000 |
| IV year students | 146 | 4.17 | .0604 | 66.749 | .000 |

Source: Authors.

The results show that the students who studied in senior years have an ambition of getting an entrepreneur. T test applied to identify whether the courses offered for developing entrepreneurship knowledge and skills such as 'Business and Society', 'Business Planning', Project management', and 'Small and Medium Enterprise Management' have an impact in shaping entrepreneur intention results are summarized in table 4.

**Table 4.** Correlation between the courses the students enrolled and entrepreneur intention

| | Enrolled the courses | | | Did not enroll the courses | | | t | Sig. (2-tailed) |
|---|---|---|---|---|---|---|---|---|
| | N | Mean | Std. Error | N | Mean | Std. Error | | |
| Business& Society | 463 | 3.63 | .0380 | 103 | 3.63 | .0931 | 38.776 | .000 |
| Business Planning | 101 | 3.86 | .0833 | 465 | 3.46 | .0409 | 48.774 | .000 |
| Project management | 500 | 3.86 | .0863 | 66 | 3.57 | .0395 | 45.998 | .000 |
| Small & Medium business management | 235 | 3.69 | .0520 | 331 | 3.41 | .0487 | 67.233 | .000 |

Source: Authors.

The students who completed 'Business and Society', 'Project Management', 'Small and Medium Enterprise Management' were highly motivated in becoming entrepreneurs compared to the ones who did not enrol these courses, however 'Business and Society' course had no impact in an entrepreneur intention of the students as there was no difference in the intention if they completed the course or not.

Moreover, test on whether the family members or close people engaged in the business had an influence in students' entrepreneur intention (refer to table 5).

**Table 5.** Whether the family members or close people engaged in the business have an impact in students' entrepreneur intention

| | Family members or close people engaged in business | | | Family members or close people not engaged in business | | | t | Sig. (2-tailed) |
|---|---|---|---|---|---|---|---|---|
| | N | Mean | Std. Error | N | Mean | Std. Error | | |
| EI | 384 | 3.85 | .0353 | 182 | 3.18 | .0601 | 87.466 | .000 |

Source: Authors.

The test results show that the students whose family is engaged in a business had a high intention of becoming entrepreneurs (mean value 3.85) compared to the ones who no family members engaged in the business had low intention (mean value 3.18). From the results it can be concluded that the business family background has an influence in students' entrepreneur intention. Factor analysis presented in table 6.

https://doi.org/10.1051/shsconf/20207301034

**Table 6.** Pearson Correlation Analysis result

| Correlations | | | | | | |
|---|---|---|---|---|---|---|
| | | EI | EA | SN | TPB | EE |
| EI | Pearson Correlation | 1 | | | | |
| | Sig. (2-tailed) | | | | | |
| EA | Pearson Correlation | .563** | 1 | | | |
| | Sig. (2-tailed) | .000 | | | | |
| SN | Pearson Correlation | .426** | .498** | 1 | | |
| | Sig. (2-tailed) | .000 | .000 | | | |
| TPB | Pearson Correlation | .473** | .417** | .324** | 1 | |
| | Sig. (2-tailed) | .000 | .000 | .000 | | |
| EE | Pearson Correlation | .286** | .305** | .385** | .370** | 1 |
| | Sig. (2-tailed) | .000 | .000 | .000 | .000 | |

**. Correlation is significant at the 0.01 level (2-tailed).

Source: Authors.

The results demonstrate that an attitude toward to entrepreneurial intention of business school students [r = .563, p < 0.01], perceived behavioral control [r = .473, p < 0.01], subjective norm [r = .426, p < 0.01] embrace a positive strong correlation, but entrepreneurial education [r = .286, p < 0.01] revealed a weak correlation.

In the entrepreneurial intention in the study subjective norm presented [r = .498, p < 0.01], and perceived behavioral control [r = .473, p < 0.01] showed a positive strong correlation, but an entrepreneurial education had moderate correlation.

Attitude toward to entrepreneur to subjective norm showed a strong positive correlation [r = .498, p < 0.01], but entrepreneur education [r = .385, p < 0.01], perceived behavioral control [r = .324, p < 0.01] revealed moderate correlation.

Perceived behavioral control to the attitude to entrepreneur had a positive and strong correlation [r = .417, p < 0.01], and an entrepreneur education [r = .370, p < 0.01] and subjective norm [r = .324, p < 0.01] presented a moderate correlation.

For the confirmation of the hypotheses regression analysis is carried out (refer to table 7).

**Table 7.** Regression analysis result

**Model Summary**

| Model | R | R Square | Adjusted R Square | Std. Error of the Estimate |
|---|---|---|---|---|
| 1 | .590[a] | .552 | .546 | 2.5958 |

a. Predictors: (Constant) , EE, EA, TPB, SN

**ANOVA[a]**

| Model | | Sum of Squares | df | Mean Square | F | Sig. |
|---|---|---|---|---|---|---|
| 1 | Regression | 681.542 | 4 | 170.386 | 25.287 | .000[b] |
| | Residual | 3793.577 | 563 | 6.738 | | |
| | Total | 4475.120 | 567 | | | |

a. Dependent Variable: EI

b. Predictors: (Constant), EE, EA, TPB, SN

https://doi.org/10.1051/shsconf/20207301034

**Coefficients[a]**

| Model | | Unstandardized Coefficients | | Standardized Coefficients | t | Sig. |
|---|---|---|---|---|---|---|
| | | B | Std. Error | Beta | | |
| 1 | (Constant) | 8.765 | .739 | | 11.865 | .000 |
| | EA | .553 | .042 | .512 | 5.987 | .000 |
| | SN | .331 | .053 | .327 | 2.587 | .000 |
| | TPB | .389 | .030 | .331 | 2.948 | .000 |
| | EE | .234 | .037 | .211 | 1.932 | .000 |

a. Dependent Variable: EI

Source: Authors.

As results of the analysis the correlation between entrepreneurial intention and other determinants such as attitude toward behaviour, perceived behavioural control, subjective norm and entrepreneurial education demonstrated .546%.

Attitude toward to the entrepreneur [$\beta$ = .512, $p < 0.01$] had a moderate impact, and perceived behavioural control [$\beta$ = .331, $p < 0.01$], subjective norm [$\beta$ = .327 $p < 0.01$], and entrepreneurial education [$\beta$ = .211, $p < 0.01$] demonstrated that they had a low impact (see figure 3).

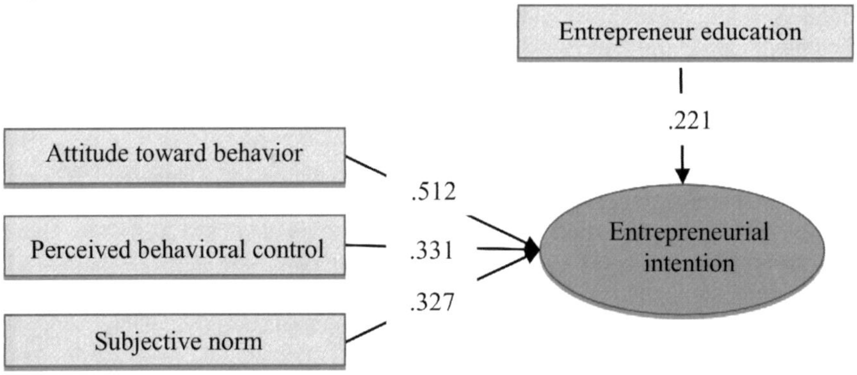


**Fig. 3.** Research result

Source: Authors.

The hypotheses are all supported as all effects of associated regulator variables (Beta coefficients) are both positive and statistically significant ($p < 0.01$).

## 4 Conclusion and Discussion

Establishing a new business venture is related with the social skills of the younger generation and their knowledge, skills and attitude are decisive in shaping future of all sectors. Accordingly, numerous studies have been conducted globally in recent years in youth entrepreneur intention, the factors influencing it, entrepreneur education and attitude.

The current study aimed at researching factors influencing in an entrepreneur intention to contribute reinforcement of application of theories and concepts in the case of business students.

An attitude toward to entrepreneur, subjective norm, perceived behavior control and entrepreneurial education are taken as independent variables, and an entrepreneur intention considered as a dependent variable and to support the proposed hypothesizes various analyses such as reliability test, correlation and regression analysis are carried out.

As correlation analysis show the all factors were correlated with the entrepreneur intention except the entrepreneurial education which indicated low correlation.

According to the results, senior year students' intention to become entrepreneurs was comparatively high. Increase in attitude toward to entrepreneur by one point resulted in the increase of entrepreneur intention by 0.512 (beta) point. In other word, the students are ready to take an opportunity and start their own businesses as they can create numerous advantages.

In overall, the students participated in the survey have a positive attitude toward the entrepreneurship.

Subjective norm also has a positive influence in entrepreneur intention and the survey results show that when family or close people's influence increases by one point, the entrepreneur intention increases by 0.327 (beta) point. The students who are from families who engaged in business had more positive attitude toward entrepreneur intention as they have support and sympathy.

The study revealed that perceived behavioral control has a positive impact in the entrepreneur intention. As the knowledge and skills in performing a private business upgrades the behavior control increases. In the survey, one-point increase of perceived behavioral control resulted in the 0.331 (beta) point in entrepreneur intention. It is clearly seen from the study that despite the potential challenges in performing business, the confidence build upon education experiences contributes to the increase of the entrepreneurial education. According to the survey, the respondent felt confident in overcoming challenges in developing their business ideas.

The survey results demonstrated that more the students complete the courses in entrepreneurship, the entrepreneur intention increases by 0.221(beta) point. Although the entrepreneur education had a slight impact to the entrepreneur education, the knowledge and education offered at the university influenced positively.

## References

1. W.J. Baumol, R.E. Litan, C.J. Schramm, *Good capitalism, bad capitalism and the economics of growth and prosperity*. New Haven, CT: Yale University Press (2007)
2. M.R. Uddin, T.K. Bose, Determinants of entrepreneurial intention of business students in Bangladesh. *International Journal of Business and Management*, **7**(24), 128-137 (2012)
3. I. Ajzen, Perceived behavioral control, self-efficacy, locus of control, and the theory of planned behavior. *Journal of Applied Social Psychology*, **32**(4), 665-683 (2002)
4. V.K. Gupta, N.M. Bhawe, The influence of proactive personality and stereotype threat on women's entrepreneurial intentions. *Journal of Leadership and Organizational Studies*, **13**, 73-85 (2007)
5. A. Shapero, L. Sokol, The Social Dimensions of Entrepreneurship. In: *C.A. Kent, D.L. Sexton, & K.H. Vesper (Eds.). Encyclopedia of Entrepreneurship. Englewood Cliffs,* NJ: Prentice-Hall, 72-90 (1982)
6. B. Bird, Implementing entrepreneurial ideas: The case for intention. *Academy of Management Review*, **1**(3), 442-453 (1988)

7. P.B. Robinson, D.V. Stimpson, J.C. Huefner, H. K. Hunt, An attitude approach to the prediction of entrepreneurship. *Entrepreneurship Theory and Practice*, 13-31 (1991)
8. N.F. Krueger, A.L. Carsrud, Entrepreneurial intentions: Applying the theory of planned behavior. *Entrepreneurship and Regional Development*, **5**, 315-330 (1993)
9. N.G. Boyd, G.S. Vozikis, The Influence of Self-efficacy on the Development of Entrepreneurial Intentions and Actions. *Entrepreneurship Theory and Practice*, **18**(4), 63-77 (1994)
10. L. Kolvereid, Prediction of employment status choice intentions. *Entrepreneurship Theory Practice,* **21**(1), 47-57 (1996)
11. C.W. Scheiner, *Fundamental Determinants of Entrepreneurial Behavior.* Wiesbaden: Gabler research (2009)
12. A. Bakri, A. Mehrez, Factors influencing entrepreneurial intentions among Arab students. *International Journal of Entrepreneurship*, **21**(3), 1-17 (2017)
13. C. del Rio-Rama et al., Entrepreneurial intentions and Entrepreneurship Education to University students in Portugal. *ETechnology, Innovation and Education Journal,* **2**(7), 1-11 (2019)
14. T.J. Bae et al., The relationship between Entrepreneurship Education and Entrepreneurial intentions: A Meta-Analytical Review. *Entrepreneurship Theory and Practice*, **38**(2), 217-254 (2019)
15. G. Gorman, D. Hanlon, W. King, Some research perspectives on entrepreneurship education and education for small business management: a ten-year literature review. *International Small Business Journal: Researching Entrepreneurship*, **15**(3), 56-77 (1997)
16. F. Linan, Intention-based models of entrepreneurship education. *Piccola Impresa / Small Business*, **2004**(3), 11-35 (2004)
17. J.C. Nunnally, *Psychometric theory.* 2nd ed. New York: McGraw-Hill (1978)

SHS Web of Conferences **73**, 01035 (2020) https://doi.org/10.1051/shsconf/20207301035
*IES2019*

# Controlling of Strategic Organizational Changes at the University

*Anna* Zotova[1*], *Valentina* Mantoulenko[1] and *Irina* Plaksina[1]

[1]Samara State University of Economics, Sovetskoi Armii Street 141, 443090, Samara, Russia

**Abstract.** Since modern organizations are characterized by a variety of goals, the definition of the effectiveness of organizational changes can be specified as a degree of results achievement (under the conditions that these results are adequate to the established and intended goals of organizational changes and the changes themselves meet certain needs of stakeholders and create conditions for constant development of an organization). This research is based on general scientific research methods: system analysis methods, scientific abstraction and generalization, comparative and categorical analysis, empirical description methods, as well as behavioral, systematic, process specifying scientific methodological approaches of the theory of management. The authors classified strategic interests of the main stakeholders in universities activities in Russian Federation; the element base and the system of interactions between them was performed in the model of interests coordination during the changes implementation process in the system of higher education and control methods when making changes were analyzed. In conclusion various forms of controlling were suggested and this variation was derived from various university stakeholders' interests.
**Key words:** strategic changes, organizational changes, universities, stakeholders, methods of controlling

## 1 Introduction

In accordance with the systems theory, the content of the results evaluation system can be formed by identifying its elements and determining links between them [1]. Obviously, the need to assess the effectiveness of organizational changes in the management process arises at least twice: at the stage of implementing organizational changes and after their completion.

An important component of this mechanism is the outline of the control system at the implementation level, consistently covering all stages of the management cycle, starting from the stage of formulation of goals and strategic planning, and ending with the stage of results evaluation and development of future recommendations. The most important functions are the control of the achievement of the set goals, the identification of deviations from the plan, the analysis of causes of these deviations and participation in the development of corrective actions. The ability of an organization to achieve its goals depends on the degree of mutual

* Corresponding author: azotova@mail.ru

integration and coordination of all elements of the considered complex mechanism of change management at the university.

The higher education is a specific economic activity studied recently more thoroughly [2, 3] and aimed at the realization of economic interests of a wide range of contact audiences. These interests determine, directly or indirectly, objectives, content and results of activities of higher education institutions. Therefore, it is necessary to assess the degree of acceptability of results of ongoing organizational changes not only from the standpoint of the organization itself, but also from the standpoint of those contact audiences whose interests led to the implementation of these changes, as well as all others whose interests may potentially be affected.

Effective implementation of changes was described in a number of works [4] in the system of higher education should take into account the mechanism of interests harmonization by carrying out changes in the higher education. Mainly the aspects of interests coordination for various types of organizations were studied by a number of researchers [4-7] If the control over the effectiveness of changes is based on indicators that assess the potential of the university in its readiness for changes and improvement of its competitiveness in the inter-university space as well as among students, it largely involves the participation of contact audiences at the stage of developing strategic and current plans, as well as in the development of targets and the vision of the university and is mandatory for the effective functioning of the educational institution. The impact of changes is controlled at the micro level (university level), and in the system of the external environment of the university (that rigidly fixes the degree of freedom by decision-making of all participants of market interactions) and is reflected at the meso-and macro-level.

## 2 Research methodology

The purpose of the study is structuring the changes in higher education into detailed typology, considering new trends in the development of higher education in the Russian Federation and in the world. The typology and substantiation of the element base and system interactions in the model of interests' coordination while implementing the changes are theoretical and based on such methods as scientific abstraction and generalization, comparative and categorical analysis.

The applied part of the research was in identification of strategic "Gap map" between the marketed system of higher education products and services and the interests of the main contact audiences. Identifying gaps in needs and the interests of key stakeholders regarding the development, provision, obtaining and using higher education was performed on the basis of empirical description methods, as well as behavioral and systematic approaches and results in analytical modules studies formed at this stage. Results obtained in the framework of the presented research modules are necessary basis for obtaining further logically linked scientific and practical results: these modules allow create a management decision support system for implementation of changes in the system of higher education carried out in order to increase its sustainability and competitiveness.

## 3 Results

During the first research phase the authors analyzed the dual character of educational services of higher education, and highlighted a few factors requiring consideration when assessing the efficiency and effectiveness of investment in higher education – both public and private:

- the result is formed both at the micro and macro levels, and the result at the micro level differs quantitatively and qualitatively from the sum of the results obtained at the micro

level, which is determined by the dual nature of the service itself, as well as the existence of external effects;
- the result should be evaluated not only in the coordinates of an economic, but also social effect;
- the result is formed on a very long horizon.

These factors should be considered when assessing the effectiveness and efficiency of measures related to organizational changes in the higher education system (systemic changes) and in the activities of individual higher education institutions (substantive changes).

Let us consider some specific features of the choice of ways to harmonize interests in the system of higher education. To do this, a matrix of ways to coordinate interests should be built, one axis of which represents the realization degree of universities' interests, the other - the realization degree of stakeholders' interests. Since the choice of a method for the coordination of interests is determined by the degree of dependence of the interacting subjects on each other, it is necessary to identify groups of universities' stakeholders by the degree of their interdependence. As a basis for such a selection can serve as a model of Aubrey L. Mendelow [8-10], where two criteria are used:

- "power": the degree of influence (high/low) on the activities of the university ("power" characterizes the ability of a stakeholder to get the desired result from the university on the basis of instruments of coercion, financial tools or emotional resources);
- "interest": the level of interest (high/low) in the results of the university activities (the interest characterizes needs of stakeholders and their dependence on the results of the university activities, as well as the desire of stakeholders to influence the university).

The Power-interest grid of Aubrey L. Mendelow allows us to distinguish four groups of stakeholders:

- group I: stakeholders with a high degree of influence and low interest;
- group II: stakeholders with a high degree of influence and high interest;
- group III: stakeholders with low level of influence and low interest;
- group IV: stakeholders with low level of influence and high interest.

Consider the distribution of stakeholders in the higher education system according to these groups:

The first group of stakeholders includes:

- the state, which is the founder (for the majority of Russian (state) universities) and the Supervisory authority (for all universities), and accordingly, has a strong impact on the activities of higher education institutions (through mechanisms of budget funding, licensing and accreditation procedures); at the same time, the level of state interest in universities should be considered as low, since the state is interested only in compliance with the law and does not depend on the results of their activities;
- state and non-state funds supporting scientific, technical and innovation activities: they have a strong impact on the activities of universities through financial instruments (grant support system) and are not directly interested in the results of their activities.

The second group of stakeholders includes:

- households (applicants and their parents): they have a strong influence on the activities of universities on the basis of financial instruments (fees for educational services, which becomes the main source of funds for universities in the conditions of reduction of budget funding from the state) and at the same time they are highly dependent on the results of educational activities of universities (interested in high quality of the educational process);
- employers: they have a strong influence on the activities of universities through financial instruments (fees for educational services, R&D results) and emotional resources (employment of university graduates), they are also strongly dependent on the results of

university activities (interested in the quality of educational services, R&D, highly qualified specialists);

- partner universities: they influence the activities of universities mainly through financial instruments and emotional resources (as part of the implementation of joint educational, innovative and other projects), they are strongly dependent on the results of this interaction (it should be noted that strong influence and high interest are observed, as a rule, only in the sphere of network interaction between universities);
- universities-competitors: they have a strong impact on the activities of universities on the basis of financial instruments (competition for budget funding, household funds, employers) and thus they are interested in the negative results of other educational institutions (unlike other groups of stakeholders universities-competitors are not interested in the results of university activities, but in their absence);
- the university management: it has a strong influence on the activities of the university through the instruments of coercion (it forms the management policy) and has a high interest in the results of university activities (increase in budget and extra-budgetary funding, compliance with the legislation, the business reputation of the university).

The third group of stakeholders includes mass media, public organizations, professional communities, regional innovation infrastructure (business incubators, etc.), university graduates, secondary schools. These groups of stakeholders are characterized by a low level of influence (mainly through emotional resources) and low interest (they do not depend directly on the activities of universities).

The fourth group of stakeholders should include employees of the university (teaching staff and other categories of employees) who have a high interest in the results of the university activities (they are interested in decent pay and working conditions, stable functioning of the educational organization), but have little leverage over the university.

Thus, based on the Mendelow's model, we can identify the most important stakeholder groups (groups II, I and IV), which shall be agreed in the first place.

The development of modern management concepts such as value-based management (VBM), M. Porter's [11] Value Chain Analysis, supply chain management (SCM) and further - entrepreneurial networks has moved away from the post-war approach to the organization as a closed system. The operation of an open system involves its interaction with the external environment; it is not required from a closed system. In the classical and post-war scientific management, organizations were usually considered as closed systems, which allowed to simplify the problem and effectively use the methods of quantitative analysis (there were no external influences). It is obvious that in reality every organization is an open system, and ignoring environmental factors can be one of the reasons for its failure. This approach formed the basis of the strategic management of the organization.

In this case, in order to assess the success of organizational changes, it should be taken into account that the management of organizational changes as a process of purposeful impact on the activities of the organization is carried out for results that the organization achieves in the external environment and to achieve which it mobilizes certain resources. Since the results of the organization's activities exist only outside its borders, in the external environment, therefore, it is the reaction of the external environment that should be considered as the basis for the final assessment of the effectiveness of organizational changes. Obviously, those organizational changes should be considered as more successful and effective ones that lead to an improvement in the position of the organization in its operation environment.

However, the task monitoring the effectiveness is carried out primarily at the micro level, i.e. monitoring the effectiveness of changes is implemented mainly at the university level. The service of providing higher education is a paid service according to the legislation of the Russian Federation. There are fundamental differences in the mechanism of providing services in the higher education system: public interests are realized through the use of the budget financing system (currently in the Russian Federation the mechanisms of state assignment and subsidies are used, the transition to the state order is discussed), private - through the mechanism of contractual financing at the expenses of individuals and legal entities. These services, provided within the framework of the state assignment on a competitive basis, are "free" for a particular consumer who does not pay for them directly, but, nevertheless, they are paid to the producer at the expense of the budget. However, the declared amount of state guarantees for higher education opportunities in a direction and level of training may deviate from the volume of demand both in one and the other direction [12]. Thus, the university itself is primarily interested in the formation of an effective system of higher education from the point of view of the rational use of both allocated budget and "contractual" resources.

Control is a legal category (activity of specially created regulatory bodies to monitor compliance with the legislation by all economic entities) and managerial category (an integral element of management at both macro-and micro-levels in order to ensure the expediency and efficiency of operations). Accordingly, there are different control forms, depending on the grounds on which it can be carried out. For the purpose of this study, the authors considered the existing forms of control from the standpoint of the implementation of the interests of different stakeholders groups and their interest level in the results of university activities and the influence degree on the interests of the universities themselves.

**Table 1.** The classification of controlling forms in accordance to stakeholders' interests grid

| Controlling forms | | Stakeholders with high level of interdependence | Stakeholders with middle level of interdependence | Stakeholders with low level of interdependence |
|---|---|---|---|---|
| **Terms of changes** | **strategic** | University authorities | State<br>State and non-profit funds | Media<br>Public organizations<br>Professional associations |
| | **tactical** | Households<br>Employers<br>Partner universities<br>Competitor universities | - | Regional innovation infrastructure<br>Graduates<br>Schools |
| | **operational** | - | Employees | |
| **Way** | **financial** | University authorities | State<br>State and non-profit funds | |
| | **administrative** | University authorities | State<br>State and non-profit funds | |
| **Functional subsystems** | **Innovative and entrepreneurial activities** | Partner universities<br>Competitor universities | Employees | Media<br>Public organizations<br>Professional associations<br>Graduates<br>Schools |
| | **R&D** | Employers<br>Partner universities | State<br>State and non-profit funds | Regional innovative infrastructure |
| | **The process of giving the educational services** | Households<br>Employers | State<br>State and non-profit funds | Schools |

| | | | | |
|---|---|---|---|---|
| Scale of control | **total** | University authorities | State | Media<br>Public organizations<br>Professional associations<br>Graduates<br>Schools |
| | **ccollegial** | Employers<br>Partner universities<br>Competitor universities | State and non-profit funds | - |
| | **individual** | Households | Employees | - |
| Terms of controlling procedures | **preliminary** | Households<br>Employers<br>University authorities | State and non-profit funds<br>Employees | Media<br>Public organizations<br>Professional associations<br>Schools |
| | **current** | Partner universities<br>Competitor universities<br>University authorities | Employees | Media<br>Public organizations<br>Professional associations<br>Regional innovation infrastructure |
| | **final** | Households<br>Employers<br>University authorities | State and non-profit funds | Media<br>Public organizations<br>Professional associations<br>Graduates<br>Schools |
| Directions of organizational sphere | **external** | Households<br>Employers<br>Partner universities<br>Competitor universities<br>University authorities | State and non-profit funds | Media<br>Public organizations<br>Professional associations<br>Graduates<br>Schools<br>Regional innovation infrastructure |
| | **internal** | University authorities | Employees | - |
| Method of obtaining and processing information | **statistical** | Households<br>Employers<br>Partner universities<br>Competitor universities<br>University authorities | State and non-profit funds | Media<br>Public organizations<br>Professional associations<br>Schools<br>Regional innovation infrastructure |
| | **calculated and analytical** | University authorities | Employees | Graduates |
| | **registration** | - | State | - |

Source: Authors.

## 4 Discussion

The transition from the conceptual level description of the assessment system of the organizational changes effectiveness to the physical level involves the formation of a set of indicators that are a means of measuring the achieved results [13]

Reducing the indicators of sustainable development of a university only to a financial and economic assessment is impossible due to, first, the dual nature of educational services producing external effects of both a private (for third parties) and public (for society as a whole) nature. Secondly, as have already been noted, the result of the activities of higher educational institutions is formed both at the micro and macro levels, where the result both quantitatively and qualitatively differs from the sum of the results obtained at the micro level and is subject to assessment in coordinates not only economic but also social effect.

## 5 Conclusion

As a result of the study, the HE changes typology was developed, taking into account new development trends of HE in the Russian Federation and in the world, based on the identification of many interconnected systemic and substantial changes and sources of their occurrence both at the level of individual universities, and the entire system of higher education. The continuous nature of changes in higher education requires developing fundamentally new approaches to monitoring and evaluating expected results of these changes. These results should consider bridging the gaps between needs (expectations)of interest groups of higher education system and university ability to satisfy these needs by providing educational services, research work, graduate training.

Based on the distribution of university contact audiences by the degree and nature of influence, the level and nature of interests the matrix was built. It shows the ways to coordinate the interests of the university and its stakeholders, which allowed to draw conclusions about the distribution of interest coordination methods among stakeholder groups. Formed approach to these choice helps to align the interests of the university with stakeholders taking into account the degree of their interdependence, the market influence and administrative mechanisms for the process of reconciling the interests.

Depending on the degree of universities adaptability to changing conditions of functioning, which is determined by such characteristics as necessity, ability, readiness for changes under the influence of a variety of objective and subjective factors, organizational changes can have different levels. The nature of change causes determines their different focus. Ultimately, this leads to a different content of organizational changes, which requires differentiated approach to managing them.

When planning changes in the university, it is necessary to take into account that the effectiveness of these changes is not only the achievement of the target indicators of the quality of educational services and the number of scientific developments (expressed in financial and quality indicators), it is the possibility of creating innovations, and the effectiveness of communication with contact audiences.

## Acknowledgment

The research is done in the frame of the state task of the Ministry of Education and Science of the Russian Federation №26.9402017/PC "Change management in high education system on the basis of sustainable development and interest agreement".

Also, the project 574060-EPP-1-2016-KZ-EPPKA2-CBHE-SP "Enhancement of Higher Education and Corporate Sectors Integration in Accordance with New Social Environment" (ENINEDU) co-funded by the Erasmus+ Program helped in analysis of modern educational trends.

## References

1. V. Mogilevskiy, *System methodology: Verbal approach*. Moscow: Economics (1999)
2. L. Armstrong, The real crisis in higher education? [online], Available at: https://www.changinghighereducation.com/2011/02/three-articlesfrom-among-many-that-contain-important-questions-for-higher-education-californiavisual-effects-firms-faci.html (2011)
3. J. Salmi, Global View on Tertiary Education. Welcome to the University of the Future [online], Available at: http://tertiaryeducation.org (2013)
4. W. Barnet, G. Carroll, Modelling internal organizational change. *Annual review of*

*Sociology*, **21**(1), 217-236 (1995)

5. J. Storberg-Walker, R. Torraco, Change and Higher Education: A Multidisciplinary Approach. *Academy of Human Resource Development International Conference (AHRD) Proceeding,* pp. 811-818 (2004)
6. G. Halasz, Organisational Change and Development in Higher Education. *Higher Education Management and Development - Compendium for Managers / Eds. J. Huisman*, pp. 51-63 (2010)
7. R. Torraco, Organization Development and Change in Universities: Implications for research and practice. *Advances in Developing Human Resources*, **7** (3), 422-437 (2005)
8. A. Mendelow, Stakeholder Mapping. *Proceedings of the 2nd International Conference on Information Systems* (1991)
9. N. Botten, *Management accounting - business strategy*. CIMA Learning System 2007. Elsevier Science (2006)
10. L. Bourne, *Stakeholder relationship management: A maturity model for organizational implementation*. Gower Publishing (2012)
11. M.E. Porter, Towards a dynamic theory of strategy. *Strategic Management Journal*, **12,** 95-117 (1991)
12. V. Ziryanov, National education system in Russia: Problems and perspectives in the context of reforms and globalization. *Moscow University Vestnik. Series Sociology and Politology,* **3** *(*2015)
13. V. Varfalovskaya, The forming of assessment system of organizational changes efficiency. *Young Scientist,* **8**, 87-90 (2012)

# Human Resource Management in SMEs in the Czech Republic – Specifics and Key Requirements

*Zdeněk* Caha[1*], *Jan* Urban[1]

[1]Institute of Technology and Business in České Budějovice, Faculty of Corporate Strategy, Okružní 517/10, 37001 České Budějovice, Czech Republic

**Abstract.** A frequent problem regarding human resources management in SMEs is that established rules are not followed. This is because managers either do not fully understand the rules or the importance thereof, or because of a lack of professional support or rigorous control by a human resources department. Another reason is that not all SMEs have personnel specialists or HR managers at their disposal, as a consequence of which the control and advisory role falls under the responsibility of line management. The solution to this conundrum may be the adoption of a personnel manual, a document that summarises best practice regarding human resources management based on the experience of professionally managed and financially prosperous small companies. This can be understood to represent a user-friendly methodological guide for the successful implementation of personnel management activities in SMEs by line managers, a guide which includes clearly defined methodological, control and advisory tasks and functions.
**Key words:** human resource, management, requirements, SMEs

## 1 Introduction

Personnel management processes in SMEs are either performed formally, but inconsistently [1-3], and often subject to the individual approach of a line manager, or not performed adequately at all [4-7]. The potential to exploit personnel management as a tool with which to support corporate performance is therefore lacking [8], as a result of which its role and control functions within an organisation do not correspond to the role of other staff functions, such as financial management.

This fact, which is supported by practical/empirical observations, became the initial hypothesis for this article, which was later verified on the basis of the results of a questionnaire survey.

The solution to this conundrum may be the adoption of a "personnel manual" [9-11], a document that summarises best practice regarding human resources management based on the experience of professionally managed and financially prosperous small companies. This can be understood to represent a user-friendly (certified) methodological guide to the

* Corresponding author: caha@mail.vstecb.cz

successful implementation of personnel management activities in SMEs, a guide which includes clearly defined methodological, control and advisory tasks and functions.

People are a key component of a successful business operation, and therefore the role of human resources management in each organization is to ensure the highest possible efficiency within the human potential usage [12]. Additionally, the importance of human capital is also discussed in a number of studies, e.g. [13, 14], which emphasize the necessity of developing entrepreneurial and managerial competencies for university students as potential future managers.

The principles of personnel management set out in such a manual should be sufficiently simple and user-friendly, but at the same time unambiguous and binding for all line managers within an organisation. Its contents must not only correspond to the general principles of professional human resources management, but also to the goals, current situation and main priorities of the organisation, as dictated by its personnel and business strategies.

The specific conditions an SME operates under means that it is highly recommended that such a personnel manual be created at all times in cooperation with line management and personnel management. Nevertheless, within this context, appropriate management training may also be needed to guarantee and/or support its role.

## 2 Data and Methods

To verify the hypothesis concerning the specifics of human resources management in SMEs and the requirements for improvement, a questionnaire survey was conducted among 220 SMEs in the Czech Republic. The questionnaire consisted of 30 questions on personnel management as a whole (its position and role within the organisation), personnel costs, the recruitment and selection of employees, employee motivation, employee education and development and the implementation of corporate culture and ethics.

The questions focused mainly on how the company created its personnel management methods/procedures for individual areas, to what extent these methods/procedures are uniform and the importance of the role of personnel management (and the individual areas it covers) within the company.

In order to better explain any possible differences, additional information on the participating SMEs was collated, namely legal status, respectively ownership structure, the market in which the company operates, its size, financial success, including the outlook for the future, the development of average personnel/labour costs and business plans for the future. The survey also sought to establish whether the companies have a personnel department or not.

The data were subsequently subjected to single and multiple statistical analyses.

## 3 Results

The results of the research revealed that just over one-third of the SMEs do not have uniform personnel management tools in place, or if they do, these tools are not used consistently. It was also found that this fact is linked to the size of the company and the presence or absence of a personnel department. In addition, in two-thirds of companies, it was found that the principles underpinning personnel management are set by top management and in the remaining one-third in conjunction with other managers or senior employees.

According to the research, SMEs have problems hiring new employees, especially qualified personnel (one-third of them). However, less than one-third indicated that these problems apply to all employees.

Just under two-thirds of the SMEs reported that their staff turnover rate was below 5%, with only 11% stating that it was above 10%. This result indicates that staff stability is not an issue at this time. However, the increasing problems associated with the retention of qualified employees may be a more worrying trend, as reported by more than one-third of the participating companies.

The findings on the incentive effect of employees´ variable wage component, which less than a quarter of the companies considered insufficient, is also linked to the previous result. The results concerning non-financial motivation revealed marked differences between companies, with less than one-third of the SMEs viewing their work as being of special interest or prestigious, and over 30% seeing no room for such considerations.

With regards to perceptions on the (further) education needs of staff, relatively large differences were found among the SMEs, with just under 60% seeing a need and almost 30% not perceiving any need at all.

The proportion of employees that undergo thorough initial training was found to be relatively high in the participating SMEs (68%), whereas only 20% of the companies indicated that such training is limited to certain categories of employees. The further education needs of employees were found to be organised randomly in about 50% of the companies, with only a third of them indicating that this is organised according to a plan.

The surveyed SMEs regard a code of ethics as a crucial tool for ethical management. However, only 27% of the companies indicated that they actively work with one, another 27% indicated they have it, but don't apply it much, and another third that they don´t have it and are not considering the creation of one.

### 3.1 Principles underpinning a personnel management manual for SMEs

On the basis of the results obtained through the questionnaire, a personnel management manual for SMEs should clearly state that:

- professional performance of personnel management activities serve to increase the performance of SMEs;
- performance of personnel-management activities forms part of the duties of all executives and is one of the criteria for their evaluation and remuneration;
- personnel management activities are carried out in cooperation with managers and the personnel department;
- personnel management activities fall under the responsibility of individual managers, which cannot be delegated to subordinates.

At the same time, it should focus specifically on the main areas of personnel management, i.e. on methods for the planning and management of personnel costs, including methods for the continuous analysis of the causes of deviations between planned and actual costs; recruitment and selection methods and tools suitable for SMEs, both in terms of cost and professional assessment, and where appropriate, the adaptation of candidates and/or staff/ new employees; the potential introduction of more efficient tools for motivating staff and rewarding performance, including non-financial motivational tools that will reduce staff turnover (improve staff stability); methods for identifying the training needs of staff, as well as methods for increasing the efficiency and effectiveness of corporate training, including its organisation; the rules and tools for ethical management in SMEs, including the drafting and composition of a code of ethics to underpin such procedures.

The following rules covering the main HR management areas can serve as a brief example.

### 3.1.1 Employee selection and hiring

The rules in a personnel manual for hiring new employees should serve primarily to professionalise the selection and reduction of unwanted staff departures. Once again, the basic principles should be summarised in simple, clear rules, for example, as follows:

- recruitment is based on a uniform recruitment process (methodology); the basis of the assessment of candidates is an interview conducted on the basis of a single format (structured interview forms), focusing on the criteria set out in the job description. The results of the individual interviews are recorded for the final evaluation of the selected candidates;
- the assessment of candidates is carried out by the chief executive officer or a senior manager in conjunction with a human resources officer;
- each assessor has the right to comment negatively on a candidate's admission; in such cases the candidate should not be accepted;
- in the case of a large number of candidates, especially for managerial and commercial positions, the selection process is carried out using the assessment centre method, with the participation of a senior manager and a human resources officer;
- the strengths and weaknesses of a candidate identified during the recruitment process (as reflected in the candidate's assessment sheet) serve as the basis for the development of an individual adaptation plan for the newly recruited employee;
- the HR department must prepare an annual report for the management of the organisation on the causes behind the departure of employees and the level(s) of undesirable staff turnover in individual departments/sections;
- the level of undesirable staff turnover in individual departments/sections of the organization is an important criterion that is taken into consideration within the framework of management´s regular evaluations.

### 3.1.2 Employee remuneration and motivation

The aim of the personnel manual in this area is to anchor the basic rules in order to produce cost-effective remuneration that contributes to an employee´s motivation, performance and job satisfaction. These rules may vary considerably from one organisation to another, but the following principles can be given as examples:

- the remuneration is based on a fixed wage component corresponding to the tasks and qualifications set out in the job description. The share of the fixed wage component for workers and those in technical-economic positions should not fall below 70% of the total wage, for specialised and managerial positions (not in top management) below 50% of the total wage;
- performance rewards (bonuses) are provided on the basis of a regular annual performance evaluation according to the criteria set for each job group. Performance remuneration for workers and those in technical-economic positions is provided up to an amount equivalent to a maximum of 15%, resp. 20% of their basic wage, for specialists and managers (not in top management) up to a maximum of 30% of their basic wage;
- extra ordinary performance, in particular the achievement of a specific objective, is remunerated through extraordinary remuneration;
- employee benefits are used to increase the motivation and job satisfaction of employees and are determined differently for individual groups (levels) or positions;

- intangible forms of remuneration for excellent performance (work success, initiatives, savings, improvement proposals) are used to motivate employees. These include verbal praise, praise at a meeting, a thank-you letter, company diploma, presentation in a company magazine, etc.

### 3.1.3. Employee development

The principles for employee development can be divided according to the nature of the reasons and needs for education. For example, the education may be required under legislation, as a result of amendments to laws and decrees, or may concern technical education, education aimed at acquiring and developing managerial, business and/or language skills, or may involve participation in professional conferences. The main principles for employee development can be summarised as follows:

- the basis for effective training events, especially managerial training events, is annual planning, ex-post evaluation and support for the use of the newly acquired knowledge;
- the training plan is prepared by the HR department based on managers' suggestions;
- the precondition for sending an employee to a training event is the agreed objective for the event, as determined by a senior manager, and the agreed success criteria between the employee and their supervisor;
- the evaluation of the training event takes place immediately after the event. This focuses on quickly capturing the main findings of the event, which, if applicable, provides valuable information for the analysis of the causes of its failure to achieve. The evaluation of educational event after one month, respectively three months by senior manager is very important for assessing the effectiveness of the educational event and internal information transfer. The appointment of internal lecturers for individual areas of corporate training will be carried out by the personnel management of the organisation on the basis of proposals put forward by top management.

### 3.1.4. Employee development

The creation and application of a code of ethics, i.e. a set of values, principles and rules by which an organisation abides, is an effective tool with which to enforce them. The development of a code of ethics should be based primarily on the following steps:

- the identification of key ethical issues or uncertainties faced by employees in different business areas;
- a review of corporate or managerial practices with respect to the implications thereof for employee ethics. An example could be to verify that the company's remuneration system does not encourage unethical behaviour towards customers; and
- an analysis of other factors or pressures that may lead to the unethical behaviour of managers and employees.

The incorporation of ethical standards into an organisation's practices relies on a number of tools and measures. The main ones are:

- the publication of a code of ethics;
- workshops that highlight the application of ethical principles and standards in the day-to-day decision-making of managers and/or other employees. A part of these workshops should be dedicated to emphasising the importance of ethical leadership, i.e. examples of managers adhering to the principles enshrined in a code of ethics, which is usually one of the most effective tools for the implementation of a code of

ethics. The 360-degree evaluation method can be used to verify how such examples work;
- training programs that inform employees of the importance of and the principles enshrined in the code of ethics, including explanations of ethical standards and dilemmas, as well as of the disciplinary consequences of failing to adhere to or violating the provisions therein;
- the establishment of a corporate hotline to offer assistance or advice to employees who find themselves in ethically complicated or contentious situations. These are measures that mainly affect larger companies;
- the projection of the principles enshrined in the code of ethics into the personnel policies and personnel management tools (guidelines and rules) of an organisation, especially with regards to recruitment, remuneration and the evaluation and promotion of employees.

By implementing such measures, a code of ethics, respectively the observance there of, becomes a key criterion for the continuous and regular evaluation of employees as well as for the implementation of other personnel activities and decisions.

## 4 Conclusions and recommendations

Personnel management processes in SMEs are either performed formally, but inconsistently, often subject to the individual approach of a line manager, or not performed adequately at all. The solution to this conundrum may be the adoption of a "personnel manual", a certified document that summarizes best practice regarding human resources management based on the experience of professionally managed and financially prosperous small companies.

Based on the above findings, a certified personnel management methodology for SMEs should, subject to company size, focus on and bind executives to basic rules, procedures and methods for individual personnel management processes. The general principles that underpin personnel management activities should reflect the mission and basic rules that apply to all personnel processes/activities in an organization. They should clearly state that the:
- professional performance of personnel management activities serve to increase the performance of SMEs;
- performance of personnel-management activities form part of the duties of all executives and is one of the criteria for their evaluation and remuneration;
- personnel management activities are carried out in cooperation with managers and the personnel department;
- personnel management activities fall under the responsibility of individual managers, which cannot be delegated to subordinates.

## Acknowledgements

This article was supported by the project "Stabilization and development of SME in rural areas", reg. No. TL01000349, the TACR Éta programme.

## References

1. R.E. Lerchster, M. Ukowitz, Strategy and organisational development in small and medium-sized enterprises – shifting from abstraction to practical work. *Gruppendynamik und Organisationsberatung*, **40**(1), 69-84 (2009)

2. L.E.S. Vega, O. Gonzáles-Morales, L. Feliciano García, Perception of entrepreneurs about relevant competences and characteristics for employment. *Revista Espanola de Orientacion y Psicopedagogia*, **27**(1), 29-46 (2016)
3. D.T. Matt, E. Rauch, P. Dallasega, Knowledge work and knowledge management in small and medium sized engineer-to-order enterprises. *2014 International conference on production research – regional conference Africa, Europe and the Middle East and 3rd international conference on quality and innovation in engineering and management (ICPR-AEM 2014)*, pp. 316-321 (2014)
4. R. Woll, S. Peplowsky, F. Steinberg, P. Tursch, Six Sigma education for small and medium sized enterprises. *11th IMEKO TC14 Symposium on Laser Metrology for Precision Measurement and Inspection in Industry, LMPMI 2014*, pp. 345-348 (2014)
5. K. Zimermanová, Selected actual aspects of employees remuneration in small and medium-sized companies. *Ekonomie a Management*, **13**(3), 33-44 (2010)
6. N. Ruettgens, K. Mayrberger, Holistic systematisation and individualisation of on-the-job training in small and medium sized enterprises with the help of the demonstrator essyst education support system. INTED2017: 11th International Technology, Education and Development Conference, 4446–4452, (2017)
7. A. Hilbig, A. Proske, G. Damnik, F. Faselt, H. Korndle, Designing workplace learning and knowledge exchange: A postgraduate training program for professionals in SME. *CSEDU 2013 – Proceedings of the 5th International Conference on Computer Supported Education 2013*, pp. 635-638 (2013)
8. J.A. Comesana-Benavides, J.C. Prado-Prado, Implementing Key Performance Indicators in Small and Medium-sized Enterprises through personnel participation. A case study. *Direccion y Organizacion,* **48**, 17-22 (2012)
9. C. Ipsen, L. Gish, Organizational-level interventions in small and medium-sized enterprises: Enabling and inhibiting factors in the PoWRS program. *Safety Science*, **71**(1), 264-274 (2015)
10. H.C. Kim, J.Y. Min, K.B. Min, S.G. Park, Job Strain and the Risk for Occupational Injury in Small- to Medium-Sized Manufacturing Enterprises: A Prospective Study of 1,209 Korean Employees. *American journal of industrial medicine*, **52**(4), 322-330 (2009).
11. Y. Li, T. Li, H.F. Zhao, H.R. Dong, W.W. Jiao, Back Propagation Employee Selection Model for Small and Medium-sized High-tech Enterprises. *Proceedings of 2009 International Conference of Management Science and Information System*, **1-4**, pp. 1557–1560 (2009)
12. K. Haviernikova, M. Kordos, The Perception of Selected Risks Focused on Human Resources as Results of Cluster Cooperation: The Comparison Between Reality and Expectations of Small and Medium Enterprises. *Proceedings of Third International Conference on Economic and Business Management (FEBM 2018)*, pp. 437-440 (2018).
13. I. Kmecová, D. Smetanová, Business Aptitude of University Students and Its Analysis. In Khalid S. Soliman. *Proceedings of the 30th International Business Information Management Association Conference Vision 2020: Sustainable Economic development, Innovation Management, and Global Growth*, pp. 683-690 (2017).
14. I. Kmecová, D. Smetanová, Investigating the Dependencies of Business Potential of University Students. *Proceedings of the 30th International Business Information Management Association Conference Vision 2020: Sustainable Economic development, Innovation Management, and Global Growth*, pp. 733-739 (2017)

# Innovation activities of Czech businesses: differences between urban and rural businesses

*Iveta* Kmecová[1*], *Marek* Vokoun[1]

[1]The Institute of Technology and Business in České Budějovice, Faculty of Corporate Strategy, Okružní 517/10, 37001 České Budějovice, Czech Republic

**Abstract.** Innovative activities of small and medium-size enterprises (SMEs) in rural areas and in Czech economy are relatively unexplored areas. Socioeconomic and environmental contributions of SMEs activities can be gained not only by means of ecological innovative activities of rural SMEs. Rural areas are also a suitable localization strategy for obtaining public money from subsidies, as there is no great direct competition and lower costs. SMEs economic contributions are irreplaceable both for urban and rural areas. Socioeconomic impact of ecological innovations for urban and rural areas is connected with the advantages of circular economy, such as reduced energy consumption, less pollution and waste, and well-organized product recycling options after use. The hypotheses are focused on the localization of businesses into rural areas, size of enterprises introducing innovations in the market, and ecological product and processes innovations. Estimation is based on 4 innovation phases as in the case of the general innovation business process. The 2014 data set is provided by the Czech Statistical Office contains the Community Innovation Survey. The results for 2014 indicate that localization does not affect the first phases of the innovation process, and that the innovative activities increase the costs of the R&D activities.
**Keywords:** Rural areas, sustainable innovation, innovative approach, socioeconomic development

## 1 Introduction

Small and medium-size enterprises (SMEs) in rural areas are more likely to use Open Innovation Framework for obtaining and application of technologies. Internal research and development activities focused on creating new products are traditionally associated with large enterprises [1]. Rural area is more often the domain of small and medium-size enterprises, which cooperate on innovations to a large extent, enter open innovation projects, and acquire and purchase predominantly external knowledge and technologies.

The research which includes innovative activities of small and medium-size enterprises is quite extensive and shows ambiguous results in the area of rural development and innovative activities. The results indicate that the historical and cultural context of the rural development and national innovation policy [2].

* Corresponding author: kmecova@mail.vstecb.cz

There are many theoretical frameworks for SMEs development, where the new strategies are created according to large manufacturing enterprises (see [3, p. 899]. The product portfolio of SMEs is usually not as large as that of large companies; SMEs also tend to fail in strategic planning, although they are more innovation-intensive compared to large enterprises [4].

The consequences of failing strategies are more likely to appear for SMEs in urban areas than for those in urban areas, since there are more manufacturing companies that kind of "mirror" each other. Sustainability ideas and strategies can help enterprises to start and develop their own continuous innovation projects. However, there is still a problem unsolved: How to persuade strategic managers to adopt general academic sustainability frameworks and consider also ecologic issues?

Eco-innovators are considered long-term sustainable entrepreneurs since they contribute significantly to solving environmental and social problems. Business is associated with profit maximizing, where a part of it is used for the company social responsibility policies (CSR). Such a sustainable entrepreneur has own ecological and social preferences [5].

Hallstedt, Thompson and Lindahl [6] propose key elements for the perspective of strategic sustainability leading to product ecological innovation. The main element is the commitment of the managers. The mission of the enterprise must consider sustainability and environmental care. Another issue to take into account is the supply chain sustainability, building long-term company image as a sustainable company, etc. The final part deals with the management audit, that is, evaluation, long term plans and reverse planning. This is hard to apply in rural sector, since the research describes 6 large European manufacturing product innovators. However, the idea of joining a sustainable supply chain is applicable for rural sector and ecological innovations.

The quality of the network in the territory is crucial for stimulating innovation activities. The network of suppliers, universities, and other organizations is important for dissemination of ideas and technologies of sustainability [7, 8]. However, there can be a problem with the fact that the networks are to large extent a product of market mechanism where competition is monitored, and the business strategies depend on costs and benefits. If the sustainability ideas are too costly for the members of the network, they will not be disseminated. Public support can be a solution that is provided either in a traditional way or by means of experimental and "fail-proof" adaptive urban design and pilot testing [9].

## 2 Data and methods

Czech innovation surveys (CSI) in the Czech Republic in 2014 contain issues concerning innovations with environmental benefits for an enterprise or final product user, and the information of the enterprise localization and its size [10]. Apart from the innovation activities discussed, effective and systematic corporate education (as part of the overall company strategy) also appears to play a certain role affecting sustainability and competitiveness in the present economy of knowledge [11, 12].

The dependent variables (Table 1) describe innovative activities of enterprises. Innovative variable represents all companies with zero research and development costs (that is, internal research and development expenditures, obtaining external knowledge, purchasing machinery, equipment, software, and facilities, trainings focused on innovation costs and costs related to introducing innovations in the market).

This paper deals with small and medium enterprises (SMEs). Size is determined only by the number of full-time employees. Urban space is determined by the 3-digit NUTS code (criterium was based on districts cities with less than 25,000 inhabitants).

**Table 1.** Dependent variables

| Variable | Obs. | Mean Value | Standard Deviation | Min | Max |
|---|---|---|---|---|---|
| R&D innovator (%) | 5198 | 0.40 | 0.49 | 0 | 1 |
| New-to-the-market innovator (%) | 1761 | 0.55 | 0.49 | 0 | 1 |
| R&D per one employee (1,000 CZK) | 5198 | 65.97 | 261.22 | 0 | 7018 |
| Environmental innovation (%) | 5198 | 0.36 | 0.48 | 0 | 1 |
| Rural enterprise (%) | 5198 | 0.19 | 0.39 | 0 | 1 |

Source: Own processing, based on CIS data.

The quality of innovation can be distinguished, but is based on self-reporting. There are three possible innovator variables. The most restrictive is the definition of a patent, which consists of observations by businesses that have applied for a patent at the European Patent Office. Innovators who have introduced a new or significantly improved product to the market before the competition are called New-to-the-market innovators. Innovators who have introduced a new or significantly improved product that is already available on the market are more like "imitators" and are adapting to the technological boundary of the industry, the so-called new-to-the-firm innovator.

**Table 2.** Summary statistics of eco-innovation activities

| Environmental benefits obtained within your enterprise | Number of obs. | Mean Value | Standard Deviation |
|---|---|---|---|
| Reduced material or water use per unit of output | 5198 | 0.15 | 0.36 |
| Reduced energy use or CO2 'footprint' (reduce total CO2 production) | 5198 | 0.20 | 0.40 |
| Reduced air, water, noise or soil pollution | 5198 | 0.15 | 0.35 |
| Replaced a share of materials with less polluting or hazardous substitutes | 5198 | 0.10 | 0.30 |
| Recycled waste, water, or materials for own use or sale | 5198 | 0.18 | 0.39 |
| Environmental benefits obtained during the consumption or use of a good or service by the end user | Number of obs. | Mean Value | Standard Deviation |
| Reduced energy use or CO2 'footprint' | 5198 | 0.15 | 0.35 |
| Reduced air, water, noise or soil pollution | 5198 | 0.11 | 0.31 |
| Facilitated recycling of product after use | 5198 | 0.10 | 0.30 |

Source: Own processing.

Ecological innovative activities are dominated by innovations focused on the production process in the enterprise, namely Reduced energy use or CO2 'footprint' (reduce total CO2 production), Recycled waste, water, or materials for own use or sale and Reduced material or water use per unit of output. This results in eco-friendly production, which brings benefits in the form of Reduced energy use or CO2 'footprint' (Table 2) for the consumer and the environment.

**Table 3:** Estimation method – based on Heckman procedure

| | |
|---|---|
| Innovation decision ($r_i^*$) | $\begin{cases} r_i^* = 1\ if\ r_i = (X_{1i}\beta_1 + \varepsilon_{i_1}) > 0 \\ r_i^* = 0 \quad if\ r_i \leq 0 \end{cases}$ |
| R&D intensity variable ($k_i^*$) | $k_i^* = \ln(k_i)\,|(r_i > 0) = X_{2i}\beta_2 + \varepsilon_{i_2}$ |

Source: Own processing.

The vector of parameters to be estimated is denoted by βn (with n = 1 and 2). In case of Probit probability, marginal effects at means are reported. The first innovation-decision equation (ri*) accounts for selection into all R&D activities, it deals with the probability of an enterprise i to engage in non-zero R&D. This is specified as a Probit model, i.e. P(ri*>0) = Φ(X1iβ1), where ri* equals 1 if enterprise i has non-zero total R&D expenditures. Cluster robust standard errors were used. In the next equation only non-zero R&D expenditures innovators are considered. The second linear equation (ki*) describes the log of total R&D expenditures to the number of employees in enterprise i. This equation is uniquely dependent on market orientation and public funding variables (Table 3). In relation to that, the calculations were based on the so-called Heckamn coorection (procedure) which is possible to use when modifications of non-randomly selected units should occur.

## 3 Results

There is no evidence about different patterns of general R&D activities between the rural and urban space at the beginning of the innovation process (table 4). The decision to innovate and innovation intensity of rural enterprises is similar to urban enterprises. In other words, there are more important variables. The decision to innovate (non-zero R&D expenditures, no restrictions about the innovation level) is positively dependent on market orientation and size. SMEs are less likely to engage innovation projects regardless of the localization. On the other hand, they are more intensive than large companies in terms of R&D spending per employee.

Eco-innovation activities contribute to the higher innovation intensity. Intuitively, the public funding (central government and EU), the high technological and knowledge level contribute to the higher R&D intensity. Interestingly, prestigious Horizon 2020 programs did not contribute to the higher R&D intensity. Cooperation increased the intensity as well as foreign ownership.

**Table 4:** Innovation decision and R&D intensity

| Varaibles | Innovator (0,1) | R&D exp. per employee (ln) |
|---|---|---|
| Number of employees (ln) | 0.230*** | -0.362*** |
| | (0.02) | (0.06) |
| Rural company (<25 thousand) | -0.045 | -0.103 |
| | (0.05) | (0.10) |
| Eco-innovation activity | | 0.365*** |
| | | (0.08) |
| Part of a group | 0.206*** | 0.074 |
| | (0.04) | (0.11) |
| Technological level | 0.176*** | 0.007 |
| Low-Medium | (0.06) | (0.13) |
| Technological level | 0.432*** | 0.236 |
| Medium | (0.06) | (0.15) |
| Technological level | 0.375*** | 0.620*** |
| High | (0.11) | (0.22) |
| Knowledge level | 0.071 | 0.362*** |
| High | (0.05) | (0.11) |
| Foreign ownership | | 0.300*** |
| | | (0.10) |
| Funding (0/1) | | -0.036 |
| Local or regional authorities | | (0.13) |
| Funding (0/1) | | 0.714*** |
| Central government | | (0.09) |
| Funding (0/1) | | 0.718*** |
| The European Union | | (0.10) |
| Funding (0/1) | | 0.298 |
| Horizon 2020 | | (0.18) |
| Co-operated on innovation activities | | 0.471*** |
| | | (0.08) |
| Largest market | 0.447*** | |
| National | (0.05) | |
| Largest market | 0.299*** | |
| Europe | (0.05) | |
| Largest market | 0.402*** | |
| World/Other | (0.09) | |
| Mill's ratio (hazard) | | -0.259 |
| | | (0.32) |
| Constant | -1.656*** | 4.355*** |
| | (0.07) | (0.57) |
| Observations | 5198 | 2130 |
| Adjusted $R^2$ | | 0.145 |

* $p < 0.10$, ** $p < 0.05$, *** $p < 0.01$

Source: Own processing.

## 4 Conclusion

Non-urban localization is an innovation friendly strategy. Results suggest that innovation networks are available for companies located in the rural areas. This paper explored general innovation activity in terms of non-zero R&D expenditures as the criteria for being an innovator. Higher-level innovation activities are more likely to differ, yet this is a matter of further research.

Additionally, it may be argued that SMEs are more active in terms of R&D expenditures per one employee. Eco-innovation activities aiming at new products and production processes contributed to the higher R&D intensity. The behavior and innovation activities of the SMEs in the rural space can be explored and compared to the large enterprises' activities, but this is also suggested as further research, since panel data structure would be more suitable for capturing the dynamics of both areas.

## 5 Final reflection

Small and medium-size enterprises (SMEs) represent an important element in the European economy. SMEs make up a substantial part of economy as such, and contribute significantly to overall employment. The role of SMEs in the individual regions consists in certain social dimension. Maintaining traditional and local production in the region influences its character. For these enterprises it is now increasingly more and more difficult to be competitive.

The European Commission appeals to member states to support SMEs, since they can be considered a key element of the economic growth, job opportunities, and innovations.

## Acknowledgements

This article was supported by the project "Stabilization and development of SME in rural areas", reg. No. TL01000349, the TACR Éta programme.

## References

1. V. van de Vrande, J.P.J. de Jong, W. Vanhaverbeke, M. de Rochemont, Open innovation in SMEs: trends, motives and management challenges. *Technovation*, **29**(6-7), 423-437 (2008)
2. N. Rosenbusch, J. Brinckmann, A. Bausch, Is innovation always beneficial? A meta-analysis of the relationship between innovation and performance in SMEs. *Journal of Business Venturing*, **26**(4), 441-457 (2011)
3. M. Terziovski, Innovation Practice and Its Performance Implications in Small and Medium Enterprises (SMEs) in the Manufacturing Sector: A Resource-Based View. *Strategic Management Journal*, **31**(8), 892-902 (2010)
4. J.W. Brook, F. Pagnanelli, Engineering and Technology Management for Sustainable Business Development. *Journal of Engineering and Technology Management*, **34**(C), 1-8 (2014)
5. S. Schaltegger, M. Wagner, Sustainable Entrepreneurship and Sustainability Innovation: Categories and Interactions. *Business Strategy and the Environment*, **20**(4), 222-237 (2011)
6. S.I. Hallstedt, A.W. Thompson, P. Lindahl, Key elements for implementing a strategic sustainability perspective in the product innovation process. *Journal of Cleaner Production*, **51**, 277-288 (2013)
7. D. Hart, K. Bell, L. Lindenfeld, S. Jain, T. Johnson, D. Ranco, B. McGill, Strengthening the role of universities in addressing sustainability challenges: the Mitchell Center for Sustainability Solutions as an institutional experiment. *Ecology and Society,* **20**(2) (2015)

8. W. Kanda, O. Hjelm, J. Clausen, D. Bienkowska, Roles of intermediaries in supporting eco-innovation. *Journal of Cleaner Production*, **205**, 1006-1016 (2018)
9. J. Ahern, S. Cilliers, J. Niemelä, Handbook of Research on Sustainable Development and Economics. *Landscape and Urban Planning*, 125, 254-259 (2014)
10. Czech Statistical Office. *Vzory Statistických Výkazů [Patterns of Statistical Statements]* (2015)
11. Z. Caha, M. Ruschak, Differences in the Organization and Planning of Corporate Education in Medium-Sized and Large Companies in the Czech Republic. *Vision 2020: Sustainable Economic development, Innovation Management, and Global Growth*, pp. 2526-2533 (2017)
12. Z. Caha, Factors Affecting the Selection of an External Education Provider among Companies in the Czech Republic. *Innovation Management and Education Excellence through Vision 2020*, pp. 6009-6018 (2018).

# Value generators in forestry and logging

*Veronika* Machová[1*], *Jakub* Horák[2]

[1]Institute of Technology and Business, School of Expertness and Valuation, Okružní 517/10, 37001 České Budějovice, Czech Republic

[2]University of Žilina, The Faculty of Operation and Economics of Transport and Communications, Univerzitná 8215/1, 01026 Žilina, Slovakia

**Abstract.** The aim of business is the company value which depends on its ability to generate future profits. It means that all the future decisions must be[1] adopted with regard to the influence on the company value. The subject of interest in this paper concerns forestry and logging companies, particularly financial statements of companies that were active on the Czech market in 2016 and thus showed a profit. Aim of this paper is to identify particular value generators of forestry and logging companies in the Czech Republic in 2016. For this reason, the best possible methodology was developed to determine them and the objective was eventually achieved. Using artificial neural networks, a total of 14 indicators, entering a company value creation process most frequently, were identified and measured by the EVA equity indicator. The following key variables were identified as company value generators: external sources, long-term liabilities, material and energy consumption, other operating expenses and sales of goods.
**Key words:** value generators, financial statements, artificial neural networks, profit

## 1 Introduction

Companies must be strong and efficient to be able to exist for a long time because only in such a condition they are able to cope with tough and hard competition, which is the consequence of the globalization of world economy. The desire of every company is to achieve its goals and successfully fulfil its tasks [1].

Brealey, Mayers and Allen [2] claim the company value is a function of benefits which are generated by the company's assets, assets lifetime and the expected pace of growth of future benefits. Nowadays the company value is influenced by several key indicators. One of them is an expected growth [3]. Nenkov [4] states that the aim of business is the company value which depends on its ability to generate future profits. It means that all the future decisions must be adopted with regard to the influence on the company value. It is necessary for a company to concentrate mainly on its position in the market, uniqueness and quality of offered products, innovative abilities of the company employees, effective cost management, ability to effectively use available sources and abilities to search for new investment opportunities [5].

* Corresponding author: machova@mail.vstecb.cz

The term of value generator began to appear in academic literature in relation to the concept of shareholder value. Bluszcz and Kijewska [6] maintain that the value generator is a set of several basic economic values which define a company in their summary. The indicator measuring the creation of company value in an annual time horizon is an economic value added EVA [7].

Issham et al. [8] acknowledge that EVA is a useful and significant instrument for the evaluation of company performance and enables to combine such factors as accountancy, economy and market information. Morard and Balu [9] point to the fact that EVA tries to express the real economic company profit. The significant advantage of this indicator is an understandable evaluation of the economic performance of company and the consideration of the degree of investment risks [10]. The economic value added is the most progressive instrument of the measuring of company's performance based on the principle of value management [11].

Linna and Jaakkola [12] claim the methods of the complex evaluation of company are able to record the influence of all the undesirable factors, which influence the whole company. The instruments for the evaluation of company are crucial in all the processes across the whole firm. The evaluation of company is an instrument how to discover the strengths and weaknesses, opportunities and threads and how the company is successful in the market in comparison with its competitors [13].

The industry of forestry involves wood-processing industry, printing and publishing industry, pulp, paper and cardboard industries. At present time the sectors of forestry are becoming significantly more important [14]. Forestry çompanies of the Czech Republic increasingly depend on the environment. It is possible to claim Czech forestry is characterized by the biological characteristics of production. It means that there is a high dependence on the environmental conditions, long production cycles, time difference between production and relating work processes, demands for large deposits of private production and seasonality. All such factors significantly influence human resources in forestry [15]. Low productivity of this sector is a key factor for the economic backwardness of the Czech Republic. Vnoučková, Urbancová and Smolová [16] admits that the main source of the enhancement of low productivity is mainly the intensification of the activities of the sector. Burivalova, Şekercioğlu and Koh [17] define logging as the processing of woody plants for the production of timber for the purpose of the supplying of world furniture, paper, construction and other markets. However, the consequence of the rise of logging is the tendency to decrease the diversity of species.

It is possible to use bonus and bankruptcy models, economic value added EVA, etc. for the purpose of the measurement of company value. Moreover, it is possible to even use artificial neural networks the inspiration of which is biological neurons [18]. One of their abilities is learning which enables them to even avoid strongly non-linear, hidden dependencies [19]. At presence, artificial neural networks are used in many fields, e.g. agriculture, economy or healthcare, and for various applications. It is obvious that they are used in applications related to the identification of the value generators of companies [20].

Neural networks have been broadly used for the evaluation of companies' performance since 1990's. Joshi and Lam [21] examine the complex prognosis of business, the measurement of performance in particular. They have attempted to model a system of forecast with the help of regressive neural networks and the methods of time series forecasting. Their research deals with the preliminary results of the forecasting of performance in business. The authors state that neural networks are a right instrument for the dealing with such problems. Dameri, Garelli and Resta [22] use neural networks for the analysis of company performance and for the support of the policies of the development of small regions. They focus on the application of unmodified neural network in order to discover the clusters of companies in the micro-regions within urban borders [22]. Li and Wu

[23] argue that not only can the reliable prognosis of commercial performances in the shop with commodities (CVS) help with making a right decision in local circumstances but they can also help with reducing the storage costs and therefore to significantly increase the profit. The main objective of their work was to design a forecasting system based on an extended fuzzy neural network (EFNN) in order to be able to forecast the business performance of CVS. EFNN can provide complex and accurate activating functions and adapt a wider range of non-linear models. The experiments using their methods involving several benchmarking methods have proved that the designed methods have optimal accuracy when they forecast the permanent positive performance of CVS.

## 2 Data and Methods

Data for analysis are included in the Albertina database of Bisnode Czech Republic, a.s. The subject of interest in this paper concerns forestry and logging companies, particularly financial statements of companies that were active on the Czech market in 2016 and thus showed a profit. The data set involves companies classified in the CZ NACE classification of economic activities under Section A: Agriculture, forestry, fishing, Division 02 – Forestry and logging, namely sub-sections 021 – Forestry and other forestry activities and 022 – Timber harvesting. The examined data set contains records of 316 companies and their complete financial statements (except for annexes). Data from balance sheets, profit and loss statements and cash flow statements will therefore be used. The data will then be put into a table using an Excel spreadsheet. Each row of the table will represent one of the companies that are listed alphabetically. Individual columns contain information from the financial statements.

In the next step, an EVA value will be calculated for shareholders (owners) of each company in a given year of its activities on the market, i.e. an EVA equity parameter.

Firstly, it is necessary to calculate a weighted average cost of capital according to formula No. 1 [24]:

$$WACC = r_f + r_{LA} + r_{podnikatelské} + r_{FinStab} \quad (1)$$

Where the following applies:
$WACC$ = Weighted Average Cost of Capital,
$r_f$ = Risk Free,
$r_{LA}$ = A function of indicators characterizing a company size,
$r_{podnikatelské}$ = A function of indicators characterizing production performance,
$r_{FinStab}$ = A function of indicators characterizing relationships between a company's assets and sources of their coverage.

Then, it is necessary to determine an alternative cost of equity value using formula No. 2 according to Neumaierová and Neumaier [24]:

$$r_e = \frac{WACC * \frac{UZ}{A} - (1-d) * \frac{U}{BU+O} * (\frac{UZ}{A} * \frac{VK}{A})}{\frac{VK}{A}} \quad (2)$$

Where the following applies:
$r_e$ = Rate of Equity,
$WACC$ = Weighted Average Cost of Capital,
$UZ$ = Remunerated Sources (equity and interest-bearing foreign capital),
$A$ = Assets,
$VK$ = Equity,

$BU$ = Bank Loans,

$O$ = Bonds,

$\frac{U}{BU+O}$ = Interest Rate, $i$ (Interest) may also be used,

$d$ = Income Tax Rate, t (Tax) may also be used.

Economic value added for shareholders is derived from a relationship given in formula No. 3 [24]:

$$EVA\ Equity = (ROE - r_e) * VK \tag{3}$$

Where the following applies:

$ROE$ = Return on Equity.

The set will then be cleared of companies where a problem with the EVA equity calculation arises, which may be caused by unknown or zero values of items necessary for the calculation. The final (resulting) spreadsheet will be imported into the Statistica software, version 12, to determine a certain extent to which the EVA equity indicator depends on individual items in the financial statements.

Basic data statistics (correlation matrix) will be performed first. Considering a correlation between two variables, their close connection is evident. Identifying these items in the financial statements is necessary for subsequent procedures. Subsequently, tools for automated neural networks using non-linear regression (while assuming at least one hidden layer) will be employed. The EVA equity indicator shall be considered as a dependent variable in the neural networks, but variables will be selected with respect to the business theory of production factors (for more details, see [25]). The data set will then be divided into three individual sets, i.e. training, test and validation. There will be 70% of input data in the training set, whereas the remaining 30% will be divided equally into the test set and the validation set, i.e. 15% of all input data, with neural structures being created with the use of the training data set. To verify reliability of neuronal structures found, the test and validation data sets are used. In total, there will be 10,000 neural networks generated, of which the 5 best performing networks will be preserved. However, the artificial neural networking process may be terminated prematurely if neural networks being created no longer show signs of improvement. Improvement of neural structures is considered to be a state where a particular value of the sum of squares, or their disorder, decreases as the least squares method and entropy will serve as the basis in this respect. Two types of neural networks, MLP (multi-layer perceptron neural networks) and RBF (radial basic function neural networks), shall be applied. In relation to hidden and output layers, one will consider all distribution functions that the above software allows, i.e. linear function, logistic function, atanh (hyperbolic tangent) function, exponential function and sine function.

The above will then result in neural structures having abilities to predict the EVA equity based on input data, from which it will be possible to deduce its likely value. The model will only consider variables with real significance for the EVA equity indicator´s final value. A neural network that can describe the relationship as correctly as possible will be selected. Such network will have the best performance in the training, test and validation data sets, minimum errors in each data set, and the ability for a clear economic interpretation. Sensitivity analysis shall also be carried out in order to determine which variables enter the calculation while significantly affecting the result. Eventually, value generators of the forestry and logging companies will be identified.

## 3 Results

Having cleared the input data of companies without the possibility of calculating the EVA equity indicator, financial statements data of 316 forestry and logging companies in the Czech Republic remain for their analysis. Based on the methodology and the found mutual correlation of data and economic interpretation, the following independent variables were determined to enter the calculation: total assets, intangible fixed assets, inventories, long-term receivables, profit / loss of previous years, external sources, long-term liabilities, bank loans and aids, short-term liabilities, sales of goods, material and energy consumption, outputs, other operating expenses and financial result.

Table 1 shows the top five generated and preserved neural networks

**Table 1**: Preserved neural structures

| Network Name | Training Perform. | Test Perform. | Validation Perform. | Training Error | Test Error | Validation Error | Training Algorithm | Error Function | Hidden Layer Activation | Output Activation Function |
|---|---|---|---|---|---|---|---|---|---|---|
| MLP 14-26-1 | 0.983301 | 0.651976 | 0.940207 | 9250127 | 2965921 | 17553151 | BFGS 50 | Sum of squares | Hyperbolic tangent | Hyperbolic tangent |
| MLP 14-7-1 | 0.986775 | 0.805951 | 0.946692 | 7337913 | 1158857 | 5330874 | BFGS 102 | Sum of squares | Logistic | Logistic |
| MLP 14-6-1 | 0.988330 | 0.810602 | 0.954352 | 6480130 | 2376759 | 6183490 | BFGS 78 | Sum of squares | Hyperbolic tangent | Logistic |
| MLP 14-6-1 | 0.982432 | 0.519603 | 0.945845 | 9895705 | 2194911 | 5692542 | BFGS 49 | Sum of squares | Logistic | Identity |
| MLP 14-6-1 | 0.981778 | 0.727762 | 0.946615 | 10200325 | 1248142 | 6144683 | BFGS 54 | Sum of squares | Logistic | Identity |

Source: Authors.

The table clearly shows that all 5 preserved neural networks are of the MLP type, so the best characteristics are assumed here. The training algorithm included different Quasi-Newton variants in each case. The least squares method was determined as an error function for each of the preserved networks. In three cases, hidden layers of neurons were activated by the logistic function, and as for the other two, the hyperbolic tangent was the hidden layer activation function. The output activation function coincided with the activation function only in two cases – twice the logistic function and the identity function, and once the hyperbolic tangent function. In the first layer of all preserved neural networks, there is the same number of neurons, i.e. 14. In the third to the fifth preserved neural networks, there is also the same number of hidden layers, i.e. 6.
Table 2 shows relevance of the networks generated.

**Table 2**: Performance of generated networks

| Neural Network | Training | Test | Validation |
|---|---|---|---|
| MLP 14-26-1 | 0.983301 | 0.651976 | 0.940207 |
| MLP 14-7-1 | 0.986775 | 0.805951 | 0.946692 |
| MLP 14-6-1 | 0.988330 | 0.810602 | 0.954352 |
| MLP 14-6-1 | 0.982432 | 0.519603 | 0.945845 |
| MLP 14-6-1 | 0.981778 | 0.727762 | 0.946615 |

Source: Authors.

Regarding the table above, performance of the data sets in the preserved networks may be compared, where it follows that the higher the performance value is (i.e. the correlation coefficient), the better, with a value of 1.00 representing 100% performance. In the best possible manner, the values should be at their highest or the same in all three data sets [26]. It may be seen that MLP 14-6-1 achieves the highest performance and is also one of the preserved networks with the least number of hidden layers. The test data set also shows lower values in all preserved neural networks in comparison with the training and validation sets whose data group performance is of a very good standard in all preserved networks.

In order to better estimate the correct result, Table 3 gives parameters of predictions made using the individual networks.

**Table 3**: Parameters of predictions

| Prediction Parameter | 1.MLP 14-26-1 | 2.MLP 14-7-1 | 3.MLP 14-6-1 | 4.MLP 14-6-1 | 5.MLP 14-6-1 |
|---|---|---|---|---|---|
| Minimum predictions (Training) | -54976.2 | -51168.9 | -50011.8 | -45851.6 | -45985.0 |
| Maximum predictions (Training) | 336735.3 | 336815.3 | 336806.5 | 326929.8 | 329309.7 |
| Minimum predictions (Test) | -17787.8 | -14718.8 | -22380.4 | -8574.2 | -11077.9 |
| Maximum predictions (Test) | 11845.0 | 5832.6 | 5165.3 | 9894.7 | 6933.4 |
| Minimum predictions (Validation) | -94052.6 | -51359.0 | -50346.1 | -67637.4 | -49941.0 |
| Maximum predictions (Validation) | 8527.9 | 18913.4 | 7802.9 | 8957.1 | 7261.0 |
| Minimum residuals (Training) | -23745.1 | -24830.4 | -21073.2 | -27310.1 | -25763.3 |
| Maximum residuals (Training) | 20387.5 | 14329.2 | 16071.0 | 17147.8 | 16440.1 |
| Minimum residuals (Test) | -10263.1 | -4793.6 | -3822.5 | -9522.3 | -7198.4 |
| Maximum residuals (Test) | 8355.2 | 5286.2 | 12947.8 | 4006.9 | 2900.5 |
| Minimum residuals (Validation) | -8400.4 | -9390.8 | -10403.7 | -9208.3 | -10808.7 |
| Maximum residuals (Validation) | 33302.9 | 11365.1 | 12315.5 | 10489.0 | 12185.2 |
| Minimum standard residuals (Training) | -7.8 | -9.2 | -8.3 | -8.7 | -8.1 |
| Maximum standard residuals(Training) | 6.7 | 5.3 | 6.3 | 5.5 | 5.1 |
| Minimum standard residuals (Test) | -6.0 | -4.5 | -2.5 | -6.4 | -6.4 |
| Maximum standard residuals (Test) | 4.9 | 4.9 | 8.4 | 2.7 | 2.6 |
| Minimum standard residuals (Validation) | -2.0 | -4.1 | -4.2 | -3.9 | -4.4 |
| Maximum standard residuals (Validation) | 7.9 | 4.9 | 5.0 | 4.4 | 4.9 |

Source: Authors.

It is apparent from Table 3 that the second and the third preserved neural networks have the most similar results between the minimum values of the training and the validation data group predictions. In the training and the test data sets, the two networks also have very similar maximum values and very low minimum standard residual values. In the validation data group, however, the two networks are almost identical, both in the minimum and the maximum values.

Finally, sensitivity analysis was conducted to identify value generators affecting the paper´s industry of interest. The analysis is summarized in Table 4.

**Table 4**: Sensitivity Analysis

| Indicator | 1.MLP 14-26-1 | 2.MLP 14-7-1 | 3.MLP 14-6-1 | 4.MLP 14-6-1 | 5.MLP 14-6-1 | Average |
|---|---|---|---|---|---|---|
| External sources | 5.13075 | 1.56081 | 4.43047 | 72.29672 | 82.27947 | 33.13965 |
| Long-term liabilities | 15.78779 | 23.33231 | 25.29810 | 37.55004 | 26.79865 | 25.75338 |
| Material and energy consumption | 1.33732 | 2.42928 | 62.03859 | 2.06980 | 1.95573 | 13.96614 |
| Other operating expenses | 1.75454 | 3.71286 | 45.21382 | 1.43566 | 1.23112 | 10.66960 |
| Sales of goods | 1.18108 | 3.03160 | 37.32229 | 2.28484 | 2.26545 | 9.21705 |
| Short-term liabilities | 2.27391 | 6.11197 | 17.78270 | 9.09170 | 7.10810 | 8.47368 |
| Financial result | 27.15702 | 1.94822 | 4.88490 | 3.47984 | 2.43826 | 7.98165 |
| Output | 1.322852 | 2.435458 | 2.664672 | 5.572817 | 9.468574 | 4.292875 |
| Inventories | 2.895362 | 3.023984 | 1.675177 | 3.437309 | 3.246568 | 2.855680 |
| Long-term receivables | 1.022961 | 1.047101 | 1.085227 | 4.296865 | 6.120598 | 2.714550 |
| Profit / Loss of previous years | 6.183940 | 1.839675 | 2.100683 | 1.163658 | 1.415109 | 2.540613 |
| Bank loans and aids | 1.501360 | 1.066261 | 3.978169 | 3.034107 | 2.565586 | 2.429096 |
| Total assets | 1.301109 | 1.435446 | 2.466732 | 3.897952 | 1.635627 | 2.147373 |
| Intangible fixed assets | 1.003957 | 1.023130 | 1.056619 | 1.478891 | 2.925599 | 1.497640 |

Source: Authors.

Even without a more careful examination of the above sensitivity analysis table, it is obvious that in the fourth and the fifth preserved networks, External resources are of the highest importance with high significance indices. Long-term liabilities of the above companies also reach high values in all preserved networks. Interestingly, the last three networks with at least one indicator value greater than 25 are also networks with the same as well as lowest number of hidden layers (6). In the first neural network, Financial result and Long-term liabilities are the most important generators of company values (MLP 14-7-1). The same applies to Material and energy consumption in the third network (3. MLP 14-6-1) and External sources in the fourth and the fifth network (4. and 5. MLP 14-6-1). The remaining selected items also play a significant but not so fundamentally influencing role in creating company values.

## 4 Conclusion

The paper´s objective was to identify particular value generators of forestry and logging companies in the Czech Republic in 2016. For this reason, the best possible methodology was developed to determine them and the objective was eventually achieved. Using artificial neural networks, a total of 14 indicators, entering a company value creation process most frequently, were identified and measured by the EVA equity indicator. The following key variables were identified as company value generators: external sources, long-term liabilities, material and energy consumption, other operating expenses and sales of goods. Thus, forestry and logging companies operating in the Czech Republic should primarily focus on the previously mentioned items of financial statement. However, other, though not so important, value generators, whose ranking of importance is shown in Table 4, should not be neglected. The results presented have great potential and it is appropriate to follow them up with further research. As part of a new trend related to the increasing popularity of the EVA equity indicator, it would therefore be advisable to determine what impact the above variables have on the indicator and what their relationship is to it. Another possible course of action may reside in decomposition of partial indicators and their incorporation into operational and tactical goals of the companies, since, as may be argued, all businesses worldwide have one common objective, i.e. creating values for business shareholders (owners).

## Acknowledgements

This article was supported by the project "Stabilization and development of SME in rural areas", reg. No. TL01000349, the TACR Éta programme.

## References

1. T. Kliestik, M. Mišánková, K. Valášková, L. Švábová, Bankruptcy prevention: New effort to reflect on legal and social changes. *Science and Engineering Ethics*, **24**(2), 791-803 (2018)
2. R.A. Brealey, S.C. Myers, F. Allen, *Principles of corporate finance.* 10th ed., 914 p. (2013)
3. J. Vrbka, Z. Rowland, Assessing the financial health of companies engaged in mining and extraction using methods of complex evaluation of enterprises. *Contributions to Economics,* 321-333 (2019)
4. D. Nenkov, Growth policy and value creation in companies. *Ikonomicheski Izsledvania*, **25**(4), 36-65 (2016)
5. J. Kliestiková, M. Mišánková, T. Kliestik, Bankruptcy in Slovakia: International comparison of the creditor´s position. *Oeconomia Copernicana*, **8**(2), 221-237 (2017)
6. A. Bluszcz, A. Kijewska, A. Sojda, Economic value added in metallurgy and mining sector in Poland. *Metalurgija*, **54**(2), 437-440 (2015)
7. J. Vrbka, Modeling the future development of top company indicator EVA equity using indicator breakdown and sensitivity analysis on an example of a specific company. *Proceedings of the 5th International Conference Innovation Management, Entrepreneurship and Sustainability (IMES2017)*, pp. 1095-1106 (2017)
8. I. Issham, A.S.M. Fazilah, Y.S. Hwa, A.A. Kamil, Economic value added (EVA) as a performance measurement for GLCs vs Non-GLCs: Evidence in Bursa Malaysia. *Prague Economic Papers*, **17**(2), 168-179 (2008)
9. B. Morard, F.O. Balu, Developing a practical model for calculating the economic value added. *Economic Computation and Economic Cybernetic Studies and Research*, **43**(3), 107-122 (2009)
10. V. Stehel, M. Vochozka, The Analysis of the Economical Value Added in Transport. *Nase More*, **63**(3), 185-188 (2016)
11. B. Kollar, T. Kliestik, Simulation approach in credit risk models. *4th International Conference on Applied Social Science. Information Engineering Research Institute, Advances in Education Research*, **51**, 150-155 (2014)
12. P. Linna, H. Jaakkola, Toward Finding Culture Assessment Tools for SE Companies. *PICMET 2010: Technology Management for Global Economic Growth* (2010)
13. M. Vochozka, Z. Rowland, J. Vrbka, Financial analysis of an average transport company in the Czech Republic. *Nase More*, **63**(3), 227-236 (2016)
14. M. Dieter, C. Thoroe, Forestry and forest industry sector in the Federal Republic of Germany according to the new European-Wide sector definition. *Forstwissenschaftliches Centralblatt*, **122**(2), 138-151 (2003)
15. M. Levá, H. Čermáková, H. Vostrovská, M. Stárová, The Profitability Assessment of Selected Companies Providing Services to Forestry in the Czech Republic. *Reports of Forestry Research-Zpravy lesnickeho vyzkumu*, **61**(2), 145-150 (2016)

16. L. Vnoučková, H. Urbancová, H. Smolová, Strategic talent management in agricultural and forestry companies. *Agricultural Economics (Zemědělská ekonomika)*, **62**(8), 345-355 (2016)
17. Z. Burivalova, Ç.H. Şekercioğlu, L.P. Koh, Thresholds of Logging Intensity to Maintain Tropical Forest Biodiversity. *Current Biology*, **24**(16), 1893-1898 (2014)
18. M.S. Hossain, Z.CH. Ong, Z. Ismail, S. Noroozi, S.Y. Khoo, Artificial neural networks for vibration based inverse parametric identifications: A review. *Applied Soft Computing*, **52**, 203-219 (2017)
19. J. Vrbka, Z. Rowland. Stock price development forecasting using neural networks. *SHS Web of Conferences: Innovative Economic Symposium 2017 – Strategic Partnership in International Trade*, **39** (2017)
20. P. Horváthová, Methodology of Introduction and Utilization of Talent Management in the Organization. *Littera Scripta*, **6**(2), 35-46 (2013)
21. A. Joshi, S.S.Y. Lam, Business performance prediction using recurrent neural networks. *IIE Annual Conference and Exhibition*, pp. 1-6 (2006)
22. R.P. Dameri, R. Garelli, M. Resta, Unsupervised Neural Networks for the Analysis of Business Performance at Infra-City Level. *Organizational Innovation and Change*, 203-215 (2016)
23. S.G. Li, Z.M. Wu, Business performance forecasting of convenience store based on enhanced fuzzy neural network. *Neural Computing and Applications*, **17**(5-6), 569-578 (2008)
24. I. Neumaierová, I. Neumaier, Finanční analýza průmyslu a stavebnictví za rok 2007 [Financial analysis of industry and construction for the year 2007]. *Analýzy MPO [Analyzes of MIT]*, 187 p. (2008)
25. G. Wöhe, E. Kislingerová, *Úvod do podnikového hospodářství [Introduction to business economy]*. 2$^{nd}$ ed. Prague: C.H. Beck, 928 p. (2007)
26. J. Vrbka, Z. Rowland, P. Šuleř, Comparison of neural networks and regression time series in estimating the development of the EU and the PRC trade balance. *SHS Web of Conferences: Innovative Economic Symposium 2018 – Milestones and Trends of World Economy*, **61** (2019)

# Condition Factor Analysis in the Region of South Bohemia

*Petra* Pártlová[1*]

[1]Institute of Technology and Business in České Budějovice, Faculty of Corporate Strategy, Okružní 517/10, 37001 České Budějovice, Czech Republic

**Abstract.** Socio-economic transformation in many post-communist states has influenced the dynamic process in such states with the differentiation trends. Such process often causes disparities in different regions, creates relations and ties, often contradictory, and spatial socio-economic polarization in the regions. The above-mentioned processes and their impacts are demonstrated on the example of the South Bohemian region, in particular the LAG territory. Using the methods of correlation analysis, the author demonstrates the conditionality of the indicators in the area of interest. Different links and relations of socio-economic, infrastructural and natural factors to the emergence of disparities in different LAGs are analyzed. The most important factors of the emerging disparities within the region and the LAGs include the relations between the progressiveness of the economic structure and population density and between the demographic structure of the region and the infrastructure, as revealed by the analysis.
**Key words:** conditionality factors, South Bohemia Region, correlation analysis

## 1 Introduction

In connection with the changes in the theory of regional sciences [1], regional studies and the related issues of rural areas have developed significantly in recent [2]. Similarly, in space polarization research localization theories were applied significantly [3], based on the work of TJH von Thünen, A. Weber.

Efforts to deal with regional disparities followed, especially within the theory of the core-periphery of Myrdal [4], Perroux [5], quoted in Havlíček et al. [6] and in particular Friedmann [7] and his theory of polarized development.

It was Friedmann who introduced the term of core-periphery. Leimgruber [8] then explains the relations and links that affect the differentiation of the territory, arguing that peripheral areas should be assessed from an economic perspective, however also social, political, cultural and environmental aspects and their impact on these peripheral regions should be taken into account.

A significant change is related to the shift from the study of general issues of regional system development to the study of problem areas, respectively the shift from extensive forms of regional research to a combination of extensive and intensive research [3].

* Corresponding author: partlova@email.vstecb.cz

Similarly, Havlíček and Chromý [9] note that the periphery is a specific area with a disturbance of functional-spatial and socio-spatial relations, as the result of uneven influence of mutually conditioned social, economic, political, cultural and physical-geographical factors. They refer to peripheral territories as such areas located outside economically intensively exploited areas, characterized as remote from residential centers. Such areas are also identified by poor traffic conditions, poor transport accessibility, and low population density. The assessment of the factors affecting regional development of post-communist states and regions is addressed by many authors.

Havlíček and Chromý [9] worth mentioning in relation to the development of the regional structure of the Czech Republic. The authors divide the conditioning factors affecting the development of regions into three categories: (1) geographical factors (natural conditions and resources, settlement hierarchy, geographical location) (2) socio-economic factors (capital, human resources and specific economic factors – technological maturity, technical and social infrastructure and socio-economic intensity of land use), (3) socio-cultural factors (tradition, cultural capital, etc.).

Most authors focus on a particular area only, influenced by conditioning factors. Pileček, and Müller [10] deal with the conditionality of the factor on differentiation of unemployment. Netrdová, and Nosek [11] focus on the area of population differentiation. Blažek and Netrdová, [12] investigate the conditionality of several factors. In particular, they deal with the unemployment rate, the intensity of housing construction and voter turnout. Novotný, Janská and Čermáková [13] study the placement of foreigners in the Czech Republic and its conditioning factors. Socio-cultural conditionality and evaluation of natural and cultural and landscape heritage for the development of tourism are discussed by Chromý, Semian and Kučera [14].

The conditionality of transport opportunities is discussed by Květoň and Marada [15]. Most of the studies are focused mainly on the regions and districts. Studies are much less concerned with research at municipal level. However, there are several studies dealing with the differentiation of lower territorial units. These include the following authors: Musil, Müller; Blazek, Netrdova; Perlin, Kucerova, Kucera; Novak, Netrdova and Maceskova, Ourednicek, Temelova [16], [12], [17-19]. In Slovakia, the issue is discussed by Buček, Korec, Ira et al., Korec, Ondoš. [20-24].

## 2 Data and methods

The research used secondary data from Ekotoxa. The data were divided into four pillars (economic, social, infrastructure-institutional and environmental). In total, 602 municipalities in 17 local action groups of the South Bohemian Region were analyzed, except for towns with a population of over 25,000, i.e. all municipalities that may be part of a regional LAG.

A total of 29 indicators were selected. The proposed and subsequently selected indicators were consulted by CzechInvest and some adjustments of the indicators were made by Ekotoxa.

Finally, four indicators were selected in the economic pillar (Progressivity Index of the Economic Structure, Trend of Progressivity Index of the Economic Structure, Rate of Business Activity, Natural Presumptions of Recreation); seven indicators in the social pillar (Population Density, Aging Population, Economic Load Index Trend, Natives, Higher Education Trend, Unemployment Trend, Czech Citizenship); eleven indicators in infrastructure (Accessibility of the area by the public transport on Saturday, Accessibility of the area by A and major roads, Accessibility of railway stations, Residents living in permanently occupied dwellings connected to water supply, gas and public sewerage, Nursery school facilities, Primary school facilities, Accessibility of secondary schools with

Maturita, Accessibility of senior care homes, Adult general practice , Availability of hospitals and outpatient facilities); and seven indicators in the environmental pillar (Production capacity of land resources, Ecological fragmentation, Share of water areas, Chemical status of surface water, Soil visualization trend, Forest hazard zones, T specific territorial emissions from stationary sources). The indicators were assigned codes from K101 to K407.

Subsequently, the data were compared in order to determine the conditionality of different factors, both within different pillars and within the same pillars, and to determine the spatial differentiation of conditional factors on four pillars in the South Bohemian region. A correlation analysis method was used. The correlation analysis is focused on finding the intensity of potential dependencies between random variables of the sample. The result of the analysis is to determine the degree of dependence, by calculating the values of different types of correlation coefficients (such as Pearson and Spearman correlation coefficient).

Using the Pearson correlation analysis is conditioned [25] by normal probability distribution of the data with an expected linear relation.

The Pearson correlation coefficient of r is the most important measure of the strength of the relation of two random continuous variables X and Y [26]. Correlation coefficients (rP) determine the degree of tightness (force) of the relation between two variables, taking values $-1 \leq r \leq +1$ and at $r = 0$ the variables X and Y are completely independent (discontinuous, unrelated).

The higher the value of the correlation coefficient, the higher is the dependency (in causal relation and the association (in association relation) between X and Y variables. If r is positive, then the value of Y increases with increasing value of X.

If r is negative, the value of Y decreases with an increasing value of X. There is a causal relation between X and Y if X unequivocally causes and affects Y (a change in X causes a change in Y). In the association relation in influencing, changing X (or Y), Y (or X) also changes, but we do not know whether this change is caused by X or Y or another variable. An independent variable is a property (phenomenon) that is the cause or condition of the occurrence of another property (phenomenon). A dependent variable is a property that is the result (consequence) of the action of an independent variable. Pearson's correlation coefficient should not be used if one of the variables X, Y is not random (its values are fixed), the linearity of the expected relation is not assumed, nor the normal distribution of the variables X and Y.

Person correlation coefficient is calculated directly from the paired values of the X and Y variables. Similarly, to the arithmetic mean and standard deviation, it is greatly influenced by outliers. Person rp is defined as the mean value of the product of standardized random variables:

$$r_p = \frac{n \cdot \sum_{i=l}^{n} x_i y_i - \sum_{i=l}^{n} x_i \cdot \sum_{i=l}^{n} y_i}{\sqrt{[n \cdot \sum_{i=l}^{n} x_i^2 - (\sum_{i=l}^{n} x_i)^2] \cdot [n \cdot \sum_{i=l}^{n} y_i^2 - (\sum_{i=l}^{n} y_i)^2]}} \quad (1)$$

Explanatory notes:
$r_p$ - Pearson correlation coefficient
n - Number of pairs compared
x - Dependent variable
y - Independent variable

Hendl defines the strength of the relation as [26]:
Weak if $| r_p | = 0.1$ to 0.3.
Medium if $|r_p| = 0.3$ to 0.7.
Strong if $| r_p| = 0.7$ to 1.

Interpretation of the force of dependence of the variables by the correlation coefficient is influenced by the number of the analyzed data (and measured values). Therefore, the statistical test, and the tables of statistical significance for Person correlation coefficient are used. The test significance of the statistical significance test for the correlation coefficient is H0: $r = 0$. If the test hypothesis is rejected at a significance level (such as $\alpha = 0.05$), it is possible to accept the result of calculating the correlation coefficient as statistically significant.

## 2.1 Analysis of conditionality factors in the pillars of the region of South Bohemia

Person correlation was used to find out all combinations of dependence between the proposed indicators for the rural area of the region of South Bohemia. The resulting matrix shows that the dependence manifested across the pillars.

| index | K101 | K102 | K103 | K104 | K201 | K202 | K203 | K204 | K205 | K206 | K207 | K301 | K302 | K303 | K304 | K305 | K306 | K307 | K308 | K309 | K310 | K311 | K401 | K402 | K403 | K404 | K405 | K406 | K407 |
|---|---|---|---|---|---|---|---|---|---|---|---|---|---|---|---|---|---|---|---|---|---|---|---|---|---|---|---|---|---|
| K101 | 1,00 | 0,1 | 0,14 | 0,21 | 0,53 | -0,01 | -0,02 | -0,36 | 0,18 | -0,05 | -0,25 | 0,44 | 0,47 | -0,14 | -0,22 | 0,27 | 0,26 | 0,3 | -0,24 | -0,17 | 0,25 | -0,17 | 0,07 | 0,26 | 0,11 | -0,15 | -0,13 | 0,01 | -0,09 |
| K102 | 0,10 | 1 | -0 | -0,06 | -0,23 | -0,06 | -0,08 | 0,02 | 0,02 | -0,08 | 0,08 | -0,17 | -0,19 | 0,11 | 0,13 | -0,09 | -0,15 | -0,17 | 0,01 | 0,06 | -0,17 | 0,02 | -0,05 | -0,15 | 0,03 | 0,01 | 0,07 | -0 | 0,04 |
| K103 | 0,14 | -0 | 1 | 0,1 | -0,04 | -0,1 | -0,07 | -0,12 | 0,24 | 0,02 | -0,03 | -0,05 | 0,05 | -0,01 | 0,01 | 0,03 | -0,05 | -0,11 | -0,05 | -0,08 | -0,11 | -0,07 | 0,03 | -0,03 | -0,06 | 0,08 | 0 | -0 | 0,1 |
| K104 | 0,21 | -0,06 | 0,1 | 1 | 0,03 | 0,16 | 0,04 | -0,18 | -0,01 | 0,08 | -0,16 | 0,19 | 0,2 | 0,15 | 0,02 | 0,12 | 0,26 | 0,31 | 0,14 | 0,15 | 0,23 | 0,05 | 0,12 | -0,04 | -0,07 | -0,15 | -0,28 | -0,46 | 0,06 |
| K201 | 0,53 | -0,23 | -0,04 | 0,03 | 1 | -0,03 | 0,05 | -0,11 | 0,14 | -0,01 | -0,14 | 0,58 | 0,5 | -0,25 | -0,27 | 0,39 | 0,41 | 0,37 | -0,39 | -0,29 | 0,3 | -0,34 | 0,02 | 0,38 | 0,06 | -0,06 | -0,11 | 0,11 | -0,44 |
| K202 | -0,01 | -0,06 | -0,1 | 0,16 | -0,03 | 1 | 0,4 | -0,11 | -0,19 | -0 | -0,16 | 0,08 | 0,07 | 0,01 | -0,01 | 0,12 | 0,07 | 0,13 | 0,13 | 0,13 | 0,09 | 0,12 | 0,05 | 0,02 | -0,03 | -0,06 | -0,04 | -0,13 | 0 |
| K203 | -0,02 | -0,08 | -0,07 | 0,04 | 0,05 | 0,4 | 1 | -0,01 | -0,05 | 0,03 | -0,08 | 0,04 | 0,06 | -0,07 | -0,02 | 0,09 | 0,04 | 0,04 | 0,01 | 0,05 | 0,06 | 0,07 | 0,02 | 0,03 | -0 | 0 | -0,02 | -0,07 | 0 |
| K204 | -0,36 | 0,02 | -0,12 | -0,18 | -0,11 | -0,11 | -0,01 | 1 | -0,09 | -0 | 0,35 | -0,08 | -0,17 | 0,13 | 0,14 | -0,19 | -0,04 | -0,02 | 0,13 | 0,07 | -0,03 | 0,07 | -0,1 | -0,11 | -0,06 | 0,06 | 0,05 | 0,08 | -0,11 |
| K205 | 0,18 | 0,02 | 0,24 | -0,01 | 0,14 | -0,19 | -0,05 | -0,09 | 1 | -0,02 | 0,05 | 0,09 | 0,14 | -0,01 | 0,02 | 0,07 | 0,02 | -0,03 | -0,11 | -0,07 | -0,06 | -0,17 | 0,01 | 0,07 | -0,01 | -0,02 | -0,04 | 0,05 | -0,02 |
| K206 | -0,05 | -0,08 | 0,02 | 0,08 | -0,01 | -0 | 0,03 | -0 | -0,02 | 1 | 0,02 | 0,04 | 0,03 | -0,02 | 0 | 0,03 | 0,02 | -0,01 | -0,04 | -0,08 | -0,02 | -0,01 | 0,03 | 0,06 | -0,11 | 0,04 | -0,02 | -0,05 | 0,04 |
| K207 | -0,25 | 0,08 | -0,03 | -0,16 | -0,14 | -0,16 | -0,08 | 0,35 | 0,05 | 0,02 | 1 | -0,22 | -0,28 | 0,15 | 0,19 | -0,14 | -0,17 | -0,17 | 0,06 | 0,02 | -0,18 | 0,07 | -0,14 | -0,17 | -0,04 | 0,1 | 0,13 | 0,07 | 0,06 |
| K301 | 0,44 | -0,17 | -0,05 | 0,19 | 0,58 | 0,08 | 0,04 | -0,08 | 0,09 | 0,04 | -0,22 | 1 | 0,81 | -0,2 | -0,31 | 0,3 | 0,48 | 0,55 | -0,17 | -0,11 | 0,45 | -0,2 | 0,09 | 0,56 | 0,13 | -0,27 | -0,19 | 0,01 | -0,28 |
| K302 | 0,47 | -0,19 | 0,05 | 0,2 | 0,5 | 0,07 | 0,06 | -0,17 | 0,14 | 0,03 | -0,28 | 0,81 | 1 | -0,2 | -0,4 | 0,28 | 0,39 | 0,44 | -0,17 | -0,08 | 0,36 | -0,16 | 0,1 | 0,54 | 0,09 | -0,25 | -0,2 | -0,03 | -0,21 |
| K303 | -0,14 | 0,11 | -0,01 | 0,15 | -0,25 | 0,01 | -0,07 | 0,13 | -0,01 | -0,02 | 0,15 | -0,2 | -0,2 | 1 | 0,46 | -0,1 | -0,01 | -0,02 | 0,29 | 0,25 | -0,04 | 0,18 | 0,02 | -0,4 | -0,22 | 0,04 | -0,01 | -0,19 | 0,1 |
| K304 | -0,22 | 0,13 | 0,01 | 0,02 | -0,27 | -0,01 | -0,02 | 0,14 | 0,02 | 0 | 0,19 | -0,31 | -0,4 | 0,46 | 1 | -0,17 | -0,11 | -0,13 | 0,3 | 0,23 | -0,09 | 0,2 | -0,02 | -0,36 | -0,09 | 0,1 | -0,03 | -0,1 | 0,07 |
| K305 | 0,27 | -0,09 | 0,03 | 0,12 | 0,39 | 0,12 | 0,09 | -0,19 | 0,07 | 0,03 | -0,14 | 0,3 | 0,28 | -0,1 | -0,17 | 1 | 0,3 | 0,31 | -0,15 | -0,11 | 0,23 | -0,2 | 0,1 | 0,19 | 0,03 | -0,09 | -0,01 | -0,03 | -0,14 |
| K306 | 0,26 | -0,15 | -0,05 | 0,26 | 0,41 | 0,07 | 0,04 | -0,04 | 0,02 | 0,02 | -0,17 | 0,48 | 0,39 | -0,01 | -0,11 | 0,3 | 1 | 0,67 | -0,04 | -0,01 | 0,51 | -0,07 | 0,09 | 0,18 | 0,09 | -0,22 | -0,15 | -0,12 | -0,24 |
| K307 | 0,30 | -0,17 | -0,11 | 0,31 | 0,37 | 0,13 | 0,04 | -0,02 | -0,03 | -0,01 | -0,17 | 0,55 | 0,44 | -0,02 | -0,13 | 0,31 | 0,67 | 1 | 0,04 | 0,06 | 0,68 | -0,04 | 0,1 | 0,2 | 0,07 | -0,29 | -0,17 | -0,13 | -0,19 |
| K308 | -0,24 | 0,01 | -0,05 | 0,14 | -0,39 | 0,13 | 0,01 | 0,13 | -0,11 | -0,04 | 0,06 | -0,17 | -0,17 | 0,29 | 0,3 | -0,15 | -0,04 | 0,04 | 1 | 0,66 | 0,07 | 0,51 | 0,07 | -0,17 | -0,03 | 0,01 | 0,01 | -0,18 | 0,21 |
| K309 | -0,17 | 0,06 | -0,08 | 0,15 | -0,29 | 0,13 | 0,05 | 0,07 | -0,07 | -0,08 | 0,02 | -0,11 | -0,08 | 0,25 | 0,23 | -0,11 | -0,01 | 0,06 | 0,66 | 1 | 0,06 | 0,32 | 0,01 | -0,16 | -0,04 | -0,05 | -0,06 | -0,15 | 0,16 |
| K310 | 0,25 | -0,17 | -0,11 | 0,23 | 0,3 | 0,09 | 0,06 | -0,03 | -0,06 | -0,02 | -0,18 | 0,45 | 0,36 | -0,04 | -0,09 | 0,23 | 0,51 | 0,68 | 0,07 | 0,06 | 1 | 0,05 | 0,08 | 0,17 | 0,14 | -0,26 | -0,14 | -0,06 | -0,15 |
| K311 | -0,17 | 0,02 | -0,07 | 0,05 | -0,34 | 0,12 | 0,07 | 0,07 | -0,17 | -0,01 | 0,07 | -0,2 | -0,16 | 0,18 | 0,2 | -0,2 | -0,07 | -0,04 | 0,51 | 0,32 | 0,05 | 1 | 0,08 | -0,14 | 0,07 | -0,03 | 0,07 | -0,24 | 0,13 |
| K401 | 0,07 | -0,05 | 0,03 | 0,12 | 0,02 | 0,05 | 0,02 | -0,1 | 0,01 | 0,03 | -0,14 | 0,09 | 0,1 | 0,02 | -0,02 | 0,1 | 0,09 | 0,1 | 0,07 | 0,01 | 0,08 | 0,08 | 1 | 0,04 | -0,06 | -0,04 | -0,04 | -0,34 | 0,04 |
| K402 | 0,26 | -0,15 | -0,03 | -0,04 | 0,38 | 0,02 | 0,03 | -0,11 | 0,07 | 0,06 | -0,17 | 0,56 | 0,54 | -0,4 | -0,36 | 0,19 | 0,18 | 0,2 | -0,17 | -0,16 | 0,17 | -0,14 | 0,04 | 1 | 0,14 | -0,23 | -0,06 | 0,06 | -0,11 |
| K403 | 0,11 | 0,03 | -0,06 | -0,07 | 0,06 | -0,03 | -0 | -0,06 | -0,01 | -0,11 | -0,04 | 0,13 | 0,09 | -0,22 | -0,09 | 0,03 | 0,09 | 0,07 | -0,03 | -0,04 | 0,14 | 0,07 | -0,06 | 0,14 | 1 | -0,32 | 0,02 | 0,13 | 0 |
| K404 | -0,15 | 0,01 | 0,08 | -0,15 | -0,06 | -0,06 | 0 | 0,06 | -0,02 | 0,04 | 0,1 | -0,27 | -0,25 | 0,04 | 0,1 | -0,09 | -0,22 | -0,29 | 0,01 | -0,05 | -0,26 | -0,03 | -0,04 | -0,23 | -0,32 | 1 | 0,09 | 0,12 | -0,04 |
| K405 | -0,13 | 0,07 | 0 | -0,28 | -0,11 | -0,04 | -0,02 | 0,05 | -0,04 | -0,02 | 0,13 | -0,19 | -0,2 | -0,01 | -0,03 | -0,01 | -0,15 | -0,17 | 0,01 | -0,06 | -0,14 | 0,07 | -0,04 | -0,06 | 0,02 | 0,09 | 1 | 0,15 | 0,03 |
| K406 | 0,01 | -0 | -0 | -0,46 | 0,11 | -0,13 | -0,07 | 0,08 | 0,05 | -0,05 | 0,07 | 0,01 | -0,03 | -0,19 | -0,1 | -0,03 | -0,12 | -0,13 | -0,18 | -0,15 | -0,06 | -0,24 | -0,34 | 0,06 | 0,13 | 0,12 | 0,15 | 1 | -0,01 |
| K407 | -0,09 | 0,04 | 0,1 | 0,06 | -0,44 | 0 | 0 | -0,11 | -0,02 | 0,04 | 0,06 | -0,28 | -0,21 | 0,1 | 0,07 | -0,14 | -0,24 | -0,19 | 0,21 | 0,16 | -0,15 | 0,13 | 0,04 | -0,11 | 0 | -0,04 | 0,03 | -0,01 | 1 |

**Fig. 1:** Correlation matrix

Source: Own processing.

Legend:

K101 Progressivity Index of the Economic Structure

K102 Trend of Progressivity Index of the Economic Structure

K103 Rate of Business Activity

K104 Natural Presumptions of Recreation

K201 Population Density

K202 Aging Population

K203 Economic Load Index Trend

K204 Natives

K205 Higher Education Trend

K305 Residents living in permanently occupied dwellings connected to water supply, gas and public sewerage

K306 Nursery school facilities

K307 Primary school facilities

K308 Accessibility of secondary schools with Maturita

K309 Accessibility of senior care homes

K310 Adult general practice

K311 Availability of hospitals and outpatient facilities

K401 Production capacity of land resources

K402 Ecological fragmentation

K206 Unemployment Trend

K207 Czech Citizenship

K301 Availability by public transport on weekdays

K302 Accessibility of the area by the public transport on Saturday

K303 Accessibility of the area by A and major roads

K304 Accessibility of railway stations

K403 Share of water areas

K404 Chemical status of surface water

K405 , Soil visualization trend

K406 Forest hazard zones

K407 T specific territorial emissions from stationary sources

# 3 Results

## 3.1 The economic pillar versus the social pillar

In the economic pillar, the dependence between the indicator of progressiveness of the economic structure and population density was revealed. Significance between these factors was confirmed and the low value of the indicator of progressivity of economic structure and low value of population density is confirmed mostly in economically weak or rural municipalities, some of them characterized as peripheral, in the long-term (such as the rural area near the Central-South Bohemian administrative boundary – LAG Blatensko, LAG Brána Písecka and LAG Krajina Srdce (Landscape of the Heart), LAS Střední Povltaví, LAG Vltava), defined by low population density and a permanent outflow of population. The above-mentioned indicator is also related to the Natives indicator, defined as share of people born in the municipality of their permanent residence. The results proved the dependence in the LAG Landscape of the Heart, the LAG Střední Povltaví and the LAG Blatensko and also in the peripheral region of the LAG Czech Canada, administratively adjacent to the Vysočina region. The dependence was not found in in the LAG of Brána Písecka area, the reason may be an increase in new business locations in this area due to higher relocation to this region, due to job opportunities.

## 3.2 The economic pillar versus the infrastructure pillar

Another indicator that affects the progressiveness of the economic structure is the Accessibility of the area by the public transport on weekdays and on Saturdays. The worst situation is in the Lužnice LAG, whose average number of connections per day is 23; and 4 on Saturdays, followed by the Blatensko LAG with 29 (4) connections; with the same result, surprisingly in the Třeboňsko LAG 29 (7) and Krajina Srdce 29 (7) and LAG Strakonicko 33 (4). On the other hand, the LAGs of Lužnicko, Hlubocko-Lišovsko, Sdružení Růže and Pomalší are the best. From the results, it is possible to point out the possibility of direct connection of business structure, concentration of the enterprises and job opportunities, higher level of services to the attractiveness of settlements and regions. It can also be verified from the map data obtained that the size of settlements closely correlates with the size of the enterprises. And it is also possible to observe that the development of smaller settlements arises around large settlements by the process of suburbanization, where there is an apparent increase in connections mainly on weekdays.

## 3.3 The economic pillar versus the environmental pillar

Natural potential for recreation and forest zoning is discussed. The functionality of forests (Vyskot, 2000), and changes in these functions in the forest ecosystem are proved. Forest zoning affects the socio-recreational potential of the forest. Such potential is quantified.

The highest socio – recreational potential is, for example, in forests of national parks and spa forests. On the contrary, the lowest value of forests is related to those anthropogenically degraded. The analysis revealed the highest natural potential for recreation in the regions of the Šumavsko LAG, the Rozkvět LAG, the Růže LAG and the Česká Kanada LAG. On the contrary, the worst situation appears in the LAGS of LAG Blatensko, Strakonicko and Vodňanská Ryba.

### 3.4 The social pillar versus the infrastructure pillar

In the social pillar, a significant dependence between demographic structure and infrastructure was found [28-32]. Demographic factors are undoubtedly among the indicators affecting the accessibility of public transport in regions and municipalities. As an example, there are the remote locations, both poor and peripheral rural areas, mostly small settlements with low population density, very often accompanied by inaccessibility. It is the peripheral location of the area that is one of the external causes that indirectly affects the location of services and job opportunities. Such significance applies in the LAGs of Třeboňsko 29 (7) and Krajina Srdce with 33 (7) public transport connections, both on weekdays and Saturdays.

There are also significant dependencies between the number of public bus services and the population size of the settlement. There is a direct dependence and a direct rate between a higher rate of public transport service and a higher population in settlements. This dependence was not confirmed for rail transport, probably explained by the fact that there is no mass construction of railways, as is the case with roads, the railways are only modernized. It is interesting to note that the dependence of unemployment on transport infrastructure was not proved [33]. This may be the result of a relatively largely motorized countryside in response to a reduction in the number of public bus and rail services in the 1990s. It is therefore clear that peripheral rural regions, and structurally affected areas suffering from high unemployment, are less motorized and public transport is also of social importance.

### 3.5 Dependencies of the infrastructure pillar indicators on the environmental pillar

Based on the analysis of the results, the impact of ecological fragmentation on transport services was proved to be a very important factor. It was proved that higher fragmentation of the settlement system (fragmentation of municipalities, i.e. more local parts) negatively affects the provision of transport services by public transport and vice versa. It was also shown that at the level of basic settlement units there is a traffic exclusion of the population, which is caused by the unavailability of public transport and hence the absence of a transport alternative [34].

### 3.6 Reciprocal dependencies in the infrastructure pillar

The results of the analysis showed a mutual interaction between transport services and the number of primary and nursery schools [35]. The relation of spatial exclusion is confirmed, i.e. low number of public transport services to remote places with low population density per capita to density of the socially excluded persons (for example children and adolescents). Spatial exclusion is a specific case where the distance of localities in the territory is in direct relation very often accompanied by traffic unavailability. It is the peripheral location of the settlement that is one of the external causes that cannot be influenced by the availability of public services (nursery and primary schools). Children and adolescents who do not have a driving license and who cannot directly influence the location of basic public services (number of connections). The age of the population, which determines their transport possibilities, is a factor beyond control.

### 3.7 Reciprocal dependence in the social pillar

Interdependencies were also proved for the social pillar indicators, in particular population density with the trend of the economic burden index. It can be inferred from this dependence that the lower the value of the economic load index, the more favorable is the ratio between the economically inactive and the active population. The population is not aging in the region [36]. Indications of this dependence, with lower population density and lower value of the economic burden index, occur in the regions of the Rozkvět and Sdružení Růže LAGs. Conversely, this is not true in the LAGs of Lužnicko, Strakonicko, Blanský les and Česká Kanada (these are typical peripheral regions where this trend is very significant).

A lower dependency was also found for the indicator of "the natives" and the share of the population with Czech citizenship. Such signification was proved to be very dependent in the LAGs of Krajina Srdce and Střední Povltaví regions, where there is a high proportion of inhabitants born in the municipality of mother's permanent residence and at the same time these regions have a high proportion of inhabitants with Czech citizenship. On the contrary, there is the LAG of Pomalší with the largest share of the population is from Slovakia and Ukraine. It can be assumed that the inhabitants of these countries moved to the region to take advantage of the labor market opportunities.

### 3.8 Reciprocal dependencies in the environmental pillar

The research showed that there is interdependence between the variables in the environmental pillar. It proved that there is a degree of dependence between forest zoning and soil production capacity. The regions with lower soil production potential report a higher zone of forest threat, such as in the regions of LAGs of Blatensko, Střední Povltaví, Brána Písecka, Krajina srdce, Vodňanská ryba, and Strakonicko. To the contrary, the with higher soil production potential have lower forest threat zones, such as the LAGs of Šumavsko, Sdružení Růže, Rozkvět, and Pomalší. It should be noted here that mountain forests are currently endangered by spruce bark beetle. This disrupts the forest zonation and there has been a greater threat to these forests in the last decade. However, it should be noted that in Šumava, where the largest forest zoning in the South Bohemian Forest Region has occurred, repeated large-scale disturbance of spruce stands has occurred in the last at least 3,000 years, with a period of approximately 180 years [37]. Then spontaneous forest ecosystem recovery is successful [38].

## 4 Conclusion

The study of spatial differentiation based on the analysis of four pillars (economic, social infrastructure and environmental) is an important perspective on the identification of conditional dependencies in the LAGs of the research, within two different pillars, and the reciprocal dependence of one pillar. It is also interesting to find out the individual regional dependencies, which are reflected in the differentiation of regions in the South Bohemian region and it is possible to distinguish between developing and lagging regions and rural environmental regions and other rural urban regions, such as suburbia.

The aim of this research was to assess the impact of different factors on the territorial disparities of LAG regions in the South Bohemian Region. As revealed by the research, worse results in the economic pillar also correlate with indifferent and worse results in the social pillar as revealed in the LAGs of Střední Povltaví, Vltava, Krajina Srdce and Blatensko.

In the LAGs of Krajina srdce and Blatensko the dependencies of infrastructural pillars and the availability of the area by public transport on weekdays were further confirmed. All the above-mentioned LAGs also occurred in assessing the dependence of the social pillar on infrastructure. These regions are largely represented by poor and peripheral communities with low population density. Here, it is interesting to point out the fact that a relation of unemployment in the transport infrastructure was not proved. The indicator of lower population density showed another dependence on the economic burden index - the lower the population density in the regions, the lower the value of the economic burden index. Such dependency was manifested, for example, in the regions of Rozkvět LAG and Sdružení Růže LAG. The analysis of the correlation of the economic and environmental pillar revealed that the regions of the LAGS of Šumava, Rozvět and Sdružení Růže had the greatest potential for recreation. The worst situation was in the LAG of Blatensko. Similarly, these regions (LAGs of Šumavsko, Rozvět and Sdružení Růže) have higher soil production potential and lower forest threat zones. The results of the correlation showed also other dependencies, although not so significant. However, common characteristics are seen in some LAGs. Certainly, it would be appropriate in further research to create more common characteristics from the pillars under review to reveal similarity patterns according to differentiated territory.

## Acknowledgements

This article was supported by the project "Stabilization and development of SME in rural areas", reg. No. TL01000349, the TACR Éta programme.

## References

1. J. Blažek, D. Uhlíř, *Teorie regionálniho rozvoje: nástin, kritika, implikace [Theory of regional development: outline, criticism, implication]*. Prague: Karolinum, 343 (2011)
2. M. Hampl, Regional differentiation of current socio-economic development in the Czech Republic. *Czech Sociological Review*, **43**(5), 889-910 (2007)
3. J. Blažek, Regionální inovační systémy a globální produkční sítě: dvojí optika na zdroje konkurenceschopnosti v současném světě [Regional innovation systems and global production networks: Two views on the source of competitiveness in the present-day world?]. *Geografie*, **117**(2), 209-233 (2012)
4. G. Myrdal, *Economic Theory and Underdeveloped Regions*. London: Duckworth, (1957)
5. F. Perroux, Note sur la notion de pole de croissance [Note on the concept of growth pole]. *Economique Applique*, **8**(1-2), 307-320 (1955)
6. T. Havlíček, V. Jančák, P. Chromý, M. Marada, *Vybrané teoreticko-metodologické aspekty a trendy geografického výzkumu periferních oblastí [Selected theoretical-methodological aspects and trends of geographic research of peripheral areas]*. In: M. Novotná (ed.): Problémy periferních oblastí [Problems of peripheral areas], KSGRR Faculty of Science, Charles University, Prague, 6-24 (2005)

7. J. Friedmann, Regional Development Policy: A Case Study of Venezuela. M.I.T. Press (1966)
8. S. Leimgruber, Investigation of the Rubber-Brass Adhesion Layer Using the Olefin-Metathesis Method. *Rubber Chemistry and Technology*, **88**, 219-233 (2015)
9. T. Havlíček, P. Chromý, Příspěvek k teorii polarizovaného vývoje území se zaměřením na periferní oblasti [Contribution to the theory of polarized development of a territory, with a special attention paid to peripheral regions]. *Geografie*, **106**(1), 1-11 (1966)
10. J. Pileček, J. Müller, Regionální disparity ve vývoji nezaměstnanosti v období hospodářské krize [Regional disparities in the development of unemployment in times of economic crisis]. Regionální disparity a hospodářské subjekty v regionálním rozvoji [Regional disparities and economic operators in regional development], p. 74-87 (2010)
11. P. Netrdová, V. Nosek, Přístupy k měření významu geografického rozměru společenských nerovností [Approaches to measuring the importance of the geographical dimension of social inequalities]. *Geografie*, **114**(1), 52-65 (2009)
12. J. Blažek, P. Netrdová, Regional unemployment impacts of the global financial crisis on the new member states of the EU in Central and Eastern Europe. *European Urban and Regional Studies*, **19**(1), 42-61 (2012)
13. J. Novotný, E. Janská, D. Čermáková, Rozmístění cizinců v Česku a jeho podmiňující faktory: pokus o kvantitativní analýzu [Distribution of foreigners in the Czech Republic and its conditioning factors: an attempt at quantitative analysis]. *Geografie*, **112**(2), 204-220 (2007)
14. P. Chromý, M. Semian, Z. Kučera, Regionální vědomí a regionální identita v Česku: případová studie Českého ráje [Regional awareness and regional identity in the Czech Republic: a case study of the Bohemian Paradise]. *Geografie*, **119**(3), 259-277 (2014)
15. V. Květoň, M. Marada, Diferenciace nabídky dopravních příležitostí v českých obcích a sociogeografických mikroregionech [Differentiation of transport opportunities in Czech municipalities and socio-geographic micro-regions]. *Geografie*, **115**(1), 21-43 (2010)
16. J. Musil, J. Müller, Vnitřní periferie v České republice jako mechanizmus sociální exkluze [Internal periphery in the Czech Republic as a mechanism of social exclusion]. *Sociologický časopis*, **44**(2), 321-348 (2008)
17. R. Perlín, S. Kučerová, Z. Kučera, Typologie venkovského prostoru Česka [Typology of rural space in the Czech Republic]. *Geografie*, **115**(2), 161-187 (2010)
18. J. Novák, P. Netrdová, Prostorové vzorce sociálně-ekonomické diferenciace obcí v České republice [Spatial patterns of socio-economic differentiation of municipalities in the Czech Republic]. *Sociologický časopis*, **47**(4), 717-744 (2011)
19. M. Macešková, M. Ouředníček, J. Temelová. Sociálně prostorová diferenciace v České republice: implikace pro veřejnou (regionální) politiku [Socio-spatial differentiation in the Czech Republic: implications for public (regional) policy]. *Ekonomický časopis*, **57**(7), 700-715 (2009)
20. M. Buček, Regionálna integrácia Slovenskej republiky: stratégia využitia podpory Európskej únie [Regional integration of the Slovak Republic: strategy of using the support of the European Union]. *Ekonomický časopis,* **51**(8), 982-996 (2003)
21. P. Korec, Regionálny rozvoj Slovenska v rokoch 1989-2004 [Regional development of Slovakia in 1989-2004]. *Geo-grafika* (2005)

22. V. Ira, L. Falťan, J. Pašiak, P. Gladoš, Podoby regionálnych odlišností na Slovensku [Forms of regional differences in Slovakia]. *Sociologický ústav* (2005)
23. P. Korec, S. Ondoš, Less developed regions of Slovakia: identification and characteristics. *Acta Geographica Universitatis Comenianae*, **48**, 55-74 (2006).
24. P. Korec, S. Ondoš, *Historické a geografické aspekty regionálnej divergencie na Slovensku [Historical and geographical aspects of regional divergence in Slovakia]*. In S. Kraft, K. Mičková, J. Rypl, P. Švec, M. Vančura. eds.: Česká geografie v evropském prostoru [Czech geography in European space], 626-631 (2007)
25. M. Meloun, J. Militký, *Interaktivní statistická analýza dat [Interactive statistical data analysis]*. Prague: Karolinum (2012)
26. J. Hendl, *Přehled statistických metod zpracování dat: analýza a metaanalýza dat [Overview of statistical methods of data processing: data analysis and meta-analysis]*. Prague: Portál (2004)
27. A. Vyskot, *Kvantifikace a hodnocení funkcí lesů České republiky [Quantification and evaluation of forest functions in the Czech Republic]*. Prague: Ministry of Environment (2003)
28. S. Nutley, Rural Areas Accessibility Problem. In: B. Hoyle, R. Knowles eds.: *Modern Transport Geography*, 2nd rev. ed., Wiley and sons, Chichester, 185-215 (1998)
29. M.J. Dargay, Determinants of car ownership in rural and urban areas: a pseudo-panel analysis. *Transportation Research Part*, **38**, 351-366 (2002)
30. J. Farrington, C. Farrington. Rural accessibility, social inclusion and social justice: towards conceptualisation. *Journal of Transport Geography*, **13**, 1-12 (2005).
31. J. Mcdonagh. Transport policy instruments and transport-related social exclusion in rural Republic of Ireland. *Journal of Transport Geography*, **14**, 355-366 (2006)
32. M. Marada, T. Hudeček, Accessibility of Peripheral Regins: A Case of Czechia. EUROPA XXI, Institute of Geography and Spatial Organization, Warszawa, p. 43-49, (2006)
33. K. Inhlanfeldt, D. Sjoquist, The spatial mismatch hypothesis: a review of recent studies and their implications for welfare reform. *Housing Policy Debate*, **9**, 849-892 (1998)
34. V. Květoň, J. Chmelík, P. Vondráčková, M. Marada, Developments in the public transport serviceability of rural settlements with examples from various types of micro-regions. *Acta Universitatis Carolinae – Geographica*, 1/2012,51-63 (2012)
35. V. Hulík, K. Tesárková, K., Dopady demografického vývoje na vzdělávací soustavu v České republice [Impacts of demographic development on the educational system in the Czech Republic]. *Orbis Scholae*, **3**(3), 7-23 (2009)
36. S. Kučerová, Z. Kučera, Vztah periferiality a vzdělávání: Lze definovat periferní oblasti na základě vývoje sítě základních škol? [Relationship between peripheral and education: Is it possible to define peripheral areas based on the development of the network of primary schools?] *Acta geographica Universitatis Comenianae*, (53), 59-73 (2009)
37. H. Svobodová, Vývoj horských lesů na Šumavě [Development of mountain forests in the Bohemian Forest]. In: Z. Neuhäuslová et al., *Mapa potenciální přirozené vegetace NP Šumava [Map of potential natural vegetation of Šumava National Park]*. Silva Gabreta, Suppl. 1/2001 (2002)
38. M. Jonášová, K. Prach. Central-European mountain spruce (Picea abies (L.) Karst.) forests: regeneration of tree species after a bark beetle outbreak. *Ecological Engineering*, **23**(1), 15-27 (2004)

# Strategic Management and Decision Making of Small and Medium-Sized Enterprises in the Czech Republic

*Jarmila* Straková[1*], *Milan* Talíř[1]

[1]Institute of Technology and Business in České Budějovice, Faculty of Corporate Strategy, Okružní 517/10, 37001 České Budějovice, Czech Republic

**Abstract.** The topic of strategic management and decision making was underestimated by SMEs for a long term. Managers believed that experience, intuition and operability are the key success factors of their management work. These opinions have recently been abandoned in the management practice. This paper is focused on the analysis of strategic management and decision making and presents the outputs from research based on a test sample of 381 enterprises, of which 114 microenterprises, 141 small enterprises and 126 medium-sized enterprises. The company size categorization within this sample is corresponding to the SME structure in the South Bohemian Region. Lack of awareness of procedures for development of essential strategic documents such as vision, mission statements and a corporate strategy was observed. More than 60 % SMEs do not have such documents. are Similar results were obtained with respect to the development of a corporate strategy using the methods of a strategic situation analysis. This finding is alarming and its adverse impact can primarily be expected in a period of economic recession or downturn.
**Key words:** strategic management, vision, mission, corporate strategy, situation analysis

## 1 Introduction

Small and medium-sized enterprises (SMEs) in the Czech Republic have gone through a number of development stages. Year 1989 was a "crunch year" for implementation of strategic management and decision making elements into SME management systems. Transformation of the national economy was underway in that period of time, consisting in transition from centrally planned and closed economy to market oriented and open national economy. SMEs gradually changed to private ownership based businesses. The application of strategic management and decision making processes to SME corporate practice is limited by development of the strategic management science. It began to form at the end of the 60s and at the beginning of the 70s of the last century, in particular in the world advanced economies with developed market economy.

[*] Corresponding author: strakova@mail.vstecb.cz

Porter [1], the guru of the global management science with Czech roots defines new categories in strategic management such as competitiveness, competitive advantage, competitive environment, and recently also the value chain. The key areas of successful small and medium-sized enterprises are characterized by the Peterson and Rondstadt concepts [2]. They define business success and prerequisites for business development using two factors – business know-how and know-who. Scientific analyses of local and foreign authors demonstrate that SMEs are unique as well as highly flexible [3-5]. Some authors concentrated on research of business as an activity using specific features of SME strategic potential and exploring the position of a business depending on its life cycle [6]. Many authors analyse the behaviour of entrepreneurs and the importance of personality traits of a manager with respect to SME, e.g. Birch [7] and Gunderson [8].

In the late 20th century, the issue of business success models in the SME environment was brought to the forefront of the international agenda [9]. The 90s of the last century are considered as a significant economic milestone in Europe. Both the world as well as European markets were saturated and businesses had to change the existing methods of management and planning. In the Czech Republic, the last decade of the last century is characterized by establishing SMEs in a new market environment at the national as well as international level. Their stabilization and transformation role, but also the high level of absorption in terms of engagement of workers gradually laid off by dissolved or ineffective operations are unique in Europe. More than 1 mil. workers found their jobs with SMEs, of which more than 60% in services.

The beginning of the 21st century is characterized by dynamic and significant changes in the corporate environment. It is the period of time when the efficiency, stability and relationship with the external environment have become the priority for SMEs [10-11]. The dynamics of corporate resources then foster the business growth and sustainability and last but not least, they help to create competitive advantages in the business environment [12-15]. All these authors apply the concept of competitive strategy by M. E. Porter and his five force analysis and the related publications of Gibbs [6], Lundström a Stevenson [16]. The current stage of development is characterized by the growing need for advanced management methods in SMEs and it is correct to say that the scientific platform does not always respond adequately.

Small and medium-sized enterprises in the Czech Republic mainly operate in rural areas, covering more than 70% of the Czech territory. The analysis of the SME and Sole Trader Association of the Czech Republic [17] shows a highly negative trend of business development in rural areas where more than 45% businesses are categorized as less developed. One of the tools that can considerably help to improve the economic situation of SMEs in general (and particularly for rural areas) is a modern management and administration concept where strategic management plays a key role.

## 2 Data and Methods

The research was conducted on a test sample of SMEs, in particular from the South Bohemian Region. The start of research dates back to 2016 and it is still ongoing. The sample includes 381 enterprises, of which 114 microenterprises, 141 small enterprises and 126 medium-sized enterprises. The size structure of this sample is corresponding to the SME structure in the South Bohemian Region.

Two hypotheses were created:

H1: Creating a SME vision and mission statement and corporate strategy is fully respected by the business sphere as a prerequisite for business stability and development.

H1: Situation analysis is widely used in the Czech business practice for projection of a corporate strategy and definition of development priorities.

The Pearson's chi-squared test as a statistical method was used to measure dependence / independence of selected category variables (internal and external strategic analyses) in order to formulate a corporate strategy.

The formula for Pearson's chi-squared test of independence is as follows:

$$\chi 2 = \frac{\sum_{i=1}^{k}(X_i - Np_i)^2}{Np_i} \quad (1)$$

| | |
|---|---|
| X2 | chi-square |
| Xi | empirical frequency (actual) |
| Npi | theoretical frequency (expected) |

Our tests apply two hypotheses, null (H0) and alternative (HA). H0 (no dependence between selected variables), HA (dependence between selected variables). The evaluation criterion for verification or rejection of a hypothesis is p-value within the interval <0.1>. If the p-value is $< \alpha$ (at the set significance level $\alpha = 0.05$), HA will be confirmed. Statistical calculations were supported by the statistical program R.

## 3 Results

The following conclusions are derived from the chart. The fact that the majority of enterprises/respondents developed their vision and mission statements and a long-term development plan (corporate strategy) was an unexpected but positive finding It is mainly surprising in case of microenterprises where given elements are created by the company owner or manager. However, a high rate (app 50%) of them are not concerned with a long-term business plan which is a negative finding, especially for small and medium-size enterprises in view of the expected economic development. It was observed during a partial analysis and interviews that a strategic plan was elaborated on the tactical and operative level. As regards decomposition of strategic goals into tactical and operative objectives, strategic goals were found as highly prevailing in micro and small enterprises unlike small and medium-sized enterprises where strategic goals are slightly more decomposed into tactical objectives It can be explained by the characteristics of the relevant size categories of businesses.

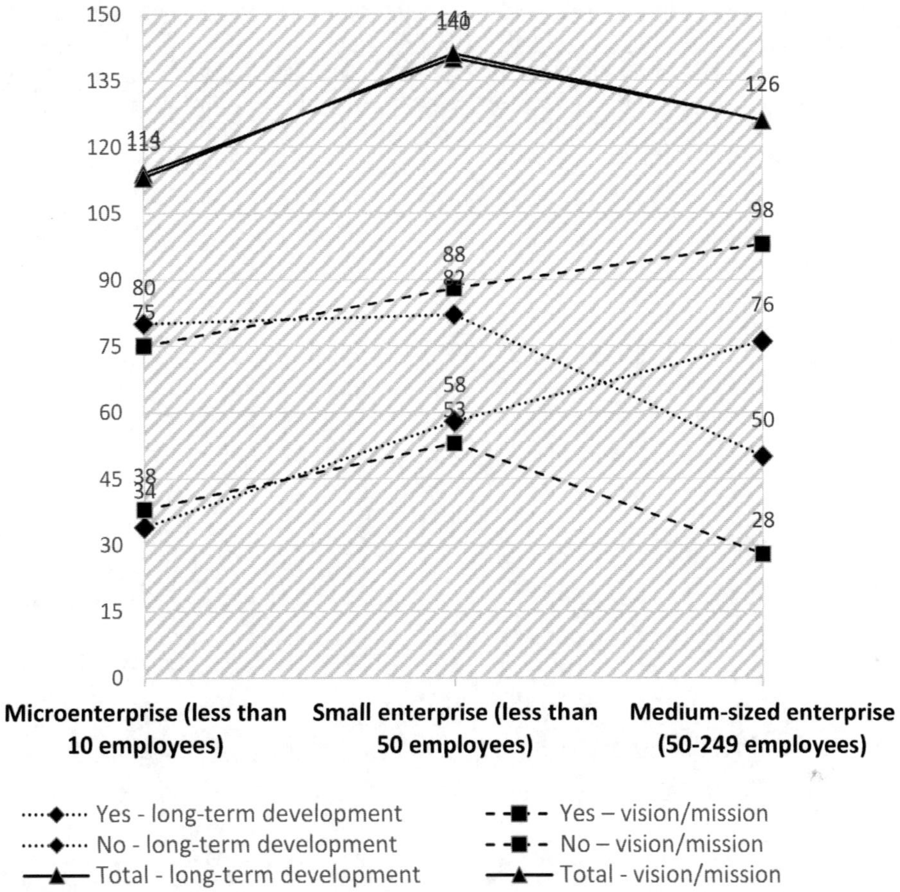


**Fig. 1.** Comparison of long-term development plan and vision/mission statements depending on the size of SME

Source: Own research.

Another research analysis was aimed at staffing for development of a strategic plan and partial departmental strategies in SMEs (see Chart 2). The development of a corporate strategy is mostly contributed by owners, namely for micro and small enterprises, or in some cases working together with the top management of medium-sized enterprises.

In the case of microenterprises, the owner is involved both in development of a corporate strategy and naturally also elaboration (decomposition) of such strategy into partial objectives.

In the case of small enterprises, owners are still highly involved in development of a corporate strategy while the entire top management team is more involved in development of a corporate strategy and the contribution of senior executives is slightly increasing. The decomposition of a corporate strategy into partial departmental strategies is still mostly contributed by the company owner followed by the entire top management team with slightly increased involvement of senior executives.

In the case of medium-sized enterprises, the situation regarding the development of a corporate strategy is quite different from the two categories above mentioned. A corporate strategy is mostly contributed by the entire top management team followed by the company owner and senior executives. For decomposition of a corporate strategy into partial departmental strategies, the key role is played by the entire top management, and the owner and senior executives then in equal parts. A significant dependence between the development of a corporate strategy and the company size category ($\alpha$ = 0.05, p = 0.003973, chi-square = 11.0564, V = 0.2492) was confirmed across the test sample.

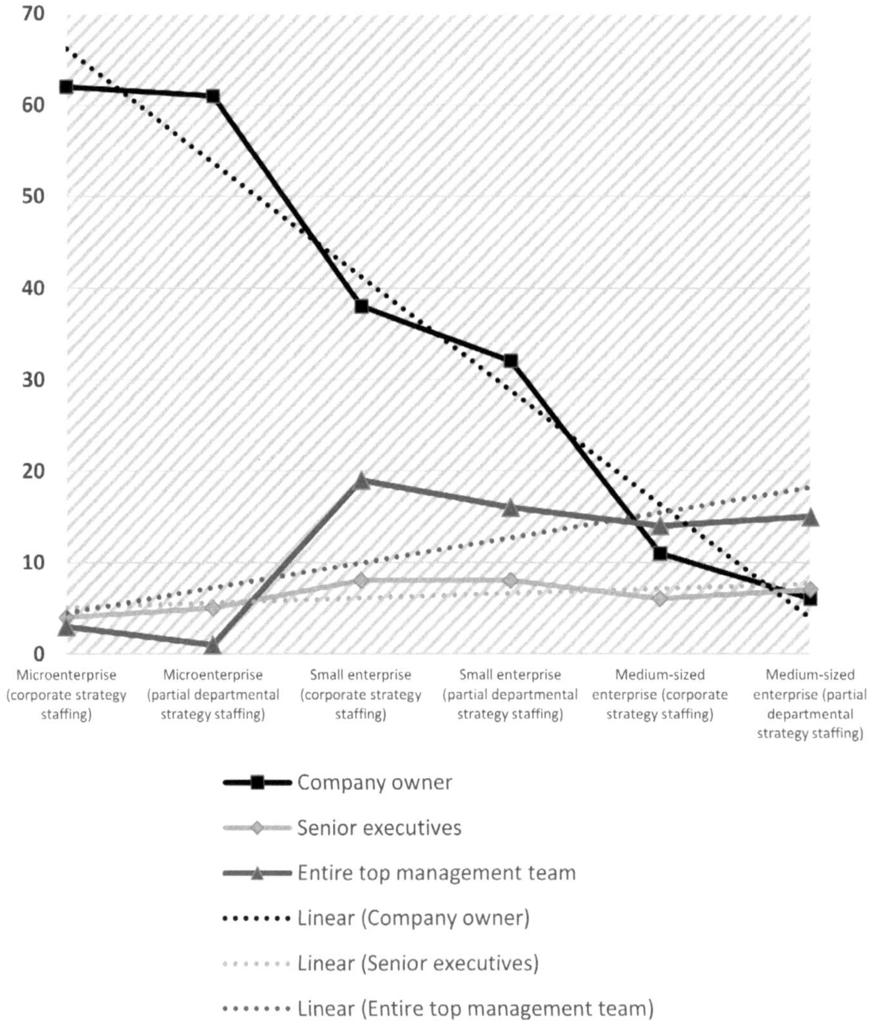


**Fig. 2** Staffing for development of a corporate strategy and partial departmental strategies in SMEs

Source: Own research.

The outputs as presented in the previous part are related to the solution of hypothesis 1 concerning the frequency of developing vision and mission statements and a corporate strategy and adherence to those documents by the company managers. This hypothesis is rejected based on the summary evaluation of results. The reason may be that more than half of the tested organizations do not create such documents and the supporting reasons are often based on very erroneous arguments contradictory to management principles. As an example, superiority of intuitive decision making to strategic thinking, in particular for microenterprises, or the declining importance of strategy in the turbulent and discontinuous world, etc.

The next part of this paper will explore the knowledge/awareness and use of situation analysis methods by the individual size categories and their application. The internal and external environment groups were analysed separately.

The following Chart 3 illustrates the use of internal analyses for the business environment. In the ascending order: GE model is least used, followed by an exposure analysis and SPACE analysis. On the contrary, a SWOT analysis, portfolio analysis and the analysis of internal resources of an organization analyses are most frequently used. The negative result is the fact that more than 60% companies do not have sufficient knowledge of the above analyses; it is a very negative finding especially for the category of medium-sized enterprises.

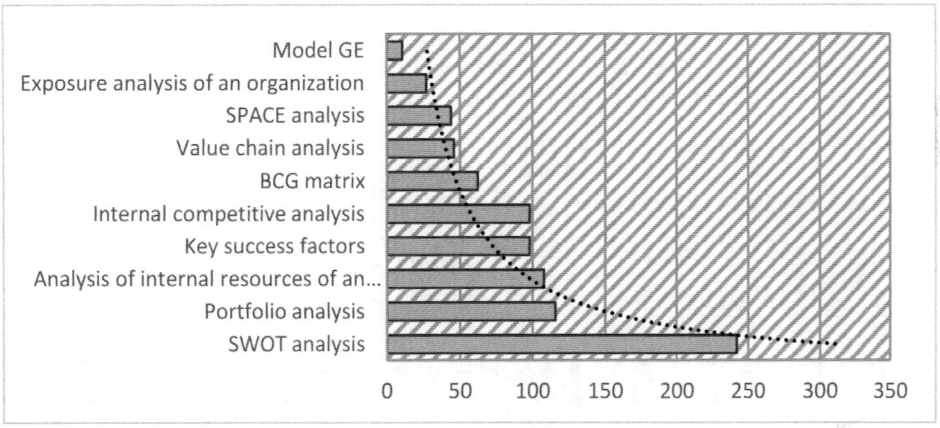


**Fig. 3.** Frequency of strategic analyses in the internal business environment

Source: Own research.

Frequency of strategic analyses in the external business environment illustrates the figure 4.

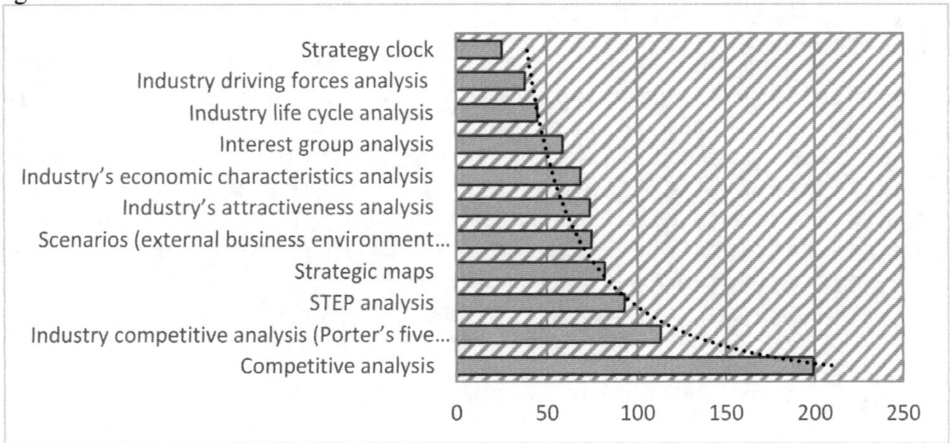


**Fig. 4.** Frequency of strategic analyses in the external business environment

Source: Own research.

As regards the knowledge and use of external methods of a situation analysis (Fig. 4), the least used methods include: Strategy clock, industry driving forces analysis and industry life cycle analysis. On the opposite side, we find a competitive analysis – Porter's model which is a positive finding especially in the period when domestic and foreign competition is increasing. These analyses also include a STEP analysis which is an adequate supplement to the two above methods. The negative fact is that more than more than 60% companies are not aware of these methods, including their practical application in the business environment.

As regards internal analyses, dependence of development of a corporate strategy on the sector focus, size and scope / sphere of influence of an organization (Tab.1) was analysed together with the frequency of strategic methods. Dependence was only established in a value chain analysis with respect to sector differentiation of tested companies. This finding is fully consistent with the theory for development of a corporate strategy that states that value chain will be among the key tools for development of a corporate strategy in future. No statistically significant values were established in other internal analyses both in terms of size categorization of tested companies and their scope / sphere of influence. In accordance with the outputs of the solution, the second hypothesis is also rejected.

**Table 1.** Dependence of development of a corporate strategy on the sector focus, size and scope / sphere of influence – internal analyses

| Analysis | Company focus p.value | Company size p.value | Scope / sphere of influence p.value |
|---|---|---|---|
| SWOT analysis | 0.8921 | 0.4663 | 0.4305 |
| Portfolio analysis | 0.3586 | 0.8461 | 0.1377 |
| Analysis of internal resources of an organization | 0,3162 | 0.4976 | 0.5966 |
| Key success factors | 0.9874 | 0.1927 | 0.2976 |
| Internal competitive analysis | 0.3086 | 0.4195 | 0.9636 |
| BCG matrix | 0.871 | 0.4078 | 0.9106 |
| Value chain analysis | **0.008428** | 0.1927 | 0.6386 |
| SPACE analysis | 0.913 | 0.3004 | 0.887 |
| Company's exposure analysis | 0.6613 | 0.9451 | 0.3936 |
| GE model | 0.6139 | 0.7167 | 0.5219 |

Source: Own research.

Table 2 illustrates dependence of development of a corporate strategy on the sector focus, size and scope of an organization in external analyses. An interest group analysis and industry driving forces analysis indicated statistically significant values only in terms of company size categorization. A lower level of dependence was identified in an industry driving forces analysis, namely the category of scope/sphere of influence of tested companies. The negative finding was that the Porter's model was absent, or an internal industry structure analysis considered as very important on a theoretical level and essential for creating a corporate strategy.

**Table 2** Dependence of development of a corporate strategy on the sector focus, size and scope / sphere of influence – external analyses

| Analysis | Company focus p.value | Company size p.value | Scope / sphere of influence p.value |
|---|---|---|---|
| Competitive analysis | 0.6514 | 0.1242 | 0.5069 |
| Industry competitive analysis (Porter's five forces model) | 0.7713 | 0.671 | 0.6984 |
| STEP analysis | 0.8112 | 0.3615 | 0.1917 |
| Strategic maps | 0.198 | 0.437 | 0.7463 |
| Scenarios (external business environment development prediction) | 0.429 | 0.6221 | 0.8786 |
| Industry's attractiveness analysis | 0.1391 | 0.8971 | 0.903 |
| Industry's economic characteristics analysis | 0.935 | 0.7879 | 0.2913 |
| Interest group analysis | 0.8808 | **0.04944** | 0.9808 |
| Industry life cycle analysis | 0.5083 | 0.4595 | 0.1912 |
| Industry driving forces analysis | 0.4083 | **0.01219** | 0.09801 |
| Strategic clock | 0.8145 | 0.3031 | 0.2803 |

Source: Own research.

It is right to say that the research was conducted in a period of strong economic growth which could affect the approach and mind-set of managers. This fact does not necessarily indicate that tested companies do not address the key factors determining the internal structure of an industry, but the importance of those is decreased in a period promoting growth.

Unlike the evaluation results of each individual factor, the group evaluation of factors (Table 3) confirmed dependence of the technological and political and legal aspects in the manufacturing and industrial sector, and all aspects in the service sector.

**Table 3.** Expected impact of the business macro environment on the development of a corporate strategy and profitability

| ***Manufacturing and industrial sector*** | **p value** | **chi-square** |
|---|---|---|
| Social and cultural | 0.5791 | 4.7287 |
| **Technological** | 0.04612 | 12.812 |
| Economic | 0.06491 | 11.8708 |
| **Political and legal** | 0.03092 | 13.887 |
| ***Service sector*** | | |
| **Social and cultural** | 0.02671 | 14.2751 |
| **Technological** | 0.00000003647 | 45.5452 |
| **Economic** | 0.000001417 | 37.4832 |
| **Political** | 0.000003536 | 35.4432 |

Source: Own research.

It can be concluded that the technical and technological aspect, including the innovation process, and the political and legal aspect are crucial for the manufacturing sector; a higher level of cohesion with the external environment was observed in the service sector as well as a direct impact of those factors on changes in the corporate strategy.

## 4 Conclusion

In accordance with the outputs from the solution, evidencing that more than 60% companies do not create vision and mission statements and a corporate strategy, the first hypothesis was rejected. It is alarming, but it can be rectified in the medium term. The participation of research organizations, including universities and public administration and regulatory authorities will be important to rectify this negative situation.

A prerequisite for efficient and effective strategic management is knowledge of situation analysis methods and tools where similar lack of knowledge and implementation in business practice was observed. Therefore, the second hypothesis expecting such knowledge and use in business practice by the majority of businesses, or managers was also rejected. Some internal as well as external methods show absolute lack of awareness. The average knowledge of those methods is app 50%. There the managerial literacy is at a very low level and again, the key role must be played by research institutions and transfer of their knowledge into business practice.

## 5 Suggestions for Practice

In the last decade, the business environment has undergone some fundamental changes affecting all of its elements. This actual and ever more intensive trend involves system changes in the concept of strategic management. The most important changes are:

- Corporate strategy should not only be consistent with requirements of the local market, but it should consider the regional, national markets, and recently also international aspects of the globalized, or globalized market.
- Standard concept of strategy as a long-term fixed document loses its original purpose due to turbulent changes in the external business environment.
- Corporate strategy has recently become a document on various alternatives of approach and processes according to the specific business situation and this statement applies more to SMEs than larger businesses.
- Every organization should create a vision and mission statement as well as strategic goals. These fundamentals are important for establishing a company and relationships and expectations for behaviour in the workplace and a corporate culture, and for defining the position where the company should be visible in a certain period of time and what performance and financial figures should be achieved by the company. Their projection must ensure a flexible response to external environmental effects.
- The existence of a long-term development plan, or a corporate strategy provides the basis for business sustainability and efficiency and it can be referred to as the precondition for existence of a business. If absent, the business can lose its competitive advantage.
- Human resources are the essential attribute for competitive advantage and profitability of an organization; continuous and systematic training of senior managers on strategic management and decision making processes is a prerequisite for existence of a viable business.

## Acknowledgements

This article was supported by the project "Stabilization and development of SME in rural areas", reg. No. TL01000349, the TACR Éta programme.

## References

1. M.E. Porter, *Konkurenční strategie: metody pro analýzu odvětví a konkurentů [Competitive stratégy: methods for analysis of industry and competitors]*. Prague: Victoria Publishing (1994)
2. R. Peterson, R. Rondstadt, A silent strength: Entrepreneurial know who. *In paper presented to the 16th European Small Business Seminar* (1987)
3. R. Sanchez, Strategic flexibility in product competition. *Strategic Management Journal,* **16**, 135-159 (1995)
4. G. Rajdeep, P. Tansuhaj, Building Organizational Capabilities for Managing Economic Crisis: The Role of Market Orientation and Strategic Flexibilit. *Journal of Marketing*, **65**(2), 67-80 (2001)
5. P.R. Krugman, M. Obstfeld, M.J. Melitz, *International Economics: Theory and Policy*, Harlow: Pearson Education Limited (2015)
6. G. Gibbs, *Learning by doing a guide to teaching and learning methods*. EMU Oxford Brookes University, UK (1988)
7. D.G.W. Birch, *The Job Generation Process*. MIT Program on Neighborhood and Regional Change (1979)
8. G. Gunderson, *The Wealth Creators: An Entrepreneurial History of the United States*. New York: Truman Talley Books, E. P. Dutton (1989)
9. J. Mugler, *Podnikové hospodářství malých a středních podniků I [Business administration of small and medium enterprises I]*. Plzeň: University of West Bohemia (2011)
10. C.E. Helfat, *Dynamic capabilities: understanding strategic change in organizations*. Malden: Blackwell (2007)
11. R.E. Miles, C.C. Snow, A.D. Meyer, *Organizational strategy, structure, and process*. Stanford, Calif: Stanford University Press (2008)
12. M. Augier, T.J. Teece, Understanding complex organization: the role of know-how, internal structure, and human behavior in the evolution of capabilities. *Industrial and Corporate Change,* **15**(2), 395-416 (2006)
13. D.J. Teece, *Strategy, Innovation and the Theory of the Firm*. Northampton: Edward Elgar Publishing (2012)
14. V. Ambrosini, C. Bowman, What are Dynamic Capabilities and are They a Useful Construct in Strategic Management? *International Journal of Management Review,* **11**(1), 29-49 (2009)
15. A. Kuuluvainen, *Dynamic capabilitiesin the International Growth of Small and Medium-Sized Firms*. Turku School of Economics: Uniprint (2011)
16. A. Lundström, L.A. Stevenson, *Entrepreneurship Policy. Theory and Practice*. New York: Springer (2005)
17. AMSP ČR [online], Available at: http://www.amsp.cz/10-pruzkum-amsp-cr-nazory-podnikatelu-na-moderni-metody (2017)

# The Development of Organizational and Management Structures in Small-scale and Mid-scale Entrepreneurship in the Czech Republic

*Jan* Váchal[1*], *Milan* Talíř[1]

[1]Institute of Technology and Business, Faculty of Corporate Strategy, Okružní 517/10, 37001 České Budějovice, Czech Republic

**Abstract.** The business architecture, whose integral part is an organizational and management company structure, will be playing more important role in the future while pursuing its business goals. As a matter of fact, the organization and management of business processes that constitute its organizational and management structure demonstrate significant changes as a result of the changeable business environment. These issues are discussed in this article. The testing data set 381 SMEs (from which 114 were micro-enterprises, 141 small-scaled enterprises and 126 mid-scale enterprises, including without limitations enterprises from South Bohemian Region) analyzed applied organizational and management structures in separate scale categories of enterprises including degrees and stages of management. Pearson's chi-squared test was used as a statistical method; p-value was employed to test zero and alternative hypotheses. SMEs manifested a transition from linear and line-staff structures to matrix and functional ones; mid-scale enterprises used strategic business units. Micro-enterprises and partly small-scale enterprises employed degree of management, mid-scale 3 and more degrees of management. The conclusion of the article specifies the estimated development of the organizational architecture in SMEs in the Czech Republic in relation to expected changes in the external business environment in particular.
**Key words**: organizational structure, management structure, levels of management, scope management

## 1 Introduction

In order to deal with the issue of organizational and management structures in small-scale and mid-scale enterprises (SMES), this topic needs to be anchored in the overall development of small-scale and mid-scale entrepreneurship in the world and the Czech Republic. Small-scale and mid-scale entrepreneurship thrived in the most of the ancient cultures. Until the Second World War, the then ČSR (Czechoslovak Republic) belonged to the most industrially

* Corresponding author: vachal@mail.vstecb.cz

developed countries in the world. The vast majority of trade subjects, where vocational and business trades dominated, were no bigger than five workers. The social significance of small-scale entrepreneurship and its positive influence on the economy and welfare of the country was acknowledged no sooner than after the Second World War within the programme of the post war restoration of national economies. The development in the Czech Republic and other countries of the Central and Eastern Europe in particular was profoundly different from that of their western counterparts. The period after the Second World War saw a considerable decline. The winding-up of private SMES resulted in gaining a dominant position of large and state-owned enterprises that were established in an administrative way.

The important milestone in the development of the business environment on the market was marked with year 1989. Constitutional Act No. 100/1991 Sb. which granted equal rights to all kinds of ownership was enacted in 1991. The business sphere demonstrated an exponential increase. Only 19,000 enterprises were registered in 1989; on the other hand, year 2018 saw a tenfold increment. SMES got stable in the Czech Republic and became stabilizers of the national economy as it had been in other EU member states, and, they not only got into new sales markets, but also had to face competitors; all the same, they were at the same time supported by both – the state and the EU.

In regard to the development of organizational and management structures, these before the Second World War may be nowadays termed as linear, i.e. line-staff structures when applying directive and centrist management. It is mostly line-staff structures with department and profession specifications that developed after the Second World War. This, then innovative, development was discontinued by the decline and winding up of the predominating part of SMES in the Czech Republic. On the other hand, the world economy sees a rapid development of SMES mostly as a result of changes in their macro environment. Burns and Stalker [1] and other authors emphasize the relation between the organizational structure, changes in the external environment and efficiency of the organization. After 1989 the Czech Republic goes through a rapid development of all units of small-scale and mid-scale entrepreneurship (micro enterprises, small-scale enterprises and mid-scale enterprises) which means that the development of new organizational and management structures also takes place. Changes in the external environment of SMES occurring from the end of the last century require substantial changes in their current architecture [2]. Banner [3] argues that these changes mostly refer to organizational and management structures [4]. Friesen [5] specifies these factors that cause changes in organizational and management structures as general variables of organizational structures. Hodge, Anthony and Gales [6] define organizational structures as a means for classifying and coordinating the organization according to the kind of work, tasks, goals and processes.

O´Neal, Beauvais and Scholl [7] refer to conventional definitions of the following authors: Mintzberg [8-9], Katz and Kahn [10] and Burns and Stalker [1]. These authors define organizational structures as a stage of centralization of the decision-making process, establishment of rules, authority, communication and compensation, standardization of workflows and skills including control activities of the enterprise. Mintzberg [8-9] analyses the relations between the organizational structure and the age of the company, size, adopted strategy, available technologies, external environment and culture of the organization. Moreover, he, as one of the first authors, also points out the relationship between the organizational structure and corporate strategy. A similar opinion is expressed by Miller in his articles [11] where, apart from the relationship between the business organization and corporate strategy, he speaks about strategic goals. On the other hand, Friesen [5] entertain a different opinion on the definition of the organizational structure. They argue that the organizational structure is determined by the current life-cycle period in which the specific enterprise currently exists. As a matter of fact, some structures are laid down by the relevant legislation and local conventions such as Améba, virtual enterprise etc.

The formal part of the organizational structure is usually implemented by organizational schemes, policies and procedures [8-9], assignment of competences, roles and responsibilities, management and control of formal communication mechanisms and human resources management [12]. Wang and Ahmed [13] refer to these elements as 'hard'. On the other hand, there are also 'soft' elements often referred to as informal structure including e.g. interpersonal relationships and relationships on the diagonal communication stage, which are not explicitly expressed in the organizational code of the enterprise. Hodge, Anthony and Gales [6] declare that the informal structure is not only a logical consequence of social networks and relationships between employees, which develop within the work process, but it can also develop as a result of a defective or ineffective formal structure. Wang and Ahmed [13] argue that what is tolerated is an inconsistency between the informal structure and formal organizational rules, relationships and standards and competency relations. A large number of authors [4, 5, 2] nowadays point out that SMESs will have to be more flexible, adaptable, agile and integrating in regard to their competitiveness and effectiveness in the near future. Hunter [12] emphasizes the importance of innovations and he further argues that enterprises should continuously improve their technical equipment, work processes and inputs and outputs. He declares that the organizational structure needs to be sometimes changed in order to achieve new and progressive goals. As applied by analogy, Cooper [14] and Hunter [12] point out the importance of the implementation of new innovative business structures. In relation to ongoing changes in organizational structures, term de-structured forms of organizational structures including new, perspective structures such as network, process, cellular, quantum structures and organized design.

## 2 Data and Methods

The research was carried out on a selected data set of SMESs including without limitations enterprises from South Bohemian Region. The research started in 2016 and has been conducted until now. The data set contains 381 enterprises from which 114 are micro enterprises, 141 small-scale enterprises and 126 mid-scale enterprises. Test sample was set up in cooperation with the Czech Statistical Office to reflect the actual structure of the companies within South Bohemian region.

Two hypotheses were formulated:

H1: Linear and line-staff organizational and management structures predominate in SMES.

H2: Single stage management system predominates in micro enterprises and small-scale enterprises; three and more management stages predominate in mid-scale enterprises.

Pearson's chi-squared test was used as a statistical method in order to determine the dependence / independence of the selected category variables (internal and external strategic analyses) for formulating corporate strategy.

The equation for Pearson's chi-square test of independence is as follows:

$$\chi 2 = \frac{\sum_{i=1}^{k}(X_i - Np_i)^2}{Np_i} \qquad (1)$$

Where:

X2 = chi-square

Xi = empirical frequency (actual)

Npi = theoretical frequency (expected)

The tests suggest two hypotheses – zero ($H_0$) and alternative ($H_A$). $H_0$ (there is no correlation between selected variables), $H_A$ (there is a correlation between selected variables).

The assessment criterion for accepting or rejecting the hypothesis is p-value which oscillates on interval <0.1>. In case that the resulting p-value is $< \alpha$ (0.05 confidence interval), $H_A$ will be confirmed. Statistical calculations were carried out using R statistic program.

## 3 Results

The outcomes are ordered in the result section according to the defined hypothesis. It is possible to confirm the validity of the first hypothesis in accordance with the outcomes in the graphs 1 to 3. The hypothesis deals with the analysis of the organizational and managerial structures in relation to the size categorization of SMESs.

It applies a departmental managerial structure in the size which is more than 70.6% in case of microenterprises (Fig. 1). 72.7% of which generates profit, 23.4% of companies have a balanced business and 3.9% of companies with a departmental managerial structure is at loss. The other organizational structures are not found in an increased quantity. It is a very surprising finding, especially in terms of linear structure.

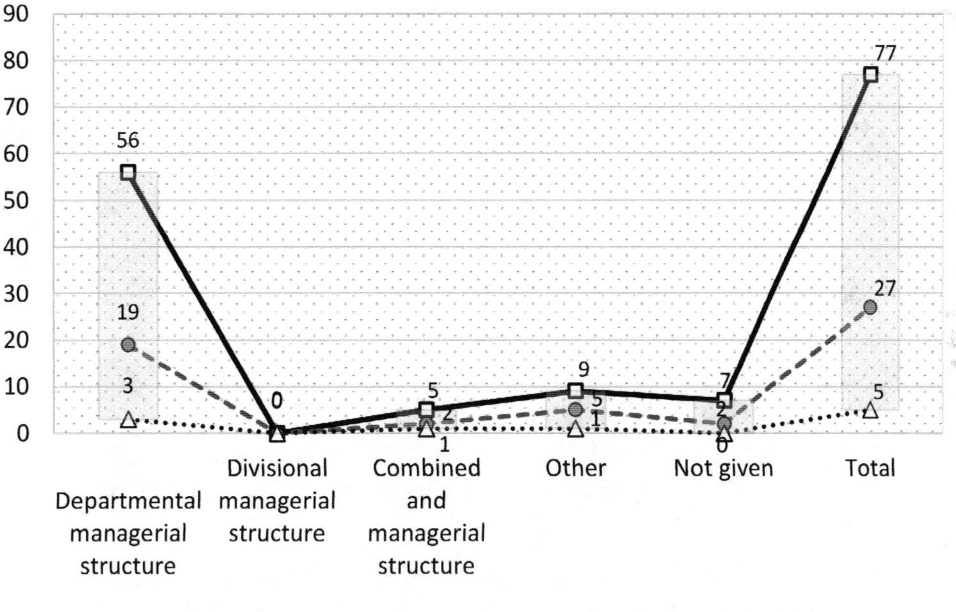


**Fig. 1.** The dependence of the type of organizational structure, size category, on the profitability of company in case of microenterprises

Source: Own research.

The departmental management structure is applied on small enterprises (Fig. 2) in agreement with micro-enterprises. There are namely 86.4% of enterprises. The profit in their economy is 63.6%, 17.4% of enterprises manage in a balanced way and at loss is 2.5% enterprises with departmental management structure. The second with a significantly lower frequency of use in business practice is the combined management structure, with 82.4% of enterprises generating profits, 11.8% operating in a balanced way and 5.9% of enterprises are at loss. Other structures such as divisional and combined structures are insignificant.

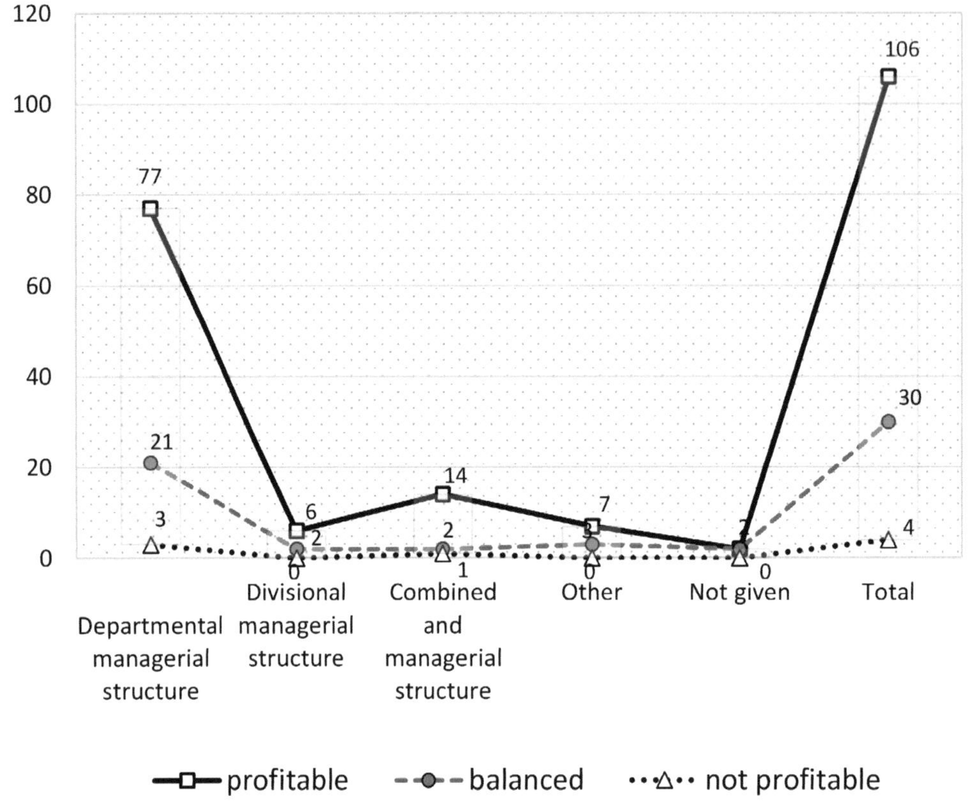


**Fig. 2.** The dependence of the type of organizational structure, size category, on the profitability of company in case of small enterprises

Source: Own research

Medium-sized enterprises (Fig. 3) also apply the departmental management structure the most of all, i.e. 78.9%. 75.3% of them generate profits, 18.6% of enterprises are in a balanced way, 6.2% show a loss. The second in terms of the frequency of use is the divisional management structure followed by the combined structure. In case of divisional structure, 66.7% of enterprises report a profit, 25% of enterprises manage their balance and 8.3% are at loss. In case of combined structure, 50% of enterprises are profitable, 30% are balanced and 20% show a loss.

In summary, departmental managerial structures predominate in all size categories of SE, most of all in micro and small enterprises, followed by medium-sized enterprises. The findings from controlled interviews have been confirmed. The transition from linear and line-staff structures to departmental structures was confirmed by the leading SMES managers. For medium-sized enterprises, the implementation of process management elements can be expected within departmental (functional) structures. However, for specialized departments, the process management process will be predominantly applied. During contact interviews, especially during structured interviews, professional and terminological illiteracy was clearly manifested. This finding is a challenge especially for educational institutions (especially relevant universities).

The trend of transition from classical structures to departmental structures was also confirmed by statistical analysis (P.value = 0.000008287). At the same time, the dependence between the type of organizational structure and the size category was proved (P.value = 1,649e-06, V = 0,2999).

It can be stated that the production and industry sector is significantly less dependent (p.value = 0.01509) from the point of view of the dependence of the organizational structure on the sectoriality of the company, in contrast to the dependence indicated in the service sector (p.value = 0.00013).

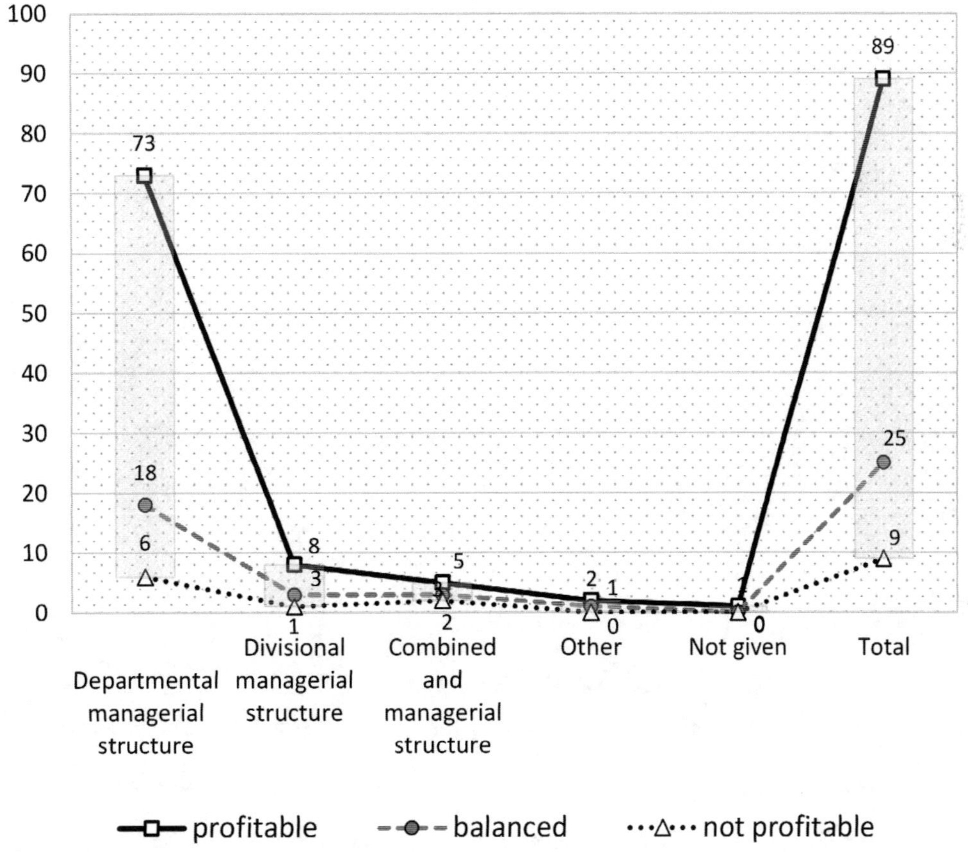


**Fig. 3.** The dependence of the type of organizational structure, size category, on the profitability of company in case of middle-sized enterprises

Source: Own research.

Another part of the research was focused on the analysis of the number of the levels of management depending on the size category of enterprises. Regarding the steepness of managerial structures (Fig. 4) it can be stated that micro and small enterprises dominate one level of management, medium-sized enterprises are predominantly characterized by two levels of management. The size of the company itself predetermines the number of management levels, and statistical dependence was demonstrated during the research. Especially in case of small enterprises, to a lesser extent micro enterprises, this is a consequence of gradually prevailing departmental, functional organizational and managerial structures. In case of medium enterprises this process has been running for several years. Functional structures reflect the changes that are currently taking place in the business sector. It is mainly the specialization of production, increasing innovative activities in these companies, introduction of information technology at the time of growing competitive pressure. Recently, the start of the digitizing of corporate processes has started to make an influence. In terms of management satisfaction with the current organizational level, 90% satisfaction was recorded.

In terms of hypothesis 2, which focuses on the number of control steps for individual size categories of SMESs, its validity can be confirmed. The prevailing share of one level of management was proved in micro enterprises. Similarly, in case of small enterprises, the gradual transformation to the departmental structure leads to an extension of the levels of management to 2 or rather 3 levels. In case of medium-sized enterprises, 3 or more levels of management are reported, i.e. in agreement with the research hypothesis.

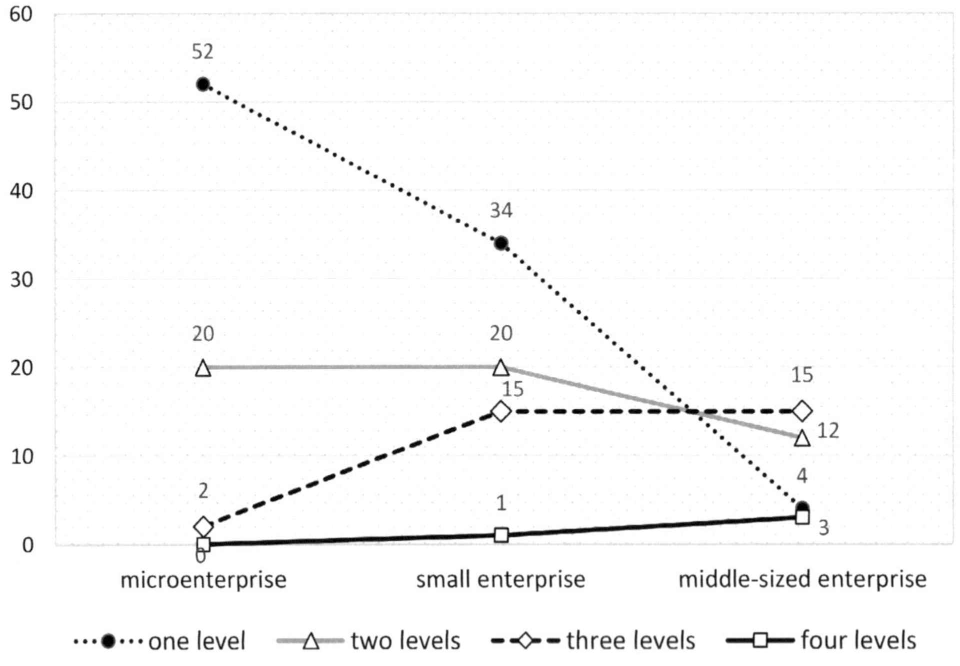


**Fig. 4**. The number of the levels of management in the individual size categories of SMSEs

Source: Own research.

The number of staff (scope management) on the individual the levels of management of SMESs was analyzed as a partial research question with the following outcomes:

6 or fewer employees were identified at the level of TOP management. For medium-sized companies, it is recommended to reallocate competencies and create functional units such as Sales Director, CFO or HR Director.

In the middle management position, the number of managers at the tactical level is in the range of twenty-five or less employees. Exceptionally, there is a range of twenty-five or more employees, especially in manufacturing companies.

At the operational level the degree of range is naturally the largest and reaches 30-45 employees in the size category of micro and small enterprises depending on the focus of the company.

The analyzed numbers of employees at individual management levels were statistically tested in terms of the influence of this number on the quality of management work of managers. The obtained statistical values ($\alpha = 0.05$, $p = 0.00000001239$, Chi-square = 47.8964, Number of respondents = 381, V = 0.5421) prove a very high degree of dependence between the number of the levels of management and the number of managers especially on TOP and tactical level. The degree of dependence is 0.5421, which can be considered to revealing a high degree of dependence.

Based on the obtained results ($\alpha = 0.05$, $p = 0.000000008665$, Chi-square = 48.6737, Number of respondents = 153, V = 0.5259) it can be stated (P value is significantly lower than parameter $\alpha$), that there is a very high degree of dependence between the level of management and the size of the business. The strength of the contingency is also high: 0.5259. Microenterprises with no fixed organizational structure (subject to exceptions) were not included in the testing of this dependence.

The last tested dependency was focused on the dependence of the organizational structure type and the number of the levels of management. The results ($\alpha = 0.05$, $p = 0.719$) show that the P value is 0.719, therefore it is significantly higher than the parameter $\alpha$. On the basis of the null hypothesis, there is no dependency between the organizational structure of the company and the number of management levels.

## 4 Conclusion

The fundamental decision for all business entities for the coming period is to move towards a change in the corporate architecture. In particular, it will be the projection of a new, progressive, innovative organizational and managerial framework with a digitization aspect. If this applies in general, it is a limiting factor for the further development of SMESs. In particular, this size category will face new challenges in the near future, mainly as a result of increasing competition, permanent technological innovation, or, for example, changes in principles, forms and rules in dealing with customers. The requirements for the provision of products and services according to the specific requirements of the customer will increase. The speed of this development has allowed the emergence of new organizational structures and the abandonment of classical structures such as linear, line-staff structures, etc. in the course of the last ten years. A new bridging term of the de-structured form was created, which describes the newly created structures. The development of the use of modern production as well as information technologies and the development of the Internet mean that the potential of the dynamics of the decision-making process increases and the number of alternatives that must be identified and taken into account increases as well. The newly created organizational structures should reflect this reality and enable companies to be developed in the form of innovation and increasing flexibility.

The results of the research which are the subject of this paper are summarized in the following way:

- SMESs are undergoing intensive rebuilding of their organizational and managerial framework, but the restructuring process will be intensified if the business cycle changes.
- SMESs are already intensively approaching the transformation of their entire corporate architecture, especially organizational and managerial structures on a functional and departmental basis of organizational structures. At the same time, there is a growing need for highly qualified specialists to professionally support this organizational transformation.
- Size categorization has proved to be the decisive factor in generating the new organizational framework. The sectoriality of the company is not crucial in this case,

but the stronger dependence between rebuilding and profitability has been proved in the field of services.

- In terms of the steepness (number of the levels of management) of organizational structures, there is a significant dependence on size categorization, less on sectoriality, which is a different outcome than that was obtained in the survey 10 years ago.
- The number of managers at individual the levels of management is highly differentiated and it is not advisable to recommend generally applicable principles. Nevertheless, it can be assumed that top managers will be subject to increasing pressure on their managerial competence, including knowledge of modern forms of management. At the tactical and operational level, demands are increasing especially in the area of human resources management.

## 5 Suggestions for practice

- It is advisable to carry out the transformation of corporate architecture in a period of economic growth.
- The change in the organization and management of the company is the result of a regular process of developing corporate structures, not a fashionable matter.
- Before the actual transformation of the organizational and managerial structure, it is necessary to process, incorporate, discuss and enforce the principles of changes in organization and management in the company and to define responsibilities of the implementation team.
- TOP management should not use the transformation of corporate architecture to deal with personal issues arising from personal antipathy towards employees.
- The existing functional elements of the corporate architecture should be preserved, while the rebuilding targets only non-functional and outdated elements.
- Every business is a living organism and a highly structured system. Therefore, its conversion must be specific and original and respect the general principles of modern management.
- One of the prerequisites for a successful transformation of business structures is to create space for an innovative process in the area of organization and management through the projection of new, innovative structures.
- The success of the proposed changes depends on the functioning communication corporate system.
- One of the limiting factors of the successful implementation of changes is to provide entire, objective and updated information for all the corporate employees because it is the only way how to secure their full support.
- Teamwork, decentralized management and flexible relations between the managerial entities should dominate in the innovative models of management.
- It is necessary to provide an objective and relevant feedback in the course of the whole transformation of corporate architecture.
- Control is an ally, not an enemy. It is necessary to create a control system including all the problematic areas which are subject to the transformation.

## Acknowledgements

This article was supported by the project "Stabilization and development of SME in rural areas", reg. No. TL01000349, the TACR Éta programme.

## References

1. T. Burns, G.M. Stalker, *The management of innovation*. Tavistock Publications (1961)
2. J.K. Galbraith, E. Lawler, The challenge of change: organizing for competitive advantage. In: *Tomorrow's organization: craft winning capabilities in a dynamic world*, eds. S. A. Mohrman, J. R. Galbraith, E. E. Lawler, Jossey-Bass, San Francisco, Calif, 1-20 (1998)
3. D.K. Banner, *Designing effective organizations: traditional & transformational views*. Sage: Thousand Oaks, CA (1995)
4. R. Ashkenas, D. Ulrich, T. Jick, S. Kerr, *The Boundaryless organization: breaking the chains of organizational structure*. Jossey-Bass: San Francisco, CA (1995)
5. G.B. Friesen, Organization design for the 21st Century. *Consulting to management*, **16**(3), 32-51 (2005)
6. B.J. Hodge, W.P. Anthony, L.M. Gales, *Organization theory: a strategic approach*, 5th ed. Prentice Hall: Upper Saddle River, New Jersey (1996)
7. J.W. O'Neill, L.L. Beauvais, R.W, Scholl, The use of organizational culture and structure to guide strategic behaviour: an information processing perspective. *The Journal of Behavioural and Applied Management*, **2**(2), 131-150 (2001)
8. H. Mintzberg, *H. Mintzberg on management: inside our strange world of organizations*. Free Press: New York (1989)
9. H. Mintzberg, *The structuring of organizations: a synthesis of the research*. Prentice-Hall: Englewood Cliffs, NJ (1979)
10. D. Katz, R.L. Kahn, *The social psychology of organizations*. John Wiley & Sons: New York (1966)
11. D. Miller, Configurations of strategy and structure: towards a synthesis. In: *Readings in strategic management*, eds. D. Asch, C. Bowman. Macmillan Education: London (1989)
12. J. Hunter, Improving organizational performance through the use of effective elements of organizational structure. *International Journal of Health Care Quality Assurance*, **15**(4/5), 12-21 (2002)
13. C.L. Wang, P.K. Ahmed, *The informal structure: hidden energies within an organisation*, Management Research Centre, Wolverhampton Business School Working Paper Series 2002, University of Wolverhampton, Telford, Shropshire (2002)
14. D. Cooper, Organizational change: From public to private sector a UK based reflective case study. *Journal of American Academy of Business*, **5**(1/2), 474-481 (2004)

# AUTHOR INDEX